AF539946

Biochemical Aspects of Plant Physiology

Biochemical Aspects of Plant Physiology Technology & Methodology

Amitav Bhattacharya
Former Principal Scientist (Plant Physiology)
Indian Institute of Pulses Research
Kanpur – 208 024, U.P.

Vijaylaxmi
Senior Scientist (Plant Physiology)
Indian Institute of Pulses Research
Kanpur – 208 024, U.P.

NEW INDIA PUBLISHING AGENCY
New Delhi – 110 034

NEW INDIA PUBLISHING AGENCY
101, Vikas Surya Plaza, CU Block, LSC Market
Pitam Pura, New Delhi 110 034, India
Phone: + 91 (11)27 34 17 17 Fax: + 91(11) 27 34 16 16
Email: info@nipabooks.com
Web: www.nipabooks.com

Feedback at feedbacks@nipabooks.com

ISBN: 978-93-83305-90-2

Composed, Designed and Printed in India

Preface

Physiology is one of the fundamental subjects of life science and knowledge of its practical aspects, particularly the biochemical aspect(s) is absolutely essential for the post graduate students as well as for researcher. We have always felt the necessity for a manual on orientation towards biochemical aspect of practical plant physiology. The present book is written in simple language covering the various aspects of basic and a clear-cut practical concept has been offered in the manual in respect of all quantitative experiments.

Biochemical methods are used in all branches of biological science including agriculture. Biochemical aspect is an integral part of plant physiology and this aspect is used to explain nearly all the phenomenon of physiological aspect of plant and/or crop. Technology and Methods for Biochemical Aspects of Plant Physiology is mainly intended for Post Graduate students and Researchers of Universities and of different Research Institutes. As it covers a broad range of subjects on the basic as well as the practical aspects of biochemical part of Plant Physiology, it is likely that it will be also useful for any student attending different theoretical or practical Plant Physiology as well as Biochemistry courses. The book builds on pre-existing knowledge obtained at under graduate classes for different aspects of Biochemistry and Plant Physiology. It assumes a solid background and experience in handling of chemical as well as bio-chemicals. This book discuss the theoretical principles and the practical. This book has the description of different biochemical estimations, and it contains detailed experimental protocol(s) to perform experiments. It rather contains a collection and description of principles that will help the students perform successful experiments on biochemical aspect of plant physiology on their own during their future carrier. The experience of the author gathered during several years of service at the Indian Institute of Pulses Research, Kanpur resulted in a material that enables students to efficiently use practical on Biochemical Aspects of Plant Physiology later during their training. Moreover, the present book also provides a solid background on basic aspects biochemical estimations, a prerequisite for successful experimental design.

By going through this book students will acquire practical knowledge regarding the techniques used and methods applied to investigate the properties of macromolecules. Students will become familiar with commonly used laboratory ware and instrumentation of a biochemical and molecular biological laboratory, learn the requirements for biochemical aspect(s). They learn how to correctly use biochemical units of measure and to make solutions and buffers for basic biochemical experiments. They will be able to determine the charge of weak acids, bases and macromolecules by taking into account their chemical environment. They learn how to measure the macromolecular concentration of solutions by spectrophotometry. They will be able to design protocols for the purification of proteins from cell cultures or tissues. They will become familiar with the physico-chemical background of enzyme action and the fundamentals of enzyme kinetics.

The method(s) given in this book will be useful for conducting practical classes of undergraduate and post graduate students in Plant Physiology, Biochemistry, Biotechnology, Microbiology, Agricultural science, Environmental science, Nutrition, Pharmaceutical science and other biology-related subjects. This book will be a boon for research workers as it covers procedures from the classical micro-Kjeldhal nitrogen assimilation to modern technologies such as PCR and Real Time PCR. We do not claim current furnishings of technology and methodology in any way exhaustive, but we do hope that it may certainly be to value to our readers. Technologies related to biochemical aspect of plant physiological are subject of tremendous interest and importance to plant physiologist, agronomist, horticulturist, ecologist, and biochemists. This book is intended to provide recognised methods related various biochemical processes involved in life processes in a comprehensive form.

Further, one can visualize that technologies and methods used for biochemical basis of plant physiology such as photosynthesis, photorespiration, plant pigments, carbon and nitrogen assimilation, plant nutrients, phenols, secondary metabolites, nucleic acid and vitamins should be very useful to not only post graduate student, but to research workers also. Thus this book is meant to be used in conjunction with a standard text of plant physiology though elementary principles relating to the technologies and methodologies are briefed.

Suggestions and advice are most welcome for the improvements in structural and for additions of current technologies as well as methodologies. We shall feel successful in our efforts publishing this book if we know how this text has become meaningful to our readers in planning their research strategies. We would appreciate suggestions from teachers and students who use the book. Hope this book shall serve the need of students, teachers and researchers. The suggestions and advice are most welcome for the improvements in structural and for additions of current techniques. We shall feel successful in our efforts publishing this book if we know how this text has become meaningful to our readers in planning their research strategies. We would appreciate suggestions from teachers and students who use the book on how we could improve upon it.

We sincerely hope that the manual will meet the requirement of undergraduate and post graduate students studying plant physiology and biochemistry and will be glad to accept constructive criticism and suggestions from the faculty, students and readers to make this manual a better one in future.

A. Bhattacharya
Vijaylaxmi

Contents

Preface v

1. Plant Physiology and its Biochemical Aspect **1**

2. Biological Samples and Chemical Substances **5**

2.1. Analytical Balances 5
2.2. Centrifuges 6
2.3. Other Laboratory Techniques 9
2.4. The UV-VIS spectrophotometer 11
2.5. Flow Injection Analysis 14
2.6. Near Infra-red Spectroscopy & Near Infra-red Analyser 15
2.7. Photoelectric Flame Photometer 17
2.8. Flame Emission Spectroscopy 20
2.9. Fluorimetry 22
2.10. Gas Chromatography 27
2.11. Atomic Absorption Spectrophotometer 39
2.12. High Performance (high pressure) Liquid Chromatography (HPLC) 42
2.13. Electrophoresis 46
2.14. Polyacrylamide Gel Electrophoresis (PAGE) 53
2.15. Isoelectric Focusing 60
2.16. Polymerase Chain Reaction (PCR) Machine 62
2.17. Real-time Polymerase Chain Reaction 72
2.18. Radio Tracer Technique 84
2.19. Scintillation Counter 90
2.20. Liquid Scintillation Counter 92
2.21. Geiger Counter 94

3. About Units 97

3.1. Numeric Expression of Quantities 98

3.2. Units of Measurement 100

4. About Solutions 103

4.1. Quantitative Description of Solutions, Concentration Units .. 107

4.2. Preparation of Solutions 108

5. Buffer and Buffer Solution 111

5.1. Universal Buffer Mixtures 113

5.2. Common Buffer Compounds Used in Biology 114

5.3. Composition of Different Buffer Systems 119

6. Carbon Assimilation 131

6.1. Photosynthesis and Photorespiration 132

6.2. Photosynthesis 132

6.3. Photorespiration 135

6.4. Estimation of Ribulose-1,5-diphosphate Carboxylase (RuBPcase EC. 4.1.1.39.) activity 136

6.5. Estimation of Phospho-enol Pyruvate Carboxylase (EC 4.1.1.31) activity 138

6.6. Estimation of Glycolate Oxidase (EC. 1.1.3.15.) Activity 142

6.7. Estimation of Glutamate Glyoxylate Aminotransferase (EC.2.6.1.4) Activity 144

6.8. Estimation for Activity of Isocitrate lyase (EC. 4.1.3.1) 147

6.9. Estimation of Glycolate Content 149

6.10. Estimtion of Starch Synthase (EC. 2.4.1.21) Activity 150

6.11. Estimation of Starch Phosphorylase (EC. 2.4.1.1) Activity .. 153

6.12. Estimation of Glucose Pyrophosphorylase (EC 2.4.2.13) Activity 157

6.13. Estimation of Malate Synthase (EC. 2.3.3.9) Activity 159

6.14. Estimation of Sucrose Synthase (EC 2.4.1.13) Activity 162

6.15. Estimaion of Invertase (EC 3.2.1.26) Activity 163

6.16. Estimation of Keto Organic Acid 165

7. Carbohydrates 167

7.1. Qualitative Reaction of Carbohydrates 169

7.2. Estimation of Reducing Sugars by Nelson-Somogyi Method.... 172

7.3. Estimation of Reducing Sugar by Dinitrosasylic Acid Method .. 174

7.4. Estimation of Total Carbohydrate by Anthrone Method 176
7.5. Estimation of Total Carbohydrate by Phenol Sulphuric Acid ... 177
7.6. Estimation of Starch by Anthrone 178
7.7. Estimation of Amylose 179
7.8. Estimation of Cellulose 182
7.9. Estimation of Hemicellulose 184
7.10. Estimation of Fructose and Inulin 186
7.11. Estimation of Pectic Substances 190
7.12. Estimation of Crude Fibre 195
7.13. Estimation of Pyruvic Acid 197
7.14. Estimation of α-Amylase Activity (EC. 3.2.1.1) 199
7.15. Estimation of β-Amylase Activity (EC. 3.2.1.2) 201
7.16. Estimation of Cellulase (E.C. 3.2.1.4) Activity 202
7.17. Estimation of Phosphorylase (EC. 2.4.1.1) Activity 205
7.18. Estimation of Phosphatase (EC. 3.1.3.1, & EC. 3.1.3.2) Activity 207
7.19. Assay of Phosphatase Using Glycerophosphate 210
7.20. Estimation of Malate Dehydrogenase (L-malate: NAD+ Oxidore-ductase EC 1.1.1.37) Activity 212
7.21. Estimating Alcohol Dehydrogenase (EC. 1.1.1.1) Activity 214
7.22. Estimation of Lactate Dehydrogenase (E.C. 1.1.1.27) Activity 219

8. Nitrogen Assimilation 221

8.1. Estimation of Nitrogenase (EC. 1.18.6.1 or 1.19.6.1) Activity 224
8.2. Estimation of Nitrate Reductase (EC. 1.6.6.1-3.) Activity 228
8.3. Estimation of Nitrite Reductase (NADPH): Nitrite Oxidoreductase (EC. 1.6.6.4) Activity 231
8.4. Estimation of Glutamate Dehydrogenase (L-glutamate: NADP Oxidoreductase (deaminase) EC. 1.4.1.4 and L-glutamate NAD Oxidoreductase (deaminating) EC 1.4.1.2) Activity 233
8.5. Estimation of Glutamine Synthatase (EC. 6.3.1.2) Activity .. 235
8.6. Estimation of Glutamate Synthase (EC. 1.4.7.1.) Activity 237
8.7. Estimation of Asparagine Synthetase (L-Aspartate; Ammonia Ligase EC. 6.3.1.4) Activity 239

8.8. Estimation of Glutathion Synthatase (E.C. 6.3.1.2) Activity . 241
8.9. Estimation of Nitrate Content 242
8.10. Estimation of Protease (EC. 3.4.21.40) Activity 244
8.11. Estimation of Aspartic Proteinase (EC. 3.4.23.23) Activity .. 247

9. Proteins and Enzymes 251
9.1. Protein Purification 254
9.2. Determination of Protein Concentration 262
9.3. Cell Disruption, Cell Fractionation and Protein Isolation 264
9.4. Protein Fractionation Based on Particle Size 274
9.5. Important Parameters for Ion Exchange-based Separation ... 283
9.6. Polyacrylamide-sodium dodecyl Sulphate Electrophoresis (SDS-PAGE) of Proteins 291
9.7. Estimation of Crude Protein Content 294
9.8. Enzymes 304
9.9. Enzyme Commission Nomenclature 307
9.10. Enzyme Kinetics 309
9.11. Isoenzymes 315
9.12. Isoenzyme Analysis 316
9.13. Native PAGE Separation and Detection of Lactate Dehydrogenase Isoenzymes 317

10. Amino Acid 321
10.1. Essential Amino Acids 324
10.2. Estimation of Glycine, Alanine and Isoleucine 326
10.3. Estimation of Lysine 327
10.4. Estimation of Methionine 330
10.5. Estimation of Arginine 332
10.6. Estimation on Tryptophan 334
10.7. Estimation of Cystine 337
10.8. Estimation of Proline 339
10.9. Estimation of Total Amino Acid using Cyanide-acetate Buffer .. 341
10.10. Estimation of Free Amino Acid through Paper Chromatography 343
10.11. Estimation of Aspartate Aminotransferase (Glutamate: Oxaloacetate Aminotrasferase EC. 2.6.1.1) Activity 344
10.12. Estimation of Alanine Aminotrasferase (Glutamate Pyruvate Aminotransferase EC. 2.6.1.2) Activity 347

11. Lipids 351
- 11.1. Categories of Lipids 352
- 11.2. Biological Functions of Lipids 355
- 11.3. Estimation of Oil in Oilseeds 358
- 11.4. Estimation of Free Fatty Acids 359
- 11.5. Estimation of Iodine Value of an Oil 361
- 11.6. Estimation of Saponification Value of Oils 363
- 11.7. Estimation of Peroxide Value of an Oil 366
- 11.8. Estimation of Lipase (E.C. 3.1.1.3) Activity 367

12. Plant Nutrition 373
- 12.1. Macronutrients (Primary) 375
- 12.2. Macronutrients (Secondary and Tertiary) 377
- 12.3. Micro-nutrients 378
- 12.4. Colorimetric Estimation of Phosphorus, Iron, Manganese, Aluminium and Crude silica in Plant Tissue 380
- 12.5. EDTA method for Estimation of Calcium and Magnesium 385
- 12.6. Determination of Zinc, Copper, Manganese, Calcium, Magnesium, Potassium and Sodium in Plant 389
- 12.7. Estimation of Boron 395
- 12.8. Estimation of Extractable Chloride 396
- 12.9. Estimation of Total Sulfur 397
- 12.10. Estimation of Molybdenum 399

13. Plant Pigments 401
- 13.1. Pigments in Plants 402
- 13.2. Estimation of Chlorophyll Contents 403
- 13.3. Chlorophyll Fluorescence 411
- 13.4. Isolation of Chloroplasts 415
- 13.5. Anthocyanin 418
- 13.6. Analysis of the Flavonoid Content in Plant Extracts 422
- 13.7. Estimation of Carotenes and Xanthophylls 423
- 13.8. Estimation of Lycopene 428
- 13.9. Curcumin 431

14. Phenols 435
- 14.1. Estimation of Total Phenolic Compounds 436
- 14.2. Estimation of Chlorogenic Acid Content 438

14.3. Estimation of Tannin Content 439
14.4. Estimation of Lignin Content 443
14.5. Estimation of Capsaicin Content 447
14.6. Estimation of Solanine Content 451
14.7. Estimation of Catalase (EC 1.11.1.6.) Activity 454
14.8. Estimation of Polyphenol Oxidase (Monophenol, Dihydroxyphe -nylalanine: Oxygen Axidoreductase EC 1.14.18.1) activity 456
14.9. Estimation of Peroxidase (Donor: H_2O_2 Oxidoreductase EC. 1.11.1.7) Activity 458
14.10. Estimation of Glutathion Peroxidaes (EC. 1.11.1.9) Activity 461
14.11. Estimation of Super Oxide Dismutase (EC. 1.15.1.1) Activity 463
14.12. Estimation of Phenyl Ammonia Lyase (L-Phenylalanine Ammonia lyase EC. 4.3.1.24) Activity 468
14.13. Estimation of Pectin Methyl Esterase (Pectin pectyl hydrolase EC. 3.1.1.11) Activity 470
14.14. Estimation of Cinnamic Acid Content 473

15. Secondary Metabolites 477
15.1. Estimation of Flavonoids in Plant Samples 480
15.2. Cyanogenic Glycoside 482
15.3. Phytic Acid 484
15.4. Gossypol 488
15.5. Phytoestrogens 490
15.6. Carotenoids 492
15.7. Saponins 498

16. Plant Growth Hormones 501
16.1. Classes of Hormones 503
16.2. Colorimetric Estimation of Indole Acetic Acid 508
16.3. Colorimetric Estimation of Gibberellic Acid 508
16.4. Estimation of Abscisic Acid 509
16.5. Estimation of Ascorbic Acid Oxidases (E.C. 1.10.3.3) Activity . 511
16.6. Estimation of Indole Acetic Acid Oxidase Activity 512

17. Nucleic Acids 515

17.1. Types of Nucleic Acids 517

17.2. Extraction of Plant DNA 523

17.3. Extraction of Plant Total DNA 525

17.4. A Simple Method for DNA Isolation from Plant Sample 527

17.5. Extraction of RNA from Plant Samples 528

17.6. Isolation of Plasmid DNA 538

17.7. Quantitative Estimation of DNA from Plant Leaves 539

17.8. Quantitative Estimation of RNA from Plant Samples 541

17.9. Separation of DNA and RNA from Plant Samples and Their Colorimetric Estimation 543

18. Vitamins 545

18.1. Vitamin A (Retinol) 546

18.2. Thiamine (Vitamin B_1) 549

18.3. Riboflavin (Vitamin B_2) 553

18.4. Niacin (Vitamin B_3) 558

18.5. Pantothanic acid (Vitamin B_5) 560

18.6. Pyridoxine (Vitamin B_6) 561

18.7. Estimation of Vitamins B-Complex (B_2, B_3, B_5 and B_6) by HPLC Method 561

18.8. Folic acid (Vitamin B_9) 563

18.9. Cobalamine (Vitamin B_{12}) 567

18.10. Estimation of Vitamins B_1, B_3, B_6, B_9 and B_{12} by HPLC 569

18.11. Biotin 571

18.12. Ascorbic Acid (Vitamin C) 572

18.13. Cholecalciferol (Vitamin D) 577

18.14. Tocopherol (Vitamin E) 578

18.15. Phylloquinone (Vitamin K) 581

18.16. Estimation of Water and Fat Soluble Vitamins by HPLC 584

CHAPTER 1

Plant Physiology and its Biochemical Aspect

Plant physiology is a sub-discipline of botany concerned with the functioning, or physiology of plants. Closely related fields include plant morphology (structure of plants), plant ecology (interactions with the environment), phytochemistry (biochemistry of plants), cell biology, genetic, biophysics and molecular biology. Fundamental processes such as photosynthesis, respiration, plant nutrition, plant hormone functions, tropism, nastic movements, photoperiodism, photo-morphogenesis, circadian rhythms, environmental stress physiology, seed germination, dormancy, stomata function and transpiration, both parts of plant water relations, are studied by plant physiologists. The field of plant physiology includes the study of all the internal activities of plants those chemical and physical processes associated with life as they occur in plants. This includes study at many levels of scale of size and time. At the smallest scale are molecular interactions of photo-synthesis and internal diffusion of water, minerals, and nutrients. Major sub disciplines of plant physiology include phytochemistry (the study of the biochemistry of plants) and phytopathology (the study of disease in plants). The scope of

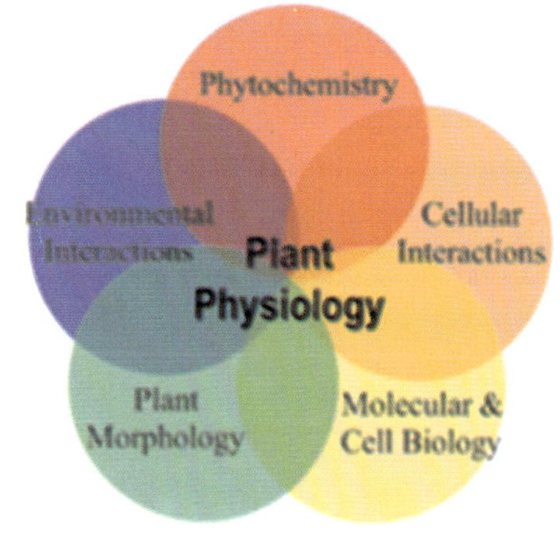

Five key areas of study within plant physiology

plant physiology as a discipline may be divided into several major areas of research.

First, the study of phytochemistry (plant chemistry) is included within the domain of plant physiology. To function and survive, plants produce a wide array of chemical compounds not found in other organisms. Photosynthesis requires a large array of pigment, enzymes, and other compounds to function. Because they cannot move, plants must also defend themselves chemically from herbivores, pathogens and competition from other plants. They do this by producing toxins and foul-tasting or smelling chemicals. Other compounds defend plants against disease, permit survival during drought, and prepare plants for dormancy. While other compounds are used to attract pollinators or herbivores to spread ripe seeds.

Secondly, plant physiology includes the study of biological and chemical processes of individual plant cells. Plant cells have a number of features that distinguish them from cells of animals, and which lead to major differences in the way that plant life behaves and responds differently from animal life. For example, plant cells have a cell wall which restricts the shape of plant cells and thereby limits the flexibility and mobility of plants. Plant cells also contain chlorophyll, a chemical compound that interacts with light in a way that enables plants to manufacture their own nutrients rather than consuming other living things as animals do.

Thirdly, plant physiology deals with interactions between cells, tissues and organs within a plant. Different cells and tissues are physically and chemically specialized to perform different functions. Roots and rhizoids function to anchor the plant and acquire minerals in the soil. Leaves catch light in order to manufacture nutrients. For both of these organs to remain living, minerals that the roots acquire must be transported to the leaves, and the nutrients manufactured in the leaves must be transported to the roots. Plants have developed a number of ways to achieve this transport, such as vascular tissue, and the functioning of the various modes of transport is studied by plant physiologists.

Fourthly, plant physiologists study the ways that plants control or regulate internal functions. Like animals, plants produce chemicals called hormones which are produced in one part of the plant to signal cells in another part of the plant to respond. Many flowering plants bloom at the appropriate time because of light-sensitive compounds that respond to the length of the night, a phenomenon known as photoperiodism. The ripening of fruit and loss of leaves in the winter are controlled in part by the production of the gas ethylene by the plant.

Finally, plant physiology includes the study of plant response to environmental conditions and their variation, a field known as environmental physiology. Stress from water loss, changes in air chemistry, or crowding by other plants can lead to changes in the way a plant functions. These changes may be affected by genetic, chemical, and physical factors.

The chemical elements of which plants are constructed principally carbon, oxygen, hydrogen, nitrogen, phosphorus, sulphur, *etc*, are the same as for all other iife forms animals, fungi, bacteria and even viruses. Only the details of the molecules into which they are assembled differ.

Biochemical aspect(s) of plant physiology involves study of the chemical processes that occur in living plant cells with ultimate aim of understanding the nature of life in molecular terms. Biochemical studies rely on the availability of appropriate analytical techniques and on the application of these techniques to the advancement of knowledge. In recent years huge advances have been made in our understanding in gene structure, and expression and in the application of techniques such as mass spectrophotometry to the study of protein structure and function. The discipline of molecular biology overlap with that of biochemistry and in many respects the aim of the two disciplines compliment each other. Molecular biology is focused on the molecular understanding of the processes of replication, transcription and translation of genetic material whereas biochemistry exploits the techniques and findings of molecular biology to advance our understanding of cellular processes.

Biochemical aspect of plant physiology is the branch of life science which deals with the study of chemical reactions occurring in living cells and organisms. Life is a chemical process involving thousands of different reactions occurring in an organized manner. These are called metabolic reactions. The term 'Biochemistry' was first introduced by the German Chemist Carl Neuberg in 1903. It takes into account of the studies related to the nature of the chemical constituents of living matter, their transformations in biological systems and the energy changes associated with these transformations. Plant physiology may thus be treated as a discipline in which biological phenomena are analyzed in terms of chemistry. The biochemical aspect of physiology for the same reason, has been variously named as 'Biological Chemistry' or 'Chemical Biology'. Modern biochemistry has two branches, descriptive biochemistry and dynamic biochemistry. Descriptive biochemistry deals with the qualitative and quantitative characterization of the various cell components and the dynamic biochemistry deals with the elucidation of the nature and the mechanism of the reactions involving these cell components. Biochemistry is related to almost all the life sciences and without biochemistry background and knowledge, a thorough understanding of health and well being is not possible. Those who acquire a sound knowledge of biochemical of plant physiology can tackle the two central concerns of the biomedical sciences (1) the understanding and maintenance of health, and (2) the understanding and treatment of diseases.

Objectives of Biochemical aspect of Plant Physiology

The major objective of biochemical aspect of plant physiology is the complete understanding of all the chemical processes associated with living cells at the molecular level. To achieve this objective, plant physiologists have attempted to isolate numerous molecules found in cells, to determine their structures and to analyze how they function.

Biochemcal aspect is the study and application of substances, reactions and processes in animals, plants, bacteria and viruses. The aims of biochemical and molecular biological aspects of plant physiological research are complex and diverse. Investigation of the network of chemical reactions taking place in living organisms and representing the most fundamental phenomena of life, identification of the molecules playing roles in biochemical processes, determination of their structure, function and interactions, examination of the molecular background of metabolism, the flow of energy and information within organisms are all among the common goals of biochemists and molecular biologists. In accordance with this diversity of problems, a high variety of tools, instruments and methods are required to answer scientific questions effectively. This book reviews the most common tools and instruments used very frequently in almost every laboratory. The appropriate handling and storage of biological samples and other chemical substances required for research will also be discussed.

CHAPTER 2

Biological Samples and Chemical Substances

Tissue or cell samples from a living organism, different cell cultures grown in a laboratory incubator under controlled conditions, homogenates or extracts of cells and tissues, solutions of isolated and purified components (*e.g.* proteins, nucleic acids) can all be referred to as biological samples. As the medium of life is water, the majority of biological samples can be defined as aqueous solutions with one or more components, colloidal systems, or water-based suspensions (*e.g.* bacterial cells dispersed in a liquid medium). Consequently, most biochemical experiments also take place in aqueous environments. Different solids (*e.g.* chemical substances obtained from different companies, synthetic oligonucleotides or peptides) are also often necessary for biochemical research. In most cases, solids are dissolved in water (or sometimes in other solvents) prior to the experiments. Therefore, the methods of preparing solutions and measuring accurately the weight of the required solids will also be discussed below. Sometimes we use gases in the laboratory. These can be stored in gas cylinders (*e.g.* O_2), in Dewar flasks in liquid state (*e.g.* liquid nitrogen), or dissolved in water (*e.g.* HCl or NH_3). By working with gases it is very important to follow all safety instructions to avoid fire, explosions, frostbite or (in case of inhalation) asphyxia or poisoning.

2.1. Analytical Balances

An analytical balance is a class of balance designed to measure small mass in the sub-milligram range. The measuring pan of an analytical balance (0.1 mg or better) is inside a transparent enclosure with doors so that dust does not collect and so any air currents in the room do not affect the balance's operation. This enclosure is often called a draft shield. The use of a mechanically vented balance safety enclosure, which has uniquely designed acrylic airfoils, allows a smooth turbulence-free airflow that prevents balance fluctuation and the measure of mass down to 1 μg without fluctuations or loss of product. Also, the sample must be at room temperature to prevent natural convection from forming air currents inside the enclosure from causing an error in reading. Single pan mechanical substitution balance maintains consistent response throughout and the useful capacity is achieved by maintaining a constant load on the balance beam, thus the fulcrum, by subtracting mass on the same side of the beam to which the sample is added.

Electronic analytical scales measure the force needed to counter the mass being measured rather than using actual masses. As such they must have calibration adjustments made to compensate for gravitational differences. They use an electromagnet to generate a force to counter the sample being measured and outputs the result by measuring the force needed to achieve balance. Such measurement device is called electromagnetic force restoration sensor. Digital laboratory balances with of different capacity and accuracy are available for the measurements of the weight of samples and chemicals. Top loading balances typically work with 0.1 g accuracy in the range from several grams to several hundred grams. For the precise measurement of small weights, analytical balances are used. These devices usually work with 0.01 mg accuracy in the range from 0.1 mg to several grams. A vibration free, flat and horizontal surface is required for proper operation of such balances usually a free standing table with a marble surface plate. Moreover, the convection of the air around the balance should be avoided. (Air currents can exert forces on the pan of the balance comparable to the force exerted by the sample being measured). For this purpose, analytical balances have a built - in enclosure with doors made of transparent plastic sheets. Measurements can be performed within this protective enclosure. Disposable weighing dishes and laboratory spatulas with spooned ends are important accessories of weight quantification. It is also possible to dispense solids directly into a beaker or a laboratory tube instead of a weighing dish - however, heavy vessels (with a mass above the upper limit of the balance) must not be applied. The mass of the empty dish must be determined first. Then, the display of the scale must be set to show zero by pushing the, tare button. Next, the substance is added into the weighing dish. The display will thus show the pure mass of the substance.

2.2. Centrifuges

Cells can readily be harvested from liquid cultures by using different centrifuges. Similarly, any suspension or floating colloidal particles (*e.g.* precipitated proteins) in a solution can be separated into fractions by spinning the sample in a centrifuge. The resulting fractions are referred to as, supernatant (*i.e.* the solution) and pellet (*i.e.* the particles collected at the bottom of the centrifuge tube, pressed together into a compact mass). Centrifuges are relatively simple devices having stationary and rotary parts. The rotation generated by the electric motor of the centrifuge is transmitted to the rotor harbouring the samples contained within appropriate centrifuge tubes. Many biochemical samples are heat sensitive; for instance, proteins denature at elevated temperatures.

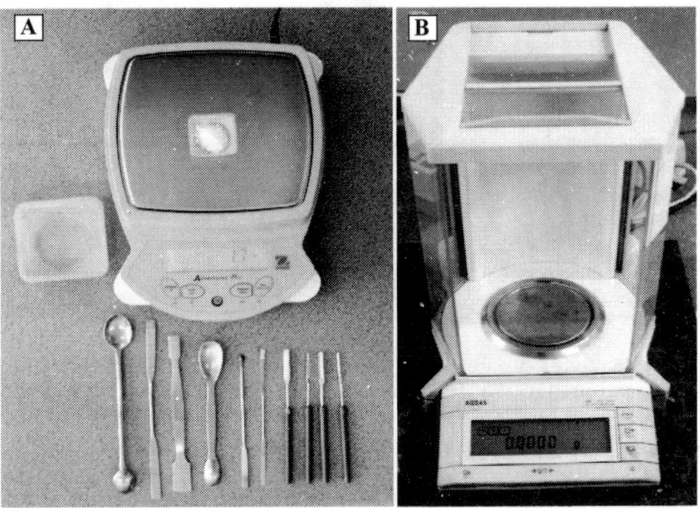

A. Top loading balance with disposable weighing dishes, laboratory spatulas and spoons B. Analytical balance on a vibration-free surface

Such samples require refrigerated centrifuges in which low temperature can be maintained during centrifugation. Centrifuges are available in different sizes ranging from simple bench-top centrifuges to preparative devices with much higher capacities (volumes up to several litres). Eppendorf tubes or Falcon tubes fit into some rotors of bench-top centri-fuges, thereby simplifying the processing of many samples. Most centrifuges can be used in conjunction with several different rotors. This versatility allows users to adapt their centrifuges easily according to the actual requirements. Centrifuge tubes must be chosen according to the manufacturer‘s instructions. Importantly, the filled rotor must be counter-balanced during operation. To achieve this, tubes with equal weights must be placed into opposite buckets or holes of the rotor. The weight balance should always be checked by simple two-armed or digital scales. If the weights of the sample tubes are different, counter-balances must be prepared by filling similar tubes with water. Unbalanced rotors are subject to extremely high forces during rotation. Even a small asymmetry of the weights around the axis of rotation can result in the breakage of the rotor shaft at high angular speeds. In such cases, the rotor may also damage the whole device and even cause serious personal injuries.

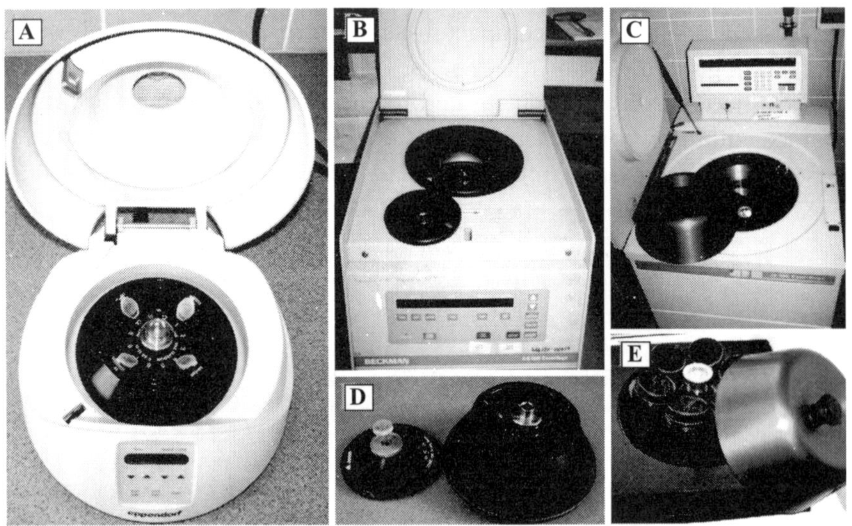

A) Bench-top Eppendorf centrifuge with samples arranged symmetrically around the axis of the rotor. B) Semi-preparative, refrigerated centrifuge with a rotor for Eppendorf tubes. C) Preparative, refrigerated centrifuge with a rotor for the handling of volumes up to 6 litres. D) Semi-preparative rotor with lid. E) Preparative rotor with lid.

The rotation speed of centrifuges is often specified as the number of revolutions per minute (RPM). However, as the force applied to the sample depends not only on the actual RPM value but also on the radius of the rotor, the relative centrifugal force (RCF) is more informative about a particular experiment. This defines acceleration according to the mass of particles floating in the sample. Therefore, RCF values are given as relative acceleration values (the centrifugal acceleration compared to *g*, ~9.8 m/s^2, the acceleration due to gravity on the surface of Earth). Thus, if the same sample is spun at equal RPM values in two centrifuges with different rotor geometries (different rotor radii), the results will be different. However, if equal RCF values are applied, sedimentation forces will be identical. Therefore, RPM values are only informative when specified together with the rotor type or radius. Simple centrifuges can provide maximal accelerations around 104 *g*. **Ultracentrifuges** are capable of operating at maximal accelerations in the range of 105-1,00,000*g*. To reach high angular speeds, ultracentrifuges generate vacuum around the rotor to decrease aerodynamic drag. Ultracentrifuges can be used for preparative (*e.g.* removal of cell debris from a lysate prior to the isolation of recombinant proteins) or analytical purposes (*e.g.* investigation of the interactions or the oligomerisation of different biomolecules).

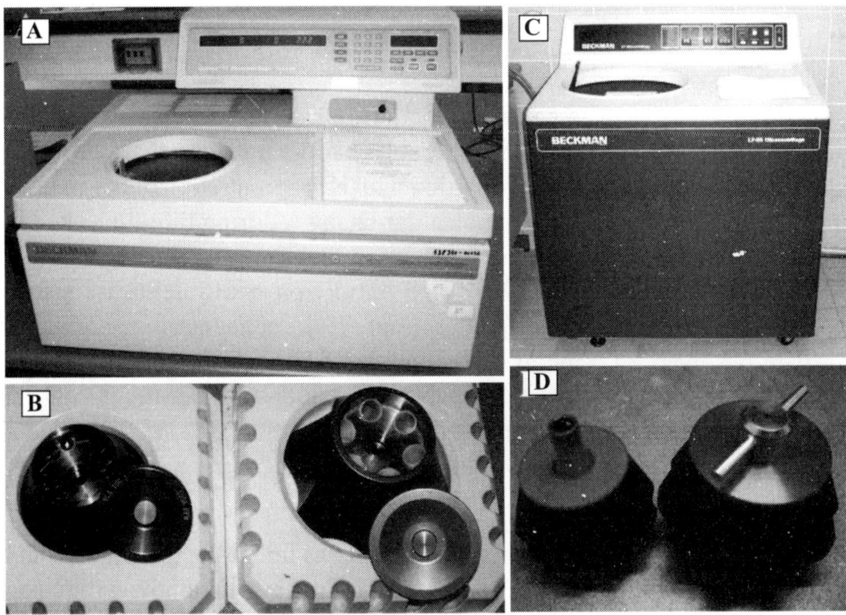

A) Micro-ultracentrifuge (RPMmax = 100000 min^{-1}, RCFmax = 300000 *g*). B) Rotors compatible with the micro-ultracentrifuge. C) Semi-preparative ultracentrifuge (RP Mmax = 70000 min^{-1}, RCFmax = 350000 *g*). D) Rotors compatible with the semi-preparative ultracentrifuge.

2.3. Other Laboratory Techniques

Spectrophotometry is one of the most widely used analytical procedures in biochemistry. The technique is well suited for simple routine determination of small quantities of materials. These measurements require that the examined material have an absorption maximum at one point of the spectrum. If the absorption maximum falls into the visible part of the spectrum, the material is coloured. Analyses can be performed in the visible spectrum also if the material in question has no colour by itself, but a chemical reaction can be performed that leads to the formation of a coloured product. Analyses in the ultraviolet spectrum are widespread, too, since many colourless materials exhibit intense absorption in this range (190-320 nm). Quantitative measurements in spectrophotometry are based on the Lambert-Beer law of light absorption by solutions. The amount of light absorbed by the sample is defined as the ratio of the intensities of the incident and the transmitted light.

From the degree of absorption, the concentration of the absorbing component can be deduced by using the following equation. If the intensity of light, I_0, diminishes to I after passing through a path of length L in a medium, then, according to the Lambert-Beer law:

$$E = \log\frac{I_0}{I} = \varepsilon \cdot c \cdot L$$

The expression log I_0 / I is called extinction (E), absorbance (A) or optical density (OD). At a wavelength where the solvent does not absorb, according to the Lambert-Beer law, E will be proportional to the concentration of the solution (c) if the solute does not undergo molecular changes during dilution. The law is valid only for monochromatic light of a given wavelength. The degree of the absorption is characterised by å, named extinction coefficient, a constant independent of the concentration of the solute. If the concentration is given in mol/l, then we speak of a molar extinction coefficient or molar absorptivity. Molar absorptivity gives the extinction of a solution of 1 mol/L (M) at the given wavelength and 1 cm path length; its unit is thus $M^{-1}cm^{-1}$. Its value is characteristic of the material in question, but also depends on the solvent and the temperature.

A photometer is a device measuring the intensity of light after passing through a sample (most often, a solution). Absorption of light by the sample will reduce the measured light intensity. A spectrophotometer is a photometer that can measure the absorbance of the sample at different wavelengths of the light. The absorbance depends on the presence and concentration of absorbing molecules). Hence, spectrophotometers are widely used to determine the concentration of different biomolecules that show characteristic absorption (*e.g.* proteins, nucleic acids). Sample holders used in spectrophotometers are called cuvettes. Depending on the wavelength range of the measurements, cuvettes made of transparent plastics, glass or quartz glass may be applicable.

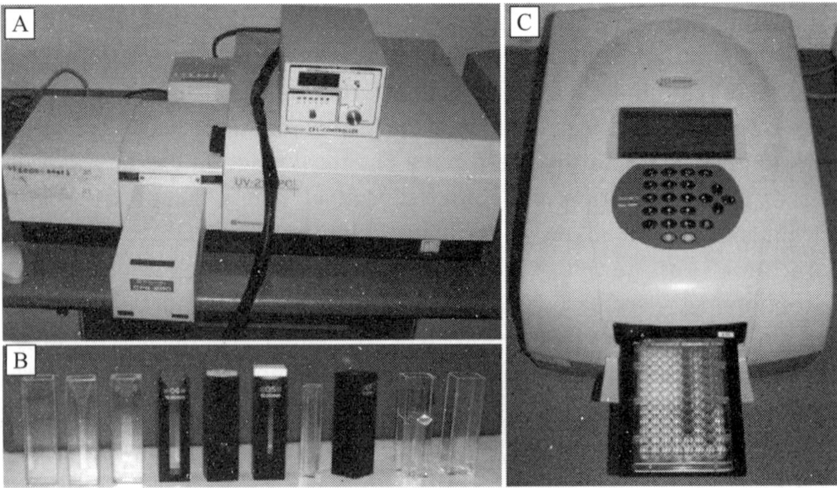

A) Spectrophotometer. B) Different types of cuvettes: photometric sample holders made of plastic, glass or quartz. C) Plate reader with samples loaded onto a 96-well microplate.

Plastic or glass cuvettes can be applied in the range of visible light (they are transparent as they do not absorb visible light).

Quartz cuvettes have low absorption both in the UV and visible range of the spectrum. If the absorbance of many samples needs to be determined simultaneously, plate readers can be used instead of spectrophotometers. These instruments can dramatically reduce the required working time. Samples are dispensed into the wells of plastic plates (*e.g.* those of a 96-well microplate with 8 rows and 12 columns of wells), and subsequently placed into the plate reader. Absorbance readings are transferred to and processed by a computer.

It is often necessary to isolate and purify one or several components from a solution. Various methods are available for the separation of biomolecules. Electrophoretic and chromatographic methods represent two major families of separation techniques.

2.4. The UV-VIS spectrophotometer

The spectrophotometer is an apparatus devised for the measurement of absorbance. It can produce a beam of light of a given wavelength, direct it at a sample (usually in solution in a cuvette) and measure the intensity of the transmitted light. For all this, a light source, a monochromator, a sample holder, a detector and a display are required.

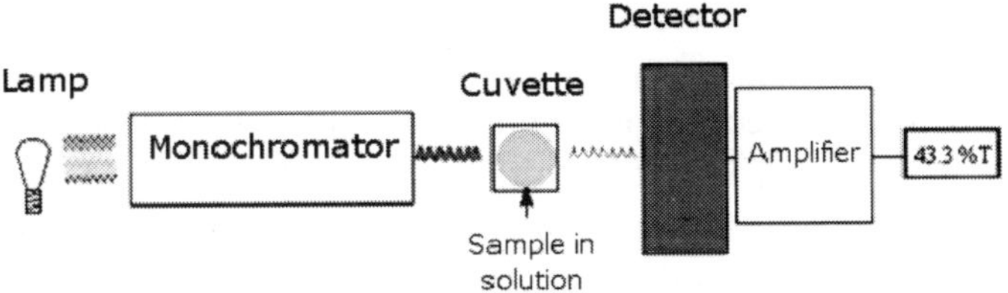

Schematic structure of a photometer

In the wavelength range of 200 to 320 nm, the light source can be either a deuterium, a high-pressure hydrogen or a high-pressure xenon lamp. To generate light in the visible or infrared spectrum, tungsten filament lamps are used. In a UV-VIS type photometer, both types of light sources can be found. The role of the monochromator is to select light of a given wavelength from the continuous spectrum of the light source. In modern devices, a diffraction grating has replaced the previously used prism. The produced light is not strictly monochromatic, but its spectral bandwidth is relatively narrow. Light entering and leaving the monochromator passes through slits whose width determines the bandwidth, the wider the slits, the wider the band-width. However, narrowing the slits reduces the intensity of the incident light. Thus, one must find a compromise between spectral purity and intensity, and set the slits accordingly. Light passes through the sample placed in a holder (cuvette) made of plastic, glass, quartz or other transparent material. Plastic and glass cuvettes are cheap but cannot be used at wavelengths under 280 and 320 nm, respectively, because their absorbance becomes too high. In the case of shorter wavelengths, quartz cuvettes are used. The quality and condition of the cuvette is critical for the measurements. Some devices are capable of keeping the temperature of the sample constant during measurement by a thermostatted cuvette holder. This is important, for example, when measuring the rate of chemical reactions.

Earler version of photometric devices are usually single-beam ones meaning that they can measure one sample at a time. More sophisticated devices have a double beam, and thus they can take two measurements simultaneously. One cuvette holds the sample, *i.e.* the solution containing the substance of interest, while the other (the reference) holds a solution that is identical to the sample in all aspects except that is lacks the substance of interest. The device automatically subtracts the absorbance of the reference from that of the sample, and thus it records a differential spectrum. This kind of spectrum not only characterises the studied molecule better, but also corrects for the error stemming from the fluctuations in the intensity of light generated by the lamp. The intensity of the transmitted light is measured by a light-sensitive electronic detector (photodiode or photomultiplier). Detector type and quality significantly affect the attributes of the system. Photomultipliers are characterised by high speed and high sensitivity in a wide spectral range. Photodiodes are small in size and have a lower sensitivity. They are used in so-called diode-array devices that measure a complete spectrum at a time. Such measurements are performed by illuminating the sample with multi-wavelength light and separating its components only after passing through the sample so that each diode measures at a given wavelength. Besides measuring at a given wavelength, one can also determine complete spectra, *i.e.* the dependence of absorbance on wavelength by changing the wavelength. Up-to-date spectrophotometers can take spectra between preset values automatically. Simpler devices can measure in the range of 0-1 absorbance units (AU), whereas higher quality ones do in a range of 0-2 AU (or even wider). However, as mentioned above, absorbance values at or higher than 2 are prone to large error.

Charged particles in a homogeneous electric field experience different forces. Positively charged particles are repelled from the positively charged electrode (*i.e.* the anode), while they are attracted by the negatively charged electrode (*i.e.* the cathode). Negatively charged particles are attracted by the anode and repelled by the cathode. This phenomenon also appears in aqueous solutions. Electrophoresis exploits the force exerted by an electric field on a charged biomolecule in solution. (For instance, DNA is negatively charged resulting from the deprotonated state of the phosphate moieties of its sugar-phosphate backbone). Electrophoretic experiments are carried out in gelatinous substances in which the charged biomolecules can be separated according to the differences in their charge, molecular size and shape. Gels are three-dimensionally cross-linked systems of macromolecules with porous structures. Pores are filled by the aqueous solution. As the pore sizes are comparable to the sizes of molecules to be separated, smaller molecules migrate faster than large molecules within the gel. This effect is called molecular sieving. **Chromatography** is another large family of separation techniques. In chromatography, the components of a mixture (*e.g.* a solution of biomolecules) are separated based on their differential partitioning between two phases: the so-called stationary and the so-called mobile phase. In biochemical experimentation, the stationary phase is generally composed of small beads of a polymeric substance filled into a column (a tube made of glass, plastic or metal with a filter at the bottom), while the mobile phase (or eluent) is the solution carrying the analyte (*i.e.* the mixture of substances to be separated) through the stationary phase. Separation can be based on the size, shape, charge, isoelectric point, hydrophobicity or specific binding affinity or biological activity of the compounds. Different stationary phases are available to exploit these molecular features. The mobile phase must be chosen according to the properties of the sample and the stationary phase. Automated or semi-automated chromatographic systems typically include vessels for different buffers and solvents, pump(s) responsible for the delivery of the mobile phase, an injector responsible for loading the samples onto the column, various chromatographic columns, detector(s) monitoring the composition of the eluate (*i.e.* the solution leaving the column), a fraction collector, and a computer for controlling the units and analyzing data.

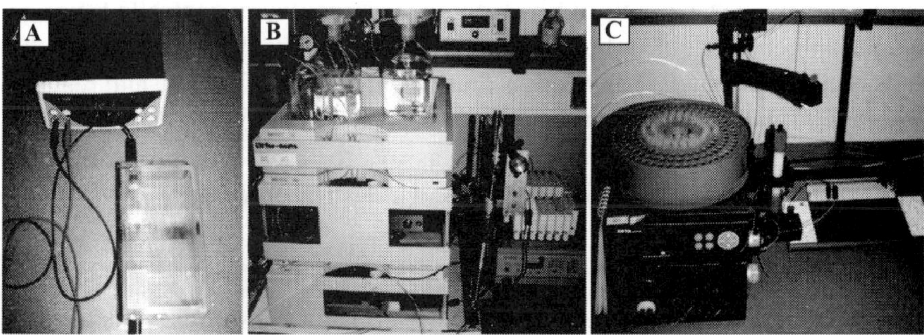

A) Power supply and buffer tank for gel electrophoresis. Gels are placed into the buffer tank. B) High-Performance Liquid Chromatography (HPLC) system. C) Fast Protein Liquid Chromatography (FPLC) system.

2.5. Flow Injection Analysis

Flow injection analysis (FIA) is an approach to chemical analysis that is accomplished by injecting a plug of sample into a flowing carrier stream. The principle is similar to that of segmented flow analysis (SFA) but no air is injected into the sample or reagent streams. FIA is an automated method in which a sample is injected into a continuous flow of a carrier solution that mixes with other continuously flowing solutions before reaching a detector. Precision is dramatically increased when FIA is used instead of manual injections and as a result very specific FIA systems have been developed for a wide array of analytical techniques. Flow Injection evolved into Sequential Injection and Bead Injection which are novel techniques based on computer controlled programmable flow.

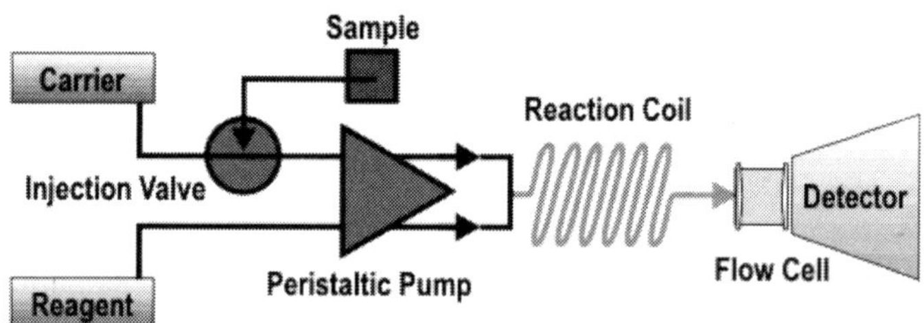

Principle of Flow Injection Analysis

Principles of Operation

A sample (analyte) is injected into a carrier solution which mixes through radial and convection diffusion with a reagent for a period of time (depends on the flow rate and the coil length and diameter) before the sample passes through a detector to waste. A Peristaltic pump is the most commonly used pump in FIA instruments but new variants of FIA technique use computer controlled syringe pumps that generate discontinuous flow that is precisely choreographed to the needs of assay protocol. The result is dramatic decrease of sample and reagent consumption and waste generation. The new technologies broadened applicability of FIA technique from laboratory serial assays to research applications and continuous monitoring of environmental and industrial processes.

Detectors

A flow through detector is located downstream from the sample injector and records a chemical physical parameter. Various different types of detector can be used for example:

- colorimeter
- fluorimeter
- ion-selective electrode
- biosensors

2.6. Near Infra-red Spectroscopy & Near Infra-red Analyser

Near-infrared spectroscopy (NIRS) is a spectroscopic method that uses the near-infrared region of the electromagnetic spectrum (from about 800 nm to 2500 nm). Typical applications include quick estimation of protein, fats *etc.*, pharmaceutical, medical diagnostics (including blood sugar and pulse oximetry), food and agrochemical quality control, and combustion research, as well as research in functional imaging and fundamental research. Near-infrared spectroscopy is based on molecular overtone and combination vibrations. Such transitions are forbidden by the selection rules of quantum mechanics. As a result, the molar absorptivity in the near-IR region is typically quite small. One advantage is that NIR can typically penetrate much farther into a sample than mid infrared radiation. Near-infrared spectroscopy is, therefore, not a particularly sensitive technique, but it can be very useful in probing bulk material with little or no sample preparation.

The molecular overtone and combination bands seen in the near-IR are typically very broad, leading to complex spectra; it can be difficult to assign specific features to specific chemical components. Multivariate (multiple variables) calibration techniques (*e.g.*, principal components analysis, partial least squares, or artificial neural networks) are often employed to extract the desired chemical information. Careful development of a set of calibration samples and application of multivariate calibration techniques is essential for near-infrared analytical methods.

Instrumentation for near-IR (NIR) spectroscopy is similar to instruments for the UV-visible and mid-IR ranges. There is a source, a detector, and a dispersive element (such as a prism, or, more commonly, a diffraction grating) to allow the intensity at different wavelengths to be recorded. Fourier transform NIR instruments using an interferometer are also common, especially for wavelengths above ~1000 nm. Depending on the sample, the spectrum can be measured in either reflection or transmission. Common incandescent or quartz halogen light bulbs are most often used as broadband sources of near-infrared radiation for analytical applications. Light-emitting diodes (LEDs) are also used; they offer greater lifetime and spectral stability and reduced power requirements.

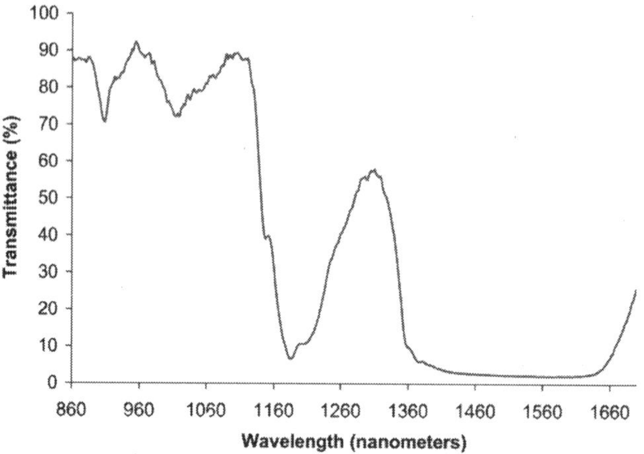

Near infra-red spectrum of liquid ethanol

The type of detector used depends primarily on the range of wavelengths to be measured. Silicon-based CCDs are suitable for the shorter end of the NIR range, but are not sufficiently sensitive over most of the range (over 1000 nm). InGaAs and PbS devices are more suitable though less sensitive than CCDs. In certain diode array (DA) NIRS instruments, both silicon-based and InGaAs detectors are employed in the same instrument. Such instruments can record both UV-visible and NIR spectra 'simultaneously'. Instruments intended for chemical imaging in the NIR may use a 2D array detector with an acousto-optic tunable filter. Multiple images may be recorded sequentially at different narrow wavelength bands.

Many commercial instruments for UV/vis spectroscopy are capable of recording spectra in the NIR range (to perhaps ~900 nm). In the same way, the range of some mid-IR instruments may extend into the NIR. In these instruments, the detector used for the NIR wavelengths is often the same detector used for the instrument's "main" range of interest. -Near-infrared spectroscopy is widely applied in agriculture for determining the quality of forages, grains, and grain products, oilseeds, coffee, tea, spices, fruits, vegetables, sugarcane, beve-rages, fats, and oils,

CropScan near infrared transmission analysers are used for measuring protein, oil and moisture in grains and oil seeds.

dairy products, eggs, meat, and other agricultural products. It is widely used to quantify the composition of agricultural products because it meets the criteria of being accurate, reliable, rapid, non-destructive, and inexpensive.

The first industrial application of near infra-red spectroscopy began in the 1950s. In the first applications, NIRS was used only as an add-on unit to other optical devices that used other wavelengths such as ultraviolet (UV), visible (Vis), or mid-infrared (MIR) spectrometers. In the 1980s, a single-unit, stand-alone NIRS system was made available, but the application of NIRS was focused more on chemical analysis. With the introduction of light-fiber optics in the mid-1980s and the monochromator-detector developments in early-1990s, NIRs became a more powerful tool for scientific research. This optical method can be used in a number of fields of science including physics, physiology, or medicine. It is only in the last few decades that NIRs began to be used as a medical tool for monitoring patients.

2.7. Photoelectric Flame Photometer

A photoelectric flame photometer is a device used in inorganic chemical analysis to determine the concentration of certain metal ions, among them sodium, potassium, lithium and calcium. Group 1 and Group 2 metals are quite sensitive to Flame Photometry due to their low excitation energies. In principle, it is a controlled flame test with the intensity of the flame colour quantified by photoelectric circuitry. The intensity of the colour will depend on the energy that had been absorbed by the atoms that was sufficient to vaporise them. The sample is introduced to the flame at a constant rate. Filters select which colours the photometer detects and exclude the influence of other ions. Before use, the device requires calibration with a series of standard solutions of the ion to be tested.

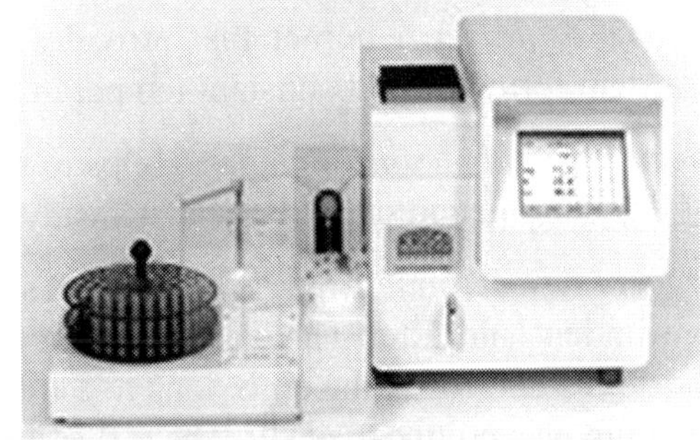

Flame photometer FP8800 for simultaneous determination of up to 4 alkali and alkali earth element concentrations in aqueous samples. Courtesy of A.KRÜSS Optronic

Photoelectric flame photometry, a branch of atomic spectroscopy is used for inorganic chemical analysis for determining the concentration of certain metal ions such as sodium, potassium, lithium, calcium, cesium, *etc*. In flame photometry the species (metal ions) used in the spectrum are in the form of atoms. The International Union of Pure and Applied Chemistry (IUPAC) Committee on Spectroscopic

Nomenclature has recommended it as flame atomic emission spectrometry (FAES). The basis of flame photometric working is that, the species of alkali metals (Group 1) and alkaline earth metals (Group II) metals are dissociated due to the thermal energy provided by the flame source. Due to this thermal excitation, some of the atoms are excited to a higher energy level where they are not stable. The absorbance of light due to the electrons excitation can be measured by using the direct absorption techniques. The subsequent loss of energy will result in the movement of excited atoms to the low energy ground state with emission of some radiations, which can be visualized in the visible region of the spectrum. The absorbance of light due to the electrons excitation can be measured by using the direct absorption techniques while the emitting radiation intensity is measured using the emission techniques. The wavelength of emitted light is specific for specific elements. Flame photometry is crude but cheap compared to flame emission spectroscopy, where the emitted light is analysed with a monochromator. Its status is similar to that of the colorimeter (which uses filters) compared to the spectrophotometer (which uses a monochromator). It also has the range of metals that could be analysed and the limit of detection are also considered

Analysis of samples by Flame photometer

Parts of a flame photometer

Source of flame: A burner that provides flame and can be maintained in a constant form and at a constant temperature.

Nebuliser and mixing chamber: Helps to transport the homogeneous solution of the substance into the flame at a steady rate.

Optical system (optical filter): The optical system comprises three parts: convex mirror, lens and filter. The convex mirror helps to transmit light emitted from the atoms and focus the emissions to the lens. The convex lens help to focus the light on a point called slit. The reflections from the mirror pass through the slit and reach the filters. This will isolate the wavelength to be measured from that of any other extraneous emissions. Hence it acts as interference type colour filters.

Photo detector: Detect the emitted light and measure the intensity of radiation emitted by the flame. That is, the emitted radiation is converted to an electrical signal with the help of photo detector. The produced electrical signals are directly proportional to the intensity of light.

A brief overview of the process:

- The solvent is first evaporated leaving fine divided solid particles.
- This solid particles move towards the flame, where the gaseous atoms and ions are produced.
- The ions absorb the energy from the flame and excited to high energy levels.
- When the atoms return to the ground state radiation of the characteristic element is emitted.
- The intensity of emitted light is related to the concentration of the element.

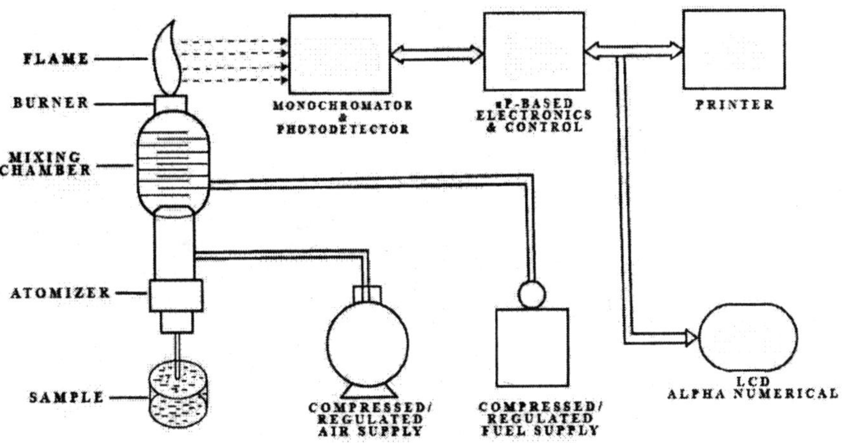

A schematic representation of flame photometer

Technique of Flame Emission Spectrophotometery

Flame photometer has both quantitative and qualitative applications. Flame photometer with monochromators emits radiations of characteristic wavelengths which help to detect the presence of a particular metal in the sample. This help to determine the availability of alkali and alkaline earth metals which are critical for soil cultivation. In agriculture, the fertilizer requirement of the soil is analyzed by flame test analysis of the soil. In clinical field, Na^+ and K^+ ions in body fluids, muscles and heart can be determined by diluting the blood serum and aspiration into the flame. Analysis of soft drinks, fruit juices and alcoholic beverages can also be analyzed by using flame photometry.

The emission spectrum of a chemical element or chemical compound is the spectrum of frequencies of electromagnetic radiation emitted due to an atom's electrons making a transition from a high energy state to a lower energy state. The energy of the emitted photon is equal to the energy difference between the two states. There are many possible electron transitions for each atom, and each transition has a specific energy difference. This collection of different transitions, leading to different radiated wavelengths, makes up an emission spectrum. Each element's emission spectrum is unique. Therefore, spectroscopy can be used to identify the elements in matter of unknown composition. Similarly, the emission spectra of molecules can be used in chemical analysis of substances.

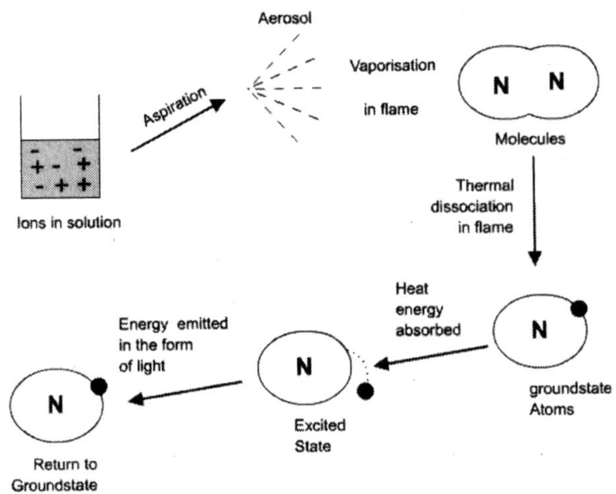

2.8. Flame Emission Spectroscopy

Emission spectroscopy is a spectroscopic technique which examines the wavelengths of photons emitted by atoms or molecules during their transition from an excited state to a lower energy state. Each element emits a characteristic set of discrete wavelengths according to its electronic structure, and by observing these wavelengths the elemental composition of the sample can be determined. Emission spectroscopy developed in the late 19th century and efforts in theoretical explanation of atomic emission spectra eventually led to quantum mechanics.

There are many ways in which atoms can be brought to an excited state. Interaction with electromagnetic radiation is used in fluorescence spectroscopy, protons or other heavier particles in Particle-induced X-ray Emission and electrons or X-ray photons in Energy-dispersive X-ray spectroscopy or X-ray fluorescence. The simplest method is to heat the sample to a high temperature, after which the excitations are produced by collisions between the sample atoms. This method is used in flame emission spectroscopy.

Although the emission lines are caused by a transition between quantized energy states and may at first look very sharp, they do have a finite width, *i.e.* they are composed of more than one wavelength of light. This spectral line broadening has many different causes. The emission spectrum can be used to determine the composition of a material, since it is different for each element of the periodic table. One example is astronomical spectroscopy: identifying the composition of stars by analyzing the received light. The emission spectrum characteristics of some elements are plainly visible to the naked eye when these elements are heated. For example, when platinum wire is dipped into a strontium nitrate solution and then inserted into a flame, the strontium atoms emit a red colour. Similarly, when copper is inserted into a flame, the flame becomes green. These definite characteristics allow elements to be identified by their atomic emission spectrum. Not all emitted lights are perceptible to the naked eye, as the spectrum also includes ultraviolet rays and infrared lighting. An emission is formed when an excited gas is viewed directly through a spectroscope.

In flame emission spectrometry, the sample solution is nebulised (converted in to a fine aerosol) and introduced into the flame where it is desolated, vaporized, and atomized, all in rapid succession. Subsequently, atoms and molecules are raised to excited states via thermal collisions with the constituents of the partially burned flame gases. Upon their return to a lower or ground electronic state, the excited atoms and molecules emit radiation characteristic of the sample components. The emitted radiation passes through a monochromator that isolates the specific wavelength for the desired analysis. A photodetector measures the radiant power of the selected radiation, which is then amplified and sent to a readout device, meter, recorder, or microcomputer system. Combustion flames provides a means of converting analytes in solution to atoms in the vapor phase freed of their chemical surroundings. These free atoms are then transformed into excited electronic states by one of two methods: absorption of additional thermal energy from the flame or absorption of radiant energy from an external source of radiation. In the first method, known as flame emission spectroscopy (FES), the energy from the flame also supplies the energy necessary to move the electrons of the free atoms from the ground state to excited states. The intensity of radiation emitted by these excited atoms returning to the ground state provides the basis for analytical determinations in FES.

The basic components of flame spectrometric instruments provide the following functions required in each method: (1) deliver the analyte to the flame, (2) induce the spectral transitions (absorption or emission) necessary for the determination of the analyte, (3) isolate the spectral lines required for the analysis, (4) detect the increase or decrease in intensity of radiation at the isolated lines(s), and (5) record these intensity data. Pretreatment of Sample Flame FES requires that the analy l e be dissolved ina solution in order to undergo nebulization (see the next section). The wet chemistry necessary to dissolve the sample in a matrix suitable for either flame method is often an important component of the analytical process. The analyst must be aware of substances that interfere with the emission measurement. When these substances are in the sample, they must be removed or masked (complexes). Reagents used to dissolve samples must not contain substances that lead to interference problems.

Applications

Most applications of FES have been the determination of trace metals, especially in liquid samples. It should be remembered that FES offers a simple, inexpensive, and sensitive method for detecting common metals, including the alkali and alkaline earths, as well as several transition metals such as Fe, Mn, Cu, and Zn. FES has been extended to include a number of non-metals: H, B, C, N, P, As, O, S, Se, Te, halogens, and noble gases. FES detectors for P and S are commercially available for use in gas chromatography. FES has found wide application in agricultural and environmental analysis, industrial analyses of ferrous metals and alloys as well as glasses and ceramic materials, and clinical analyses of body fluids. FES can be easily automated to handle a large number of samples. Array detectors interfaced to a microcomputer system permit simultaneous analyses of several elements in a single sample.

2.9. Fluorimetry

Fluorimetry is the quantitative study of the fluorescence of fluorescent molecules. Many biomolecules are fluorescent or can be labelled with fluorescent molecules, making fluorimetry a widely used tool in analytical and imaging methods. As the available photon-detecting devices are highly sensitive—even a single photon can be detected and one fluorophore can emit millions of photons in a second, fluorimetry is suitable for and is often used in single-molecule experiments.

Physical basis of fluorescence

Photons of a given wavelength are absorbed by the fluorophore and excite some of its electrons. The system remains in this excited state for only a few nanoseconds and then relaxes into its ground state. (Note that light travels 30 centimetres in a single nanosecond.) When returning from the excited state to the ground state, the electron may emit a photon. This is known as fluorescent emission. The wavelength of the absorbed photon is always shorter than that of the emitted photon (*i.e.* the energy of the emitted light is lower than that of the absorbed one). This phenomenon, the so-called Stokes shift, is an important attribute of fluorescence both in theory and practice. The relations between the wavelength (λ, nm), frequency (ν, 1/s) and energy (E, J) of light are the following: $\lambda = c/\nu$, where c is the speed of light (approximately 300 000 km/s).

The fluorimeter

The Stokes shift facilitates the creation of highly sensitive methods of detection of fluorescence. As the wavelengths of the exciting and detected (emitted) light differ, the background created by the exciting light can be minimised by using a proper setup. There are two ways to avoid that the exciting light gets into the detector: (1) Measurements are often carried out in a geometric arrangement in which the detection of emission is perpendicular to the exciting beam of light. (2) Light filters are placed between the light source and the sample and also between the sample and the detector. Light of only a certain wavelength range can pass through these filters. Photons of the exciting light leaving the sample will not reach the detector as they are absorbed by the emission filter. In many cases, monochromators are used instead of filters. Their advantage is that the selected wavelength can be set rather freely and more precisely compared to filters that are set to a given interval and adjustments can only be made by replacing them.

This double protection of the detector from the exciting light is necessary due to the fact that the intensity of fluorescent light is usually two or three orders of magnitude smaller than that of the exciting light. This means that even if only 1 or 0.1 % of the exciting light reaches the detector, half of the detected signal intensity would arise from the exciting light and only the other half from the emission of the sample. This would result in a 50% background signal level, as the detector is unable to distinguish photons based on their wavelength.

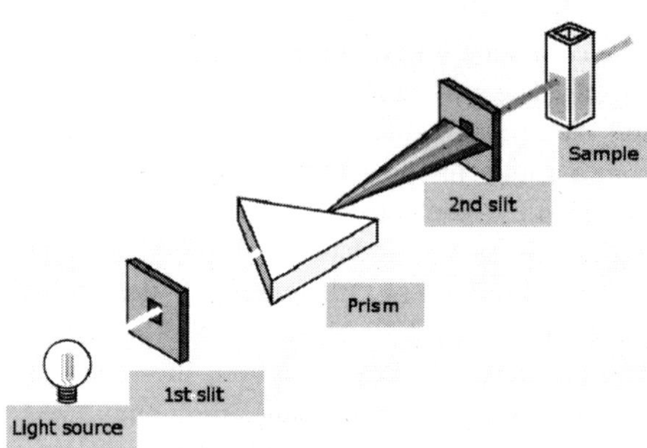

Scheme of a monochromator. From white (wide-spectrum) light, the monochromator is able to select light within a given narrow spectrum. White light is projected onto a prism splitting it to its components, effectively creating a rainbow behind it. On its way to the sample, light must pass through a small slit and therefore only a small part of the spectrum (a practically homogenous light beam) reaches it. Wavelength of the light leaving the monochromator can be changed by rotating the prism as this will let a different part of the rainbow through the slit.

Fluorophores

Fluorophores are characterised by specific fluorescence spectra, namely their excitation (absorption) spectrum and emission spectrum. The excitation spectrum is recorded by measuring the intensity of emission at a given wavelength while the wavelength of excitation is continuously changed. The emission spectrum is recorded by measuring the intensity of the emitted light as a function of its wavelength while the wavelength of the exciting light is kept constant. The shape of the excitation spectrum is usually the same as the shape of the emission spectrum. However, due to the Stokes shift, the emission spectrum is shifted towards red compared to the excitation spectrum, and usually the shape of the two spectra are mirror images of each other.

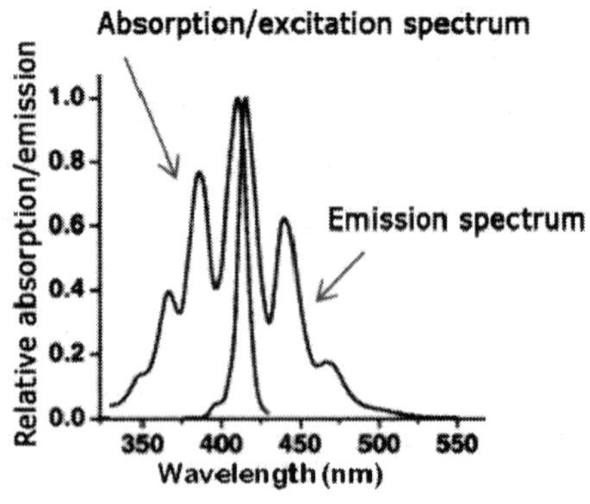

Absorption (excitation) and emission spectra of pyrene

The intensity of fluorescence of a molecule is sensitive to its environment. Emission intensity is significantly affected by the pH and the polarity of the solvent as well as the temperature. Usually, an apolar solvent and a decrease in temperature will increase the intensity. The immediate environment of the fluorophore is an important factor, too. Another molecule or group moving close to the fluorophore can change the intensity of fluorescence. Due to these attributes, fluorimetry is well suited to the study of different chemical reactions and/or conformational changes, aggregation and dissociation. In proteins, two amino acids have side chains with significant fluorescence: tryptophan and tyrosine. The fluorescence of these groups in a protein is called the intrinsic fluorescence of the protein. Tryptophan is a relatively rare amino acid; most proteins contain only one or a few tryptophans. Tyrosine is much more frequent; there are usually five to ten times more tyrosines in a protein than tryptophans. On the other hand, the fluorescence intensity of tryptophan is much higher than that of tyrosine.

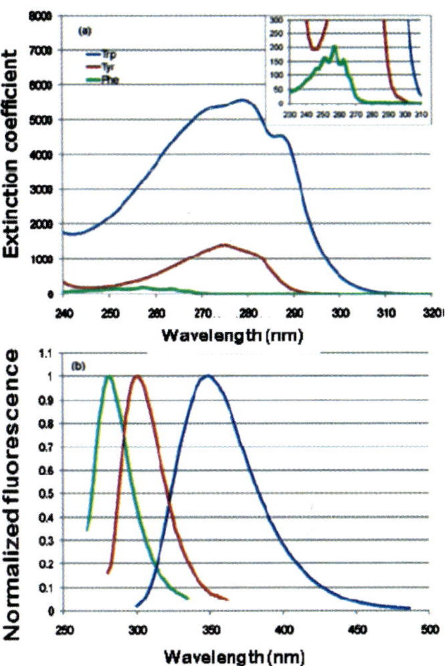

Extinction (A) and emission (B) spectra of tryptophan, tyrosine and phenylalanine. (Note that the three amino acids shown display markedly different fluorescence intensities. For visibility, emission spectra shown in panel B were normalised to their individual maxima.)

The above shown spectra clearly show that the fluorescence of tryptophan can be studied specifically even in the presence of tyrosines, since if the excitation is set to 295 nm and the detection of emission is set to 350 nm, the fluorescence of tyrosine can be neglected. Both the intensity of the fluorescence and the shape of the emission spectrum are sensitive to the surroundings of the side chain, which often changes upon conformational changes of the protein. Tryptophan fluorimetry is therefore suitable to detect conformational changes of enzymes and other proteins. It can also be applied to detect the binding of ligands to proteins as well as the di- or multimerisation of proteins, provided that the reaction results in a change in the surroundings of a tryptophan side chain. The environment of tryptophans obviously changes on unfolding of proteins. Consequently, fluorescence is well suited also for following denaturation of proteins.

Tryptophan and tyrosine fluorescence is not the only way to detect and investigate proteins using fluorescence. There are proteins that undergo post-translational modifications including the covalent isomerisation of three amino acids that makes them fluorescent. The first such protein discovered was the green fluorescent protein (GFP), which is expressed naturally in the jellyfish *Aequorea victoria* (phylum *Cnidaria*). Since then, fluorescent proteins were isolated from many other species. A large number of recombinantly modified forms of GFP were created in the last 20 years, all different in their fluorescence and colour. The intrinsic fluorescence of GFP can be used to label proteins. If we create a chimera from the genes of GFP and another protein of interest—in other words, we attach the gene of GFP to the 5‘ or 3‘ end of the gene encoding the other protein—this construct will be transcribed and translated into a protein that will have GFP fused to it at its N- or C-terminus. Thus, if using an appropriate vector we transform an organism and introduce this new gene into it, its product will show a green fluorescence when excited. Through this phenomenon, we can easily locate proteins on the tissue, cellular or subcellular levels. As a variety of differently coloured fluorescent proteins are at our disposal, we can even measure colocalisation of labelled proteins *in vivo*. The application of fluorescent proteins in biology was such a significant technological breakthrough that its pioneers were awarded a Nobel prize in 2008.

Fluorescence, phosphorescence and chemiluminescence

Even though the phenomena mentioned in the title are similar in many ways, it is important to make distinctions. In all three cases, the source of the light is an excited electron that, while returning to its ground state, can emit part of its excitation energy as a photon with some probability. In the case of fluorescence, excitation is performed using light of a wavelength that is optimally absorbed by the molecule. The transition between the excited and ground states is direct and fast (occurs on the nanosecond time scale). In the case of phosphorescence, the difference lies in the manner and rate of emission. In this case, the electron does not, immediately fall back from its excited state into its ground state, but is able to enter a particular alternative excited state. All transitions leading from this state to the ground state are so-called forbidden transitions. This does not mean that they do not happen at all, but their probability is rather low. This way, the lifetime of the excited state can increase from nanoseconds to milliseconds, minutes or even hours. The difference between fluorescence and chemiluminescence is not in the way the emission occurs, but in how the excitation is achieved. In the latter case the energy necessary to excite the electron comes not from light but from a chemical reaction. This is the way some living organisms can produce light. We call this bioluminescence. Fireflies and anglerfishes of deep seas are well known examples of the occurrence of this phenomenon.

2.10. Gas Chromatography

Gas chromatography (GC), is a common type of chromatography used in analytical chemistry for separating and analyzing compounds that can be vaporized without decomposition. Typical uses of GC include testing the purity of a particular substance, or separating the different components of a mixture (the relative amounts of such components can also be determined). In some situations, GC may help in identifying a compound. In preparative chromatography, GC can be used to prepare pure compounds from a mixture. In gas chromatography, the *mobile phase* (or "moving phase") is a carrier gas, usually an inert gas such as helium or an unreactive gas such as nitrogen. The *stationary phase* is a microscopic layer of liquid or polymer on an inert solid support, inside a piece of glass or metal tubing called a column (a homage to the fractionating column used in distillation). The instrument used to perform gas chromatography is called a *gas chromatograph* (or "aerograph", "gas separator").

The gaseous compounds being analyzed interact with the walls of the column, which is coated with a stationary phase. This causes each compound to elute at a different time, known as the *retention time* of the compound. The comparison of retention times is what gives GC its analytical usefulness.

Gas chromatography is in principle similar to column chromatography (as well as other forms of chromatography, such as HPLC, TLC), but has several notable differences. Firstly, the process of separating the compounds in a mixture is carried out between a liquid stationary phase and a gas mobile phase, whereas in column chromatography the stationary phase is a solid and the mobile phase is a liquid. (Hence the full name of the procedure is "Gas–liquid chromatography", referring to the mobile and stationary phases, respectively.) Secondly, the column through which the gas phase passes is located in an oven where the temperature of the gas can be controlled, whereas column chromatography (typically) has no such temperature control. Thirdly, the concentration of a compound in the gas phase is solely a function of the vapour pressure of the gas.

Gas chromatography is also similar to fractional distillation, since both processes separate the components of a mixture primarily based on boiling point (or vapour pressure) differences. However, fractional distillation is typically used to separate components of a mixture on a large scale, whereas GC can be used on a much smaller scale (*i.e.* micro scale).

Gas chromatography is also sometimes known as vapour-phase chromatography (VPC), or gas–liquid partition chromatography (GLPC). These alternative names, as well as their respective abbreviations, are frequently used in scientific literature. Strictly speaking, GLPC is the most correct terminology, and is thus preferred by many authors.

A gas chromatograph is a chemical analysis instrument for separating chemicals in a complex sample. A gas chromatograph uses a flow-through narrow tube through which different chemical constituents of a sample pass in a gas stream (carrier gas, *mobile phase*) at different rates depending on their various chemical and physical properties and their interaction with a specific column filling. As the chemicals exit the end of the column, they are detected and identified electronically. The function of the stationary phase in the column is to separate different components, causing each one to exit the column at a different time. Other parameters that can be used to alter the order or time of retention are the carrier gas flow rate, column length and the temperature.

A gas chromatograph with a headspace sampler

In a GC analysis, a known volume of gaseous or liquid analyte is injected into the "entrance" (head) of the column, usually using a micro syringe (or, solid phase micro extraction fibers, or a gas source switching system). As the carrier gas sweeps the analyte molecules through the column, this motion is inhibited by the adsorption of the analyte molecules either onto the column walls or onto packing materials in the column. The rate at which the molecules progress along the column depends on the strength of adsorption, which in turn depends on the type of molecule and on the stationary phase materials. Since each type of molecule has a different rate of progression, the various components of the analyte mixture are separated as they progress along the column and reach the end of the column at different times (retention time). A detector is used to monitor the outlet stream from the column; thus, the time at which each component reaches the outlet and the amount of that component can be determined. Generally, substances are identified (qualitatively) by the order in which they emerge (elute) from the column and by the retention time of the analyte in the column.

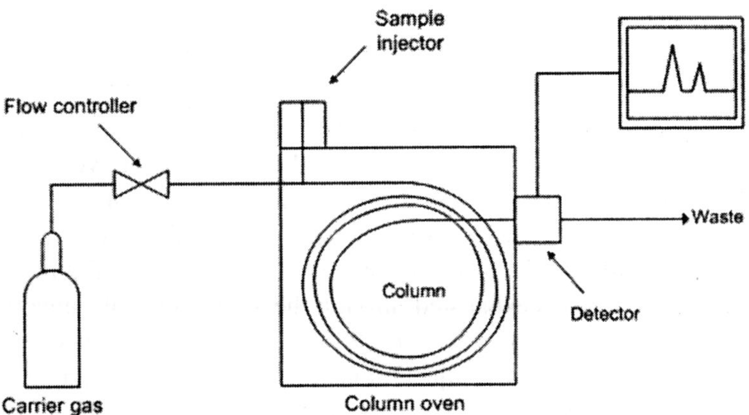

Flow diagram of gas chromatograph

Physical components

Auto samplers

The auto sampler provides the means to introduce a sample automatically into the inlets. Manual insertion of the sample is possible but is no longer common. Automatic insertion provides better reproducibility and time-optimization.

Different kinds of auto samplers exist. Auto samplers can be classified in relation to sample capacity (auto-injectors vs. auto samplers, where auto-injectors can work a small number of samples), to robotic technologies (XYZ robot vs. rotating robot – the most common), or to analysis:

- Liquid
- Static head-space by syringe technology
- Dynamic head-space by transfer-line technology
- Solid phase micro extraction (SPME)

Traditionally auto sampler manufacturers are different from GC manufacturers and currently no GC manufacturer offers a complete range of auto samplers.

Inlets

The column inlet (or injector) provides the means to introduce a sample into a continuous flow of carrier gas. The inlet is a piece of hardware attached to the column head.

Common inlet types are:

- S/SL (split/split less) injector; a sample is introduced into a heated small chamber via a syringe through a septum – the heat facilitates volatilization of the sample and sample matrix. The carrier gas then either sweeps the entirety (split less mode) or a portion (split mode) of the sample into the column. In split mode, a part of the sample/carrier gas mixture in the injection chamber is exhausted through the split vent. Split injection is preferred when working with samples with high analyte concentrations (>0.1%) whereas splitless injection is best suited for trace analysis with low amounts of analytes (<0.01%). In splitless mode, the split valve opens after a pre-set amount of time to purge heavier elements that would otherwise contaminate the system. This pre-set (split less) time should be optimized, the shorter time (*e.g.*, 0.2 min) ensures less tailing but loss in response, the longer time (2 min) increases tailing but also signal.
- On-column inlet; the sample is here introduced directly into the column in its entirety without heat, or at a temperature below the boiling point of the solvent. The low temperature condenses the sample into a narrow zone. The column and inlet can then be heated, releasing the sample into the gas phase. This ensures the lowest possible temperature for chromatography and keeps samples from decomposing above their boiling point.
- PTV injector; Temperature-programmed sample introduction was first described by Vogt in 1979. Originally Vogt developed the technique as a method for the introduction of large sample volumes (up to 250 μl) in capillary GC. Vogt introduced the sample into the liner at a controlled injection rate. The temperature of the liner was chosen slightly below the boiling point of the solvent. The low-boiling solvent was continuously evaporated and vented through the split line. Based on this technique, Poy developed the programmed temperature vaporising injector; PTV. By introducing the sample at a low initial liner temperature many of the disadvantages of the classic hot injection techniques could be circumvented.
- Gas source inlet or gas switching valve; gaseous samples in collection bottles are connected to what is most commonly a six-port switching

valve. The carrier gas flow is not interrupted while a sample can be expanded into a previously evacuated sample loop. Upon switching, the contents of the sample loop are inserted into the carrier gas stream.

- P/T (Purge-and-Trap) system; An inert gas is bubbled through an aqueous sample causing insoluble volatile chemicals to be purged from the matrix. The volatiles are 'trapped' on an absorbent column (known as a trap or concentrator) at ambient temperature. The trap is then heated and the volatiles are directed into the carrier gas stream. Samples requiring preconcentration or purification can be introduced via such a system, usually hooked up to the S/SL port.

The choice of carrier gas (mobile phase) is important. Hydrogen has a range of flow rates that are comparable to helium in efficiency. However, helium may be more efficient and provide the best separation if flow rates are optimized. Helium is non-flammable and works with a greater number of detectors and older instruments. Therefore, helium is the most common carrier gas used. However, the price of helium has gone up considerably over recent years, causing an increasing number of chromatographers to switch to hydrogen gas. Historical use, rather than rational consideration, may contribute to the continued preferential use of helium.

Detectors

The most commonly used detectors are the flame ionization detector (FID) and the thermal conductivity detector (TCD). Both are sensitive to a wide range of components, and both work over a wide range of concentrations. While TCDs are essentially universal and can be used to detect any component other than the carrier gas (as long as their thermal conductivities are different from that of the carrier gas, at detector temperature), FIDs are sensitive primarily to hydrocarbons, and are more sensitive to them than TCD. However, a FID cannot detect water. Both detectors are also quite robust. Since TCD is non-destructive, it can be operated in-series before a FID (destructive), thus providing complementary detection of the same analytes.

Other detectors are sensitive only to specific types of substances, or work well only in narrower ranges of concentrations. They include:

- Thermal Conductivity detector (TCD), this common detector relies on the thermal conductivity of matter passing around a tungsten-rhenium filament with a current travelling through it. In this set up helium or nitrogen serve as the carrier gas because of their relatively high thermal conductivity which keep the filament cool and maintain uniform resistivity and electrical efficiency of the filament. However, when analyte molecules elute from the column, mixed with carrier gas, the thermal conductivity decreases and this causes a detector response. The response is due to the decreased thermal conductivity causing an increase in filament temperature and resistivity resulting in fluctuations in voltage. Detector sensitivity is proportional to filament current while it's inversely proportional to the immediate environmental temperature of that detector as well as flow rate of the carrier gas.

- Flame Ionization detector (FID), in this common detector electrodes are placed adjacent to a flame fuelled by hydrogen / air near the exit of the column, and when carbon containing compounds exit the column they are pyrolyzed by the flame. This detector works only for organic / hydrocarbon containing compounds due to the ability of the carbons to form cations and electrons upon pyrolysis which generates a current between the electrodes. The increase in current is translated and appears as a peak in a chromatogram. FIDs have low detection limits (a few picograms per second, but they are unable to generate ions from carbonyl containing carbons. FID compatible carrier gasses include nitrogen, helium, and argon.
- Catalytic combustion detector (CCD), which measures combustible hydrocarbons and hydrogen.
- Discharge ionization detector (DID), which uses a high-voltage electric discharge to produce ions.
- Dry electrolytic conductivity detector (DELCD), which uses an air phase and high temperature (v. Coulsen) to measure chlorinated compounds.
- Electron capture detector (ECD), which uses a radioactive beta particles (electron) source to measure the degree of electron capture. ECD are used for the detection of molecules containing electronegative / withdrawing elements and functional groups like halogens, carbonyl, nitriles, nitro groups, and organometalics. In this type of detector either nitrogen or 5% methane in argon is used as the mobile phase carrier gas. The carrier gas passes between two electrodes placed at the end of the column, and adjacent to the anode (negative electrode) resides a radioactive foil such as 63_{Ni}. The radioactive foil emits a beta particle (electron) which collides with and ionizes the carrier gas to generate more ions resulting in a current. When analyte molecules with electronegative / withdrawing elements or functional groups electrons are captured which results in a decrease in current generating a detector response.

- Flame photometric detector (FPD), which uses a photomultiplier tube to detect spectral lines of the compounds as they are burned in a flame. Compounds eluting off the column are carried into a hydrogen fueled flame which excites specific elements in the molecules, and the excited elements (P, S, halogens, some metals) emit light of specific characteristic wavelengths. The emitted light is filtered and detected by a photomultiplier tube. In particular, phosphorus emission is around 510-536nm and sulfur emission at 394nm.
- Atomic Emission Detector (AED), a sample eluting from a column enters a chamber which is energized by microwaves that induce a plasma. The plasma causes the analyte sample to decompose and certain elements generate an atomic emission spectra. The atomic emission spectra is diffracted by a diffraction gradient and detected by a series of photomultiplier tubes.
- Hall electrolytic conductivity detector (ElCD)
- Helium ionization detector (HID)
- Nitrogen-phosphorus detector (NPD), a form of thermionic detector where nitrogen and phosphorus alter the work function on a specially coated bead and a resulting current is measured.
- Infrared detector (IRD)
- Mass spectrometer (MS) – also called (GC-MS) highly effective and sensitive, even in a small quantity of sample.
- Photon ionization detector (PID)
- Pulsed discharge ionization detector (PDD)
- Thermionic ionization detector (TID)

Some gas chromatographs are connected to a mass spectrometer which acts as the detector. The combination is known as GC-MS. Some GC-MS are connected to an NMR spectrometer which acts as a backup detector. This combination is known as GC-MS-NMR. Some GC-MS-NMR are connected to an infrared spectrophotometer which acts as a backup detector. This combination is known as GC-MS-NMR-IR. It must, however, be stressed this is very rare as most analyses needed can be concluded via purely GC-MS.

Methods

Two valves are used to switch the test gas into the sample loop. After filling the sample loop with test gas, the valves are switched again applying carrier gas pressure to the sample loop and forcing the sample through the Column for separation.

The method is the collection of conditions in which the GC operates for a given analysis. Method development is the process of determining what conditions are adequate and/or ideal for the analysis required.

Conditions which can be varied to accommodate a required analysis include inlet temperature, detector temperature, column temperature and temperature program, carrier gas and carrier gas flow rates, the column's stationary phase, diameter and length, inlet type and flow rates, sample size and injection technique. Depending on the detector(s) (see below) installed on the GC, there may be a number of detector conditions that can also be varied. Some GCs also include valves which can change the route of sample and carrier flow. The timing of the opening and closing of these valves can be important to method development.

Carrier gas selection and flow rates

Typical carrier gases include helium, nitrogen, argon, hydrogen and air. Which gas to use is usually determined by the detector being used, for example, a DID requires helium as the carrier gas. When analyzing gas samples, however, the carrier is sometimes selected based on the sample's matrix, for example, when analyzing a mixture in argon, an argon carrier is preferred, because the argon in the sample does not show up on the chromatogram. Safety and availability can also influence carrier selection, for example, hydrogen is flammable, and high-purity helium can be difficult to obtain in some areas of the world. As a result of helium becoming scarcer, hydrogen is often being substituted for helium as a carrier gas in several applications.

The purity of the carrier gas is also frequently determined by the detector, though the level of sensitivity needed can also play a significant role. Typically, purities of 99.995% or higher are used. The most common purity grades required by modern instruments for the majority of sensitivities are 5.0 grades, or 99.999% pure meaning that there is a total of 10ppm of impurities in the carrier gas that could affect the results. The highest purity grades in common use are 6.0 grades, but the need for detection at very low levels in some forensic and environmental applications has driven the need for carrier gases at 7.0 grade purity and these are now commercially available. Trade names for typical purities include "Zero Grade," "Ultra-High Purity (UHP) Grade," "4.5 Grade" and "5.0 Grade."

The carrier gas linear velocity affects the analysis in the same way that temperature does (see above). The higher the linear velocity the faster the analysis, but the lower the separation between analytes. Selecting the linear velocity is therefore the same compromise between the level of separation and length of analysis as selecting the column temperature. The linear velocity will be implemented by means of the carrier gas flow rate, with regards to the inner diameter of the column.

With GCs made before the 1990s, carrier flow rate was controlled indirectly by controlling the carrier inlet pressure, or "column head pressure." The actual flow rate was measured at the outlet of the column or the detector with an electronic flow meter, or a bubble flow meter, and could be an involved, time consuming, and frustrating process. The pressure setting was not able to be varied during the run, and thus the flow was essentially constant during the analysis. The relation between flow rate and inlet pressure is calculated with Poiseuille's equation for compressible fluids. Many modern GCs, however, electronically measure the flow rate, and electronically control the carrier gas pressure to set the flow rate. Consequently, carrier pressures and flow rates can be adjusted during the run, creating pressure/flow programs similar to temperature programs.

Stationary compound selection

The polarity of the solute is crucial for the choice of stationary compound, which in an optimal case would have a similar polarity as the solute. Common stationary phases in open tubular columns are cyanopropylphenyl dimethyl polysiloxane, carbowax polyethyleneglycol, biscyanopropyl cyanopropylphenyl polysiloxane and diphenyl dimethyl polysiloxane. For packed columns more options are available.

Inlet types and flow rates

The choice of inlet type and injection technique depends on if the sample is in liquid, gas, adsorbed, or solid form, and on whether a solvent matrix is present that has to be vaporized. Dissolved samples can be introduced directly onto the column via a COC injector, if the conditions are well known; if a solvent matrix has to be vaporized and partially removed, a S/SL injector is used (most common injection technique); gaseous samples (*e.g.*, air cylinders) are usually injected using a gas switching valve system; adsorbed samples (*e.g.*, on adsorbent tubes) are introduced using either an external (on-line or off-line) desorption apparatus such as a purge-and-trap system, or are desorbed in the injector (SPME applications).

Sample size and injection technique

The real chromatographic analysis starts with the introduction of the sample onto the column. The development of capillary gas chromatography resulted in many practical problems with the injection technique. The technique of on-column injection, often used with packed columns, is usually not possible with capillary columns. The injection system in the capillary gas chromatograph should fulfil the following two requirements:

1. The amount injected should not overload the column.
2. The width of the injected plug should be small compared to the spreading due to the chromatographic process. Failure to comply with this requirement will reduce the separation capability of the column. As a general rule, the volume injected, V_{inj}, and the volume of the detector cell, V_{det}, should be about 1/10 of the volume occupied by the portion of sample containing the molecules of interest (analytes) when they exit the column.

Some general requirements which a good injection technique should fulfill are:

- It should be possible to obtain the column's optimum separation efficiency.
- It should allow accurate and reproducible injections of small amounts of representative samples.
- It should induce no change in sample composition. It should not exhibit discrimination based on differences in boiling point, polarity, concentration or thermal/catalytic stability.
- It should be applicable for trace analysis as well as for undiluted samples.

However, there are a number of problems inherent in the use of syringes for injection, even when they are not damaged:

- Even the best syringes claim an accuracy of only 3%, and in unskilled hands, errors are much larger
- The needle may cut small pieces of rubber from the septum as it injects sample through it. These can block the needle and prevent the syringe filling the next time it is used. It may not be obvious of what happened.
- A fraction of the sample may get trapped in the rubber, to be released during subsequent injections. This can give rise to ghost peaks in the chromatogram.

- There may be selective loss of the more volatile components of the sample by evaporation from the tip of the needle.

Column selection

The choice of column depends on the sample and the active measured. The main chemical attribute regarded when choosing a column is the polarity of the mixture, but functional groups can play a large part in column selection. The polarity of the sample must closely match the polarity of the column stationary phase to increase resolution and separation while reducing run time. The separation and run time also depends on the film thickness (of the stationary phase), the column diameter and the column length.

Column temperature and temperature program

The column(s) in a GC are contained in an oven, the temperature of which is precisely controlled electronically. (When discussing the "temperature of the column," an analyst is technically referring to the temperature of the column oven. The distinction, however, is not important and will not subsequently be made in this article).

The rate at which a sample passes through the column is directly proportional to the temperature of the column. The higher the column temperature, the faster the sample moves through the column. However, the faster a sample moves through the column, the less it interacts with the stationary phase, and the less the analytes are separated. In general, the column temperature is selected to compromise between the length of the analysis and the level of separation.

A method which holds the column at the same temperature for the entire analysis is called "isothermal." Most methods, however, increase the column temperature during the analysis, the initial temperature, rate of temperature increase (the temperature "ramp"), and final temperature are called the "temperature program." A temperature program allows analytes that elute early in the analysis to separate adequately, while shortening the time it takes for late-eluting analytes to pass through the column.

Qualitative analysis

Generally chromatographic data is presented as a graph of detector response (y-axis) against retention time (x-axis), which is called a chromatogram. This provides a spectrum of peaks for a sample representing the analytes present in a sample eluting from the column at different times. Retention time can be used to identify analytes if the method conditions are constant. Also, the pattern of peaks will be constant for a sample under constant conditions and can identify complex mixtures of analytes. In most modern applications however the GC is connected to a mass spectrometer or similar detector that is capable of identifying the analytes represented by the peaks.

Quantitative analysis

The area under a peak is proportional to the amount of analyte present in the chromatogram. By calculating the area of the peak using the mathematical function of integration, the concentration of an analyte in the original sample can be determined. Concentration can be calculated using a calibration curve created by finding the response for a series of concentrations of analyte, or by determining the relative response factor of an analyte. The relative response factor is the expected ratio of an analyte to an internal standard (or external standard) and is calculated by finding the response of a known amount of analyte and a constant amount of internal standard (a chemical added to the sample at a constant concentration, with a distinct retention time to the analyte).In most modern GC-MS systems, computer software is used to draw and integrate peaks, and match MS spectra to library spectra.

In general, substances that vaporize below ca. 300° C (and therefore are stable up to that temperature) can be measured quantitatively. The samples are also required to be salt-free; they should not contain ions. Very minute amounts of a substance can be measured, but it is often required that the sample must be measured in comparison to a sample containing the pure, suspected substance known as a reference standard. Various temperature programs can be used to make the readings more meaningful; for example to differentiate between substances that behave similarly during the GC process. Professionals working with GC analyze the content of a chemical product, for example in assuring the quality of products in the chemical industry; or measuring toxic substances in soil, air or water. GC is very accurate if used properly and can measure picomoles of a substance in a 1 ml liquid sample, or parts-per-billion concentrations in gaseous samples.

The GC analyse hydrocarbons (C2-C40+). In a typical experiment, a packed column is used to separate the light gases, which are then detected with a TCD. The hydrocarbons are separated using a capillary column and detected with a FID. A complication with light gas analyses that include H_2 is that He, which is the most common and most sensitive inert carrier (sensitivity is proportional to molecular mass) has an almost identical thermal conductivity to hydrogen (it is the difference in thermal conductivity between two separate filaments in a Wheatstone Bridge type arrangement that shows when a component has been eluted). For this reason, dual TCD instruments used with a separate channel for hydrogen that uses nitrogen as a carrier are common.

Argon is often used when analysing gas phase chemistry reactions such as F-T synthesis so that a single carrier gas can be used rather than two separate ones. The sensitivity is less, but this is a trade off for simplicity in the gas supply.

2.11. Atomic Absorption Spectrophotometer

Atomic absorption spectroscopy (AAS) is a spectroanalytical procedure for the quantitative determination of chemical elements employing the absorption of optical radiation (light) by free atoms in the gaseous state. In analytical chemistry the technique is used for determining the concentration of a particular element (the analyte) in a sample to be analyzed. AAS can be used to determine over 70 different elements in solution or directly in solid samples employed in physiological, biochemical, pharmacology, biophysics and toxicology research.

Atomic absorption spectrometry has many uses in different areas of chemistry such as (1) Clinical analysis: Analyzing metals in biological fluids and tissues such as whole blood, plasma, urine, saliva, brain tissue, liver, muscle tissue, semen, (2) Pharmaceuticals: In some pharma-ceutical manufacturing processes, minute quantities of a catalyst that remain in the final drug product, (3) Water analysis: Analyzing water for its metal content.

The technique makes use of absorption spectrometry to assess the concentration of an analyte in a sample. It requires standards with known analyte content to establish the relation between the measured absorbance and the analyte concentration and relies therefore on the Beer-Lambert law.

In short, the electrons of the atoms in the atomizer can be promoted to higher orbitals (excited state) for a short period of time (nanoseconds) by absorbing a defined quantity of energy (radiation of a given wavelength). This amount of energy, *i.e.*, wavelength, is specific to a particular electron transition in a particular element. In general, each wavelength corresponds to only one element, and the width of an absorption line is only of the order of a few picometers (pm), which gives the technique its elemental selectivity. The radiation flux without a sample and with a sample in the atomizer is measured using a detector, and the ratio between the two values (the absorbance) is converted to analyte concentration or mass using the Beer-Lambert Law.

Atomic absorption spectrophotometer

Instrumentation

In order to analyze a sample for its atomic constituents, it has to be atomized. The atomizers most commonly used nowadays are flames and electrothermal (graphaite tube) atomizers. The atoms should then be irradiated by optical radiation, and the radiation source could be an element-specific line radiation source or a continuum radiation source. The radiation then passes through a monochromator in order to separate the element-specific radiation from any other radiation emitted by the radiation source, which is finally measured by a detecto.

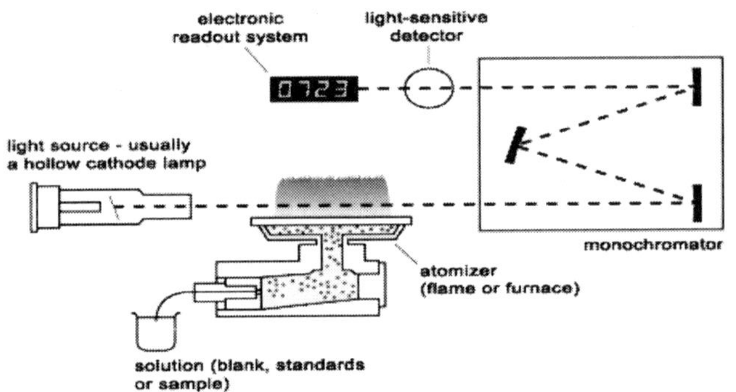

Atomic absorption spectrometer block diagram

Atomizers

The atomizers most commonly used nowadays are (spectroscopic) flames and electrothermal (graphite tube) atomizers. Other atomizers, such as glow-discharge atomization, hydride atomization, or cold-vapor atomization might be used for special purposes.

Flame atomizers

The oldest and most commonly used atomizers in AAS are flames, principally the air-acetylene flame with a temperature of about 2300 °C and the nitrous oxide system (N_2O)-acetylene flame with a temperature of about 2700 °C. The latter flame, in addition, offers a more reducing environment, being ideally suited for analytes with high affinity to oxygen.

Liquid or dissolved samples are typically used with flame atomizers. The sample solution is aspirated by a pneumatic analytical nebulizer, transformed into an aerosol, which is introduced into a spray chamber, where it is mixed with the flame gases and conditioned in a way that only the finest aerosol droplets (< 10 μm) enter the flame. This conditioning process is responsible that only about 5% of the aspirated sample solution reaches the flame, but it also guarantees a relatively high freedom from interference.

On top of the spray chamber is a burner head that produces a flame that is laterally long (usually 5–10 cm) and only a few mm deep. The radiation beam passes through this flame at its longest axis, and the flame gas flow-rates may be adjusted to produce the highest concentration of free atoms. The burner height may also be adjusted, so that the radiation beam passes through the zone of highest atom cloud density in the flame, resulting in the highest sensitivity.

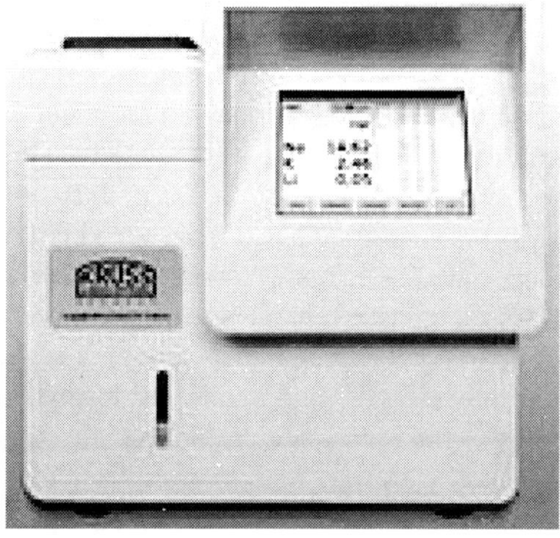

A laboratory flame photometer that uses a propane operated flame atomizer

The processes in a flame include the following stages:

- Desolvation (drying) – the solvent is evaporated and the dry sample nano-particles remain;
- Vaporization (transfer to the gaseous phase) – the solid particles are converted into gaseous molecules;
- Atomization – the molecules are dissociated into free atoms;
- Ionization – depending on the ionization potential of the analyte atoms and the energy available in a particular flame, atoms might be in part converted to gaseous ions.

Each of these stages includes the risk of interference in case the degree of phase transfer is different for the analyte in the calibration standard and in the sample. Ionization is generally undesirable, as it reduces the number of atoms that are available for measurement, *i.e.*, the sensitivity.

In flame AAS a steady-state signal is generated during the time period when the sample is aspirated. This technique is typically used for determinations in the mg L^{-1} range, and may be extended down to a few µg L^{-1} for some elements.

2.12. High Performance (high pressure) Liquid Chromatography (HPLC)

In gel filtration and ion exchange chromatography, the efficiency of chromatography increases with reducing the particle size of the gel matrix and with enhancing its size homogeneity. The efficiency reaches a new level of quality when a matrix grain size of 3-10 μm is applied. Liquid chromatography performed using such resins is called high performance liquid chromatography, abbreviated as HPLC. However, at such particle sizes, the sufficient flow of the mobile phase (eluent) can be achieved only by applying a high pressure of around 10 MPa by using special precision pumps. HPLC thus also stands for high pressure LC and according to many researchers; the abbreviation also refers to the high price of the specialised equipment.

Due to the application of high pressure, the primary requirement regarding HPLC columns is that they should be incompressible. Silica is predominantly used for this purpose. Under appropriate conditions, silica can be used to create homogeneous column media of sufficient strength and with a well-controlled particle size. For the hydrophilic silica stationary phase, only hydrophobic mobile phases can be applied. Therefore, HPLC was primarily suitable for the separation of hydrophobic organic solvent-soluble materials. Later, the chemical modification of the silica surface made possible the creation of hydrophobic silica gels. In this case, the hydrophilic-hydrophobic relation of the stationary and mobile phases became reversed, hence the term reverse-phase chromatography (RPC). Reverse-phase chromatography opened up the possibility of the separation of water-soluble substances, including the majority of molecules of biological origin. Large pore-size gels also allowed the separation of macromolecules. The hydrophobic surface is formed by long alkyl chains linked to the silica. These include octadecyl, octyl, butyl (labeled as C18, C8, C4) and also phenyl groups. Furthermore, gels containing charged groups can be used for ion exchange.

If the composition of the mobile phase is constant during chromatography, we speak of isocratic elution. Gradient elution is achieved via applying a linear or non-linear concentration gradient. In many cases, isocratic and gradient sections are combined in the elution profile. The gradient is most often created by using microprocessor-controlled, variable-speed pumps (two at least). In this case, the mobile phase components are mixed at the high-pressure side of the pumps. With the help of precision valves, the gradient can also be made by mixing the buffers at low pressure. This way, the eluent can be transmitted onto the column by using only one pump. If unmodified silica is used, the mobile phase can be created by mixing organic solvents of different hydrophobicity. In reverse-phase chromatography, a dilute aqueous solution (*e.g.* 0.1 % trifluoroacetic acid, few mM phosphate buffer) and a water-miscible organic solvent (*e.g.* acetonitrile, methanol, propanol) can be mixed to prepare a mobile phase for isocratic or gradient applications. In the biochemical practice, light absorption and fluorescence detectors are most commonly used. However, refractive index, conductivity, optical rotation and electrochemical detectors are also frequently applied. For photometric detection, the eluent must be optically clear and must not absorb light in the applied wavelength range. The use of extremely pure solvents is generally required anyway in order to avoid both the contamination of the column media and the appearance of unexpected materials during chromatography. A very important development is the appearance of mass spectrometric (MS) detectors. These are used to determine the mass of the separated components, which is a decisive parameter of a given substance. Actually, MS detectors do not determine the mass directly, but determine the mass/charge (m/z) ratio. However, the mass can be easily derived for single- and multiple-charged values. In the past two decades, ionisation techniques have been introduced in MS measurement systems. Ionisation of high molecular weight biopolymers (such as proteins) can be efficiently achieved via ESI (electrospray ionisation) or MALDI (matrix-assisted laser desorption).

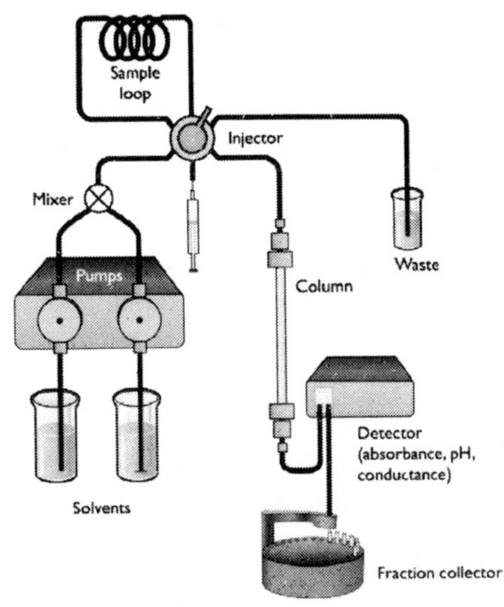

Schematic layout diagram of the HPLC equipment. The system consists of pumps that ensure the high-pressure delivery of a two-component mobile phase, a mixing unit, a sample injector, a chromatographic column and a detector.

As a result, chromatographic methods play an increasingly important role in protein research. One- or multi-dimensional high performance liquid chromatography combined with in-line mass spectrometry allows the targeted identification of all proteins expressed in a cell at a given time, *i.e.* the proteome; or a specific subset of these proteins. This is significant because, unlike the genome, the proteome is not constant: it may vary by tissue or cell type and may also depend on the physiological state or developmental stage of the individual. The detection of post-translational modifications is also of importance. For instance, the specific detection of phosphorylated proteins in the proteome aids the understanding of various biological regulatory processes. All components of the HPLC equipment must be pressure-resistant and chemically resistant. Thus, stainless steel and, in more special cases, resistant titanium alloys are applied. Moving parts, including pistons and valves, are also made of highly mechanoresistant materials (special ceramic, glass, industrial ruby *etc.*). Recently, the investigation of samples sensitive to trace metal contamination (*e.g.* some enzymes) has necessitated the application of particularly pressure-resistant plastic parts. The FPLC (fast protein liquid chromatography) system was developed for the separation of native proteins. FPLC differs from the above-described HPLC chromatography in that the column resins are specially-treated dextran-based or synthetic polymeric materials (Superdex, Superose, Sephacryl *etc.*), which are, due to their hydrophilic character and high porosity, particularly useful for the separation of biopolymers. These particles have a slightly larger size and lower pressure resistance than the silica-based HPLC media. However, FPLC media are suitable for ensuring sufficient fluid flow at pressures in the range of 0.5-1 MPa, and the efficiency of FPLC also meets most requirements of protein purification applications. FPLC columns are made of pressure-resistant borosilicate glass. Exposed metal parts are also avoided in the pumps and the piping systems. Both HPLC and FPLC provide better efficiency and sensitivity as well as lower time requirement compared to conventional chromatographic applications.

During sample preparation, one must take it seriously that the solution should be clear and free of dust or other particles. Otherwise, the apparatus may become blocked and the chromatographic column may become contaminated. The sample can be centrifuged at 20-40000 g and/or filtered through a 0.45-µm filter. The sample is preferably dissolved in the starting mobile phase. In must be ensured that the sample is fully dissolved. The volume of the sample depends on the diameter of the column used (see below).

1) Selection of the stationary phase (column medium)

In the case of nonpolar, water-insoluble materials, unmodified silica, or possibly diol media should be used. In the case of amino group-containing polar, water-soluble substances, reverse-phase C18, C8 or C4-modified silica media can be applied. Larger hydrophobic peptides and proteins bind too strongly to the C18 solid phase. In this case it is advisable to use C4 or C8 matrices.

For analytical purposes, smaller particle size matrices (around or less than 3 µm) should be used. This will increase the efficiency of separation, but has the disadvantage of increasing the pressure in the system. For semi-preparative and preparative purposes, 5-10-µm particle sizes are suitable. One must also consider the porosity of matrix particles. In the case of low molecular weight metabolites, amino acids and small peptides, the commonly-used 100-Å pore-size media are suitable. In the case of macromolecules, high porosity (wide pore), 300-Å pore-size media are to be chosen (1 Å = 0.1 nm).

2) Selection of the mobile phase

The most important characteristics of the mobile phase include purity, viscosity, UV transparency and miscibility with other solvents. HPLC techniques require special-purity ("HPLC grade") solvents, including water specially purified for this purpose. The selection of the correct column size is important in terms of the economic use of materials. High-viscosity solvents should be generally avoided as their use increases the system pressure. In terms of detectability, it is important that the mobile phase should be optically pure. Given that the most commonly used chromatographic detectors operate in the UV range, the UV absorption of the mobile phase should be considered. The most commonly used HPLC reagents (*e.g.* acetonitrile) are available in various qualities. Highest-quality reagents enable photometric detection even at a wavelength of 200 nm. In the range of 280-340 nm, it is sufficient to use less expensive grades of acetonitrile. In reverse-phase chromatogramphy, the initial mobile phase is a dilute aqueous solution. The organic component used for the reduction of solvent polarity can be *e.g.* methanol, ethanol, propanol or acetonitrile. The initial aqueous solution can be, for instance, 0.1 % formic acid, acetic acid or few mM phosphate buffer. For ion pair formation, the commonly used agent is 0.1 % tri-fluoroacetic acid. Ion pair formation enhances the retention of highly charged molecules due to charge compensation. However, if chromatography is coupled to online MS measurements, trifluoroacetic acid should be avoided as it reduces the ionisability of molecules. In this case, the use of dilute formic acid or ammonium formate is recommended. It must be taken into consideration that the mixing of the solvents during gradient elution will result in changes in the solubility of air in the mobile phase. This effect may result in air bubble formation when the solution leaves the column and the pressure is reduced. This will severely interfere with photometric detection. Mobile phase components must therefore be degassed prior to usage. Some chromatographic instruments contain a so-called degasser unit, which applies a slight vacuum to keep the concentration of dissolved air continuously low. If no degasser unit is attached to the chromatographic instrument, mobile phase components should either be degassed by vacuum, or the very poorly water-soluble helium gas should be bubbled through the solutions to expel the dissolved air. Before use, filtering of the mobile phase through a fine (0.45-µm) filter is recommended in order to get rid of fine dust contamination.

3) Selection of column size

With regard to column size selection, it is crucially important whether the chromatographic column will be used for analytical or preparative purposes. For analytical purposes, microbore or minibore columns with an internal diameter of 1-3 mm should be used. In the case of microbore/minibore columns, the applicable sample volume is 5-25 μL, which may contain 0.01-0.1 mg of material. In such applications, the amount of solvents used can be reduced significantly, which is advantageous for both financial and environmental reasons. The use of such thin columns is also recommended when only a small amount of sample is available. If a small sample is applied to a large column, the sample may disappear . The most commonly used type of column is the standard column with a 4.6-mm diameter. These columns can be used for both analytical and semi-preparative purposes. The volume of standard columns is 10-50 μL in the case of analytical uses; whereas for semi-preparative work, column volumes up to 1000 μL can be used. The amount of material that can be applied is in the range of 0.1-2 mg. Columns with a diameter of 10-20 mm can be used for preparative purposes. In such columns, the amount of material that can be applied may be 20-200 mg, and the sample volume can reach 5-10 ml. Columns are expensive. It is advisable to insert a protective, few-millimetre-long "pre-column" before the main column, with the two columns having identical column media. The pre-column can be replaced at low cost if blockage occurs.

2.13. Electrophoresis

Electrophoresis is a method used to separate charged particles from one another based on differences in their migration speed. In the course of electrophoresis, two electrodes (typically made of an inert metal, *e.g.* platinum) are immersed in two separate buffer chambers. The two chambers are not fully isolated from each other. Charged particles can migrate from one chamber to the other. By using an electric power supply, electric potential (E) is generated between the two electrodes. Due to the electric potential, electrons move by a wire between the two electrodes. More specifically, electrons move from the anode to the cathode. Hence, the anode will be positively charged, while the cathode will be negatively charged. Electrons driven to the cathode will leave the electrode and participate in a reduction reaction with water generating hydrogen gas and hydroxide ions. In the meantime, at the positive anode an oxidation reaction occurs. Electrons released from water molecules enter the electrode generating oxygen gas and free protons (which immediately form hydroxonium ions with water molecules). The amount of electrons leaving the cathode equals the amount of electrons entering the cathode. As mentioned, the two buffer chambers are interconnected such that charged particles can migrate between the two chambers. These particles are driven by the electric potential between the two electrodes. Negatively charged ions, called anions, move towards the positively charged anode, while positively charged ions, called cations, move towards the positively charged cathode.

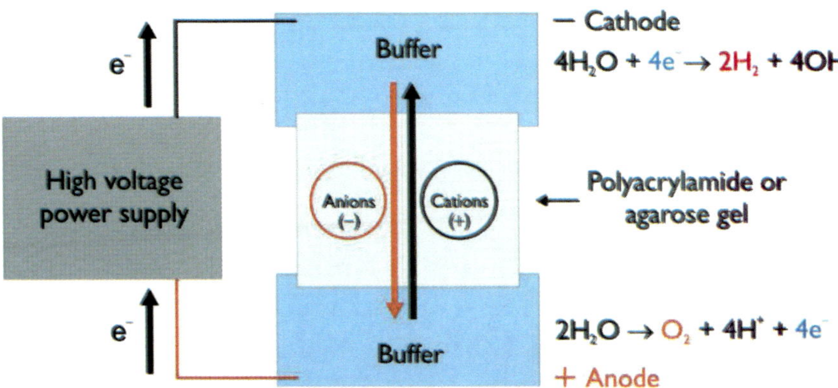

The principle of electrophoresis. In the course of electrophoresis, two electrodes are immersed in two separate buffer chambers. The two chambers are connected such that charged particles can migrate from one chamber to the other. By using a power supply, electric potential difference is generated between the two electrodes. As a result, electrons flow from one of the electrodes, the anode, towards the other electrode, the cathode. Electrons from the cathode are taken up by water molecules of the buffer, resulting in a chemical reaction which generates hydrogen gas and hydroxide ions. In the other buffer chamber, water molecules transfer electrons to the anode an in another chemical reaction that generates oxygen gas and protons. (Protons are immediately taken up by water molecules to form hydroxonium ions.) As charged particles can migrate between the two chambers due to the electric potential difference, positive ions (cations) move towards the negatively charged cathode while negatively charged ions (anions) move towards the positively charged anode.

Different ions migrate at different speeds dictated by their sizes and by the number of charges they carry. As a result, different ions can be separated from each other by electrophoresis. It is very important to understand the basic physics describing the dependence of the speed of the ion as a function of the number of charges on the ion, the size of the ion, the magnitude of the applied electric field and the nature of the medium in which the ions migrate. By understanding these basic relationships, the principles of the many different specific electrophoresis methods become comprehensible. The fundamental principle of electrophoresis is illustrated above figure. The mathematical description of the force during electrophoresis is simple. An electric force is exerted on the charged particle. The magnitude of the electric force equals the product of the charge of the particle and the electric field generated between the two electrodes proportional to the velocity of the particle. At the typically very low speed of particle migration during electrophoresis, the force is a linear function of the velocity of the particle.

The ratio of the force and the velocity is defined as the frictional coefficient. The value of this frictional coefficient is a function of the size and shape of the particle and the viscosity of the medium. The larger the particle and the more obstructing the medium, the higher the value of frictional coefficient. When electrophoresis is started, particles accelerate instantaneously to a velocity at which the magnitude of the drag force equals the magnitude of the (opposite) accelerating electric force. Once the magnitude of the two opposing forces becomes equal, the resultant force becomes zero. Therefore, each particle will move at a constant velocity characteristic of the given particle at the given accelerating potential and medium. A similar phenomenon is also there for centrifugation. There, the accelerating force is unrelated—being proportional to the mass instead of the charge of the particle—but the frictional force and the phenomenon of two opposing forces leading to a characteristic particle velocity is analogous. A useful parameter, the electrophoretic mobility of the particle, defines the velocity of the particle in a given medium when one unit of electric field is applied. (This parameter is analogous to the Svedberg units defined for centrifugation). Electrophoretic mobility is a linear function of the charge of the particle and it is a reciprocal function of the frictional coefficient, which depends on both the size of the particle and the nature of the medium, particles having different electrophoretic mobility, *i.e.* those that migrate at different speeds in the same medium and electric field, can be separated by electrophoresis. In biochemical and molecular biological studies, the most typical charged molecules that are analysed and separated by electrophoresis are proteins and nucleic acids. Electrophoresis is always performed by using a special medium, most often a gel. The corresponding methods are therefore denoted as gel electrophoresis.

About gel electrophoresis

The principle of electrophoresis does not assume any particular requirements about the nature of the liquid medium in which the ions are separated. Yet, in the great majority of currently used electrophoretic applications, the medium has a three-dimensional network structure, *i.e.* the medium is a gel. At the very beginning when the technique was invented, electrophoresis was performed without using a gel matrix. Charged particles were migrated in a homogeneous liquid phase. However, it soon became apparent that the use of a liquid medium raises at least three major difficulties. One is that the separation of different ions in an ordinary liquid is rather inefficient. It is so because a significant factor of an effective separation should be a marked size-dependent drag force exerted by the medium on the particles. Although even ordinary liquids do interfere with the migration speed of the particles in a size-dependent manner, this size dependence is quite moderate. The other big problem has a simple technical origin. In liquid phase, even very small levels of temperature inhomogeneity trigger convection that significantly compromises the resolution of the separation. Finally, in an ordinary liquid phase, the extent of diffusion is high and, in the typical timeframe of the generally slow electrophoresis experiments, diffusion decreases the efficiency of the separation. All three problems had been dealt with when, instead of ordinary liquids, gels were introduced as a medium.

The gel provides a three-dimensional molecular network structure to the liquid medium. It prevents convections and also lowers the rate of diffusion. Moreover, perhaps the most dramatic advantageous effect of the gel is that it acts as a molecular sieve: it interferes only slightly with the movement of small molecules, but drastically slows down the motion of large molecules. All gels are characterised by an average pore size. Molecules much smaller than the mean pore diameter are almost unaffected by the presence of the gel, while those that are larger than the pores practically do not migrate in the gel. When ions with sizes in the range of the pore size are migrated through the gel by electrophoresis, the gel exerts a pronounced size-dependent dragging force on them. As a consequence, the pore size distribution of the gel determines an operational size range in which different ions can be separated. Looking at this from the opposite point of view, each separation problem defines an optimal pore size to be applied. The gel has to fulfil several general criteria to be applicable for biochemical electrophoresis. It needs to be hydrophilic, chemically stable (should not participate in chemical reactions during electrophoresis), neutral (free of electric charges, otherwise it would act as an ion exchanger) and mechanically resistant (should not be too elastic or too rigid as such gels would be difficult to handle). Furthermore, as the separated ions (mostly proteins and nucleic acids) need to be visualised in the gel by some kind of staining procedure, the gel should be transparent, and should not strongly bind the dyes used for staining. Finally, and very importantly, the experimenter should be able to adjust the pore size during the preparation of the gel.

The size range of molecules (ions) studied in molecular biology is extremely broad. No single gel-forming compound is known that could cover the entire corresponding range of applicable pore sizes. Two compounds are dominantly used for gel electrophoresis: polyacrylamide and agarose. Polyacrylamide gels typically provide much smaller pores than do agarose gels. The polyacrylamide gel is formed by the radical polymerisation of acrylamide monomers. This process alone would lead to very long polymer chains instead of a three-dimensional gel. The three-dimensional network is brought about by the incorporation of *N, N'*-methylene-bis-acrylamide into the polymerising chains, which results in crosslinks between the long chains. The polyacrylamide gel is held together by covalent bonds. The pore size of polyacrylamide gels can be adjusted via the concentration of the acrylamide monomer and the ratio of the cross linking agent, *N,N'*-methylenebisacrylamide. The pore size of polyacrylamide gels corresponds to a relatively low value (compared to that of agarose gels). Polyacrylamide gels are used typically for the electrophoresis of proteins and relatively small nucleic acids. In comparison, the agarose gel is formed via non-covalent interactions between long polysaccharide chains. The pore size of agarose gels is much larger than that of acrylamide gels. Accordingly, agarose gels are used typically for the electrophoresis of large nucleic acids. The pore size of the agarose gel can be controlled via the concentration of the agarose solution. As the interaction between agarose molecules is non-covalent, the gel is formed by a physical (in contrast to a chemical) process. A suspension of agarose is heated up until the system reaches a sol state and then it is left to cool down to room temperature to reach the gel state. The following sections review the various polyacrylamide- and agarose-based electrophoresis methods.

Two-dimensional (2D) electrophoresis

The various separation methods are all aimed at separating complex systems to individual components. Separation is always based on at least one physicochemical property that shows diversity among the components. The general problem encountered in the case of complex mixtures is that not all components differ significantly from all other components when only one property is considered. Accordingly, separation based on a single property rarely results in single-component fractions. Some components will be efficiently separated from all others, while some other components will remain in the mixture. The remaining mixtures can be further fractionated by another separation technique that relies on a different physicochemical property. The most effective separation can be achieved if the combined consecutive separation steps rely on absolutely independent physicochemical properties. A good example of this is the very high-resolution two-dimensional polyacrylamide gel electrophoresis (2D-PAGE) that combines two already discussed electrophoresis methods, isoelectric focusing and SDS-PAGE.

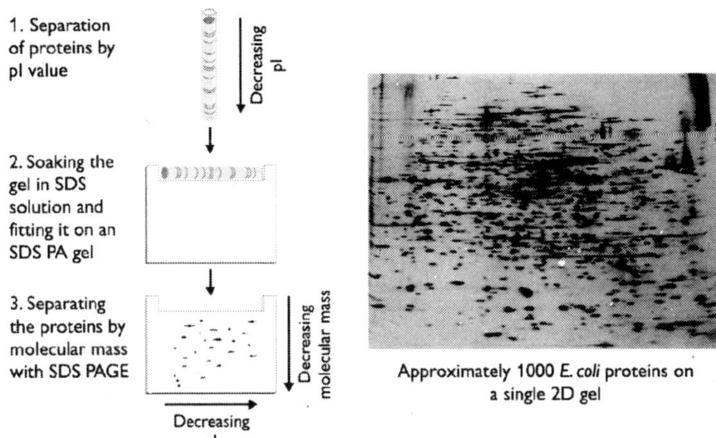

Two-dimensional (2D) electrophoresis. 2D electrophoresis is the combination of isoelectric focusing and SDS-PAGE. Proteins are first separated based on their pI values and then based on their molecular mass. As these properties are completely independent, the combination of the two separation methods provides much higher resolution than either of the two methods alone.

As the first step of 2D gel electrophoresis, isoelectric focusing is performed to separate proteins based on their pI values. Only a single sample is loaded on a gel strip in this step. The sample is separated in one dimension both in a primary and in a figurative sense. In a primary sense because the components are separated along a single line, and in a figurative sense as the separation is based on a single well-defined property, the pI value. After the first separation step has been completed in the first dimension, the gel strip is soaked in an SDS solution and is fitted tightly to one side of a classical SDS polyacrylamide gel. The second separation step is traditional SDS-PAGE, which separates proteins based on their molecular mass. This second step represents a second dimension in both a primary and a figurative sense. The second separation is performed in a second dimension in a direction rectangular to that of the first separation, and the property utilised in the second step (molecular mass) is completely independent of the one utilised in the first step (pI).

If, after the first step, some gel regions contain different proteins that coincidentally have identical pI values, these proteins will be separated from each other in the second step if their molecular mass is different. Note that every aspect discussed for SDS-PAGE also applies to the second separation step of 2D-PAGE. Van der Waals interactions that might have held protein subunits together in the course of isoelectric focusing will break and individual subunits will become separated. If disulfide bridges need to be opened up, some kind of reducing agent needs to be added. Accordingly, in the second separation step, single polypeptide chains will migrate in the gel. If isoelectric focusing collects a multimeric protein at a certain gel location, the second electrophoresis step will dissect it into individual chains. If the multimer contains subunits of different sizes, these subunits will be separated from each other in the second separation step.

Agarose gel electrophoresis

The pore size of the gel defines the operational molecular mass range in which effective separation can be achieved. The size of the DNA molecules most frequently handled in a typical recombinant DNA experiment is in the range of thousands of base pairs. Such very large molecules cannot be separated in polyacrylamide gels, as even the largest pore sizes achievable in such gels are too small for these macromolecules. Naturally, a different type of gel matrix is needed. For this purpose, agarose has become the most popular matrix. The agarose gel forms through non-covalent interactions between polysaccharide molecules. When such a gel is heated, it undergoes a phase transition from gel to sol state. The agarose powder is mixed with running buffer and the slurry is heated up to reach the sol state. The liquid is then poured in an appropriate, usually horizontally-mounted, template to solidify. The pore size is set by the concentration of agarose monomers. The agarose gel meets all of the important requirements that make a good electrophoresis matrix: it is hydrophilic, does not carry charges, chemically inert and does not absorb the dyes used for nucleic acid staining.

Staining methods

Proteins and nucleic acids absorb light only in the ultraviolet region of the spectrum and are therefore invisible for the naked eye. UV spectroscopy in the gel is not straightforward due to the high background absorbance of the gel. Therefore, following electrophoresis, proteins or nucleic acids are visualised in the gel by using various staining methods.

General protein gel stains

Several dyes have been introduced for protein gel staining. These compounds bind tightly to proteins and absorb light in the visible region of the spectrum. There is no single absolutely best compound. The choice of the dye depends mostly on the quality of the protein to be detected. The following dyes are listed in the order of increasing sensitivity: Coomassie Brilliant Blue R 250, Acidic Fast-Green, Amido black, and silver nitrate. Coomassie Brilliant Blue is the most frequently used protein gel staining dye. Depending on the thickness of the gel and the properties of the protein, as low as 0.1 microgram protein can be detected using this dye.

2.14. Polyacrylamide Gel Electrophoresis (PAGE)

Polyacrylamide gels can be used for the separation and analysis of proteins and relatively small nucleic acid molecules. For example, when it was first invented, Sanger's DNA sequencing method applied PAGE to separate linear single-stranded DNA molecules based on their length. The resolution of the PAGE method is so high that, in the size range of about 10-1000 nucleotide units, it is capable of separating DNA molecules that differ in length only by a single monomer unit. In the case of single-stranded DNA, individual molecules are separated solely based on their length. This is due to the fact that, in the case of DNA (or RNA), the number of negative charges is a simple linear function of the number of monomer units (*i.e.* the length of the molecule). In other words, the specific charge (number of charges per particle mass) is invariant, *i.e.* it is the same for all DNA molecules. It is so because each monomer unit has one phosphate moiety that carries the negative charge. When an appropriate denaturing agent, such as urea, is added to the DNA sample and the gel is heated, the shape of the varying-length linear DNA molecules becomes identical. As a consequence, denatured molecules will be separated exclusively based on their size. There are several PAGE methods (SDS-PAGE, isoelectric focusing, 2D PAGE) that can be applied mostly for the separation of proteins based on distinct molecular properties.

At a given pH, different proteins carry different amounts of electric charge. Moreover, different proteins have different shapes and sizes, too. Consequently, during electrophoresis, proteins are separated by a complex combination of their charge, shape and size. PAGE separation of proteins provides high resolution. However, as three independent molecular properties simultaneously influence electrophoretic mobility, it will provide limited room for precise interpretation. For example, when two proteins are compared, it remains hidden what makes one of them migrate faster: a larger number of electric charges, a smaller size, or a more spherical shape. Nevertheless, even the simplest PAGE method, which will be referred to as native PAGE, provides many particular advantages. In order to increase the analytical applicability of the PAGE technology, several variations of the method have been established to separate proteins based on a single molecular property. SDS-PAGE separates proteins based primarily on molecular weight, while isoelectric focusing separates proteins exclusively based on isoelectric point. In the presence of suitable initiator and catalyst compounds, acrylamide can readily polymerise in a radical process. This reaction would lead to very long polyacrylamide chains, yielding a highly viscous liquid instead of a gel. These long chains need to be cross-linked to form a three dimensional network. This is achieved by mixing *N,N'*-methylenebisacrylamide into the acrylamide solution. In essence, *N,N'*-

methylenebisacrylamide is composed of two acrylamide molecules covalently interconnected via a methylene moiety. When, during the polymerisation reaction, the acrylamide groups of *N,N'*-methylenebisacrylamide molecules become incorporated in the long polyacrylamide chains, cross-links are formed between the polyacrylamide chains leading to a gel. In the course of electrophoresis, ions (proteins or nucleic acids) are separated in this gel.

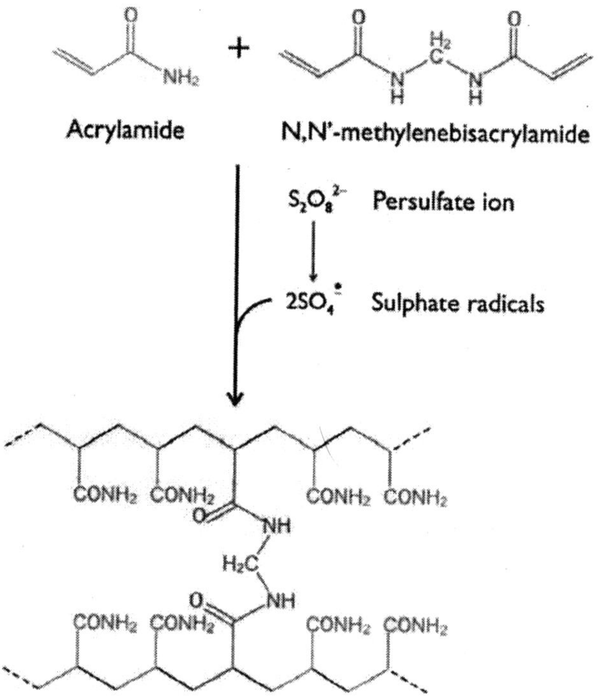

Acrylamide + N,N'-methylenebisacrylamide

$S_2O_8^{2-}$ Persulfate ion

$2SO_4^{\bullet-}$ Sulphate radicals

Molecular structure of the polyacrylamide gel. The three-dimensional molecular network comes into being by a radical polymerisation of acrylamide monomers and cross-linking *N,N'*-methylenebisacrylamide components.

Without any modification, polyacrylamide electrophoresis separates macromolecular ions based on a combination of charge, size and shape. Size (and shape) separation is due to the molecular sieving property of the gel. The size range in which molecules can be separated is dictated by the average pore size of the gel. In the case of polyacrylamide gels, this can be controlled through the concentration of the acrylamide monomer and the proportion of the cross-linking *N,N'*-methylenebisacrylamide. The acrylamide concentration can be set in the range of about 4-20 % as this is the range in which the mechanical properties of the gel are appropriate. Below this range the gel will be too soft and it will not keep its shape, while above this range it will be too rigid and prone to break. The optimal proportion of the *N,N'*-methylenebisacrylamide component is 1-3 % relative to the acrylamide component. The polyacrylamide gel possesses all advantageous properties necessary for a good electrophoresis medium, *i.e.* it is hydrophilic, free of electric charges and chemically stable. A further very important property of the polyacrylamide gel is that it does not participate in any non-specific or specific binding interaction with proteins. Furthermore, the polyacrylamide gel does not interfere with common protein staining reactions. When electrophoresis is performed under native (non-denaturing) conditions, such as near neutral pH and ambient or lower temperature, many enzymes retain their native conformation and, in turn, their enzymatic activity. This way, many enzymes can be separated and specifically detected in the gel after electrophoretic separation. In the course of creating the gel, a buffer with a properly chosen pH is mixed into the acrylamide/*N,N'*-methylenebisacrylamide solution. Radical polymerisation is subsequently triggered by suitable catalyst and initiator compounds. The catalyst is usually ammonium persulfate, which spontaneously decomposes in aqueous media, thereby generating free radicals. These free radicals in themselves cannot efficiently cleave the double bonds of the acrylamide molecule, but are able to excite the electrons of the initiator molecules. This leads to the generation of free radicals, originating from the initiator molecules, that are able to trigger radical polymerisation of acrylamide monomers. The most frequently used initiator is tetra-methylethylenediamine (TEMED). There are two types of gels according to their geometry. In early gel electrophoretic applications, gel tubes were used that allowed only a single sample to be run. Gel slabs were later introduced, allowing for many samples to be run at the same time in the same gel in parallel. Gel slabs became much more common than gel tubes. Gel slabs are created by pouring the gel-forming solution between two parallel glass sheets prior to polymerisation. Besides its higher throughput, this gel geometry provides another important advantage over gel tubes: samples are loaded side by side on such slabs and are run in the same gel at the same time. This allows for a more reliable comparison of the samples, facilitating the interpretation of experimental results.

Proper selection of pH and acrylamide concentration is instrumental for successful electrophoresis. For protein electrophoresis, the pH is set usually higher than the pI value of the proteins in the sample. At such a pH, all proteins will be negatively charged and will move towards the anode. The buffer in the medium serves two purposes. One is to set and maintain the proper pH during electrophoresis. The other function of the buffer is to establish the electric current in the medium. The majority of the electric current is carried by the ions of the buffer.

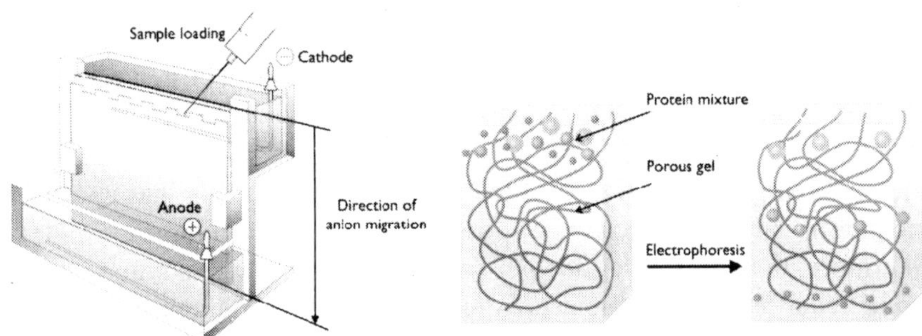

Separation of proteins in a polyacrylamide gel. As illustrated in the left panel, several samples can be run in parallel in a slab gel. Ions can move between the two electrodes only through the gel interconnecting the two chambers. The gel acts as a molecular sieve. The larger the molecule, the larger the drag force exerted on it by the gel.

Normally, the protein-ions that are separated by electrophoresis have only a negligible contribution to the current. In other words, proteins have a low ion transport number. However, if the buffer concentration is set too low, the contribution of proteins to carrying the current will increase, and the protein molecules will migrate rapidly. This usually leads to smearing of the bands of migrating proteins. On the other hand, if the buffer concentration is set too high, the mobility of the proteins will be too low. In this case the electrophoresis process would take a very long time. Unnecessary lengthening of the process provides excess time for diffusion, which lowers the resolution of separation. According to the applied buffer system, gel electrophoretic methods can be classified into two types: continuous and discontinuous. Continuous methods apply the same buffer in the gel and in the two buffer chambers containing the electrodes. The only advantage of this method lies in its simplicity. More complex discontinuous methods were introduced to provide higher resolution. SDS polyacrylamide gel electrophoresis is usually associated with such a discontinuous system. The discontinuous system applies two gels of different pore size and three different buffers. One of the gels, the resolving gel, is polymerised at a higher acrylamide concentration. The pore size of this gel is set according to the size range of the proteins to be separated. Another gel, the stacking gel is created on top of the resolving gel. The stacking gel is polymerised from a more dilute acrylamide solution to provide larger pores. This pore size does not provide a molecular sieving effect.

As mentioned above, there are three buffers: different ones in each of the two gels and a third one, the so-called "running buffer" in the buffer chambers containing the electrodes. In the gel buffers, the anion originates from a strong acid; it is usually chloride ion. Dissociation of strong acids does not depend on the pH: these acids always fully dissociate. Consequently, chloride ion is never protonated in the solution: its ionisation state is independent of the pH. On the other hand, the anion component of the running buffer is the conjugate base of a weak acid. Consequently, the ionisation state of this ion depends on the pH of the buffer. Glycinate ion is one of the most frequently used compounds for this purpose. The pH in the running buffer is set to 8.3. The protein sample is layered on the top of the stacking gel. When an electric field is generated by the power supply, the protein ions and the ions of the running buffer enter the stacking gel. The pH in the stacking gel is set to 6.8. This value is only slightly higher than the pI value of glycine (6.5). At this pH, most glycine molecules are in a neutral zwitterionic state, and only a small portion of the molecules carry a net negative charge. In this state, glycine has a low electrophoretic mobility and a corresponding low transport number. The local sparsity of ions elevates the local electric resistance of the medium. As the electric current must be of the same magnitude at any segments of the electric circuit (there is no macroscopic charge separation), the voltage will increase according to Ohm's law. Due to this effect, the migration speed of the proteins will be relatively high and the protein front will reach the chloride front in the stacking gel. The ion concentration in the chloride front is high and, therefore, here the electric resistance and the voltage are low. This slows down the protein front. This effect results in a very sharp protein front, with the protein molecules being crowded right behind the chloride ion front. The protein sample will thus enter the resolving gel in a sharp band. The pH in the resolving gel is set to about 8.8. At this pH, almost all glycinate molecules are in the anionic state. Thus, the electric mobility of glycinate increases, and the concentrating effect applied by the stacking gel ends in the resolving gel. Different proteins will be separated in the resolving gel according to their charge, size and shape. In most electrophoretic methods, a tracking dye is mixed in the sample. Usually, this dye is chosen to have a higher electrophoretic mobility than any of the components of interest (proteins or nucleic acids) in the sample. The function of the tracking dye is to visualise the running front and, in turn, the completeness of the run. The most popular tracking dye is bromophenol blue. The following sections review the various PAGE methods listed from the simplest to the most complex one.

SDS-PAGE

SDS-PAGE is an electrophoresis method to separate proteins. However, unlike in the case of native PAGE, here the proteins migrate in their denatured state. As it was mentioned in the general introduction to traditional (native) PAGE, the migration velocity of proteins is a function of their size, shape and the number of electric charges they carry. As the velocity is a complex function of these properties, native PAGE cannot be used to estimate the molecular mass of proteins. The traditional native PAGE method is similarly unable to assess whether a purified protein is composed of a single subunit or multiple subunits. Even a multi-subunit protein may migrate in a single sharp band. SDS-PAGE was introduced to analyse such cases and to allow the estimation of the molecular mass of single-subunit proteins or those of individual subunits of multi-subunit proteins. SDS-PAGE is the most prevalent PAGE method currently in use.

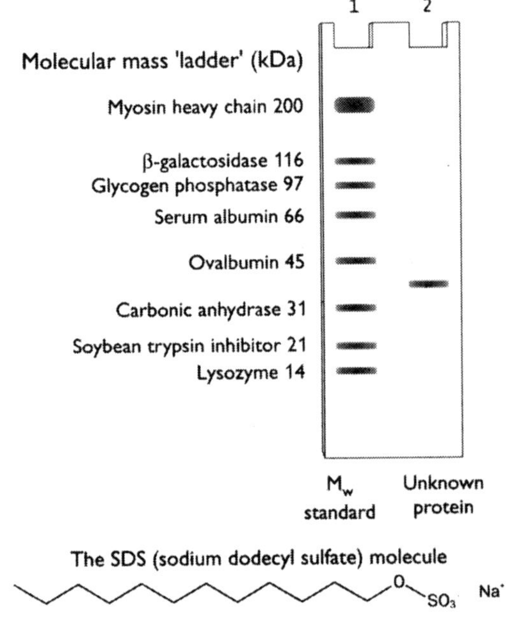

SDS polyacrylamide gel electrophoresis. SDS (sodium dodecyl sulphate) is an anionic detergent that unfolds proteins and provides them with extra negative charges. The amount of the associated SDS molecules—and therefore the number of charges—is proportional to the length of the polypeptide chain. The SDS gel separates individual polypeptide chains (monomeric proteins and subunits of multimeric proteins) according to their size. The velocity of the proteins is an inverse linear function of the logarithm of their molecular mass. Proteins of known molecular mass can be used to establish a calibration curve (a descending line) along which the unknown molecular mass of other proteins can be estimated.

SDS (sodium dodecyl sulphate) is an anionic detergent. When proteins are treated with SDS at high temperature, radical conformational changes occur. The treatment breaks all native non-covalent intermolecular (inter-subunit) and intramolecular interactions. The subunit structure of multi-subunit proteins disintegrates and the proteins unfold. If the native structure is stabilised by disulfide bridges, reducing agents are also added to open up these connections. SDS molecules bind to unfolded proteins in large excess, providing extra negative charges to the molecules.

The amount of the bound SDS molecules is largely independent of the amino acid sequence of the polypeptide chain and it is roughly a linear function of polypeptide length—*i.e.* the molecular mass of the protein. Therefore, upon SDS-treatment, the specific charge (the charge-to-mass ratio) of different proteins will become roughly identical. Another result of the treatment is that the shape of the different proteins becomes similar. The negatively charged SDS molecules repel each other, which lends a rod-like shape to the SDS-treated proteins. These factors together result in a situation analogous to the one already discussed in this chapter for the PAGE separation of linear single-stranded (denatured) DNA molecules. Instead of being separated simultaneously by charge, shape and size, SDS-treated proteins—just like denatured linear DNA molecules—will be separated solely based on their size. As size is a linear function of mass, SDS-PAGE ultimately separates proteins based on their molecular mass. SDS-PAGE is the most popular cost-effective method to estimate the molecular mass of protein subunits with considerable accuracy. The relative mobility (*i.e.* the running distance of the protein divided by the running distance of the tracking dye) of the SDS-treated protein is in inverse linear proportion to the logarithm of the molecular mass of the protein. By running several proteins of known molecular mass simultaneously alongside the protein of interest, a log molecular mass – relative mobility calibration curve (a descending linear graph) can be created. Based on the calibration curve, the estimated molecular mass of the protein in question can be easily calculated.

Table 1: The useful separating range of polyacrylamide gels as a function of acrylamide concentration. In the useful range, the log molecular mass – relative mobility relationship is linear.

Acrylamide concentration (%)	Linear range of separation (kDa)
15	12-43
10	16-68
7.5	36-94
5	57-212

SDS-PAGE is a standard method for assessing whether the sample of an isolated protein is homogeneous. Besides that, SDS-PAGE is a robust method for the analysis of large supramolecular complexes such as multi-enzyme complexes or the myofibril, as discussed below. SDS-PAGE separates and denatures individual subunits of these complexes. Thus, all polypeptide chains will migrate separately in the gel. Via various staining procedures, all subunits can be visualised and the relative amounts of these proteins (subunits) can also be determined. This allows for the identification of each subunit of a complex and provides a good estimate of the stoichiometry of subunits, too.

2.15 Isoelectric Focusing

In the course of isoelectric focusing, the conditions are set in a way that proteins will be separated exclusively based on their isoelectric point. The two termini and many side chains of proteins contain dissociable groups (weak acids or bases). The dissociation state of these groups is a function of the pH of the environment. Isoelectric focusing is based on the pH-dependent dissociation of these groups. Due to this pH-dependent phenomenon, the net electric charge of a protein molecule will be a function of the pH of the medium. If, in a given protein, the number of acidic residues (Asp, Glu) exceeds that of the basic ones (Arg, Lys, His), the protein will have a net negative charge at neutral pH. The isoelectric point (pI) of the protein *i.e.* the pH at which the net charge of the protein is zero will be in the acidic pH range. Such proteins are often denoted as acidic proteins. If the number of basic residues exceeds that of the acidic ones, the protein will be positively charged at neutral pH, and its pI value will be in the basic pH range. These proteins are often called basic proteins. Isoelectric focusing is an efficient high-resolution method because the pI values of various proteins are spread across a broad range. If the pH is lower than the pI of the protein, the protein will be positively charged and will move towards the cathode during electrophoresis. If the pH is higher than the pH of the protein, the protein will be negatively charged and will migrate towards the anode. If the pH equals the pI value, the net charge of the protein will be zero and the protein will not migrate in the gel any further. In the course of isoelectric focusing, proteins are placed in a gel representing a special medium in which the pH gradually decreases by going from the negative cathode towards the positive anode. As the protein migrates, it encounters a gradually changing pH and its net charge will also change accordingly. If it has a net negative charge and therefore moves towards the cathode, it will encounter a gradually decreasing pH, *i.e.* a more and more acidic environment. Consequently, the protein will take on more and more protons—up to a level where its net charge will be zero. This state is reached when the protein reaches a location where the pH equals its pI value. At this point, the protein will stop moving because no electric force will be exerted on it. If it spontaneously diffused further towards the anode, it would take on more protons, would become positively charged and would turn back to migrate towards the cathode.

Following the same line of thinking, if a positively charged protein moves towards the cathode, it will encounter increasing pH and lose more and more protons. It will migrate to the place where the pH equals its pI value and will thus stop. If it diffused further towards the cathode, it would become negatively charged and would turn back towards the anode. As one can see, by performing electrophoresis in a medium in which the pH decreases from the cathode towards the anode, each protein will find its place according to its pI value and will become sharply focused at that location. In addition, it does not matter where exactly the proteins were introduced in the medium between the cathode and the anode.

A decisive component of this method is the usually linear pH gradient created inside the gel. There are two methods to create such a gradient. One of them applies carrier ampholytes (ampholyte is an acronym from the words amphoteric and electrolyte). Ampholytes or zwitterions are molecules that contain both weakly acidic and weakly basic groups. Just like in the case of proteins, the net charge of ampholytes is a function of the pH. In the course of isoelectric focusing, a mixture of various ampholytes is used such that the pI of the various ampholyte components will cover a range in which the pI values of the "neighbouring" ampholytes differ only slightly. This ampholyte mixture is soaked in the gel and an appropriate electric field is generated by a power supply. This leads to a process analogous to the one already explained for proteins. Each ampholyte will migrate to the location where its net charge becomes zero. As soon as this steady-state is achieved, ampholytes will function as buffers and keep the pH of their immediate environment constant. This establishes the pH gradient in which the proteins can be separated. The other, more sophisticated method applies special ampholytes that can be covalently polymerised into the polyacrylamide gel. The appropriate ampholyte gradient is created before the gel is polymerised. This way, the gradient will be covalently fixed in the gel, providing an immobilised pH gradient. The appropriate pH range provided by the ampholyte mixture should be selected based on the pI values of the proteins to be separated. Regardless of how the pH gradient was created, once the proteins reach the location in the gel where the pH equals their pI, they finally stop moving and the system reaches a steady-state.

One of the potential technical difficulties encountered during isoelectric focusing originates from the fact that the solubility of proteins is lowest at their pI value. This can lead to the precipitation of some proteins in the gel. To prevent this unwanted process, urea is most often applied in the gel as an additive. Urea denatures proteins and keeps denatured proteins in solution. As the pI value of proteins is largely independent of their conformational state, this modification does not compromise the method. The solubility of membrane proteins can be further promoted by the addition of non-ionic detergents. Isoelectric focusing is aimed at separating proteins based exclusively on their pI value thus, independently of their size. Therefore, the molecular sieving property of the gel in this method should be avoided. The only function of the gel is to prevent free convectional flows in the medium. Accordingly, for isoelectric focusing, polyacrylamide gels are made at very low acrylamide concentrations, and sometimes even agarose gels are applied when very large pores are needed. Isoelectric focusing is usually performed in a horizontally-mounted electrophoresis apparatus and by applying intense cooling.

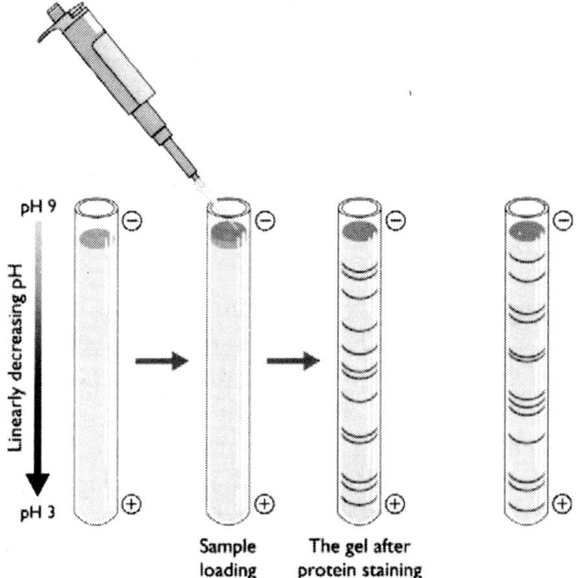

Isoelectric focusing. In the course of isoelectric focusing, a pH gradient is created in the gel (usually made of polyacrylamide, less frequently agarose). Upon electrophoresis, various proteins will accumulate in different narrow regions of the gel where the pH equals their individual pI value. At this pH, the number of positive charges equals that of the negative charges on the protein—the net charge will thus be zero. Consequently, no resultant electric force is exerted on the protein.

2.16. Polymerase Chain Reaction (PCR) Machine

The polymerase chain reaction (PCR) is a biochemical technology in molecular biology to amplify a single or a few copies of a piece of DNA across several orders of magnitude, generating thousands to millions of copies of a particular DNA sequence.

PCR is now a common and often indispensable technique used in medical and biological research labs for a variety of applications. These include DNA cloning for sequencing, DNA-based phylogeny, or functional analysis of genes; the diagnosis of hereditary diseases; the identification of generic fingerprints (used in forensic sciences and paternity testing); and the detection and diagnosis of infectious diseases.

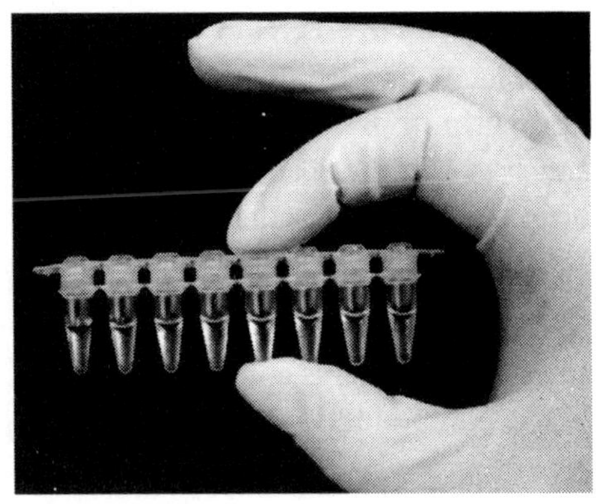

A strip of eight PCR tubes, each containing a 100 μl reaction mixture

The method relies on thermal cycling, consisting of cycles of repeated heating and cooling of the reaction for DNA melting and enzymatic replication of the DNA. Primers (short DNA fragments) containing sequences complementary to the target region along with a DNA polymerase (after which the method is named) are key components to enable selective and repeated amplification. As PCR progresses, the DNA generated is itself used as a template for replication, setting in motion a chain reaction in which the DNA template is exponentially amplified. PCR can be extensively modified to perform a wide array of genetic manipulations.

Almost all PCR applications employ a heat-stable DNA polymerase, such as Taq polymerase (an enzyme originally isolated from the bacterium *Thermus aquaticus*). This DNA polymerase enzymatically assembles a new DNA strand from DNA building-blocks, the nucleotides, by using single-stranded DNA as a template and DNA oligonucleotides (also called DNA primers), which are required for initiation of DNA synthesis. The vast majority of PCR methods use thermal cycling, *i.e.*, alternately heating and cooling the PCR sample through a defined series of temperature steps.

In the first step, the two strands of the DNA double helix are physically separated at a high temperature in a process called DNA melting. In the second step, the temperature is lowered and the two DNA strands become templates for DNA polymerase to selectively amplify the target DNA. The selectivity of PCR results from the use of primers that are complementary to the DNA region targeted for amplification under specific thermal cycling conditions.

PCR principles and procedure

PCR is used to amplify a specific region of a DNA strand (the DNA target). Most PCR methods typically amplify DNA fragments of between 0.1 and 10 kilo base pairs (kb), although some techniques allow for amplification of fragments up to 40 kb in size. The amount of amplified product is determined by the available substrates in the reaction, which become limiting as the reaction progresses.

A thermal cycler for PCR

A basic PCR set up requires several components and reagents. These components include:

- DNA template that contains the DNA region (target) to be amplified.
- Two primers that are complementary to the 3' (three prime) ends of each of the sense and anti-sense strand of the DNA target.
- Taq polymerase or another DNA polymerase with a temperature optimum at around 70° C.
- Deoxynucleoside triphosphates (dNTPs, sometimes called "deoxynucleotide triphosphates"; nucleotides containing triphosphate groups), the building-blocks from which the DNA polymerase synthesizes a new DNA strand.
- Buffer solution, providing a suitable chemical environment for optimum activity and stability of the DNA polymerase.
- Divalent cations, magnesium or manganese ions; generally Mg^{2+} is used, but Mn^{2+} can be utilized for PCR-mediated DNA mutagenesis, as higher Mn^{2+} concentration increases the error rate during DNA synthesis.
- Monovalent cation potassium ions.

The PCR is commonly carried out in a reaction volume of 10–200 μl in small reaction tubes (0.2–0.5 ml volumes) in a thermal cycler. The thermal cycler heats and cools the reaction tubes to achieve the temperatures required at each step of the reaction (see below). Many modern thermal cyclers make use of the Peltier effect, which permits both heating and cooling of the block holding the PCR tubes simply by reversing the electric current. Thin-walled reaction tubes permit favorable thermal conductivity to allow for rapid thermal equilibration. Most thermal cyclers have heated lids to prevent condensation at the top of the reaction tube. Older thermocyclers lacking a heated lid require a layer of oil on top of the reaction mixture or a ball of wax inside the tube.

Procedure

Typically, PCR consists of a series of 20-40 repeated temperature changes, called cycles, with each cycle commonly consisting of 2-3 discrete temperature steps, usually three. The cycling is often preceded by a single temperature step (called *hold*) at a high temperature (>90°C), and followed by one hold at the end for final product extension or brief storage. The temperatures used and the length of time they are applied in each cycle depend on a variety of parameters. These include the enzyme used for DNA synthesis, the concentration of divalent ions and dNTPs in the reaction, and the melting temperature (Tm) of the primers.

- *Initialization step*: This step consists of heating the reaction to a temperature of 94–96 °C (or 98 °C if extremely thermostable polymerases are used), which is held for 1–9 minutes. It is only required for polymerases that require heat activation by hot-start PCR.

- *Denaturation step*: This step is the first regular cycling event and consists of heating the reaction to 94–98 °C for 20–30 seconds. It causes melting of the DNA template by disrupting the hydrogen bonds between complementary bases, yielding single-stranded DNA molecules.

- *Annealing step*: The reaction temperature is lowered to 50–65 °C for 20–40 seconds allowing annealing of the primers to the single-stranded DNA template. Typically the annealing temperature is about 3–5 °C below the Tm of the primers used. Stable DNA–DNA hydrogen bonds are only formed when the primer sequence very closely matches the template sequence. The polymerase binds to the primer-template hybrid and begins DNA formation.

- *Extension/elongation step*: The temperature at this step depends on the DNA polymerase used; Taq polymerase has its optimum activity temperature at 75–80 °C, and commonly a temperature of 72 °C is used with this enzyme. At this step the DNA polymerase synthesizes a new DNA strand complementary to the DNA template strand by adding dNTPs that are complementary to the template in 5' to 3' direction, condensing the 5'-phosphate group of the dNTPs with the 3'-hydroxyl group at the end of the nascent (extending)] DNA strand. The extension time depends both on the DNA polymerase used and on the length of the DNA fragment to be amplified. As a rule-of-thumb, at its optimum temperature, the DNA polymerase will polymerize a thousand bases per minute. Under optimum conditions, *i.e.*, if there are no limitations due to limiting substrates or reagents, at each extension step, the amount of DNA target is doubled, leading to exponential (geometric) amplification of the specific DNA fragment.

- *Final elongation*: This single step is occasionally performed at a temperature of 70–74 °C for 5–15 minutes after the last PCR cycle to ensure that any remaining single-stranded DNA is fully extended.
- *Final hold*: This step at 4–15 °C for an indefinite time may be employed for short-term storage of the reaction.

To check whether the PCR generated the anticipated DNA fragment (also sometimes referred to as the amplimer or amplicon), agarose gel electrophoresis is employed for size separation of the PCR products. The size(s) of PCR products is determined by comparison with a DNA ladder (a molecular weight marker), which contains DNA fragments of known size, run on the gel alongside the PCR products.

PCR stages

The PCR process can be divided into three stages:

Exponential amplification: At every cycle, the amount of product is doubled (assuming 100% reaction efficiency). The reaction is very sensitive: only minute quantities of DNA need to be present.

Leveling off stage: The reaction slows as the DNA polymerase loses activity and as consumption of reagents such as dNTPs and primers causes them to become limiting.

Plateau: No more product accumulates due to exhaustion of reagents and enzyme.

PCR optimization

In practice, PCR can fail for various reasons, in part due to its sensitivity to contamination causing amplification of spurious DNA products. Because of this, a number of techniques and procedures have been developed for optimizing PCR conditions. Contamination with extraneous DNA is addressed with lab protocols and procedures that separate pre-PCR mixtures from potential DNA contaminants. This usually involves spatial separation of PCR-setup areas from areas for analysis or purification of PCR products, use of disposable plasticware, and thoroughly cleaning the work surface between reaction setups. Primer-design techniques are important in improving PCR product yield and in avoiding the formation of spurious products, and the usage of alternate buffer components or polymerase enzymes can help with amplification of long or otherwise problematic regions of DNA. Addition of reagents, such as formamide, in buffer systems may increase the specificity and yield of PCR. Computer simulations of theoretical PCR results (Electronic PCR) may be performed to assist in primer design.

Application of PCR

Selective DNA isolation

PCR allows isolation of DNA fragments from genomic DNA by selective amplification of a specific region of DNA. This use of PCR augments many methods, such as generating hybridization probe for Southern or northern hybridization and DNA cloning, which require larger amounts of DNA, representing a specific DNA region. PCR supplies these techniques with high amounts of pure DNA, enabling analysis of DNA samples even from very small amounts of starting material.

Other applications of PCR include DNA sequencing to determine unknown PCR-amplified sequences in which one of the amplification primers may be used in Sanger sequencing, isolation of a DNA sequence to expedite recombinant DNA technologies involving the insertion of a DNA sequence into a plasmid or the genetic material of another organism. Bacterial colonies (*E. coli*) can be rapidly screened by PCR for correct DNA vector constructs. PCR may also be used for genetic fingerprinting; a forensic technique used to identify a person or organism by comparing experimental DNAs through different PCR-based methods.

Some PCR 'fingerprints' methods have high discriminative power and can be used to identify genetic relationships between individuals, such as parent-child or between siblings, and are used in paternity testing. This technique may also be used to determine evolutionary relationships among organisms.

Amplification and quantification of DNA

Because PCR amplifies the regions of DNA that it targets, PCR can be used to analyze extremely small amounts of sample. This is often critical for forensic analysis, when only a trace amount of DNA is available as evidence. PCR may also be used in the analysis of ancient DNA that is tens of thousands of years old. These PCR-based techniques have been successfully used on animals, such as a forty-thousand-year-old mammoth, and also on human DNA, in applications ranging from the analysis of Egyptian mummies to the identification of a Russian tsar and the body of British king Richard III.

Quantitative PCR methods allow the estimation of the amount of a given sequence present in a sample a technique often applied to quantitatively determine levels of gene expression. Quantitative PCR is an established tool for DNA quantification that measures the accumulation of DNA product after each round of PCR amplification.

Variations on the basic PCR technique

- *Allel-specific PCR*: a diagnostic or cloning technique based on single-nucleotide variations (SNVs not to be confused with SNPs) (single-base differences in a patient). It requires prior knowledge of a DNA sequence, including differences between allels, and uses primers whose 3' ends encompass the SNV (base pair buffer around SNV usually incorporated). PCR amplification under stringent conditions is much less efficient in the presence of a mismatch between template and primer, so successful amplification with an SNP-specific primer signals presence of the specific SNP in a sequence.

- Assembly PCR or *Polymerase Cycling Assembly (PCA)*: artificial synthesis of long DNA sequences by performing PCR on a pool of long oligonucleotides with short overlapping segments. The oligonucleotides alternate between sense and antisense directions, and the overlapping segments determine the order of the PCR fragments, thereby selectively producing the final long DNA product.

- *Asymtric PCR*: preferentially amplifies one DNA strand in a double-stranded DNA template. It is used in sequencing and hybridization probing where amplification of only one of the two complementary strands is required. PCR is carried out as usual, but with a great excess of the primer for the strand targeted for amplification. Because of the slow (arithmatic) amplification later in the reaction after the limiting primer has been used up, extra cycles of PCR are required. A recent modification on this process, known as *L*inear-*A*fter-*The*-*E*xponential-PCR (LATE-PCR), uses a limiting primer with a higher melting temperature (Tm) than the excess primer to maintain reaction efficiency as the limiting primer concentration decreases mid-reaction.

- *Dial-out PCR*: a highly parallel method for retrieving accurate DNA molecules for gene synthesis. A complex library of DNA molecules is modified with unique flanking tags before massively parallel sequencing. Tag-directed primers then enable the retrieval of molecules with desired sequences by PCR.

- *Digital PCR (dPCR)*: used to measure the quantity of a target DNA sequence in a DNA sample. The DNA sample is highly diluted so that after running many PCRs in parallel, some of them will not receive a single molecule of the target DNA. The target DNA concentration is calculated using the proportion of negative outcomes. Hence the name 'digital PCR'.

- *Hellicase-dependent amplification*: similar to traditional PCR, but uses a constant temperature rather than cycling through denaturation and annealing/extension cycles. DNA helicase, an enzyme that unwinds DNA, is used in place of thermal denaturation.
- *Hot start PCR*: a technique that reduces non-specific amplification during the initial set up stages of the PCR. It may be performed manually by heating the reaction components to the denaturation temperature (*e.g.*, 95°C) before adding the polymerase. Specialized enzyme systems have been developed that inhibit the polymerase's activity at ambient temperature, either by the binding of an antibody or by the presence of covalently bound inhibitors that dissociate only after a high-temperature activation step. Hot-start/cold-finish PCR is achieved with new hybrid polymerases that are inactive at ambient temperature and are instantly activated at elongation temperature.
- *In silico PCR* (digital PCR, virtual PCR, electronic PCR, e-PCR) refers to computational tools used to calculate theoretical polymerase chain reaction results using a given set of primers (probes) to amplify DNA sequences from a sequenced genome or transcription. In silico PCR was proposed as an educational tool for molecular biology.
- *Intersequence-specific PCR* (ISSR): a PCR method for DNA fingerprinting that amplifies regions between simple sequence repeats to produce a unique fingerprint of amplified fragment lengths.
- *Inverse PCR*: is commonly used to identify the flanking sequences around genomic inserts. It involves a series of DNA digestions and self ligation, resulting in known sequences at either end of the unknown sequence.
- *Ligation-mediated PCR*: uses small DNA linkers ligated to the DNA of interest and multiple primers annealing to the DNA linkers; it has been used for DNA sequencing, genome walking, and DNA footprinting.
- *Methylation-specific PCR* (MSP): It is used to detect methylation of CpG islands in genomic DNA. DNA is first treated with sodium bisulfite, which converts unmethylated cytosine bases to uracil, which is recognized by PCR primers as thymine. Two PCRs are then carried out on the modified DNA, using primer sets identical except at any CpG islands within the primer sequences. At these points, one primer set recognizes DNA with cytosines to amplify methylated DNA, and one set recognizes DNA with uracil or thymine to amplify unmethylated DNA. MSP using qPCR can also be performed to obtain quantitative rather than qualitative information about methylation.

- *Miniprimer PCR*: uses a thermostable polymerase (S-Tbr) that can extend from short primers ("smalligos") as short as 9 or 10 nucleotides. This method permits PCR targeting to smaller primer binding regions, and is used to amplify conserved DNA sequences, such as the 16S (or eukaryotic 18S) rRNA gene.
- *Multiplex Ligation-dependent Probe Amplification* (*MLPA*): permits multiple targets to be amplified with only a single primer pair, thus avoiding the resolution limitations of multiplex PCR (see below).
- *Multiplex-PCR*: consists of multiple primer sets within a single PCR mixture to produce amplicons of varying sizes that are specific to different DNA sequences. By targeting multiple genes at once, additional information may be gained from a single test-run that otherwise would require several times the reagents and more time to perform. Annealing temperatures for each of the primer sets must be optimized to work correctly within a single reaction, and amplicon sizes. That is, their base pair length should be different enough to form distinct bands when visualized by gel electrophoresis.
- Nanoparticle-Assisted PCR (nanoPCR): In recent years, it has been reported that some nanoparticles (NPs) can enhance the efficiency of PCR (thus being called nanoPCR), and some even perform better than the original PCR enhancers. It was also found that quantum dots (QDs) can improve PCR specificity and efficiency. Single-walled carbon nanotubes (SWCNTs) and multi-walled carbon nanotubes (MWCNTs) are efficient in enhancing the amplification of long PCR. Carbon nanopowder (CNP) was reported be able to improve the efficiency of repeated PCR and long PCR. ZnO, T_iO_2, and Ag NPs were also found to increase PCR yield. Importantly, already known data has indicated that non-metallic NPs retained acceptable amplification fidelity. Given that many NPs are capable of enhancing PCR efficiency, it is clear that there is likely to be great potential for nanoPCR technology improvements and product development.
- *Nested PCR*: increases the specificity of DNA amplification, by reducing background due to non-specific amplification of DNA. Two sets of primers are used in two successive PCRs. In the first reaction, one pair of primers is used to generate DNA products, which besides the intended target, may still consist of non-specifically amplified DNA fragments. The product(s) are then used in a second PCR with a set of primers whose binding sites are completely or partially different from and located 3' of each of the primers used in the first reaction. Nested PCR is often more

successful in specifically amplifying long DNA fragments than conventional PCR, but it requires more detailed knowledge of the target sequences.

- *Overlap-extension PCR* or *Splicing by overlap extension (SOEing)* : a genetic engineering technique that is used to splice together two or more DNA fragments that contain complementary sequences. It is used to join DNA pieces containing genes, regulatory sequences, or mutations; the technique enables creation of specific and long DNA constructs. It can also introduce deletions, insertions or point mutations into a DNA sequence.
- *PAN-AC*: uses isothermal conditions for amplification, and may be used in living cells.
- *Quantitative PCR* (qPCR): used to measure the quantity of a target sequence (commonly in real-time). It quantitatively measures starting amounts of DNA, cDNA, or RNA. Quantitative PCR is commonly used to determine whether a DNA sequence is present in a sample and the number of its copies in the sample. *Quantitative PCR* has a very high degree of precision. Quantitative PCR methods use fluorescent dyes, such as Sybr Green, EvaGreen or fluorophore-containing DNA probes, such as TaqMan, to measure the amount of amplified product in real time. It is also sometimes abbreviated to RT-PCR (*real-time* PCR) but this abbreviation should be used only for reverse transcription PCR. qPCR is the appropriate contractions for quantitative PCR (real-time PCR).
- *Reverse Transcription PCR (RT-PCR)*: for amplifying DNA from RNA. Reverse transcription reverse transcribes RNA into cDNA, which is then amplified by PCR. RT-PCR is widely used in expression profiling, to determine the expression of a gene or to identify the sequence of an RNA transcript, including transcription start and termination sites. If the genomic DNA sequence of a gene is known, RT-PCR can be used to map the location of exons and introns in the gene. The 5' end of a gene (corresponding to the transcription start site) is typically identified by RACE-PCR (*Rapid Amplification of cDNA Ends*).
- *Solid Phase PCR*: encompasses multiple meanings, including Polony Amplification (where PCR colonies are derived in a gel matrix, for example), Bridge PCR (primers are covalently linked to a solid-support surface), conventional Solid Phase PCR (where Asymmetric PCR is applied in the presence of solid support bearing primer with sequence matching one of the aqueous primers) and Enhanced Solid Phase PCR (where conventional Solid Phase PCR can be improved by employing high Tm and nested solid support primer with optional application of a

thermal 'step' to favour solid support priming).

- *Thermal asymmetric interlaced PCR (TAIL-PCR)*: for isolation of an unknown sequence flanking a known sequence. Within the known sequence, TAIL-PCR uses a nested pair of primers with differing annealing temperatures; a degenerate primer is used to amplify in the other direction from the unknown sequence.
- *Touchdown PCR* (*Step-down PCR*): a variant of PCR that aims to reduce nonspecific background by gradually lowering the annealing temperature as PCR cycling progresses. The annealing temperature at the initial cycles is usually a few degrees (3-5°C) **above** the T_m of the primers used, while at the later cycles, it is a few degrees (3-5°C) **below** the primer T_m. The higher temperatures give greater specificity for primer binding, and the lower temperatures permit more efficient amplification from the specific products formed during the initial cycles.
- *Universal Fast Walking*: for genome walking and genetic fingerprinting using a more specific 'two-sided' PCR than conventional 'one-sided' approaches (using only one gene-specific primer and one general primer — which can lead to artefactual 'noise') by virtue of a mechanism involving lariat structure formation. Streamlined derivatives of UFW are LaNe RAGE (lariat-dependent nested PCR for rapid amplification of genomic DNA ends), 5'RACE LaNe and 3'RACE LaNe.

2.17. Real-time Polymerase Chain Reaction

A quantitative polymerase chain reaction (qPCR), also called real-time polymerase chain reaction, is a laboratory technique of molecular biology based on the polymerase chain reaction (PCR), which is used to amplify and simultaneously quantify a targeted DNA molecule. For one or more specific sequences in a DNA sample, quantitative PCR enables both detection and quantification. The quantity can be either an absolute number of copies or a relative amount when normalized to DNA input or additional normalizing genes.

The procedure follows the general principle of polymerase chain reaction; its key feature is that the amplified DNA is detected as the reaction progresses in "real time". This is a new approach compared to standard PCR, where the product of the reaction is detected at its end. Two common methods for the detection of products in quantitative PCR are: (1) non-specific fluorescent dyes that intercalate with any double-stranded DNA, and (2) sequence-specific DNA probes consisting of oligonucleotides that are labelled with a fluorescent reporter which permits detection only after hybridization of the probe with its complementary sequence to quantify messenger RNA (mRNA) and non-coding RNA in cells or tissues.

qPCR is the abbreviation used for quantitative PCR (real-time PCR). Real-time reverse-transcription PCR is often denoted as: qRT-PCR The acronym "RT-PCR" commonly denotes reverse transcription polymerase chain reaction and not real-time PCR, but not all authors adhere to this convention.

Cells in all organisms regulate gene expression by turnover of gene transcripts (messenger RNA, abbreviated to mRNA): The amount of an expressed gene in a cell can be measured by the number of copies of an mRNA transcript of that gene present in a sample. In order to robustly detect and quantify gene expression from small amounts of RNA, amplification of the gene transcript is necessary. The polymerase chain reaction (PCR) is a common method for amplifying DNA; for mRNA-based PCR the RNA sample is first reverse-transcribed to cDNA with reverse technique.

In order to amplify small amounts of DNA the same methodology is used as in conventional PCR using a DNA template, at least one pair of specific primers, deoxyribonucletides, a suitable buffer solution and a thermo-stable DNA polymerase. A substance marked with a fluorophore is added to this mixture in a thermal cycler that contains sensors for measuring the fluorescence of the flurophore after it has been excited at the required wavelength allowing the generation rate to be measured for one or more specific products. This allows the rate of generation of the amplified product to be measured at each PCR cycle. The data thus generated can be analysed by computer software to calculate *relative gene expression* (or *mRNA copy number*) in several samples. Quantitative PCR can also be applied to the detection and quantification of DNA in samples to determine the presence and abundance of a particular DNA sequence in these samples. This measurement is made after each amplification cycle, and this is the reason why this method is called real time PCR (that is, immediate or simultaneous PCR). In the case of RNA quantitation, the template is complementary DNA (cDNA), which is obtained by reverse transcription of ribonucleic acid

An older model three-temperature thermal cycler for PCR

(RNA). In this instance the technique used is quantitative RT-PCR or Q-RT-PCR.

- Deoxynucleoside triphosphates (dNTPs, sometimes called "deoxynucleotide triphosphates"; nucleotides containing triphosphate groups), the building-blocks from which the DNA polymerase synthesizes a new DNA strand.
- Buffer solution, providing a suitable chemical environment for optimum activity and stability of the DNA polymerase.
- Divalent cations, magnesium or manganese ions; generally Mg^{2+} is used, but Mn^{2+} can be utilized for PCR-mediated DNA mutagenesis, as higher Mn^{2+} concentration increases the error rate during DNA synthesis.
- Monovalent cation potassium ions.

The PCR is commonly carried out in a reaction volume of 10–200 μl in small reaction tubes (0.2–0.5 ml volumes) in a thermal cycler. The thermal cycler heats and cools the reaction tubes to achieve the temperatures required at each step of the reaction (see below). Many modern thermal cyclers make use of the Peltier effect, which permits both heating and cooling of the block holding the PCR tubes simply by reversing the electric current. Thin-walled reaction tubes permit favorable thermal conductivity to allow for rapid thermal equilibration. Most thermal cyclers have heated lids to prevent condensation at the top of the reaction tube. Older thermocyclers lacking a heated lid require a layer of oil on top of the reaction mixture or a ball of wax inside the tube.

Procedure

Typically, PCR consists of a series of 20-40 repeated temperature changes, called cycles, with each cycle commonly consisting of 2-3 discrete temperature steps, usually three. The cycling is often preceded by a single temperature step (called *hold*) at a high temperature (>90° C), and followed by one hold at the end for final product extension or brief storage. The temperatures used and the length of time they are applied in each cycle depend on a variety of parameters. These include the enzyme used for DNA synthesis, the concentration of divalent ions and dNTPs in the reaction, and the melting temperature (Tm) of the primers.

- *Initialization step*: This step consists of heating the reaction to a temperature of 94–96° C (or 98° C if extremely thermostable polymerases are used), which is held for 1–9 minutes. It is only required for DNA polymerases that require heat activation by hot-start PCR.

- *Denaturation step*: This step is the first regular cycling event and consists of heating the reaction to 94–98 °C for 20–30 seconds. It causes melting of the DNA template by disrupting the hydrogen bonds between complementary bases, yielding single-stranded DNA molecules.
- *Annealing step*: The reaction temperature is lowered to 50–65 °C for 20–40 seconds allowing annealing of the primers to the single-stranded DNA template. Typically the annealing temperature is about 3–5 °C below the Tm of the primers used. Stable DNA–DNA hydrogen bonds are only formed when the primer sequence very closely matches the template sequence. The polymerase binds to the primer-template hybrid and begins DNA formation.
- *Extension/elongation step*: The temperature at this step depends on the DNA polymerase used; Taq polymerase has its optimum activity temperature at 75–80 °C, and commonly a temperature of 72 °C is used with this enzyme. At this step the DNA polymerase synthesizes a new DNA strand complementary to the DNA template strand by adding dNTPs that are complementary to the template in 5' to 3' direction, condensing the 5'-phosphate group of the dNTPs with the 3'-hydroxyl group at the end of the nascent (extending) DNA strand. The extension time depends both on the DNA polymerase used and on the length of the DNA fragment to be amplified. As a rule-of-thumb, at its optimum temperature, the DNA polymerase will polymerize a thousand bases per minute. Under optimum conditions, *i.e.*, if there are no limitations due to limiting substrates or reagents, at each extension step, the amount of DNA target is doubled, leading to exponential (geometric) amplification of the specific DNA fragment.
- *Final elongation*: This single step is occasionally performed at a temperature of 70–74 °C for 5–15 minutes after the last PCR cycle to ensure that any remaining single-stranded DNA is fully extended.
- *Final hold*: This step at 4–15 °C for an indefinite time may be employed for short-term storage of the reaction.

To check whether the PCR generated the anticipated DNA fragment (also sometimes referred to as the amplimer or amplicon), agarose gel electrophoresis is employed for size separation of the PCR products. The size(s) of PCR products is determined by comparison with a DNA ladder (a molecular weight marker), which contains DNA fragments of known size, run on the gel alongside the PCR products.

PCR stages

The PCR process can be divided into three stages:

Exponential amplification: At every cycle, the amount of product is doubled (assuming 100% reaction efficiency). The reaction is very sensitive: only minute quantities of DNA need to be present.

Leveling off stage: The reaction slows as the DNA polymerase loses activity and as consumption of reagents such as dNTPs and primers causes them to become limiting.

Plateau: No more product accumulates due to exhaustion of reagents and enzyme.

PCR optimization

In practice, PCR can fail for various reasons, in part due to its sensitivity to contamination causing amplification of spurious DNA products. Because of this, a number of techniques and procedures have been developed for optimizing PCR conditions. Contamination with extraneous DNA is addressed with lab protocols and procedures that separate pre-PCR mixtures from potential DNA contaminants. This usually involves spatial separation of PCR-setup areas from areas for analysis or purification of PCR products, use of disposable plasticware, and thoroughly cleaning the work surface between reaction setups. Primer-design techniques are important in improving PCR product yield and in avoiding the formation of spurious products, and the usage of alternate buffer components or polymerase enzymes can help with amplification of long or otherwise problematic regions of DNA. Addition of reagents, such as formamide, in buffer systems may increase the specificity and yield of PCR. Computer simulations of theoretical PCR results (Electronic PCR) may be performed to assist in primer design.

Application of PCR

Selective DNA isolation

PCR allows isolation of DNA fragments from genomic DNA by selective amplification of a specific region of DNA. This use of PCR augments many methods, such as generating hybridization probes for Southern or Northern hybridization and DNA cloning, which require larger amounts of DNA, representing a specific DNA region. PCR supplies these techniques with high amounts of pure DNA, enabling analysis of DNA samples even from very small amounts of starting material.

Other applications of PCR include DNA sequencing to determine unknown PCR-amplified sequences in which one of the amplification primers may be used in Sanger sequencing, isolation of a DNA sequence to expedite recombinant DNA technologies involving the insertion of a DNA sequence into a plasmid or the genetic material of another organism. Bacterial colonies (*E. coli*) can be rapidly screened by PCR for correct DNA vector constructs. PCR may also be used for genetic fingerprinting; a forensic technique used to identify a person or organism by comparing experimental DNAs through different PCR-based methods.

Some PCR 'fingerprints' methods have high discriminative power and can be used to identify genetic relationships between individuals, such as parent-child or between siblings, and are used in paternity testing. This technique may also be used to determine evolutionary relationships among organisms.

Quantitative PCR and DNA microarray are modern methodologies for studying gene expression. Older methods were used to measure mRNA abundance: Differential display, RNAase protection assay and Northern blot. Northern blotting is often used to estimate the expression level of a gene by visualizing the abundance of its mRNA transcript in a sample. In this method, purified RNA is separated by agarose gel electrophoresis, transferred to a solid matrix (such as a nylon membrane), and probed with a specific DNA or RNA probe that is complementary to the gene of interest. Although this technique is still used to assess gene expression, it requires relatively large amounts of RNA and provides only qualitative or semi quantitative information of mRNA levels. Estimation errors arising from variations in the quantification method can be the result of DNA integrity, enzyme efficiency and many other factors. For this reason a number of standardization systems have been developed. Some have been developed for quantifying total gene expression, but the most common are aimed at quantifying the specific gene being studied in relation to another gene called a normalizing gene, which is selected for its almost constant rate of expression. These genes are also called housekeeping genes as they are usually involved in the functions related to basic cellular survival, which normally implies constitutive gene expression. This enables researchers to report a ratio for the expression of the genes of interest divided by the expression of the selected normalizer. Thereby allowing comparison of the former without actually knowing its absolute level of expression.

The most commonly used normalizing genes are those that code for the following proteins: tubulin, glyceraldehydes-3-phosphate dehydrogenase, albumin, cyclophilin, and ribosomal RNAs.

Quantitative PCR is carried out in a thermal cycler with the capacity to illuminate each sample with a beam of light of a specified wavelength and detect the fluorescence emitted by the excited fluorophore. The thermal cycler is also able to rapidly heat and chill samples, thereby taking advantage of the physicochemical properties of the nucleic acids and DNA polymerase.

Quantitative PCR with double-stranded DNA-binding dyes as reporters

A DNA-binding dye binds to all double-stranded (ds) DNA in PCR, causing fluorescence of the dye. An increase in DNA product during PCR therefore leads to an increase in fluorescence intensity and is measured at each cycle, thus allowing DNA concentrations to be quantified. However, dsDNA dyes such as SYBR Green will bind to all dsDNA PCR products, including nonspecific PCR products (such as Primer dimer). This can potentially interfere with, or prevent, accurate quantification of the intended target sequence. The SYBR Green is excited using blue light (λ_{max} = 488 nm) and it emits green light (λ_{max} = 522 nm).

1. The reaction is prepared as usual, with the addition of fluorescent dsDNA dye.
2. The reaction is run in a quantitative PCR instrument, and after each cycle, the levels of fluorescence are measured with a detector; the dye only fluoresces when bound to the dsDNA (*i.e.*, the PCR product). With reference to a standard dilution, the dsDNA concentration in the PCR can be determined.

This method has the advantage of only needing a pair of primers to carry out the amplification, which keeps costs down; however, it is only possible to amplify a product using a chain reaction.

Like other quantitative PCR methods, the values obtained do not have absolute units associated with them (*i.e.*, mRNA copies/cell). As described above, a comparison of a measured DNA/RNA sample to a standard dilution will only give a fraction or ratio of the sample relative to the standard, allowing only relative comparisons between different tissues or experimental conditions. To ensure accuracy in the quantification, it is usually necessary to normalize expression of a target gene to a stably expressed gene. This can correct possible differences in RNA quantity or quality across experimental samples.

Fluorescent reporter probe method

(1) In intact probes, reporter fluorescence is quenched. (2) Probes and the complementary DNA strand are hybridized and reporter fluorescence is still quenched. (3) During PCR, the probe is degraded by the Taq polymerase and the fluorescent reporter released.

Fluorescent reporter probes detect only the DNA containing the probe sequence; therefore, use of the reporter probe significantly increases specificity, and enables quantification even in the presence of non-specific DNA amplification. Fluorescent probes can be used in multiplex assays—for detection of several genes in the same reaction—based on specific probes with different-coloured labels, provided that all targeted genes are amplified with similar efficiency. The specificity of fluorescent reporter probes also prevents interference of measurements caused by primer dimer, which are undesirable potential by-products in PCR. However, fluorescent reporter probes do not prevent the inhibitory effect of the primer dimers, which may depress accumulation of the desired products in the reaction.

The method relies on a DNA-based probe with a fluorescent reporter at one end and a quencher of fluorescence at the opposite end of the probe. The close proximity of the reporter to the quencher prevents detection of its fluorescence; breakdown of the probe by the 5' to 3' exonuclease activity of the Taq polymerase breaks the reporter-quencher proximity and thus allows unquenched emission of fluorescence, which can be detected after excitation with a laser. An increase in the product targeted by the reporter probe at each PCR cycle therefore causes a proportional increase in fluorescence due to the breakdown of the probe and release of the reporter.

1. The PCR is prepared as usual (see PCR), and the reporter probe is added.
2. As the reaction commences, during the annealing stage of the PCR both probe and primers anneal to the DNA target.
3. Polymerisation of a new DNA strand is initiated from the primers, and

once the polymerase reaches the probe, its 5'-3'-exonuclease degrades the probe, physically separating the fluorescent reporter from the quencher, resulting in an increase in fluorescence.

4. Fluorescence is detected and measured in a real-time PCR machine, and its geometric increase corresponding to exponential increase of the product is used to determine the quantification cycle (C_q) in each reaction.

Fusion temperature analysis

Distinct fusion curves for a number of PCR products (showing distinct colours). Amplification reactions can be seen for a specific product (pink, blue) and others with a negative result (green, orange). Q-PCR permits the identification of specific, amplified DNA fragments using analysis of their melting temperature (also called T_m value). The method used is usually PCR with double-stranded DNA-binding dyes as reporters and the dye used is usually SYBR Green. The DNA melting temperature is specific to the amplified fragment. The results of this technique are obtained by comparing the dissociation curves for the analysed DNA samples.

Unlike conventional PCR, this method avoids the previous use of electrophoresis techniques to demonstrate the results of all the samples. This is because, despite being a kinetic technique, quantitative PCR is usually evaluated at a distinct end point. The technique therefore usually provides more rapid results and / or uses fewer reactants than electrophoresis. If subsequent electrophoresis is required it is only necessary to test those samples that real time PCR has shown to be doubtful and / or to ratify the results for samples that have tested positive for a specific determinant.

Quantification of gene expression

Quantifying gene expression by traditional DNA detection methods is unreliable. Detection of mRNA on a Northern blot or PCR products on a gel or Southern blot does not allow precise quantification. For example, over the 20-40 cycles of a typical PCR, the amount of DNA product reaches a plateau that is not directly correlated with the amount of target DNA in the initial PCR.

Quantitative PCR can be used to quantify nucleic acids by two common methods: relative quantification and absolute quantification. Absolute quantification gives the exact number of target DNA molecules by comparison with DNA standards using a calibration curve. It is therefore essential that the PCR of the sample and the standard have the same amplification efficiency. Relative quantification is based on internal reference genes to determine fold-differences in expression of the target gene. The quantification is expressed as the change in expression levels of mRNA interpreted as complementary DNA (cDNA, generated by reverse transcription of mRNA). Relative quantification is easier to carry out as it does not require a calibration curve as the amount of the studied gene is compared to the amount of a control housekeeping gene.

As the units used to express the results of relative quantification are unimportant the results can be compared across a number of different RT-Q-PCR. The reason for using one or more housekeeping genes is to correct non-specific variation, such as the differences in the quantity and quality of RNA used, which can affect the efficiency of reverse transcription and therefore that of the whole PCR process. However, the most crucial aspect of the process is that the reference gene must be stable.

The selection of these reference genes was traditionally carried out in molecular biology using qualitative or semi-quantitative studies such as the visual examination of RNA gels, Northern blot densitometer or semi-quantitative PCR (PCR mimics). Now, in the genome era, it is possible to carry out a more detailed estimate for many organisms using DNA microarrays. However, research has shown that amplification of the majority of reference genes used in quantifying the expression of mRNA varies according to experimental conditions. It is therefore necessary to carry out an initial statiscally sound methodological study in order to select the most suitable reference gene.

A number of statistical algorithms have been developed that can detect which gene or genes are most suitable for use under given conditions. Those like geNORM or BestKeeper can compare pairs or genometric means for a matrix of different reference genes and tissues.

Modeling

Unlike end point PCR (conventional PCR) real time PCR allows quantification of the desired product at any point in the amplification process by measuring fluorescence (in reality, measurement is made of its level over a given threshold). A commonly employed method of DNA quantification by quantitative PCR relies on plotting fluorescence against the number of cycles on a logarithmic scale. A threshold for detection of DNA-based fluorescence is set slightly above background. The number of cycles at which the fluorescence exceeds the threshold is called the threshold cycle (C_t) or, according to the MIQE guidelines, quantification cycle (C_q). During the exponential amplification phase, the sequence of the DNA target doubles every cycle. For example, a DNA sample whose C_q precedes that of another sample by 3 cycles contained $2^3 = 8$ times more template. However, the efficiency of amplification is often variable among primers and templates. Therefore, the efficiency of a primer-template combination is assessed in a titration experiment with serial dilutions of DNA template to create a standard curve of the change in C_q with each dilution. The slope of the linear regression is then used to determine the efficiency of amplification, which is 100% if a dilution of 1:2 results in a C_q difference of 1. The cycle threshold method makes several assumptions of reaction mechanism and has a reliance on data from low signal-to-noise regions of the amplification profile that can introduce substantial variance during the data analysis.

To quantify gene expression, the C_q for an RNA or DNA from the gene of interest is subtracted from the C_q of RNA/DNA from a housekeeping gene in the same sample to normalize for variation in the amount and quality of RNA between different samples. This normalization procedure is commonly called the *$\ddot{A}C_t$-method* and permits comparison of expression of a gene of interest among different samples. However, for such comparison, expression of the normalizing reference gene needs to be very similar across all the samples. Choosing a reference gene fulfilling this criterion is therefore of high importance, and often challenging, because only very few genes show equal levels of expression across a range of different conditions or tissues. Although cycle threshold analysis is integrated with many commercial software systems, there are more accurate and reliable methods of analysing amplification profile data that should be considered in cases where reproducibility is a concern.

Mechanism-based qPCR quantification methods have also been suggested, and have the advantage that they do not require a standard curve for quantification. Methods such as MAK2 have been shown to have equal or better quantitative performance to standard curve methods. These mechanism-based methods use knowledge about the polymerase amplification process to generate estimates of the original sample concentration. An extension of this approach includes an accurate model of the entire PCR reaction profile, which allows for the use of high signal-to-noise data and the ability to validate data quality prior to analysis.

Applications

There are numerous applications for quantitative polymerase chain reaction in the laboratory. It is commonly used for both diagnostic and basic research. Uses of the technique in industry include the quantification of microbial load in foods or on vegetable matter, the detection of GMOs (Genetically modified organism) and the quantification and genotyping of human viral pathogens.

Diagnostic uses

Diagnostic quantitative PCR is applied to rapidly detect nucleic acids that are diagnostic of, for example, infectious diseases, cancer and genetic abnormalities.

The introduction of quantitative PCR assays to the clinical microbiology laboratory has significantly improved the diagnosis of infectious diseases, and is deployed as a tool to detect newly emerging diseases, such as new strains of flu, in diagnostic tests.

Microbiological uses

Quantitative PCR is also used by microbiologists working in the fields of food safety, food spoilage and fermentation and for the microbial risk assessment of water quality (drinking and recreational waters) and in public health protection.

The antibacterial assay Virtual colony count utilizes a data quantification technique called Quantitative Growth Kinetics (QGK) that is mathematically identical to QPCR, except bacterial cells, rather than copies of a PCR product, increase exponentially. The QGK equivalent of the threshold cycle is referred to as the "threshold time".

Uses in research

In research settings, quantitative PCR is mainly used to provide quantitative measurements of gene transcription. The technology may be used in determining how the genetic expression of a particular gene changes over time, such as in the response of tissue and cell cultures to an administration of a pharmacological agent, progression of cell differentiation, or in response to changes in environmental conditions. It is also used for the determination of zygosity of transgenic animals used in research.

Detection of phytopathogens

The agricultural industry is constantly striving to produce plant propagules or seedlings that are free of pathogens in order to prevent economic losses and safeguard health. Systems have been developed that allow detection of small amounts of the DNA of *Phytopthora ramorum*, a fungus that kills Oaks and other species, mixed in with the DNA of the host plant. Discrimination between the DNA of the pathogen and the plant is based on the amplification of its sequences, spacers located in ribosomal RNA gene's coding area, which are characteristic for each taxon. Field-based versions of this technique have also been developed for identifying the same pathogen.

Detection of genetically modified organisms

Q-PCR (using reverse transcription) can be used to detect GMOs given its sensitivity and dynamic range in detecting DNA. Alternatives such as DNA or protein analysis are usually less sensitive. Specific primers are used that amplify not the transgene but the poromoter, terminator or even intermediate sequences used during the process of engineering the vector. As the process of creating a transgenic plant normally leads to the insertion of more than one copy of the transgene its quantity is also commonly assessed. This is often carried out by relative quantification using a control gene from the treated species that is only present as a single copy.

Clinical quantification and genotyping

Viruses can be present in humans due to direct infection or co-infections. This makes diagnosis difficult using classical techniques and can result in an incorrect prognosis and treatment. The use of Q-PCR allows both the quantification and genotyping (characterization of the strain, carried out using melting curves) of a virus such as the Hepatitis B virus. The degree of infection, quantified as the copies of the viral genome per unit of the patient's tissue, is relevant in many cases; for example, the probability that the type 1 herpes simplex virus reactivates is related to the number of infected neurons in the ganglia.

2.18. Radio Tracer Technique

A radioactive tracer, or radioactive label, is a chemical compound in which one or more atoms have been replaced by a radioisotope, so by virtue of its radioactive decay it can be used to explore the mechanism of chemical reactions by tracing the path that the radioisotope follows from reactants to products.

Radioisotopes of hydrogen, carbon, phosphorus, sulphur, and iodine have been used extensively to trace the path of biochemical reactions. A radioactive tracer can also be used to track the distribution of a substance within a natural system such as a cell or tissues. Radioactive tracers are also used to determine the location of fractures created by hydraulic functioning in natural gas production. Radioactive tracers form the basis of a variety of imaging systems, such as, PET scans, SPECT scans and technetium scans.

Methodology

Isotopes of a chemical element differ only in the mass number. For example, the isotopes of hydrogen can be written as ^{1}H, ^{2}H and ^{3}H, with the mass number at top left. When the atomic nucleus of an isotope is unstable, compounds containing this isotope are radioactive. Tritium is an example of a radioactive isotope.

The principle behind the use of radioactive tracers is that an atom in a chemical compound is replaced by another atom, of the same chemical element. The substituting atom, however, is a radioactive isotope. This process is often called radioactive labeling. The power of the technique is due to the fact that radioactive decay is much more energetic than chemical reactions. Therefore, the radioactive isotope can be present in low concentration and its presence detected by sensitive radiation detectors such as Geiger counters and scintillation counters.

There are two main ways in which radioactive tracers are used

1. When a labeled chemical compound undergoes chemical reactions one or more of the products will contain the radioactive label. Analysis of what happens to the radioactive isotope provides detailed information on the mechanism of the chemical reaction.

2. A radioactive compound is introduced into a living organism and the radio-isotope provides a means to construct an image showing the way in which that compound and its reaction products are distributed around the organism.

Production

The commonly used radioisotopes have short half lives and so do not occur in nature. They are produced by nuclear reactions. One of the most important processes is absorption of a neutron by an atomic nucleus, in which the mass number of the element concerned increases by 1 for each neutron absorbed. For example,

$$^{13}C + n \rightarrow {}^{14}C$$

In this case the atomic mass increases, but the element is unchanged. In other cases the product nucleus is unstable and decays, typically emitting protons, electrons (beta particles) or alpha particles. When a nucleus loses a proton the atomic number decreases by 1. For example,

$$^{32}S + n \rightarrow {}^{32}P + p$$

Neutron irradiation is performed in a nuclear reactor, so tracer studies are carried out close to the reactor itself. The other main method used to synthesize radioisotopes is proton bombardment. The proton are accelerated to high energy either in a cyclotron or a linear acceletor.

Tracer isotopes

Hydrogen

Tritium is produced by neutron irradiation of ^{6}Li

$$^{6}Li + n \rightarrow {}^{4}He + {}^{3}H$$

Tritium has a half-life 4,500±8 days (approximately 12.32 years), and it decays by beta decay. The electrons produced have an average energy of 5.7 keV. Because the emitted electrons have relatively low energy, the detection efficiency by scintillation counting is rather low. However, hydrogen atoms are present in all organic compounds, so tritium is frequently used as a tracer in biochemical studies.

Carbon

^{11}C decays by positron emission with a half-life of approximately 20 min. ^{11}C is one of the isotopes often used in positron emission tomography.

^{14}C decays by beta-decay, with a half-life of 5730 y. It is continuously produced in the upper atmosphere of the earth so it occurs at a trace level in the environment. However, it is not practical to use naturally-occurring ^{14}C for tracer studies. Instead it is made by neutron irradiation of the isotope ^{13}C which occurs naturally in carbon at about the 1.1% level. ^{14}C has been used extensively to trace the progress of organic molecules through metabolic pathways.

Nitrogen

^{13}N decays by positron emission with a half-life of 9.97 min. It is produced by the nuclear reaction

$$^{1}H + {}^{16}O \rightarrow {}^{13}N + {}^{4}H$$

^{13}N is used in positron emission tomography (PET scan).

Oxygen

^{15}O decays by positron emission with a half-life of 122 sec. It is used in positron emission tomography

Fluorine

^{18}F decays by positron emission with a half-life of 109 min. It is made by proton bombardment of ^{18}O in a cyclotron or linear particle accelerator. It is an important isotope in the radiopharmaceutical industry. It is used to make labelled fluorodeoxyglucose (FDG) for application in PET scans.

Phosphorus

^{32}P is made by neutron bombardment of ^{32}S

$$^{32}S + n \rightarrow {}^{32}P + p$$

It decays by beta decay with a half-life of 14.29 days. It is commonly used to study protein phosphorylation by kinases in biochemistry.

^{32}P is made in relatively low yield by neutron bombardment of ^{31}P. It is also a beta-emitter, with a half-life of 25.4 days. Though more expensive than ^{32}P, the emitted electrons are less energetic, permitting better resolution in, for example, DNA sequencing.

Both isotopes are useful for labelling nucleotides and other species that contain a phosphate group.

Sulfur

^{35}S is made by neutron bombardment of ^{35}Cl

$$^{35}Cl + n \rightarrow {}^{35}S + p$$

It decays by beta-decay with a half-life of 87.51 days. It is used to label the sulfur-containing amino-acids methionine and cysteine. When a sulfur atom replaces an oxygen atom in a phosphate group on a nucleotide a thiophosphate is produced, so ^{35}S can also be used to trace a phosphate group.

Technetium

^{99m}Tc is a very versatile radioisotope. It is easy to produce in a technetium-99m generator, by decay of ^{99}Mo.

$$^{99}Mo \rightarrow {}^{99m}Tc + e^- + ve$$

The molybdenum isotope has a half-life of approximately 66 hours (2.75 days), so the generator has a useful life of about two weeks. Most commercial ^{99m}Tc generators use column chromatography, in which ^{99}Mo in the form of molybdate, MoO_4^{2-} is adsorbed onto acid alumina (Al_2O_3). When the ^{99}Mo decays it forms pentachnetate TcO_4^-, which because of its single charge is less tightly bound to the alumina. Pulling normal saline solution through the column of immobilized ^{99}Mo elutes the soluble ^{99m}Tc, resulting in a saline solution containing the ^{99m}Tc as the dissolved sodium salt of the pertechnetate. The pertechnetate is treated with a reducing agent such as Sn^{2+} and a ligand. Different ligands form coordination complexes which give the technetium enhanced affinity for particular sites in the human body. ^{99m}Tc decays by gamma emission, with a half-life: 6.01 hours. The short half-life ensures that the body-concentration of the radioisotope falls effectively to zero in a few days.

Iodine

^{123}I is produced by proton irradiation of ^{124}Xe. The caesium isotope produced is unstable and decays to ^{123}I. The isotope is usually supplied as the iodide and hypoiodate in dilute sodium hydroxide solution, at high isotopic purity. ^{123}I has also been produced at Oak Ridge National Laboratories by proton bombardment of ^{123}Te. ^{123}I decays by electron capture with a half-life of 13.22 hours. The emitted 159 keV gamma ray is used in single photon emission captured tomography (SPECT). A 127 keV gamma ray is also emitted. ^{125}I is frequently used in radioimmunoassay because of its relatively long half-life and ability to be detected with high sensitivity by gamma counters.

^{129}I is present in the environment as a result of the testing of nuclear weapons in the atmosphere. It was also produced in the Chernobyl and Fukushima disasters. ^{129}I decays with a half-life of 15.7 million years, with low-energy beta and gamma emissions. It is not used as a tracer, though its presence in living organisms, including human beings, can be characterized by measurement of the gamma rays.

Other isotopes

Many other isotopes have been used in specialized radiopharmacological studies. The most widely used is ^{67}Ga for gallium scans. ^{67}Ga is used because, like ^{99m}Tc, it is a gamma-ray emitter and various ligands can be attached to the Ga^{3+} ion, forming a coordination complex which may have selective affinity for particular sites in the human body.

Unit of radio activity

Curie is the unit of radio activity and is abbreviated as Ci. It was originally defined as the rate at which 1.0 g of radium^{226} decays. Because of the relatively long half life of Ra^{226}, the isotope served as a convenient standard. The curie is now defined as the quantity of any radioactive substance in which the decay rate is 3.7 x 10^{10} disintegration per second (2.2 x 10^{12} DPM). Because the efficiency of most radiation detection devices are less than 100%, a given number of curies almost always yield a lower than theoretical count rate. Hence, there is the distinction between DPS and counts per minute (CPM). For example, a sample contains 1 μ Ci of radioactive material has a decay rate of 2.22 x 10^6 DPM. If only 30% of the disintegration are detected the observed count rate is 6.66 x 10^5 CPM.

Specific activity

It is not necessary for every molecule of a compound to be radioactive for us to use the radioactivity as a measure of concentration (and thereby, determine reaction rates, and so on). All that is necessary is that the sample contain enough radioactive molecules to count accurately and that we know the specific activity of the compound. Specific activity (SA) refers to the amount of radioactivity per unit amount of substance. It is, in fact, a way to designating the fraction of the total molecules present, that is radioactivity. Specific activity may be given in terms of curies per gram (Ci/g), mili curies per milligram (mCi/mg), mimicries per mili moles (mCI/mM), disintegration per minute per mill moles (DPM/mM), counts per minute per micro moles (CPM/μ M), or in any other convenient way. Once the soecific activity of a compound is known, any given count rate can be equated to the amount of the compound in a sample.

In most studies with radio active compounds, it is assumed that there are no isotopic effects, *i.e.*, it is assumed that the radioactive molecules are randomly distributed among the total molecules of the compound and behave identically to the non-radioactive molecules. This is a reasonable assumption for most of the isotopes used in biology. The one exception is H^3, which has a mass three times that of normal H. However, for most biologically important H^3-labeled compounds, the total molecular weight is not much different from that of the unlabeled compound. For example, $H^3{}_2O$ has a molecular weight only 19% greater than that of $H^1{}_2O$.

Applications

In metabolism research, Tritium and ^{14}C-labeled glucose are commonly used in glucose clamps to measure rates of glucose uptake, fatty acid synthesis, and other metabolic processes. While radioactive tracers are sometimes still used in human studies, stable isotope tracers such as ^{13}C are more commonly used in current human clamp studies. Radioactive tracers are also used to study lipoprotein metabolism in humans and experimental animals.

In medicine, tracers are applied in a number of tests, such as ^{99m}Tc in autoradiography and nuclear medicine, including simple photon emission compound tomography (SPECT), positron emission tomography (PET) and scintigraphy. The urea breath test for *Helicobacter pylori* commonly used a dose of ^{14}C labelled urea to detect *H. pylori* infection. If the labelled urea was metabolized by h. pylori in the stomach, the patient's breath would contain labelled carbon dioxide. In recent years, the use of substances enriched in the non-radioactive isotope ^{13}C has become the preferred method, avoiding patient exposure to radioactivity.

In hydraulic fracturing, radioactive tracer isotopes are injected with hydraulic fracturing fluid to determine the injection profile and location of created fractures. Tracers with different half-lives are used for each stage of hydraulic fracturing. In the United States amounts per injection of radionuclide are listed in the US Nuclear Energy Commission (NRC) guidelines. According to the NRC, some of the most commonly used tracers include antimony-124, bromine-82, iodine-125, iodine-131, iridium-192, and scandium-46. A 2003 publication by the International Atomic Energy Agency confirms the frequent use of most of the tracers above, and says that manganese-56, sodium-24, technetium-99m, silver-99m, argon-41, and xenon-133 are also used extensively because they are easily identified and measured.

2.19. Scintillation Counter

A scintillation counter is an instrument for detecting and measuring ionizing radiation. It consists of a scintilator which generates photons of light in response to incident radiation, a sensitive photomultiplier tube which converts the light to an electrical signal, and the necessary electronics to process the photomultiplier tube output.

Scintillation counters are widely used because they can be made inexpensively yet with good quantum efficiency and can measure both the intensity and the energy of incident radiation. When a charged particle strikes the scintillator, its atoms are excited and photons are emitted. These are directed at the photomultiplier tube's photocathode, which emits electrons by the photoelectric effect. These electrons are electrostatically accelerated and focused by an electrical potential so that they strike the first dynode of the tube. The impact of a single electron on the dynode releases a number of secondary electrons which are in turn accelerated to strike the second dynode. Each subsequent dynode impact releases further electrons, and so there is a current amplifying effect at each dynode stage. Each stage is at a higher potential than the previous to provide the accelerating field. The resultant output signal at the anode is in the form of a measurable pulse for each photon detected at the photocathode, and is passed to the processing electronics. The pulse carries information about the energy of the original incident radiation on the scintillator. Thus both intensity and energy of the radiation can be measured.

The scintillator must be in complete darkness so that visible light photons do not swamp the individual photon events caused by incident ionising radiation. To achieve this, a thin opaque foil, such as aluminised Mylar, is often used, though it must have a low enough mass to prevent undue attenuation of the incident radiation being measured.

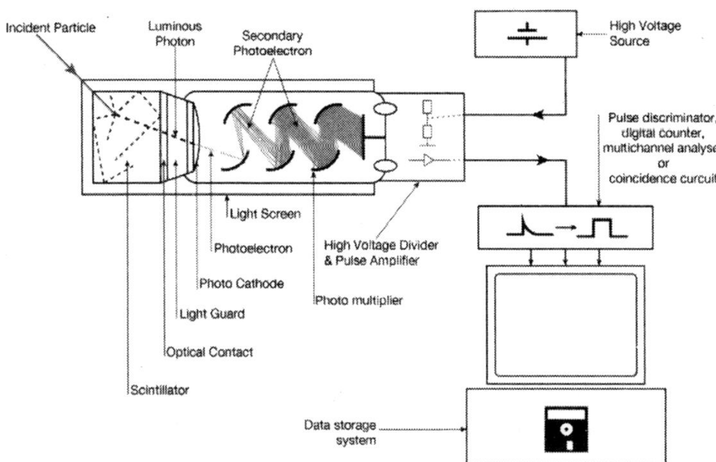

Schematic showing incident particles hitting a scintillating crystal, triggering the release of photons which are then converted into photoelectrons and multiplied in the photomultiplier.

The scintillator consists of a transparent crystal, usually a phosphor, plastic (usually containing anthracene) or organic liquid that fluoresces when struck by ionizing radiation. Cesium iodide (CsI) in crystalline form is used as the scintillator for the detection of protons and alpha particles. Sodium iodide (NaI) containing a small amount of thallium is used as a scintillator for the detection of gamma waves and Zinc Sulphide is widely used as a detector of alpha particles.

Photomultiplier tubes (photomultipliers or PMTs for short), members of the class of vaccum tubes, and more specifically vacuum phototubes, are extremely sensitive detectors of light in the ultraviolet, visible, and near-infrared ranges of the electromagnetic spectrum. These detectors multiply the current produced by incident light by as much as 100 million times (*i.e.*, 160 dB), in multiple dynode stages, enabling (for example) individual photons to be detected when the incident flux of light is very low. Unlike most vacuum tubes, they are not obsolete.

The combination of high gain, low noise, high frequency response or, equivalently, ultra-fast response, and large area of collection has earned photomultipliers an essential place in nuclear and particle physics, astronomy, medical diagonstics including blood tests, medical imaging, motion picture film scanning (telecine), radar jamming, and high-end image scanners known as drum scanners. Elements of photomultiplier technology, when integrated differently, are the basis of night vision devices.

Semiconductor devices, particularly avalanche photodiodes, are alternatives to photomultipliers; however, photomultipliers are uniquely well-suited for applications requiring low-noise, high-sensitivity detection of light that is imperfectly collimated.

Detector efficiencies

Gamma

The quantum efficiency of a gamma-ray detector (per unit volume) depends upon the density of electrons in the detector, and certain scintillating materials, such as sodium iodide and bismuth germanate, achieve high electron densities as a result of the high atomic numbers of some of the elements of which they are composed. However, detector based semiconductors, notably hyper pure germanium, have better intrinsic energy resolution than scintillators, and are preferred where feasible for gamma-ray spectrometry.

Neutron

In the case of neutron detectors, high efficiency is gained through the use of scintillating materials rich in hydrogen that scatter neutrons efficiently. Liquid scintillation counters are an efficient and practical means of quantifying beta radiation.

2.20. Liquid Scintillation Counter

Liquid scintillation counting is a standard laboratory method in the life-sciences for measuring radiation from beta-emitting radioactive isotopes. Scintillating materials are also used in differently constructed scintillation counters in many other fields.

Samples are dissolved or suspended in a "cocktail" containing a solvent (historically aromatic organics such as benzene or toluene, but more recently less hazardous solvents are used), typically some form of a surfactant, and small amounts of other additives known as "fluors" or scintillators. Scintillators can be divided into primary and secondary phosphores, differing in their luminescence properties.

Beta particles emitted from the isotopic sample transfer energy to the solvent molecules: the Π cloud of the aromatic ring absorbs the energy of the emitted particle. The energized solvent molecules typically transfer the captured energy back and forth with other solvent molecules until the energy is finally transferred to a primary scintillator. The primary phosphor will emit photons following absorption of the transferred energy. Because that light emission may be at a wavelength that does not allow efficient detection, many cocktails contain secondary phosphors that absorb the fluorescence energy of the primary phosphor and re-emit at a longer wavelength.

Liquid Scintillation Counter

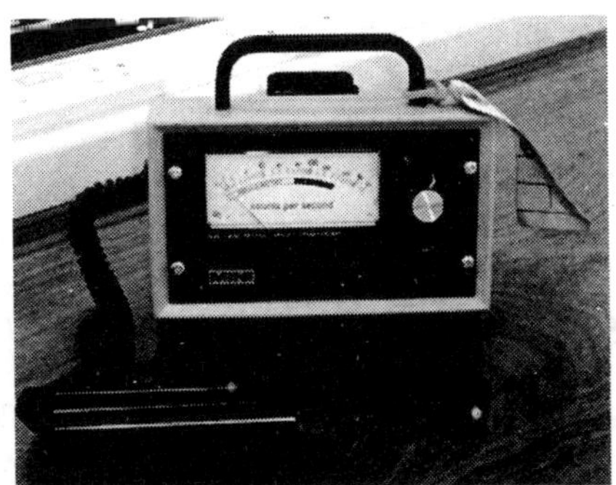

A "two-piece" bench type Geiger–Müller counter with end-window detector.

The radioactive samples and cocktail are placed in small transparent or translucent (often glass or plastic) vials that are loaded into an instrument known as a liquid scintillation counter. Newer machines may use 96-well plates with individual filters in each well. Many counters have two photomultiplier tubes connected in a coincidence circuit. The coincidence circuit assures that genuine light pulses, which reach both photomultiplier tubes, are counted, while spurious pulses (due to line noise, for example), which would only affect one of the tubes, are ignored.

Counting efficiencies under ideal conditions range from about 30% for tritium (a low-energy beta emitter) to nearly 100% for phosphorus-32, a high-energy beta emitter. Some chemical compounds (notably chlorine compounds) and highly coloured samples can interfere with the counting process. This interference, known as "quenching", can be overcome through data correction or through careful sample preparation.

High-energy beta emitters, such as phosphorus-32, can also be counted in a scintillation counter without the cocktail, instead using an aqueous solution. This technique, known as Cherenkov counting, relies on the Cherenkov radiation being detected directly by the photomultiplier tubes. Cherenkov counting in this experimental context is normally used for quick, rough measurements, since the geometry of the sample can create variations in the output.

Quenching in scintillation counting

The presence of inert material or coloured material in a sample may reduce the radioactivity observed in a scintillation counting. This quenching effect can be corrected easily by means of an internal or added standard. The first thing to be sure of is that the samples ae counted under the same condition as the standards. For example, if the samples are 0.5 ml of an enzyme assay mixture (in aquous buffer) that are added to 5.0 ml scintillation fluid, then the specific activity of the standard should be determined in 0.5 ml of the same buffer, counted in 5.0 ml of scintillation fluid.

Dilution Factor

The dilution factor if defined as:

Specific activity of precursor fed / Specfic activity of compound isolated

This factor is used frequently to espress the precursor relation of a compound in the biosynthesis of a second compound. Thus, in the sequence A → B → C → D the dilution factor for C → D would be small, whereas for A it would be large. Therefore, a small dilution factor would indicate that compound C fed to a tissue has a better precursor relationship to the final product than compound A with a large dilution factor.

2.21. Geiger Counter

A Geiger–Müller counter, also called a Geiger counter, is a type of particle detector that measures ionizing radiation. It detects the emission of nuclear radiation — alpha particles, beta particles, and gamma particles by the ionization produced in a low-pressure gas in a Geiger-Müller tube, which gives its name to the instrument. In wide and prominent use as a hand-held radiation survey instrument, it is perhaps one of the world's best-known radiation measurning instruments.

The original operating principle was discovered in 1908 in early radiation research. Since the subsequent development of the Geiger-Müller tube in 1928 the Geiger-Müller counter has been a popular instrument for use in radiation dosimetry, health physics, experimental physics, the nuclear industry, geological exploration and other fields, due to its robust sensing element and relatively low cost. However there are limitations in measuring high radiation rates and in measuring the energy of incident radiation.

Geiger counter instruments consist of two main elements; the Geiger-Müller tube, and the processing and display electronics. The radiation sensing element is an inert gas-filled Geiger-Müller tube (usually containing helium, neon, or argon with halogens added) at a low pressure which briefly conducts electrical charge when a particle or photon of radiation makes the gas conductive by ionization. The tube has the property of being able to amplify each ionization event by means of the Townsend avalanche effect and produces an easily measured current pulse which is passed to the processing electronics.

There are two main limitations of the Geiger counter. Because the output pulse from a Geiger-Müller tube is always the same magnitude regardless of the energy of the incident radiation, the tube cannot differentiate between radiation types. A further limitation is the inability to measure high radiation rates due to the "dead time" of the tube. This is an insensitive period after each ionization of the gas during which any further incident radiation will not result in a count, and the indicated rate is therefore lower than actual. Typically the dead time will reduce indicated count rates above about 10^4 to 10^5 counts per second depending on the characteristic of the tube being used. Whilst some counters have circuitry which can compensate for this, for accurate measurements ion chamber instruments are preferred for high radiation rates.

Geiger counters are widely used to detect gamma radiation, and for this the windowless tube is used. However, efficiency is generally low due to the low interaction of gamma compared with alpha and beta particles. For instance, a chrome steel G-M tube is only about 1% efficient over a wide range of energies.

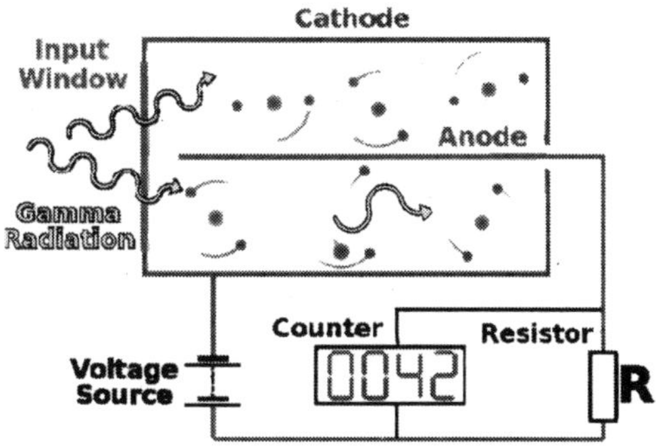

Schematic of a Geiger counter using an "end window" tube for low penetration radiation.

The article on the Geiger-Muller tube carries a more detailed account of the techniques used to detect photon radiation. But in summary, for high energy gamma this largely relies on interaction of the photon radiation with the tube wall material, usually 1–2 mm of chrome steel on a "thick-walled" tube, to produce electrons within the wall which can enter and ionize the fill gas. This is necessary as the low pressure gas in the tube has little interaction with high energy gamma photons. However, for low energy photons there is

Laboratory use of a G-M counter with end window probe to measure beta radiation from a radioactive source

greater gas interaction and the direct gas ionisation effect increases. With decreasing energy the wall effect gives way to a combination of wall effect and direct ionisation, until direct gas ionisation dominates. Due to the variance in response to different photon energies, thick walled steel tubes employ what is known as "energy compensation" which attempts to compensate for these variations over a large energy range.

A typical design for low energy photon detection for such as X-rays is a long tube with a thin wall or end window. This gives a greater gas volume, and thereby an increased chance of particle interaction.

CHAPTER 3

About Units

Units/measures are used to specify quantities. Every unit has a well-defined standard basic value, which is used as a reference in measurements. The units used in scientific practice constitute a scientifically established system, SI (Système International). Although only SI units are official , numerous non-SI units are still widely used (calorie, Ångström, Celsius *etc.*). We distinguish two types of quantities with two corresponding types of SI units. The base quantities cannot be derived from other quantities. Many of these are used in biochemistry (names of units and their symbols are given in brackets) to quantify the mass (gram, g), the length (meter, m), the time (second, s), the temperature (kelvin, K) and the charge (coulomb, C). Much larger is the group of derived quantities. Of these, the various units of concentration (to characterise the abundance of materials in solutions and other mixtures), volume and energy are the ones most frequently used in biochemistry. It is important to note that the terms mass and weight are often used interchangeably as alternatives. Technically, however, they have different meanings. The mass is the total quantity of matter in an object, which comes from the mass of all of its protons and neutrons, although not simply additively. Weight is a measure of the gravitational force exerted on an object. As the mass of protons and neutrons is the unit mass, their total mass in an object (*e.g.* the mass of a molecule, the molar mass) is a unitless number (*e.g.* the molar mass of water is 18). In other words, the molar mass is a relative number that would reflect the number of protons and neutrons in an atom or molecule, if the masses of protons and neutrons were additive. We make mass measurable by expressing it in grams because this way we can handle it as weight and measure it with a balance, the device developed for this purpose. Thus, the measurement of weight means the measurement of mass of physical objects not that of atoms or molecules, but *e.g.* that of 18 ml water. At the same time, this way we make the measured mass dependent on the place of measurement. (*e.g.*, 18 g of H_2O is not the same amount (ml or mass) of water at the poles and the equator, although the difference cannot be demonstrated by a traditional lever-arm balance due to its principle of operation). Thus the practical aspect of the relationship between mass and weight can be summarised as the following: we can measure mass only as weight, and weight is the effect of gravity exerted on mass. Weight is therefore determined by both the mass and gravitational forces, and an object can be weightless but never massless. The mass unit to express the size of large molecules is Dalton (Da), used mainly in biochemistry. This shows how many times the mass of a macromolecule (*e.g.* a protein) is larger than the mass of a hydrogen atom (more precisely, a proton or a neutron).

An important quantity is the mole, which is special to chemistry. The atomic and molar masses expressed in grams correspond to one mole (*e.g.* 18 grams of water). One mole of a substance contains Avogadro's number (6.022×10^{23}) of particles (atoms, molecules, ions or even photons). Due to the definition of the mole, and because the unit of mass (weight) is the gram, we express atomic and molecular masses (weights) in grams. These are called atomic weight and molecular weight (gram atomic weight, gram molecular weight). We get the number of moles by dividing the quantity of material present by the atomic or molar mass, both expressed in grams (number of moles (m) = mass/molar mass).

3.1. Numeric Expression of Quantities

The accuracy of numbers, significant figures

Except the numbers we obtain by counting (the number of items), all numbers are inexact because their values are determined with a certain degree of uncertainty. The source of uncertainty is the limited capacity of either the measuring device (due to the flaws in its construction or improper calibration) or the measuring person (due to improper skills). We know the degree of uncertainty only in the case of our own measurements. Without a detailed discussion of the problem, the following is useful to know. In order to estimate and give the uncertainty of data, we need two (related) pieces of information: the order of magnitude and the significant figures. The significant figures are determined by the achievable precision (exactness) of the measurement. Usually, the last figure of a measured value bears uncertainty, *i.e.* this figure should be considered as estimated (significant figures = all certain digits + one estimated digit). The value and thus the significance of zero digits in numbers is dependent on their position. Zeros are not significant if they are at the beginning of a number (leading zeros), or if they are at the end (trailing zeros) without a decimal point in the number (even though such zeros carry information about the magnitude). In contrast, zero digits are significant if they are at the end of a number containing a decimal point (because they show the exactness of measurement), as well as if they are inside the number (confined zeros, located between nonzero digits). We must consider these aspects and the achievable precision of measurements when we simplify our numbers by rounding them up or down. We usually do this when we recalculate (mathematically transform) measured data or when we obtain them by calculation. (*e.g.*, a given amount of substance should be dissolved in a calculated solvent volume of 5.4786 ml. If our measuring device is calibrated with 0.1 ml division, then the figures "6" and "8" (corresponding to 0.0006 and 0.008 ml, respectively) are immeasurable, and figure "7" (corresponding to 0.07 ml) is uncertain. In this case, the desired volume can be approximated by pipetting 5.5 ml, considering the above uncertainties.) Carefully performed rounding is always recommended because series of digits of meaningless length make calculations unnecessarily difficult and bear the danger of calculation errors.

Expression of large and small quantities: exponential and prefix forms

In most cases, measured quantities differ from the unit of the given quantity by several orders of magnitude. In these cases, the length of the number cannot be substantially reduced by rounding. Two procedures are in use to avoid the writing of many zeros in the case of very large or small numbers, which would be uncomfortable and may be a source of error. During one of these procedures, large or small numbers are converted into their exponential form while keeping the unit of quantity (*i.e.* the scale). The other procedure replaces zeros with a prefix, *i.e.* it changes the unit of quantity (the scale). For example, we can write a quantity of 0.0000043 litre (L) as 4.3×10^{-6} L (exponential form) or 4.3 µl (microlitre, prefix form) according to the first and the second procedure, respectively. In the latter expression, the "µ" sign (read as micro) is called a prefix, which reflects the degree of change in the scale relative to the basic unit of quantity (in this example, six orders of magnitude downward) this way keeping the magnitude information that was originally conveyed by the eliminated (replaced) zeros. The definition of prefixes specifies the relationship between the two procedures. Official SI prefixes are defined at steps of three orders of magnitude above or below the unit of the given quantity, *i.e.* at 10^3, 10^6, 10^9 and 10^{-3}, 10^{-6}, 10^{-9} *etc.* Other (non SI) prefixes including the deci (10^{-1}), the centi (10^{-2}) and the hecto (10^2) are not official, although widely used in civil life outside the laboratory. A mixture of prefix and exponential forms within an expression should be avoided (*e.g.* 5.2×10^{-4} µg is expressed correctly as 0.52 ng), similarly to the way in which the normal and exponential forms of numbers are used. (In exponential notation we do not write values larger than

10 or smaller than 1 such as 3100 x 10^4 or 0.12 x 10^{-2}, because these numbers are correctly expressed as 3.1 x 10^7 and 1.2 x 10^{-3}, respectively). The use of prefix or exponential forms of numbers is often insufficient to reduce their length. Therefore, rounding (see above) is also a necessary simplification.

Order of Magnitude	Prefix	Symbol
10^{12}	Tera	T
10^{9}	Glga	G
10^{6}	Mega	M
10^{3}	kilo	k
10^{0} (=1, the unit)	-	-
10^{-3}	milli	m
10^{-6}	micro	μ
10^{-9}	nano	n
10^{-12}	plco	p
10^{-15}	farnto	f
10^{-18}	atto	a

Magnitude scales, prefixes and their relationship

3.2. Units of Measurement

The French International d'Unités (the SI system) is the accepted convention for all units of measurement. Following table lists the basic and derived SI units. Next table lists numerical values for some physical constant of SI units. The third table lists the commonly used prefixs associated with quantitative terms. Fourth table gives the interconversion of non-SI units of volume.

Quantity	SI Unit	Symbol (Basic SI units)	Definition of SI unit	Equivalent of SI unit
Basic Unit				
Length	Metre	M		
Mass	Kilogram	Kg		
Time	Secnd	S		
Electric current	Ampere	A		
Temperature	Kelvin	K		
Amount of substance	Mole	Mol		
Derived Unit				
Force	Newton	N	kg m s^{-1}	Jm^{-1}
Energy, work, heat	Joule	J	kg m^2 s^{-2}	N m
Power, radiant flux	Watt	W	kg m^2 s^{-3}	J s^{-1}
Electric charge	Coulomb	C	A s	J V^{-1}
Electric potential	Volt	V	kg m^2 s^{-3} A^{-1}	JC^{-1}
Electric resistance	Ohm	Ω	kg m^2s^{-3} A^{-2}	VA^{-1}
Pressure	Pascal	Pa	kg m^{-1} s^{-2}	N m^{-2}

(Contd.)

Quantity	SI Unit	Symbol (Basic SI units)	Definition of SI unit	Equivalent of SI unit
Frequency	Hertz	Hz	s^{-1}	
Magnetic flux density	Tesla	T	$kgs^{-2}A^{-1}$	Vsm^{-2}
Other Units Based on SI				
Area	Square meter	m^2		
Volume	Cubic meter	m^3		
Density	Kilogram /unit meter	kgm^{-3}		
Concentration	Mole/cubic meter	$molm^{-3}$		

SI Units – conversion factors for non-SI units

Unit	Symbol	SI Equivalent
Avagardo constant	L or N_A	6.022×10^{23} mol^{-1}
Faraday Constant	F	9.648×10^{4} $Cmol^{-1}$
Plank constant	k	6.626×10^{-14} Js
Universal or molar gas constant	R	8.314 $JK^{-1}mol^{-1}$
Molar volume of an ideal gas at s.t.p.		22.41 dm^3mol^{-1}
Velocity of light in vacuum	c	2.997×10^{8} ms^{-1}
Energy		
Calorie	cal	4.184 J
Erg	erg	10^{-7} J
electron volt	eV	1.602×10^{-19} J
Pressure		
Atmospheric pressure	atm	101325 Pa
Bar	bar	10^5 Pa
Millimeter of Hg	mm Hg	133.322 Pa
Temperature		
Centigrade	^0C	$(t^0 C + 273.15)K$
Fahrenheit	^0F	$5/9\,(t^0 F - 32) + 273.15K$
Length		
Ångstrom	Å	10^{-10} m
Inch	in	0.0254 m
Mass		
Pound	Ib	0.4536 kg

Common unit prefixs associated with quantitative terms

Multiply	Prefix	Symbol
10^{24}	Yotta	Y
10^{21}	Zetta	Z
10^{18}	Exa	E
10^{15}	Peta	P
10^{12}	Tera	T
10^{9}	Giga	G
10^{6}	Mega	M
10^{3}	Kilo	K
10^{2}	Hecta	H
10^{1}	Deca	Da

CHAPTER 4

About Solutions

As homogeneity is a basic criterion, it is important to know the solubility of solutes, which is a temperature dependent property. Besides temperature, solubility is influenced by the acidity (pH) of the solvent due to the acid-base character of the solute. Bases tend to dissolve better in acidic solutions, whereas acids do in basic solutions. Therefore, one often needs to change the pH of solutions accordingly. As the most widely used concentration units (molarity, w/v% and v/v%) refer the amount of solute in a given volume of the solution, it is the volume of the *complete solution* that must be set to the desired (calculated) value thus, one must consider not only the solvent but the solute(s) as well. Another reason why one must add the solute(s) to an amount of solvent which is less than the final volume might be that one will need to set the pH subsequently in order to enhance solubility and/or to reach a desired pH value. The volume of acid or base solutions used for pH setting is usually not known in advance and is not negligible. If the solute was initially dissolved in the final desired volume, one would exceed this volume during pH setting (even if the volume of the solute is negligible) and, thus, the solution would not have the desired (calculated) concentration but a smaller (and mostly unknown) one. Thus, according to the correct procedure, one must take into consideration the fact that the final volume will contain the volume of the buffering component. value particularly in the case of dissolving proteins, which have a tendency to stick to the bottom of the dish, thereby making dissolution substantially slower.

Dissolution of substances can be enhanced in various ways including mixing, increasing the temperature, applying sonication or by performing dissolution in the appropriate order (*e.g.* in the case of the Coomassie solution used to stain proteins in acrylamide gels). Mixing is a çommonly used and generally harmless method. Proteins are notoriously sensitive molecules; so are a number of simpler organic compounds, too. If we need to set the pH, we must consider the pH sensitivity of the solute. Sometimes, special requirements must be met such as sterility and/or the sensitivity of the solute to light or oxygen. Prepared solutions must appear clear following the complete solution of the solute(s). If this is not the case, the solution must be filtered. The dilution of concentrated sulphuric acid is a good example of necessary precaution: it is sulphuric acid that must be added to water (and not *vice versa*) because otherwise water which boils due to the high heat of salvation would sprinkle around the hot and corrosive acidic solution, causing serious injury.

Definition of Solutions and Their Main Characteristics

True solutions are homogeneous mixtures of two or more substances. The substance present in the largest quantity is the solvent, and all other components are called solutes. Homogeneous mixtures form a single phase by eye. (For example, we cannot discern two liquids within a solution, unlike in emulsions which, therefore, are not true solutions.) In true solutions, the atoms, molecules or ions of the solute are dispersed in the solvent. Theoretically, the solvent may be a gas, a solid, or a liquid, but solutions in which the solvent is a liquid are by far the most important in living organisms. Water is practically the only solvent that functions in plants, but the solutes concerned are numerous and includes all soluble inorganic salts that may be present in the soil, as well as oxygen, CO_2, sugars, organic acids and various other soluble organic compounds. Standard solution contains a known weight of the substances dissolved in a known volume of solvent. The term concentration denotes the proportion of solutes and solvents in a solution. There are numerous ways in which concentration and composition of a solution may be expressed. Special cases are those solutions in which the dispersion of the solute is not molecular (or atomic or ionic) but we cannot visually discern the solute (*e.g.* micelles). These are not molecular dispersions; yet they do not form a distinct phase. Such solutions are called colloids. The other definition of colloid solutions is that the particle size of the solute is in the range of 1 to 1000 nanometres. The solutions of most proteins fall within this category; *i.e.* these are colloid solutions, albeit proteins in these solutions are molecular dispersions. In laboratory practice, mixtures in a liquid state are usually called solutions. In the biochemical laboratory we generally prepare aqueous solutions, *i.e.* solutions in which the solvent is water. The main reason for this is that we usually work with proteins and other biomolecules that can stably adopt their native conformation in an aqueous environment. We often prepare solutions in organic solvents in order to dissolve smaller compounds (*e.g.* substrates of enzymes) or in the case of HPLC procedures. Most of the organic solvents used in biochemical experiments are miscible with water. As a semi-quantitative characterisation of solutions, we can distinguish unsaturated, saturated and supersaturated solutions. These solutions contain less, the same, or more than the maximum possible amount of solute, respectively. (These characteristics, together with solubility, are temperature dependent).

Per cent solution: expressed in terms of percentage. In this type of a solution a known unit of the solute is dissolved in 100 units of water. When the solute is a solid and the solvent a liquid, a percent solution is prepared on a w/v basis, *i.e.*, a known wt. of a substance (solute) is added to 100 ml of the solvent.

Example: to prepare a 10 per cent of $CuSO_4$ solution we add 10.0 g of $CuSO_4$ to 100 ml of water. On the other hand, if both the solute and the solvent are liquids, percent solution is prepared on volume basis, and here, a known volume of the solute is taken and the final volume is then made up to 100 ml with the solvent.

Example: to make a 35 per cent solution of perchloric acid, 35 ml of perchloric acid is taken in a measuring cylinder and the volume is made up to 100 ml with water. This method of expression does not show the relative number of the solute that are present in the frequently in the laboratory. They are generally found in adequate precise work. When we use salts while preparing a percent solution, based on weight by volume, it will give percent of salts but not different ion in the salt.

Example: To prepare percent solution of an ion present in the salt that (for *e.g.*, Na in NaOH), the ratio of the ion to the total salt has to be determined. This is calculated based on the ion.

Mol. Wt of NaOH = 40

Na = 23

O = 16

H = 1 *i.e.*, 23g of Na is present in 40g of NaOH.

To prepare 10 percent Na solution, 10g of Na has to be dissolved in 100 ml H_2O.

23g Na is present in 40g NaOH.

17.4g of NaOH contains 10g of Na.

Normal solutions

A normal solution contains 1g equivalent weight of a dissolved substance (solute) per litre of the solution.

$$\text{1 g equivalent wt. of a substance} = \frac{\text{mol. wt. of the solution}}{\text{no. of replaceable hydrogen atoms of hydrogen equivalents}}$$

Hydrochloric acids (HCl) with mol. Wt. of 36.47 has one replaceable hydrogen atom, hence 1 liter of 1 N HCl contains 36.47g of the acid. On the other hand, sulfuric acid (H_2SO_4) having a mol. Wt. of 98, has 2 replaceable hydrogen atoms. Therefore a normal solution of H_2SO_4 contains 49g (98/2) of the acid per liter.

Similarly 1 N NaOH will contain 40g of the base per litre (as hydrogen atom is equivalent to 1 acid hydrogen atom), while 1 N $Ca(OH)_2$ contains 74.096/2=37.048 g/litre.

Salts are also considered in the same terms. 1 N K_2SO_4 contains 174.26/2=87.13 g/litre (K_2SO_4 contains 2 hydrogen equivalents).

To make 1N NaOH, 40g of NaOH is first dissolved in a small quantity of water and make up the volume of 1 litre.

To make 2N NaOH, dissolve 40 x 2 g (80g) NaOH in a small

To make 0.1 N NaOH, dissolve 40 x 0.1g (4.0g) of NaOH in a small quantity of water and make up to the volume of 1 liter.

Molar solutions

A molar substance is one containing as many grams of the solute per litre of the solution as the molecular weight of the dissolved substance, *i.e.*, one gram molecular weight of a substance dissolved in 1 litre of solution.

Example:

1.0 M sucrose: 342.2 g sucrose dissolved in a small quantity of water and the volume is made up to 1 litre.

342.2 mg dissolved in litre – 1 mM

342.2 μg dissolved in litre - 1μM

When we have to prepare the molar concentration of iron in a substance the molecular ratio of that ion to the total molecular weight of the compound has to be considered.

NaOH - Mol. Wt. = 40

40g in a litre = 1M NaOH

To get 1 M of Na the amount of Na needed is 23 g which is present in 40 g of NaOH.

K_2SO_4 – Mol.wt = 174.2

174.2g in a liter =1 M K_2SO_4

To get 1M of K the amount of K_2SO_4 needed is 87.1g (174.2/2 = 87.1)

Parts per million (ppm) solution

One mg of a substance dissolved in 1 liter of a solution will give 1ppm solution.

The quantity of substances necessary to prepare 10 ppm solution is same for substances differing in molecular weight.

Example: GA – mol wt. = 342.0

10 mg of GA in a liter - 10ppm

IAA – mol wt. =175

10mg of IAA in a litre – 10 ppm

Though the molecular weight is different, the amount necessary to prepare 10ppm is 10 mg.

The molarity of 1000 ppm of different solutions will be different depending on the molecular weight of that compound.

Eg: 1000 ppm of NaOH is 0.025 M

1000 ppm of sucrose is 0.0029 M

4.1. Quantitative Description of Solutions, Concentration Units

We specify the quantity of components in a mixture quantitatively as their ratio. The precise expression of this can be achieved via concentration units. In liquid-state solutions, it is usually a certain quantity of the solution to which the quantiti(es) of its component(s) are compared, depending on the concentration unit. For instance, in the case of molar concentration (molarity, M, mol/dm^3 or mol/l), the most widely used concentration unit in biochemistry, we refer to the number of moles of a given substance in one litre of solution. (*e.g.*, 0.2 M means 0.2 moles of a substance in one liter of solution.) Various percentage forms (%, reference to 100 units of something) are also widely used in biochemistry, although they are not SI units. In the case of weight-by-volume and volume-by-volume percentages (w/v % and v/v %, respectively) we refer to 100 ml of solution, whereas in the case of weight-by-weight percentage (w/w %) we refer to 100 g of solution. In the case of v/v % the volume of the solute, while in the case of w/v % and w/w % the weight of the solute is the basis of reference. (*e.g.*, a 15 w/v% solution contains 15 g of solute in 100 ml of solution.) During conversions from or into w/v%, one must also consider the density of the solution (unless it is one g/ml). The density of dilute solutions (up to several %) generally deviates only negligibly from that of the solvent. Therefore, the w/w % and w/v % values of such solutions are practically identical (*i.e.* 3 w/v % corresponds to approximately 3 w/w %). It is important to note that, for practical reasons, the biochemical concentration units mg/ml and μg/ml are also used in the case of protein and other macromolecular solutions in biochemical experiments.

4.2. Preparation of Solutions

As homogeneity is a basic criterion, it is important to know the solubility of solutes, which is a temperature-dependent property. Information on the solubility of compounds can be obtained from handbook tables (for inorganic compounds) or from the manufacturer (product catalogues of organic compounds). Besides temperature, solubility is influenced by the acidity (pH) of the solvent due to the acid-base character of the solute. Bases tend to dissolve better in acidic solutions, whereas acids do in basic solutions. Therefore, one often needs to change the pH of solutions accordingly. This can be most simply achieved by the addition of an inorganic acid or base (HCl, NaOH, KOH *etc.*) or by using a buffer system. In both cases, one must be aware of the fact that the prepared solution is a multi-component one, *i.e.* it contains a pH-setting substance in addition to the substance(s) to be dissolved. In order to achieve the desired concentration precisely, the measuring range of the devices used (cylinders, pipettes, balances *etc.*) must be in the range of the quantity to be measured. (For example, a 70-ml solution should not be prepared in a measuring cylinder of a maximal volume of 250 ml, but in one with a 100-ml maximal volume.) As the most widely used concentration units (molarity, w/v % and v/v %) refer the amount of solute in a given volume of the solution (see above), it is the volume of the *complete solution* that must be set to the desired (calculated) value—thus, one must consider not only the solvent but the solute(s) as well. For instance, when preparing one liter of a 4-M $CaCl_2$ solution, the required amount of $CaCl_2$ must be added not to one liter of water but to much less *(e.g.* to 500-600 ml, regardless of the condition whether it can dissolve the applied amount of $CaCl_2$), and subsequently complemented with additional water to fill up the volume to one liter. This is the only way to ascertain that the total volume of the solution ($CaCl_2$ and water together) will be one litre. During this procedure, one must take into account that the solute also contributes to the total (final) volume of the solution. Another reason why one must add the solute(s) to an amount of solvent which is less than the final volume might be that one will need to set the pH subsequently in order to enhance solubility and/or to reach a desired pH value. The volume of acid or base solutions used for pH setting is usually not known in advance—and is not negligible. If the solute was initially dissolved in the final desired volume, one would exceed this volume during pH setting (even if the volume of the solute is negligible) and, thus, the solution would not have the desired (calculated) concentration but a smaller (and mostly unknown) one. Thus, according to the correct procedure, one must take into consideration the fact that the final volume will contain the volume of the buffering component.

During mixing the components of a solution, it is often important to be aware of the appropriate and/or safe order of addition. The dilution of concentrated sulphuric acid is a good example of necessary precaution: it is sulphuric acid that must be added to water (and not *vice versa*) because otherwise water—which boils due to the high heat of solvation—would sprinkle around the hot and corrosive acidic solution, causing serious injury. The appropriate order of dissolving solid solutes is the addition of the solute into the solvent (preferably into less than the final volume aimed)—and not *vice versa*, pouring solvent onto the solute. This knowledge is of high practical value particularly in the case of dissolving proteins, which have a tendency to stick to the bottom of the dish, thereby making dissolution substantially slower. Dissolution of substances can be enhanced in various ways including mixing, increasing the temperature, applying sonication or by performing dissolution in the appropriate order (*e.g.* in the case of the Coomassie solution used to stain proteins in acrylamide gels). Mixing is a commonly used and generally harmless method. However, special care must be taken during sonication and especially during heating: one must consider the (heat) stability of the dissolved compound. Proteins are notoriously sensitive molecules; so are a number of simpler organic compounds, too. If we need to set the pH, we must consider the pH sensitivity of the solute. Undesired events and reactions can be prevented by the avoidance of the addition of large quantities or high concentrations of acids or bases. Instead, it is better to use more dilute acid or base solutions and add these in small portions. Sometimes, special requirements must be met such as sterility and/or the sensitivity of the solute to light or oxygen. If the solute is heat sensitive, the solution cannot be sterilized by heat. In these cases, filtration must be applied. Prepared solutions must appear clear following the complete solution of the solute(s). If this is not the case, the solution must be filtered. (Obviously, one should check if it is not a part of the solute that fell out of the solution.) To this end, one can use traditional tools (filter papers, glass filters) or disposable filters (or filter units) of various materials and pore sizes according to the expected purity of the solution. For example, the solutions used in FPLC, HPLC or in optical devices (photometers, fluorometers) may not contain any floating particles because these are harmful to the device (HPLC, FPLC) or may severely interfere with the measurement. Solutions are generally stored in glass or plastic containers (bottles, flasks or tubes) that must be tightly closed to prevent changes in concentration due to the slow evaporation of the solvent. (This is much more difficult to achieve in the case of more volatile organic solvents. Here, the use of glass screw-cap vials with Teflon cover is advisable.) For this reason, bakers, Erlenmeyer or other flasks and measuring cylinders (used during preparative procedures) are not useful for this purpose as they do not meet the above criterion of safe storage even in the case of aqueous solutions. Solutions of most of inorganic and organic compounds are generally stored at room temperature. However, in the biochemical laboratory, some solutions must be stored in a refrigerator or a freezer. The later way of storage is needed to achieve chemical and/or microbiological stability. Unless they are sterilised, solutions of proteins and even those of simple inorganic compounds (*e.g.* phosphate salts) provide a favourable environment for the growth of bacteria and fungi. (The appearance of these microorganisms is indicated by the increased turbidity and unpleasant smell of the solution. Algae may even appear in distilled water!) Some enzymes (*e.g.* proteases) may also slowly lose their activity. Cooling generally below freezing temperature cannot completely prevent but at least substantially slow down such processes.

Dishes and containers of solutions must be provided with appropriate labels that must remain readable throughout the whole time period of the potential usage of the solution. In the case of solutions containing one or a few different solutes, the label should include the name (or the chemical formula) and the concentration of the solute(s). In many cases, specific names are given to more complex multi-component solutions. These names must be unambiguous at least within the laboratory (*e.g.* activation solution , sonication buffe, 10x reaction mixture). Names of many solutions are international if they are widely used during a common laboratory procedure.

CHAPTER 5

Buffer and Buffer Solution

A buffer is an aqueous solution consisting of a mixture of a weak acid and its conjugate base or a weak base and its conjugate acid. Its pH changes very little when a small amount of strong acid or base is added to it and thus it is used to prevent changes in the pH of a solution. Buffer solutions are used as a means of keeping pH at a nearly constant value in a wide variety of chemical applications. Many life forms thrive only in a relatively small pH range so they utilize a buffer solution to maintain a constant pH. One example of a buffer solution found in nature is blood.

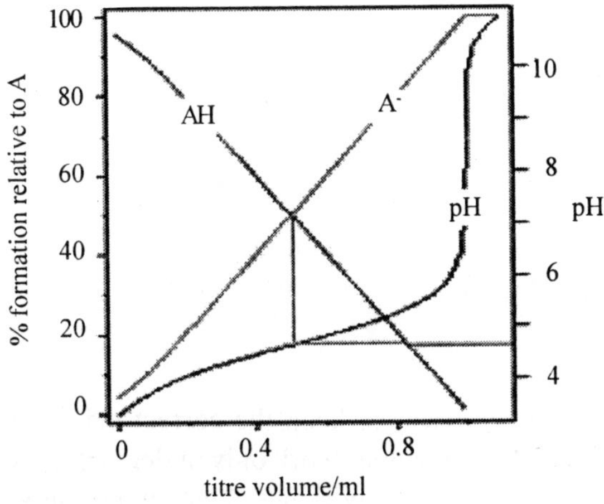

Principles of buffering

Simulated titration of an acidified solution of a weak acid (pK_a = 4.7) with alkali.

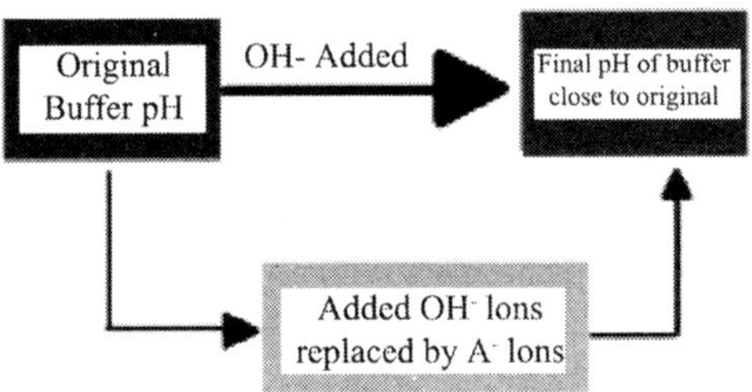

Addition of hydroxide to a mixture of a weak acid and its conjugate base

Buffer solutions achieve their resistance to pH change because of the presence of an equilibrium between the acid HA and its conjugate base A^-.

$$HA \rightleftharpoons H^+ + A^-$$

When some strong acid is added to an equilibrium mixture of the weak acid and its conjugate base, the equilibrium is shifted to the left, in accordance with Le Chatelier's principle. Because of this, the hydrogen ion concentration increases by less than the amount expected for the quantity of strong acid added. Similarly, if strong alkali is added to the mixture the hydrogen ion concentration decreases by less than the amount expected for the quantity of alkali added. The effect is illustrated by the simulated titration of a weak acid with pK_a = 4.7. The pH changes relatively slowly in the buffer region, pH = $pK_a \pm 1$, centered at pH = 4.7 where [HA] = [A^-]. The hydrogen ion concentration decreases by less than the amount expected because most of the added hydroxide ion is consumed in the reaction

$$OH^- + HA \rightarrow H_2O + A^-$$

and only a little is consumed in the neutralization reaction which results in an increase in pH.

$$OH^- + H^+ \rightarrow H_2O$$

Once the acid is more than 95% deprotonated the pH rises rapidly because most of the added alkali is consumed in the neutralization reaction.

Applications

Buffer solutions are necessary to keep the correct pH for enzymes in many organisms to work. Many enzymes work only under very precise conditions; if the pH moves outside of a narrow range, the enzymes slow or stop working and can denature. In many cases denaturation can permanently disable their catalytic activity. A buffer of carbonic acid (H_2CO_3) and bicarbonate (HCO_3^-) is present in blood plasma, to maintain a pH between 7.35 and 7.45.

The majority of biological samples that are used in research are made in buffers, especially phosphate buffered saline (PBS) at pH 7.4.

Simple buffering agents

Buffering agent	pK_a	Useful pH range
Citric acid	3.13, 4.76 – 6.40	2.1 – 7.4
Acetic acid	4.8	3.8 – 5.8
KH_2PO_4	7.2	6.2 – 8.2
CHES	9.3	8.3 – 10.3
Borate	9.24	8.25 – 10.25

For buffers in acid regions, the pH may be adjusted to a desired value by adding a strong acid such as hydrochloric acid to the buffering agent. For alkaline buffers, a strong base such as sodium hydroxide may be added. Alternatively, a buffer mixture can be made from a mixure of an acid and its conjugate base. For example, an acetate buffer can be made from a mixture of acetic acid and sodium acetate. Similarly an alkaline buffer can be made from a mixture of the base and its conjugate acid.

5.1. Universal Buffer Mixtures

By combining substances with pK_a values differing by only two or less and adjusting the pH, a wide range of buffers can be obtained. Citric acid is a useful component of a buffer mixture because it has three pK_a values, separated by less than two. The buffer range can be extended by adding other buffering agents. The following two-component mixtures (McIlvaine's buffer solutions) have a buffer range of pH 3 to 8.

0.2M Na_2HPO_4 /ml	0.1M Citric Acid /ml	pH...
20.55	79.45	3.0
38.55	61.45	4.0
51.50	48.50	5.0
63.15	36.85	6.0
82.35	17.65	7.0
97.25	2.75	8.0

A mixture containing citric acid, potassium dihydrogen phosphate, boric acid, and diethyl barbituric acid can be made to cover the pH range 2.6 to 12.

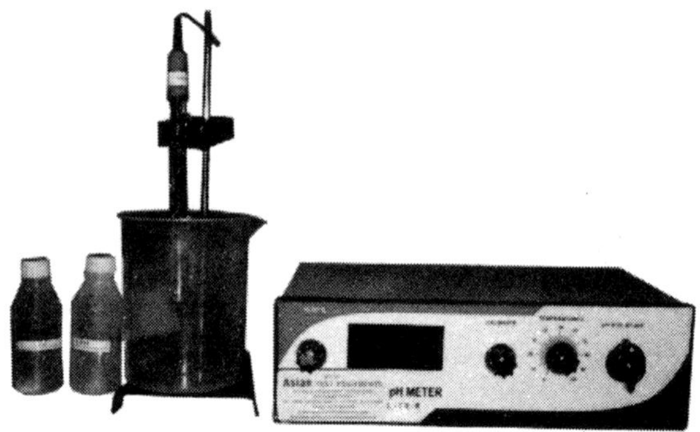

pH Meter

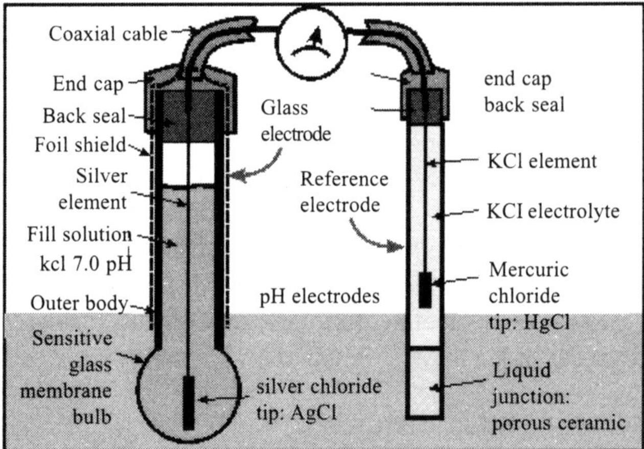

Electrodes of a pH meter

5.2. Common Buffer Compounds Used in Biology

Common Name	Pk_a at 25^0 C	Buffer range	Temp. effect dp H/dT in (1/K)	Mol. Weight	Full Compound Name
TAPS	8.43	7.7-9.1	-0.018	243.3	3-{[tris (hydroxymethyl) methyl] amino} propanesulfonic acid
Bicine	8.35	7.6-9.0	-0.018	163.2	N, N-bis (2-hydroxymethyl) glycine
Tris	8.06	7.5-9.0	-0.028	121.14	Tris (hydroxymethyl) methylamine
Tricine	8.05	7.4-8.8	-0.021	179.2	N-tris(hydroxymethyl)methylglycine
TAPSO	7.635	7.0-8.2		259.3	3-[N-Tris (hydroxy methyl) methy lamino]-2-hydroxypropansulfonic acid
HEPES	7.48	6.8-8.2	-0.014	238.3	4-2-hydroxymethyl-1-piperazinesulfonic acid

(*Contd.*)

Common Name	Pk_a at 25^0 C	Buffer range	Temp. effect dp H/dT in (1/K)	Mol. Weight	Full Compound Name
TES	7.40	6.8-8.2	-0.020	229.20	2-{[tris(hydroxymethyl) methyl] amino} ethanesulfonic acid
MOPS	7.20	6.5-7.9	-0.015	209.3	3-(N-morpholino)propanesulfonic acid
PIPES	6.76	6.1-7.5	-0.008	302.4	Piperazine-N,N'-bis(2-ethanesulfonic acid)
Cacodylate	6.27	5.0-7.4		138.0	Dimethylarsinic acid
SSC	7.0	6.5-7.5		189.1	Saline sodium citrate
MES	6.15	5.5-6.7	-0.011	195.2	2-(N-norpholino)ethanesulfonic acid
Succinic acid	7.4	7.4-7.5		181.2	2(R)-2-(methylamino)succinic acid

Buffer capacity

Buffer capacity, β, is a quantitative measure of the resistance of a buffer solution to pH change on addition of hydroxide ions. It can be defined as follows.

$$\beta \frac{dn}{d(p[H^+])}$$

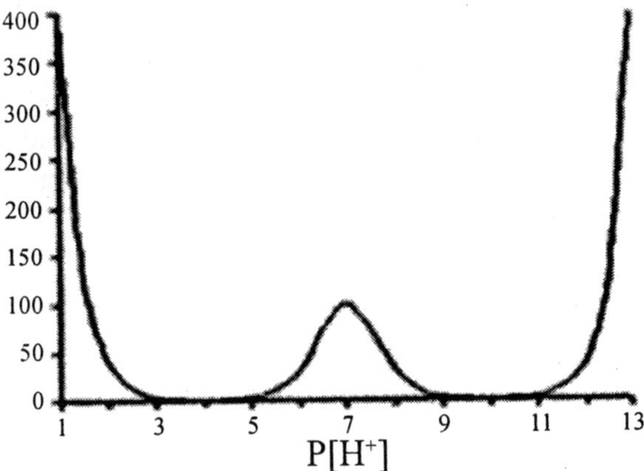

Buffer capacity for a 0.1 M solution of an acid with pK_a of 7

where **dn** is an infinitesimal amount of added base and **d(p[H$^+$])** is the resulting infinitesimal change in the cologarithm of the hydrogen ion concentration. With this definition the buffer capacity of a weak acid, with a dissociation constant K_a, can be expressed as

$$\frac{dn}{d(pH)} = 2.303\left([H^+] + \frac{C_A K_a [H^+]}{(K_a + [H^+])^2} + [OH^-]\right)$$

where C_A is the analytical concentration of the acid. pH is defined as $-\log_{10}[H^+]$.

There are three regions of high buffer capacity.

- At very low p[H$^+$] the first term predominates and β increases in proportion to the hydrogen ion concentration. This is independent of the presence or absence of buffering agents and applies to all solvents.

- In the region $p[H^+] = pK_a \pm 2$ the second term becomes important. Buffer capacity is proportional to the concentration of the buffering agent, C_A, so dilute solutions have little buffer capacity.
- At very high $p[H^+]$ the third term predominates and β increases in proportion to the hydroxide ion concentration. This is due to the self-ionization of water and is independent of the presence or absence of buffering agents.

The buffer capacity of a buffering agent is at a maximum $p[H^+] = pK_a$. It falls to 33% of the maximum value at $p[H^+] = pK_a \pm 1$ and to 10% at $p[H^+] = pK_a \pm 1.5$. For this reason the useful range is approximately $pK_a \pm 1$.

Calculating buffer pH

Monoprotic acids

First write down the equilibrium expression.

$$HA \rightleftharpoons A^- + H^+$$

This shows that when the acid dissociates equal amounts of hydrogen ion and anion are produced. The equilibrium concentrations of these three components can be calculated in an ICE table.

ICE table for a monoprotic acid

	[HA]	[A⁻]	[H⁺]
I	C_0	0	y
C	-x	x	x
E	C_0-x	x	x+y

The first row, labelled I, lists the initial conditions: the concentration of acid is C_0, initially undissociated, so the concentrations of A^- and H^+ would be zero; y is the initial concentration of *added* strong acid, such as hydrochloric acid. If strong alkali, such as sodium hydroxide, is added y will have a negative sign because alkali removes hydrogen ions from the solution. The second row, labelled C for change, specifies the changes that occur when the acid dissociates. The acid concentration decreases by an amount $-x$ and the concentrations of A^- and H^+ both increase by an amount $+x$. This follows from the equilibrium expression. The third row, labelled E for equilibrium concentrations, adds together the first two rows and shows the concentrations at equilibrium.

To find x, use the formula for the equilibrium constant in terms of concentrations:

$$K_a = \frac{[H^+][A^-]}{[HA]}$$

Substitute the concentrations with the values found in the last row of the ICE table:

$$K_a = \frac{x(x+y)}{C_0 - x}$$

Simplify to:

$$x^2 + (K_a + y)x - K_a C_0 = 0$$

With specific values for C_0, K_a and y this equation can be solved for x. Assuming that pH = $-\log_{10}[H^+]$ the pH can be calculated as pH = $-\log_{10}x$.

Polyprotic acids

Polyprotic acids are acids that can lose more than one proton. The constant for dissociation of the first proton may be denoted as K_{a1} and the constants for dissociation of successive protons as K_{a2}, *etc.* Citric acid, H_3A, is an example of a polyprotic acid as it can lose three protons.

Equilibrium	pK_a value
$H_3A \rightleftharpoons H_2A^- + H^+$	pK_{a1} = 3.13
$H_2A^- \rightleftharpoons HA^{2-} + H^+$	pK_{a2} = 4.76
$HA^{2-} \rightleftharpoons A^{3-} + H^+$	pK_{a3} = 6.40

When the difference between successive pK values is less than about three, there is overlap between the pH range of existence of the species in equilibrium. The smaller the difference, the more the overlap. In the case of citric acid, the overlap is extensive and solutions of citric acid are buffered over the whole range of pH 2.5 to 7.5.

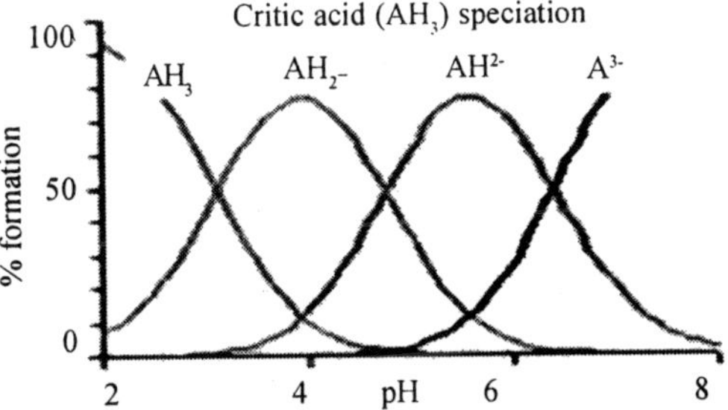

% species formation calculated for a 10 millimolar solution of citric acid.

Calculation of the pH with a polyprotic acid requires a speciation calculation to be performed. In the case of citric acid, this entails the solution of the two equations of mass balance

$$C_A = [A^{3-}] + \beta_1[A^{3-}][\mathrm{H}^+] + \beta_2[A^{3-}][H^+]^2 + \beta_3[A^{3-}][H^+]^3$$
$$C_H = [H^+] + \beta_1[A^{3-}][\mathrm{H}^+] + 2\beta_2[A^{3-}][H+]^2 + 3\beta_3[A^{3-}][H^+]^3 - K_\omega[H]^{-1}$$

C_A is the analytical concentration of the acid, C_H is the analytical concentration of added hydrogen ions, β_q are the cumulative association constants

$$\log\beta_1 = pK_{a3}, \log\beta_2 = pK_{a2} + pK_{a3}, \log\beta_3 = pK_{a1} + pK_{a2} + pK_{a3}$$

K_w is the constant for Self-ionization of water. There are two non-linear simultaneous equations in two unknown quantities $[A^{3-}]$ and $[H^+]$. Many computer programs are available to do this calculation. In general the two mass-balance equations can be written as

$$C_A = [A] + \sum p\beta q[A]^p[H^+]^q$$
$$C_H = [H^+] + \sum q\beta q[A]^p[H^+]^q - K_\omega[H]^{-1}$$

In this general expression [A] stands for the concentration of the fully deprotonated acid and the electrical charge on this species is not specified.

5.3. Composition of Different Buffer Systems

Buffer No.	A	B	Composition
1.	KCl 0.2 N (14.91 g/l)	HCl 0.2 N	25 ml A + X ml B and made to 1000 ml.
2.	Glycie 0.01M in NaCl 0.1M (7.507 g glycine + 5.844 g NaCl/l)	HCl 0.1N	X ml of A + (100 – X) ml of B
3.	Disodium citrate 0.1M (21.06 g $C_6H_8O_7.7\ H_2O$ + 200 ml NaOH 1.0 N /l)	HCl 0.1 N	X ml A + (100 – X ml) of B
4.	Potassium bipthalate 0.1 M (20.42 g $KHC_8H_4O_4$ /l)	HCl 0.1 N	50 ml A + X ml B and made to 100 ml
5.	As in No. 4	NaOH 0.1 N	50 ml A + X ml of B and made to 100 ml
6.	As in No. 3	NaOH 0.1 N	X ml A + (100 – X) ml B
7.	Monopotassium phosphate 1/15 M (9.073 g KH_2PO_4 / l)	Disodium phosphate 1/15 M (11.87 g Na_2HPO_4 / l)	X ml A + (100 – X) ml B
8.	Barbitol sodium 0.1 M (20.62 g / l)	HCl 0.1 N	X ml A + (100 – X) ml B
9.	Boric acid half neutralized 0.2 M (12.37 g / l + 100 ml NaOH 0.1 M)	HCl 0.1 N	X ml A + (100 – X ml) B
10.	As in No. 2	NaOH 0.1 N	X ml A + (100 – X ml) B
11.	As in No. 9	NaOH 0.1 N	X ml A + (100 – X ml) B
12.	Citric acid 0.1 M (21.01 g $C_6H_8O_7.\ H_2O$/ l)	Disodium phosphate 0.2 M (35.60 g $Na_2HPO_4.2H_2O$ / l)	X ml A + (100 – X ml) B
13.	To citric acid and phosphoric acid salts each equivalent to 100 ml of 1.0 N NaOH add 3.54 g crystalline ortho boric acid and 343 ml of 1.0 N NaOH and make the volume to 1.0 liter)	HCl 0.1 N	20 ml of A + X ml of B and made up to 100 ml
14.	Citric acid monopotassium phosphate, barbital, boric acid all 0.0285 N (6.004 g, 3.888 g, 5.263 g & 1.767 g /l, respectively)	NaOH 0.2 N	100 ml A + X ml B
15.	Sodium acetate 0.1 N (8.204 g $C_2H_3O_4Na$ or 13.61 g $C_2H_3O_2Na.3H_2O$ / l)	Acetic acid 0.1 N	X ml A + (100 – X)ml B

(Contd.)

Buffer No.	A	B	Composition
16.	Bβ-dimethyl glutamic acid 0.1 N (16.02 g / l)	NaOH 0.2 N	A. 100 ml A + X ml B and made to 1000 ml B. 100 ml A + X ml B + 5.844 g NaCl and made to 1000 ml.
17.	Piperazine 1.0 M (86.14 g / l)	HCl 0.1 N	5 ml A + X ml B and made up to 100 ml
18.	Tetraethyle ethylene diamine 1.0 M (172.32 g / l)	HCl 0.1 N	5 ml A + X ml B and made up to 100 ml
19.	Tris, maleic acid 0.2 N (24.23 g Tris + 23.21 g maleic acid or 19.61 g maleic anhydride / l)	NaOH 0.2 N	25 ml A + X ml B and made up to 100 ml
20.	Dimethylaminoethylamine 1.0 N (88.0 g/l)	HCl 0.1 N	5 ml A + X ml B and made up to 100 ml
21.	Imidazole 0.2 M (13.62 g/l)	HCl 0.1 N.	25 ml A + X ml B and made up to 100 ml
22.	Triethanolamine 0.5 M (76.11 g/l) containing 20 g/l ethylnetetra acetic acid disodium salt ($C_{10}H_{14}O_8Na_2.2H_2O$)	HCl 0.5 N	10 ml A + X ml B made up to 100 ml.
23.	N.dimethyl aminoleucyl glycine 0.1 M (24.33 g/l) containing NaCl 0.2 M (11.69 g/l)	NaOH 1.0 N 100 ml made up to 1 liter with A	X ml A + (100 – X) ml B
24.	Tris 0.2 N (24.23 g/l)	HCl 0.1 N	25 ml A + X ml B made up to 100 ml.
25.	2-Amino, 2-methyl propane 1-3 diol 0.1 N (10.51 g/l)	HCl 0.1 N	50 ml A + X ml B made up to 100 ml
26.	Sodium bicarbonate anhydrous 0.1 M (10.69 g/l)	Na_2CO_3 0.1 M (8.401 g/l)	X ml A + (100 – X) ml B

Note

Borate buffer: This compound forms complexes with a number of compounds such as sugars.

Citrate: This ion readily combines with Ca and cannot be used in presence of this metal ion.

The pH in many biological experiments often needs to be maintained at pH 6-8, but these are relatively few buffers which are effective over this range and particularly around pH 7.4.

Phosphate: This may act as an enzyme inhibitor, or even as a metabolite in some experiments. Heavy metals are readily precipitate from solution as the phosphate and it has a poor buffering capacity above pH 7.5.

Tris: This buffer can be used with heavy metals but may also act as an inhibitor in some system. Its major disadvantage is the temperature effect, which is often overlooked. A tris buffer of pH 7.8 at room temperature has a pH of 8.4 at 4^0 C and 7.4 at 37^0 C, so that the $[H^+]$ increases ten folds from preparation of material at 4^0 C to its measurement at 37^0 C. It has a poor buffering capacity below pH 7.5.

Table : This table gives the amount of X ml of stock solution listed above required to make a buffer of desired pH.

	Different buffer systems												
pH	1	2	3	4	5	6	7	8	9	10	11	12	13
1.0	54.2												
1.2	36.0	11.1	9.0										
1.4	23.2	26.4	17.9										
1.6	14.7	36.2	23.6										
1.8	9.3	43.9	27.6										
2.0	5.9	50.7	30.2										74.4
2.2	3.8	56.5	33.2									98.4	68.8
2.4		62.3	34.1	41.0								94.5	64.6
2.6		68.4	36.0	34.3								90.0	61.3
2.8		74.7	37.9	27.8								85.1	58.9
3.0		81.0	39.9	21.6								80.3	56.9
3.2		86.2	42.1	15.9								76.0	55.2
3.4		90.3	44.8	10.9								72.0	53.9
3.6			47.8	6.7								68.4	52.9
3.8			51.2	3.3								65.1	51.8
4.0			55.1	0.0								62.0	50.7
4.2			60.0		3.3							59.1	49.7

(*Contd.*)

pH	1	2	3	4	5	6	7	8	9	10	11	12	13
4.4			66.4		6.7							56.4	48.6
4.6			74.9		11.1							53.7	47.5
4.8			85.0		16.5							51.2	46.4
5.0			100		22.6		99.2					49.0	45.4
5.2					28.8	87.1	98.4					46.2	44.3
5.4					34.4	78.0	97.3					44.7	43.2
5.6					39.1	70.3	95.5					42.4	42.0
5.8					42.4	64.5	92.8					40.0	40.8
6.0					45.0	60.3	88.9					37.4	39.7
6.2					46.7	57.2	83.0					34.5	38.4
6.4						54.8	75.4					31.4	37.0
6.6						53.2	65.3					27.9	35.6
6.8							53.4					23.5	34.2
7.0							41.3	53.3				19.0	32.9
7.2							29.6	55.0				13.8	31.7
7.4							19.7	57.6				9.8	30.6
7.6							12.8	60.8				6.8	29.6
7.8							7.4	65.2	53.3			4.6	28.8
8.0							3.7	70.6	55.4				28.1

(Contd.)

pH	1	2	3	4	5	6	7	8	9	10	11	12	13
8.2								75.9	58.0				27.6
8.4								81.2	62.1				27.0
8.6								86.2	66.9	94.7			26.3
8.8								90.1	73.6	92.0			25.2
9.0								93.2	83.5	88.4			24.0
9.2									95.6	84.0			22.6
9.4										78.9	87.0		21.4
9.6										73.2	75.5		20.2
9.8										67.2	65.1		19.0
10.0										62.5	57.6		18.1
10.2										58.8	56.4		17.1
10.4										55.7	54.1		16.5
10.6										53.6	52.3		16.0
10.8										52.2			15.5
11.0										51.2			14.7
11.2										50.4			13.5
11.4										49.5			11.7
11.6										48.7			9.1

pH	1	2	3	4	5	6	7	8	9	10	11	12	13
11.8										47.6			5.5
12.0										46.0			1.3
12.2										43.2			
12.6										21.4			

Different Buffer systems

pH	14	15	16A	16B	17	18	19	20	21	22	23	24	25	26
1.0														
1.2														
1.4														
1.6														
1.8														
2.0														
2.2														
2.4														
2.6	1.6													
2.8	3.6													
3.0	5.7													
3.2	7.8		7.0	14.4										

(Contd.)

pH	14	15	16A	16B	17	18	19	20	21	22	23	24	25	26
3.4	9.9		13.3	20.9										
3.6	11.7		20.7	26.8										
3.8	13.5	10.9	26.3	32.4										
4.0	15.3	16.6	32.4	36.6										
4.2	17.5	23.9	36.2	40.2										
4.4	19.7	33.5	39.3	43.1										
4.6	21.9	44.9	41.3	45.7	94.3									
4.8	24.1	56.6	43.5	48.3	91.5									
5.0	26.3	67.8	45.7	51.5	87.8									
5.2	28.3	76.8	48.4	53.6	83.6									
5.4	31.0	84.0	51.3	58.2	77.6									
5.6	33.4	89.3	55.0	63.6	71.8									
5.8	39.8		58.8	68.7	66.5									
6.0	38.3		63.9	73.6	61.8	71.8	12.4	88.0						
6.2	40.8		69.5	78.5	58.2	66.5	15.2	83.3	43.4					
6.4	43.3		74.1	83.3	55.5	61.8	17.9	77.9	40.4					
6.6	45.8		83.5	87.4		58.2	20.8	72.0	36.5					
6.8	48.3		87.4	91.0		55.3	22.2	66.6	31.4					
7.0	50.9		90.1	93.2			23.7	61.9	25.4	86.2	86.4			

(Contd.)

pH	14	15	16A	16B	17	18	19	20	21	22	23	24	25	26
7.2	53.4		91.8	94.9			25.2	58.1	19.6	79.6	80.6	44.7		
7.4	55.8		93.0	95.8			26.7	55.3	14.6	71.3	72.8	42.0		
7.6	58.2		93.8	96.8			28.6		10.2	62.0	63.2	39.3		
7.8	60.5						31.2		6.6	52.0	52.1	33.7	43.9	
8.0	62.8						33.1			42.0	41.1	27.9	41.6	
8.2	65.0					46.4	36.9			31.9	31.4	17.3	38.4	
8.4	67.2					43.9	39.9			22.8	23.0	13.0	34.8	
8.6	69.3					40.9	42.7	45.4		16.0	15.9	8.8	30.7	
8.8	71.3					36.8		42.8		11.7	10.3	5.3	23.3	
9.0	73.0					31.8		39.2					17.7	
9.2	75.1					26.2		34.7					13.3	10.0
9.4	77.0					20.4		29.3					9.2	18.4
9.6	78.8					15.2		23.6					5.2	29.3
9.8	80.4					10.8		19.0					4.1	42.0
10.0	81.8					7.4		13.1					2.3	53.4
10.2	83.1							9.2						63.7
10.4	84.3							6.2						73.1
10.6	85.4													81.2

pH	14	15	16A	16B	17	18	19	20	21	22	23	24	25	26
10.8	86.5													87.9
11.0	87.8													
11.2	89.3													
11.4	91.3													
11.6	94.5													
11.8	99.0													
12.0														

Some other buffer systems

Acetate buffer (0.2M): Place X ml of 0.2 M acetic acid in 100 ml volumetric flask and make the volume with 0.2 M sodium acetate.

X ml	92	88	83	76	66	55	43	32	33	17	7
pH	3.6	3.8	4.0	4.2	4.4	4.6	4.8	5.0	5.2	5.4	5.6

Carbonate & bicarbonate buffer (0.1 M): Place X ml of 0.1 M sodium bicarbonate and make the final volume of 100 ml with 0.1 M sodium carbonate solution

X ml	93	88	82	73	61	49	38	27	18	10	5.5
pH	9.0	9.2	9.4	9.6	9.8	10.0	10.2	10.4	10.6	10.8	11.0

Citric acid – sodium citrate buffer (0.05 M): Place X ml of 0.05 M citric acid in a volumetric flask and make the volume with 0.05 M trisodium citrate

X ml	91	86	80	75	70	65	60	55
pH	3.0	3.2	3.4	3.6	3.8	4.0	4.2	4.4

Phosphate buffer (0.1 M)

Sodium phosphate buffer - Add X ml of 0.2 M sodium hydroxide to 50 ml of 0.2 M sodium dihydrogen phosphate (NaH_2PO_4) and dilute to 100 ml.

Potassium phosphate buffer: Add X ml of 0.2 M potassium hydroxide to 50 ml of 0.2 M potassium dihydrogen phosphate (KH_2PO_4) and dilute to 100 ml.

X ml	3.5	5.8	9.1	13	18	24	30	35	40	43	45	47
pH	5.8	6.0	6.2	6.4	6.6	6.8	7.0	7.2	7.4	7.6	7.8	8.0

Tris (hydroxymethyl amino methane) – HCl buffer (0.1 M): Add X ml of 0.2 M HCl to 50 ml of 0.2 M tris and make the volume up to 100 ml.

X ml	43	41	39	34	29	24	18	13	9.5	6.0
pH	7.2	7.4	7.6	7.8	8.0	8.2	8.4	8.6	8.8	9.0

Barbitone (0.07 M, pH 8.6): Dissolve 2.58 g of diethyl barbituric acid and 14.42 g of sodium diethyl barbiturate in water and make the volume up to 1.0 liter.

Borate (0.3 M, pH 8.6): Dissolve 18.55 g boric acid and add 3.65 g of NaOH and make the volume up to 1.0 liter.

Cacodylate (0.1 M, pH 6.0): Add 29.ml of 0.2 M HCl to 50 ml of sodium cacodylate and make the volume up to 100 with water.

References

Scorpio, R. (2000). *Fundamentals of Acids, Bases, Buffers & Their Application to Biochemical Systems*. ISBN 0-7872-7374-0.

Hulanicki, A. (1987). *Reactions of acids and bases in analytical chemistry*. Horwood. ISBN 0-85312-330-6. (translation editor: Mary R. Masson)

McIlvaine, T.C. (1921). "A buffer solution for colorimetric comparaison". *J. Biol.Chem.* **49**(1): 183-186.

Medham, J.; Denny, R.C.; Barnes, J.D.; Thomas, M (2000). *Vogel's textbook of quantitative chemical analysis* (5th. Ed. ed.). Harlow: Pearson Education. ISBN 0-582-22628-7. Appendix 5

Carmody, Walter R. (1961). "Easily prepared wide range buffer series". *J. Chem. Educ.* **38** (11): 559–560. Bibcode:1961JChEd..38..559C. doi:10.1021/ed038p559.

Alderighi, L.; Gans, P.; Ienco, A.; Sabatini, A.; Vacca (1999). "Hyperquad simulation and speciation (HySS): a utility program for the investigation of equilibria involving soluble and partially soluble species". *Coordination Chemistry Reviews* **184** (1): 311–318. doi:10.1016/S0010-8545(98)00260-4.

CHAPTER 6

Carbon Assimilation

Carbon assimilation or carbon fixation is the conversion of inorganic carbon (carbon dioxide) to organic compounds by living organisms. The most prominent example is photosynthesis, although chemosynthesis is another form of carbon fixation that can take place in the absence of sunlight. Organisms that grow by fixing carbon are called autotrophs. Autotrophs include photoautotrophs, which synthesize organic compounds using the energy of sunlight, and lithoautotrophs, which synthesize organic compounds using the energy of inorganic oxidation. Heterotrophs are organisms that grow using the carbon fixed by autotrophs. The organic compounds are used by heterotrophs to produce energy and to build body structures. "Fixed carbon", "reduced carbon", and "organic carbon" are equivalent terms for various organic compounds.

Net vs gross CO_2 fixation

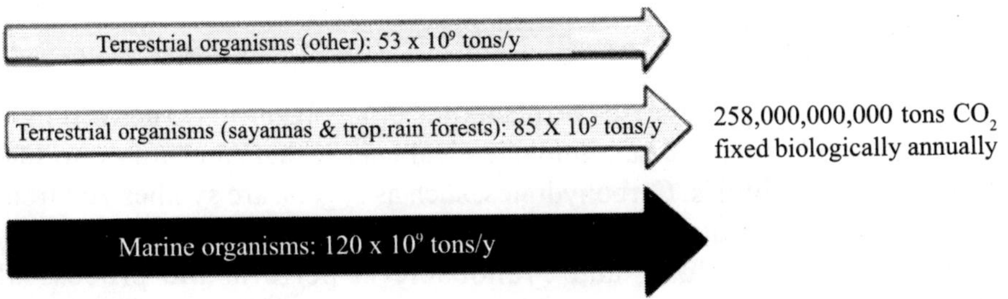

Graphic showing net annual amounts of CO_2 fixation by land and sea-based organisms

It is estimated that approximately 258 billion tons of carbon dioxide are converted by photosynthesis annually. The majority of the fixation occurs in marine environments, especially areas of high nutrients. The gross amount of carbon dioxide fixed is much larger since approximately 40% is consumed by respiration in the evenings following each day of photosynthesis. Given the scale of this process, it is understandable that RuBisCO is the most abundant protein on earth. Six autotrophic carbon fixation pathways are known. The Calvin cycle fixes carbon in the chloroplasts of plants and algae, and in the cyanobacteria. It also fixes carbon in the anoxygenic photosynthetic protobacteria called purple bacteria, and in some non-phototrophic proteobacteria.

6.1. Photosynthesis and Photorespiration

Atmospheric carbon dioxide is reduced to organic forms through photosynthesis. Among terrestrial and aquatic autotrophs, there are three photosynthetic pathways. Here we discuss the ecological and evolutionary aspects of C_3 and C_4 photosynthesis, the two most widely distributed pathways. Three photosynthetic pathways exist among terrestrial plants: C_3, C_4, and crassulacean acid metabolism (CAM) photosynthesis. C_3 photosynthesis is the ancestral pathway for carbon fixation and occurs in all taxonomic plant groups. The term C_3 photosynthesis is based on the observation that the first product of photosynthesis is a 3-carbon molecule. In C_4 photosynthesis, the initial photosynthetic product is a 4-carbon molecule. C_4 photosynthesis occurs in the more advanced plant taxa and is especially common among monocots, such as grasses and sedges, but not very common among dicots (most trees and shrubs). CAM photosynthesis, in honor of the plant family in which this pathway was first documented, occurs in many epiphytes and succulents from very arid regions. However, CAM photosynthesis is sufficiently limited in distribution that CAM plants are not an appreciable component of the global carbon cycle. This section focuses on the factors influencing the dynamics of C_3 and C_4 dominated ecosystems.

6.2. Photosynthesis

Photosynthesis is a process used by plants and other organisms to convert light energy, normally from the sun, into chemical energy that can be used to fuel the organisms' activities. Carbohydrates, such as sugars, are synthesized from carbon dioxide and water. Oxygen is also released, mostly as a waste product. Most plants, most algae, and cyanobacteria perform the process of photosynthesis. Photosynthesis maintains atmospheric oxygen levels and supplies all of the organic compounds and most of the energy necessary for all life on Earth. Although photosynthesis is performed differently by different species, the process always begins when energy from light is absorbed by proteins called reaction centres that contain green chlorophyll pigments. In plants, these proteins are held inside organelles called chloroplasts, which are most abundant in leaf cells, while in bacteria they are embedded in the plasma membrane. In these light-dependent reactions, some energy is used to strip electrons from suitable substances such as water, producing oxygen gas. Furthermore, two further compounds are generated: reduced nicotinamide adenine dinucleotide phosphate (NADPH) and adenine triphosphate (ATP), the "energy currency" of cells.

Carbon dioxide is converted into sugars in a process called carbon fixation. Carbon fixation is an endothermic redox reaction, so photosynthesis needs to supply both a source of energy to drive this process, and the electrons needed to convert carbon dioxide into a carbohydrate. This addition of the electrons is a reducing reaction. In general outline and in effect, photosynthesis is the opposite of cellular respiration, in which glucose and other compounds are oxidized to produce carbon dioxide and water, and to release exothermic chemical energy to drive the organism's metabolism. However, the two processes take place through a different sequence of chemical reactions and in different cellular compartments.

The general equation for photosynthesis is therefore:

$$2n\ CO_2 + 2n\ DH_2 + \text{photons} \rightarrow 2(CH_2O)_n + 2n\ DO$$

Carbon dioxide + electron donor + light energy → carbohydrate + oxidized electron donor

In *oxygenic* photosynthesis water is the electron donor and, since its hydrolysis releases oxygen, the equation for this process is:

$$2n\ CO_2 + 4n\ H_2O + \text{photons} \rightarrow 2(CH_2O)_n + 2n\ O_2 + 2n\ H_2O$$

carbon dioxide + water + light energy → carbohydrate + oxygen + water

Often 2n water molecules are cancelled on both sides, yielding:

$$2n\ CO_2 + 2n\ H_2O + \text{photons} \rightarrow 2(CH_2O)_n + 2n\ O_2$$

carbon dioxide + water + light energy → carbohydrate + oxygen

Other processes substitute other compounds (such as arsenite) for water in the electron-supply role; for example some microbes use sunlight to oxidize arsenite to arsenite: The equation for this reaction is:

$$CO_2 + (AsO_3^{3-}) + \text{photons} \rightarrow (AsO_4^{3-}) + CO$$

carbon dioxide + arsenite + light energy → arsenate + carbon monoxide (used to build other compounds in subsequent reactions)

Photosynthesis occurs in two stages. In the first stage, *light-dependent reactions* or *light reactions* capture the energy of light and use it to make the energy-storage molecules ATP and NADPH. During the second stage, the *light-independent reactions* use these products to capture and reduce carbon dioxide.

Most organisms that utilize photosynthesis to produce oxygen use visible light to do so, although at least three use shortwave infrared or, more specifically, far-red radiation.

In the light-independent (or "dark") reactions, the enzyme RuBisCO captures CO_2 from the atmosphere and in a process that requires the newly formed NADPH, called the Calvin-Benson Cycle, releases three-carbon sugars, which are later combined to form sucrose and starch. The overall equation for the light-independent reactions in green plants is:

$3\ CO_2 + 9\ ATP + 6\ NADPH + 6\ H^+ \rightarrow C_3H_6O_3\text{-phosphate} + 9\ ADP + 8\ P_i + 6\ NADP^+ + 3\ H_2O$

To be more specific, carbon fixation produces an intermediate product, which is then converted to the final carbohydrate products. The carbon skeletons produced by photosynthesis are then variously used to form other organic compounds, such as the building material cellulose, as precursors for lipid and amino acid biosynthesis, or as a fuel in cellular respiration. The latter occurs not only in plants but also in animals when the energy from plants gets passed through a food chain.

The fixation or reduction of carbon dioxide is a process in which carbon dioxide combines with a five-carbon sugar, ribulose 1,5-bisphosphate (RuBP), to yield two molecules of a three-carbon compound, glycerate 3-phosphate (GP), also known as 3-phosphoglycerate (PGA). GP, in the presence of ATP and NADPH from the light-dependent stages, is reduced to glyceraldehydes 3-phosphate (G3P).

C_4 plants chemically fix carbon dioxide in the cells of the mesophyll by adding it to the three-carbon molecule phosphoenolpyruvate (PEP), a reaction catalyzed by an enzyme called PEP carboxylase, creating the four-carbon organic acid oxaloacetic acid. Oxaloacetic acid or malate synthesized by this process is then translocated to specialized bundle sheath cells where the enzyme RuBisCO and other Calvin cycle enzymes are located, and where CO_2 released by decarboxylation of the four-carbon acids is then fixed by RuBisCO activity to the three-carbon sugar 3-phosphoglyceric acids. The physical separation of

RuBisCO from the oxygen-generating light reactions reduces photorespiration and increases CO_2 fixation and, thus, photosynthetic capacity of the leaf. C_4 plants can produce more sugar than C_3 plants in conditions of high light and temperature. Many important crop plants are C_4 plants, including maize, sorghum, sugarcane, and millet. Plants that do not use PEP-carboxylase in carbon fixation are called C_3 plants because the primary carboxylation reaction, catalyzed by RuBisCO, produces the three-carbon sugar 3-phosphoglyceric acids directly in the Calvin-Benson cycle. Over 90% of plants use C_3 carbon fixation, compared to 3% that use C_4 carbon fixation, however, the fact that C_4 has evolved in over 60 plant lineages makes it a striking example of convergent evolution.

Xerophytes, such as cacti and most succulents, also use PEP carboxylase to capture carbon dioxide in a process called Crassulacean acid metabolism (CAM). In contrast to C_4 metabolism, which *physically* separates the CO_2 fixation to PEP from the Calvin cycle, CAM *temporally* separates these two processes. CAM plants have a different leaf anatomy from C_3 plants, and fix the CO_2 at night, when their stomata are open. CAM plants store the CO_2 mostly in the form of malic acid via carboxylation of phosphoenolpyruvate to oxaloacetate, which is then reduced to malate. Decarboxylation of malate during the day releases CO_2 inside the leaves, thus allowing carbon fixation to 3-phosphoglycerate by RuBisCO. Sixteen thousand species of plants use CAM.

6.3. Photorespiration

Photorespiration (also known as the oxidative photosynthetic carbon cycle or C_2 photosynthesis) is a process in plant metabolism which attempts to ameliorate the consequences of a wasteful oxygenation reaction by the enzyme RuBisCO. The desired reaction is the addition of carbon dioxide to RuBP (carboxylation), a key step in the Calvin-Benson cycle, however approximately 25% of reactions by RuBisCO instead add oxygen to RuBP (oxygenation), producing a product that cannot be used within the Calvin-Benson cycle. This process reduces efficiency of photosynthesis, potentially reducing photosynthetic output by 25% in C_3 plants. Photorespiration involves a complex network of enzyme reactions that exchange metabolites between chloroplasts, leaf peroxisomes and mitochondria.

The oxygenation reaction of RuBisCO is a wasteful process because 3-Phosphoglycerate is created at a reduced rate and higher metabolic cost compared with RuBP carboxylase activity. While photorespiratory carbon cycling results in the formation of G3P eventually, there is still a net loss of carbon (around 25% of carbon fixed by photosynthesis is re-released as CO_2)

and nitrogen, as ammonia. The ammonia must be detoxified at a substantial cost to the cell. Photorespiration also incurs a direct cost of 2ATP and one NAD(P)H.

While it is common to refer to the entire process as photorespiration, technically the term refers only to the metabolic network which acts to rescue the products of the oxygenation reaction (phosphoglycolate).

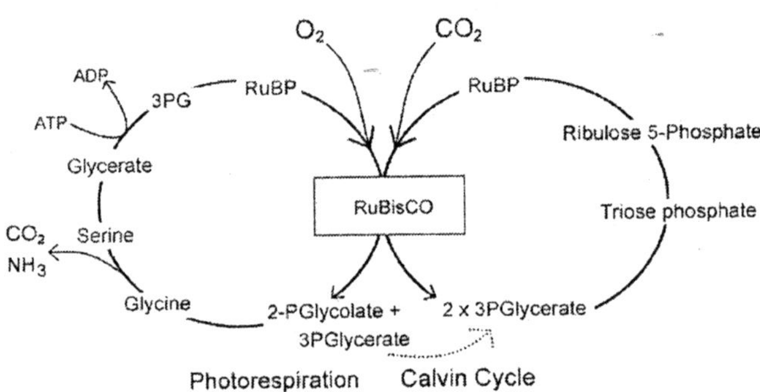

6.4. Estimation of Ribulose-1,5-diphosphate Carboxylase (RuBPcase EC. 4.1.1.39.) activity

Ribulose-1,5-bisphosphate (RuBP) is an organic substances that is involved in photosynthesis. The anion is a double phosphate ester of the ketose (ketone-containing sugar) called ribulose. Salts of this species can be isolated, but its crucial biological function involves this colourless anion in solution. The enzyme ribulose bisphosphate carboxylase/oxygenase (RuBisCO) catalyzes the reaction between RuBP with carbon dioxide. The product is the highly unstable 6-carbon intermediate known as 3-keto-2-carboxyarabinitol 1,5-bisphosphate. This six-carbon intermediate decay virtually instantaneously into two molecules of glycerate 3-phosphate (G3P). RuBisCO also catalyzes reaction of RuBP with oxygen (O_2) in a process called photorespiration, a process that is more prevalent at high temperatures. RuBP is also involved in photorespiration, in which is combines with O_2 to become PGA+ Phosphoglycolic Acid. In the Calvin cycle, RuBP is a product of the phosphorylation of ribulose-5-phosphate by ATP.

This enzyme is involved in photosynthesis. It is a double phosphate ester of ketose called ribulose. This enzyme ribulose bisphosphate carboxylase oxygenase (RuBisCO) catalyzes the reaction between RuBP with carbon dioxide. into two molecules of glycerate 3-phosphate (G3P). RuBisCO also catalyzes RuBP with oxygen (O_2) in a process called photorespiration, a process that is more prevalent at high temperatures. RuBP is also involved in photorespiration, in which is combines with O_2 to become PGA+ Phosphoglycolic Acid. In the Calvin cycle, RuBP is a product of the phosphorylation of ribulose-5-phosphate by ATP.

Reagents

- Tris buffer (0.15 M pH 7.5): 0.3 M Tris solution is prepared by dissolving 3.6342 g Tris salt in water and the final volume is made to 100 ml pH of this solution is adjusted to 7.5 with 0.2 N HCl. The final volume was made up to 200 ml.
- Grinding media: 0.112 g EDTA (0.01 M), 0.23 g KCl (0.01 M), 0.007 g $MgCl_2$ (0.001 M) and 0.046 g glutathione reduced (0.005 M) in 30 ml of phosphate buffer pH 7.5.
- NaH^*CO_3 solution: 84 mg of sodium bicarbonate was dissolved in distilled water and to this 0.5 m Ci of radioactive bicarbonate was added. The final volume of 10 ml was made with water.
- Ribulose diphosphate solution (0.04 M): 2 mg of RuDP salt of sodium is dissolved in 5 ml of distilled water.
- Magnesium chloride $MgCl_2$: 0.3 mM of $MgCl_2$ was prepared by dissolving 3.05 g in 50 ml of water.
- Ethylene diamine tetraacetic acid (EDTA): 0.01 M of EDTA was prepared by dissolving 0.186 g of EDTA in 50 ml of water.
- Scintillation Liquid: 4 g of PPO (2,5-diphenyl oxazole) and 100 mg of POPOP (1,4 bis (2,5 phenyl) oxazolyl benzene) is dissolved in 1000 ml of toluene.

Enzyme Extraction

- 100 mg of freshly cut leaf material is grinded in pre chilled mortar and pastle with 10 ml of pre chilled grinding media and acid washed sand.
- The extracts is passed through four layers of cheese cloth and the filterate is centrifuged at 10000 x g for 15 minutes at 0^0 C.
- The supernatant is used as the enzyme source.

Assay of the enzyme activity

- Enzyme assay is done in glass scintillation vials.

- In a final volume of 0.57 ml the reaction mixture consists of the following
    - 0.1 ml of enzyme extract
    - 0.2 ml of NaH*CO_3 solution (10 μ Ci / 20 μ mole $NaHCO_3$).
    - 0.05 ml ribulose diphosphate solution (0.2 μ mole)
    - 0.01 M $MgCl_2$ (3.0 μ mole)
    - 0.01 EDTA (0.1 μ mole)
    - 0.20 ml Tris-HCl buffer pH 7.5 (30.0 μ mole)
- A blank was run simultaneously having all the reagents and boiled enzyme extract.
- The reaction mixture is incubated for 10 minutes and the reaction is stopped by addition of 0.2 ml of 6.0 N acetic acid, which also remove unfixed bicarbonates.
- The solution is evaporated to dryness in an oven at 50^0 C.
- 10 ml of scintillation liquid is added to the vials and the radioactive counts are recorded by scintillation counter.
- The results are expressed as H*CO^-_3 fixed per gram fresh weight of leaf per hour.

Reference

Fair, P., Tew J. and Cresswell, C.F. (1973). Enzyme activities associated with carbon dioxide exchange in illuminated leaves of *Hordeum vulgare* L. I. Effect of light period, leaf age and position on carbon dioxide compensation point. Ann. Bot., **37:** 831-844.

6.5. Estimation of Phospho-enol Pyruvate Carboxylase (EC 4.1.1.31) activity

Phosphoenolpyruvate carboxylase (also known as PEP carboxylase, PEPCase, or PEPC; EC 4.1.1.31) is an enzyme in the family of carboxy-lyases that catalyzes the addition of bicarbonate to phosphoenolpyruvate (PEP) to form the four-carbon compound oxaloacetate:

$$PEP + HCO^-_3 \rightarrow oxaloacetate + Pi$$

This reaction is used for carbon fixation in CAM and C_4 plants where it plays a key role in photosynthesis. The enzyme is also found in some bacteria, but not in animals or fungi. After conversion of CO_2 to bicarbonate by carbonic anhydrase, PEP carboxylase assimilates the available bicarbonate into a four-carbon compound (oxaloacetate, which is further converted to malate) that can be stored or shuttled between plant cells. This allows for a separation of initial carbon fixation by contact with air and secondary carbon fixation into sugars by RuBisCO during the light-independent reactions of photosynthesis.

In succulent CAM plants adapted for growth in very dry conditions, PEP carboxylase fixes bicarbonate during the night when the plant opens its stomata to allow for gas exchange. During the day time, the plant closes the stomata to preserve water and releases CO_2 inside the leaf from the storage compounds produced during the night. This allows the plants to thrive in dry climates by conducting photosynthesis without losing water through open stomata during the day. In C4 plants, for example maize, PEP carboxylase fixes bicarbonate in the mesophyll cells of the leaf and the resulting four-carbon compound, malate, is shuttled into the bundle sheath cells where it releases CO_2 for fixation by RuBisCO. Thus, the two processes are separated spatially, allowing for RuBisCO to operate in a low-oxygen environment to circumvent photorespiration. Photorespiration occurs due to the inherent oxygenase activity of RuBisCO in which the enzyme uses oxygen instead of carbon dioxide without incorporating carbon into sugars or generating ATP. As such, it is a wasteful reaction for the plant. By comparison, C4 carbon fixation via PEP carboxylase is more efficient.

Solutions Required

- Tris·HCl / $MgCl_2$ buffer, 50 mM, pH 8.0 - Adjust to a pH of 8.0 with 20% NaOH. Add enough $MgCl_2$ for a final concentration of 1 mM.
- Tris·HCl buffer, 0.15 M pH 8.5 - Adjust to a pH of 8.5 with 20% NaOH.
- $MgCl_2$ 0.3 M - can be prepared in stock solution and stored at room temperature.
- $NaHCO_3$ 0.6 M - must be prepared fresh
- NADH 3.0 mM - must be prepared fresh
- Phosphoenolpyruvate 0.1 M - must be prepared fresh
- Acetyl CoA 15 mM - must be prepared fresh
- Malate Dehydrogenase - use porcine heart enzyme as purchased from Sigma, approximately 10000 U/ml

Preparation of Cell Extract

- Centrifuge sufficient cells so that the volume diluted down to 5 ml would give an optical density of 20-30. For example, for a broth of OD=1, use 100 ml. For a broth of OD=10, use 10 ml.
- After first pelletization of cells, resuspend in 5-15 ml of 4°C Tris/Mg (pH 8.0) buffer.
- After second pelletization of cells, resuspend in 5 ml of 4°C Tris/Mg (pH 8.0) buffer, and
- break with French Press.

Procedure

- Turn on the ultraviolet bulb on the spectrophotometer and wait 30 minutes for warm-up. Select the kinetics-time window on the instrument. Load the method "A:/nadh". This method has a run-time of 60 s, a temperature of 37°C (or another appropriate fermentation temperature), a wavelength of 340 nm and uses 2 autosamplers.
- In UVtranslucent cuvettes, take the following: Water 140 µl, Tris-HCl buffer (pH 8.5) 1000 µl, $NaHCO_3$ 25 µl, $MgCl_2$ 25 µl, NADH 50 µl, Acetyl CoA 50 µl, Malate dehydrogenase 10 µl, and PEP 50 µl. In control PEP is omitted and in place of PEP 50 µl water is added.
- Directly from the ice when ready to commence the assay, place the two cuvettes (each containing 1350 µl) into the spectrophotometer holder (position #1 for control, position #2 for experimental).
- Wait 10 minutes to allow the temperature of the solutions in the cuvettes to equilibrate.
- "Blank" and then depress "Read Samples" on the monitor.
- Simultaneously add 150 µl of the cell extract to the cuvettes.
- To mix solutions, immediately and simultaneously aspirate and dispense the contents of the cuvettes with a pipettor. Mix the solutions in this way ten times. (Count!)
- Promptly depress "start" on the monitor.
- Record the rates for the two (control and experimental) cuvettes.
- Dilution of the cell extract may be adjusted so that change in absorbance is between about 0.05 and 0.7 AU in one minute. This dilution should be

accomplished externally in a microcentrifuge tube (for example, by adding 50 μL of cell extract to 950 μL DI water to achieve a dilution of 20).

- The volume of 150 μL should always be used in the enzyme assay mixture.

Caiculatiun of Activity

One unit (U) of phosphoenolpyruvate carboxylase activity is defined as the amount of enzyme

required to produce 1.0 ìmole of oxaloacetate in one minute.

1. dA/dt (min^{-1}) = $[Rate]_{experimental} - [Rate]_{control}$ = dA/dt

2. Activity = (1000 x TV x D x (dA / dt)) / (ε x V x CF)

Activity	:	Volumetric Activity (U/L)
TV	:	Total volume in cuvette (1000 μl)
D	:	Dilution of the cell extract. (For example, if 50 ìl of cell extract were add to 950 ìl
DI	:	Water prior to using a volume of cell extract in the assay, then D=20)
V	:	Volume of cell extract used (150 μl)
ε	:	Molar extinction coefficient for NADH (6.22 l/m mol for a path length of 1.0 cm)
CF	:	Concentration Factor of cell extract (For example, if a 100 ml sample is concentrated to a 2 ml volume for the French Press, then CF=50)

Specific Activity = (Activity / Protein Concentration)

Reading

Maeba, P, Sanwal, B.D. (1969) Phosphoenolpyruvate carboxylase frm *Salmonella typhimurium* strain LT2, In "*Methods in Enzymology*", **13:** 283-288.

Cánovas, J.L. and Kornberg, H.L. (1969) Phosphoenolpyruvate carboxylase from *Escherichia coli.* In "*Methods in Enzymology,*" **13:** 288-292.

6.6. Estimation of Glycolate Oxidase (EC. 1.1.3.15.) Activity

Glycolate oxidase, (S)-2-hydroxy-acid oxidase (EC 1.1.3.15) is an enzyme that catalyses the following chemical reaction

$$\text{(S)-2-hydroxy acid} + O_2 \rightleftharpoons \text{2-oxo acid} + H_2O_2$$

Thus, the two substrates of this enzyme are (S)-2-hydroxy acid and O_2, whereas its two products are 2-oxo acid and H_2O_2.

This enzyme belongs to the family of oxidoreductases, specifically those acting on the CH-OH group of donor with oxygen as acceptor. The systematic name of this enzyme class is (S)-2-hydroxy-acid:oxygen 2-oxidoreductase. Other names in common use include glycolate oxidase, hydroxy-acid oxidase A, hydroxy-acid oxidase B, glycolate oxidase, oxidase, L-2-hydroxy acid, hydroxyacid oxidase A, L-alpha-hydroxy acid oxidase, and L-2-hydroxy acid oxidase. This enzyme participates in glyoxylate and dicarxylate metabolism. It employs one cofactor, FMN.

The first step of photosynthetic carbon fixation is catalyzed by ribulose-1,5-bisphosphate carboxylase / oxygenase, which arose when the atmosphere contained virtually no O_2. The increase of atmospheric O_2 caused by oxygenic photosynthesis gave rise to photorespiratory metabolism, a process whereby O_2 substitutes for CO_2, causing ribulose-1,5-bisphosphate carboxylase / oxygenase to produce the toxic compound 2-phosphoglycolate, which is ultimately recycled into 3-phosphoglycerate. This recycling process is vital for the success of photosynthetic organisms under the current levels of O_2 in the atmosphere but also accounts for losses of large amounts of carbon and energy. The enzyme is assayed spectrophotometrically by monitoring the formation of glyoxylate from glycolate in presence of phenylhydrazine at 30^0 C.

Materials

- Phosphate buffer: 0.1 M phosphate buffer pH 7.5 is prepared by dissolving 6.8045 g of potassium dihydrogen phosphate in 500 ml of distilled water and 8.709 g of potassium hydrogen phosphate in 500 ml of distilled water. The buffer of pH 7.5 is made by adding 16.ml of potassium dihydrogen phosphate and 84 ml of potassium hydrogen phosphate and final pH is adjusted by pH meter.

- Sodium glycolate: 0.1 M solution is prepared by dissolving 0.098g of sodium glycolate in 10 ml of distilled water.

- Phenylhydrazine hydrochloride: 0.1 M solution is prepared by dissolving 0.144 g of phenylhydrazine hydrochloride in 10 ml of distilled water.

- Cystein solution: 0.1 M solution was prepared by dissolving 0.121 g of cystein in 10 ml of distilled water.
- Hydrochloric acid: 8.62 ml of concentrated HCl (11.6 N) is diluted to 100 ml with distilled water to get 1.0 N HCl.

Extraction

- Leaves are washed and wiped with filter paper. Leaves are cut in small pieces for grinding. Weighed amount (~2.0 g) of leaf materials is taken and grounded in pre chilled mortar with 10 ml of 0.1 M phosphate buffer pH 7.5. Homogenates are centrifuged for 20 minutes at 12,000 x g at 0^0 C. Supernatant is used as source of enzyme.

Assay of enzyme activity

The assay mixture is made of following:

- Phosphate buffer pH 7.5, 2.3 ml (230 μ moles)
- Sodium glycolate 0.3 ml (30 μ moles)
- Phenyl hydrazine hydrochloride 0.2 ml (20 μ moles)
- Cystein 0.1 ml (10 μ moles)
- Enzyme extract 0.1 ml
- A blank with boiled enzyme is used as control.
- Incubation is done for 30 minutes at 35^0 C. Reaction is stopped by adding 3 ml of 1.0 N HCl. Out of this mixture 2 ml is taken and mixed with 1 ml of $K_3Fe(CN)_6$ (potassium ferricyanide) solution (10% v/v). It is kept for 15 minutes and the final volume is made to 5 ml with distilled water. Optical density is observed at 520 nm against a reagent blank.

Preparation of standard curve

A stock solution of 1 μ mole per 1 ml of sodium glyoxylate is prepared by dissolving 11.4 mg of sodium glyoxylate in 100 ml of distilled water. From this stock solution, solutions of 0.05, 0.1, 0.15, 0.2, 0.25...............0.6 μ moles are prepared by successive dilution. To these solutions, 0.2 ml of phenylhydrazine hydrochloride is added and then the mixture is shaken vigorously. Finally 1.0 each of 1.0 N HCl and potassium ferricyanide solution is added and the volume is made to 5.0 ml with distilled water. After 15 minutes. Optical density is recorded at 520 nm and standard is plotted against concentrations of glyoxylate.

Reference

Lui, NST, Roels, OA (1970). An improved method for determining glyoxylic acid. *Analytical Biochemistry*, **38:** 202-209.

6.7. Estimation of Glutamate Glyoxylate Aminotransferase (EC. 2.6.1.4) Activity

In photorespiration, peroxisomal glutamate:glyoxylate aminotransferase (GGAT) catalyzes the reaction of glutamate and glyoxylate to produce 2-oxoglutarate and glycine. Previous studies demonstrated that alanine aminotransferase-like protein functions as a photorespiratory GGAT. Photorespiratory transamination to glyoxylate, which is mediated by GGAT and serine glyoxylate aminotransferase (SGAT), is believed to play an important role in the biosynthesis and metabolism of major amino acids. To better understand its role in the regulation of amino acid levels, we produced 42 *GGAT1* overexpression lines that express different levels of *GGAT1* mRNA. The levels of free serine, glycine, and citrulline increased markedly in *GGAT1* overexpression lines compared with levels in the wild type, and levels of these amino acids were strongly correlated with levels of *GGAT1* mRNA and GGAT activity in the leaves. This accumulation began soon after exposure to light and was repressed under high levels of CO_2. Light and nutrient conditions both affected the amino acid profiles; supplementation with NH_4NO_3 increased the levels of some amino acids compared with the controls. The results suggest that the photorespiratory aminotransferase reactions catalyzed by GGAT and SGAT are both important regulators of amino acid content.

This glutamate-glyoxylate amino transferase is also known as glycine-oxoglutarate transaminase (EC 2.6.1.4) is an enzyme that catalyses the following chemical reaction.

$$\textbf{glycine + 2-oxoglutarate} \rightleftharpoons \textbf{glyoxylate + L-glutamate}$$

Thus, the two substrates of this enzyme are glycine and 2-oxoglutarate, whereas its two products are glyoxylate and L-glutamate.

This enzyme belongs to the family of transferases, specifically the transaminases, which transfer nitrogenous groups. The systematic name of this enzyme class is glycine:2-oxoglutarate aminotransferase. Other names in common use include glutamic-glyoxylic transaminase, glycine aminotransferase, glyoxylate-glutamic transaminase, L-glutamate-glyoxylate aminotransferase, and glyoxylate-glutamate aminotransferase. This enzyme participates in glycine serine and threonine metabolism. It employs one cofactor, pyridoxal phosphate.

Assay of this enzyme is done in acoordence to Rehfeld and Tolbert (1972) by incubating L-glutamic acid-u-^{14}C with with unlabeled glyoxylic acid and subsequently measuring the radio activity in the organic acid fraction of the reaction mixture, separated according to Splittstoesser (1969).

Materials

- Phosphate buffer 0.2 M pH 7.5 is prepared by dissolving 35.598 g of sodium phosphate dibasic in 1000.0 ml of water and 27.218 g of potassium dihydrogen phosphate in 1000 ml of water. The desired pH is made by mixing 80.3 ml of sodium phosphate and 19.7 ml of potassium dihydrogen phosphate. Final pH is adjusted by pH meter using 1.0 N HCl or 1.0 N NaOH.
- Phosphate buffer 0.2 M pH 7.8 is prepared by mixing 92.6 ml of sodium phosphate dibasic solution and 7.8 ml of potassium dihydrogen phosphate solution. Final pH of 7.8 is addusted by pH meter using 1.0 N HCl or 1.0 N NaOH.
- Sodium glutamate - 187.0 mg of sodium glutamate is dissolved in 10.0 ml of 0.2 phosphate buffer pH 7.8 to have 5.0 μ moles of glutamate per 0.1 ml. The final pH of 7.8 is addusted by pH meter using 1.0 N HCl or 1.0 N NaOH.
- Glyoxylic acid – 92.0 mg of monohydrated glyoxylic acid is dissolved in 10.0 ml of 0.2 phosphate buffer pH 7.8 to have 10 μ moles of glyoxylic acid per 0.1 ml. The final pH of 7.8 is addusted by pH meter using 1.0 N HCl or 1.0 N NaOH.
- Pyrodoxal-5-phosphate – 5.0 mg of pyrodoxal-5-phosphate is dissolved in 10.0 ml of water to get a solution of 50 μg of pyrodoxal-5-phosphate per 0.1 ml of solution.
- L-glutamic acid-u-^{14}C: 0.1 m Ci/ml of L-glutamic acid-u-^{14}C (sp. Activity 18.45 Ci/m mole) was appropriately diluted by water to get a solution of 1.0 μ Ci/0.5 ml.
- Sciemtillation liquid: This is prepared by dissolving 6.0 g of PPO (2,5-diphenyl oxazole) and 300 mg of POPOP (2,4-methyl 1,5-phenyl oxazole benzene) in 1000 ml of toluene.
- Dowex 50 (H^+) 200-400 mesh – This is prepared by boiling Dowex 50 in 6.0 N HCl for 5 minutes and washing them by water for chloride ions, tested against silver nitrate solution (arbitrarily prepared).

Procedure

- Approximately 2.0 g of freshly acquired leaf lets is macerated in prechilled mortar with 10.0 ml of 0.2 M phosphate buffer pH 7.5. Homogenates are centrifuged at 10,000 x g for 15 minutes at 2°C. Supernatant is used as source of enzyme.
- Following chemicals are taken for enzymatic assay:
  - Phosphate buffer 0.2 M pH 7.8 – 2.5 ml (500 μ moles)
  - Enzyme extract – 0.2 ml
  - Pyrodoxal-5-phosphate 0.1 ml (5.0 μ g)
  - Sodium glutamate 0.1 ml (5.0 μ moles)
  - L-glutamic acid-u-14_C 0.5 ml (1.0 μCi)
- The chemical mixture is first incubated at 35° C for 10 minutes and then 0.1 ml of glyoxylic acid is added and further incubated for 60 minutes at 35° C.
- Reaction is terminated by the addition of 2.0 ml of 90% (v/v) ethyl alcohol
- Boiled enzyme extract is used for determining non-enzymatic transamination.
- The assay mixture is heated over boiling water bath to remove the alcohol and volume is reduced to zero.
- The dried assy mixture is redissolved in 5.0 ml of water and passed through a column of 5.0 x 1.0 cm Dowex 50 (H^+). The column was washed by another 50.0 ml of water to elute the organic acid.
- The washing is collected and evaporated to near dryness over a boiling waterbath.
- These dried washing is redissolved in 5.0 ml of and water and out of this, 1.0 ml is taken in liquid scientillation vials.
- Liquid of vials are oven dried at 60° C and thereafter 5.0 ml of scientillation liquid is added. The vials are left overnight.
- Radio activity of the vials is counted in liquid scientillation counter.
- The radioactivity counts are corrected for non-enzymatic transamination and background counts.
- The enzyme activity is expressed as dps in organic acid/mg protein/hr.

Reading

Rehfeld, D.W. and Tolbert, N.E. (1972). Aminotransferases in peroxisomes from spinach leaves, *J. Biol. Chem.*, **247:** 4803-4811.

Splittstoesser, W.F. (1969). Arginine metabolism by pumpkin seedlings. Separation of plant extracts by ion exchange resins. *Plant Cell Physiol.*, **10 (1):** 87-94.

6.8. Estimation for Activity of Isocitrate lyase (EC. 4.1.3.1)

Isocitrate lyase (EC 4.1.3.1), or ICL, is an enzyme in the glyoxylate cycle that catalyses the cleavage of isocitrate to succinate and glyoxylate. Together with malate synthase, it bypasses the two decarboxylation steps of the tricarxylic acid cycle (TCA cycle) and is used by bacteria, fungi, and plants. The systematic name of this enzyme class is isocitrate glyoxylate-lyase (succinate-forming). Other names in common use include isocitrase, isocitritase, isocitratase, threo-Ds-isocitrate glyoxylate-lyase, and isocitrate glyoxylate-lyase. This enzyme participates in glyoxylate and dicarboxylate metabolism. This enzyme belongs to the family of lyases, specifically the oxo-acid-lyases, which cleave carbon-carbon bonds. Other enzymes also belong to this family including carboxyvinyl-carboxyphosphate phosphorylmutase (EC 2.7.8.23) which catalyses the conversion of 1-carboxyvinyl carboxyphosphonate to 3-(hydrohydroxyphosphoryl) pyruvate carbon dioxide, and phosphoenolpyruvate mutase (EC 5.4.2.9), which is involved in the biosynthesis of phosphinothricin tripeptide antibiotics.

During catalysis, isocitrate is deprotonated, and an adol clevage results in the release of succinate and glyoxylate. This reaction mechanism functions much like that of adolase in glycolysis, where a carbon-carbon bond is cleaved and an aldehyde is released. ICL competes with isicitrate dehydrogenase, an enzyme found in the TCA cycle, for isocitrate processing. Flux through these enzymes is controlled by phosphorylation of isocitrate dehydrogenase, which has a much higher affinity for isocitrate as compared to ICL. Deactivation of isocitrate dehydrogenase by phosphorylation thus leads to increased isocitrate channeling through ICL, as seen when bacteria are grown on acetate, a two-carbon compound. The ICL enzyme has been found to be functional in various archaea, bacteria, protists, plants, fungi, and nematodes. Although the gene has been found in genomes of nematodes and cnidaria, it has not been found in the genomes of placental mammals. By diverting isocitrate from the TCA cycle, the actions of ICL and malate synthase in the glyoxylate cycle result in the net assimilation of carbon from 2-carbon compounds. Thus, while the TCA cycle yields no net carbon assimilation, the glyoxylate cycle generates intermediates that can be used to synthesize glucose (via gluconogenesis), and other biosynthetic products. As a result, organisms that use ICL and malate synthase are able to synthesize glucose and metabolic intermediates from acetyl-CoA derived from acetate or from the degradation of ethanol, fatty acids or poly-â-hydroxybutyrate. This function is especially important for higher plants which use oilseeds. In these germinating seeds, the breakdown of oils generates acetyl-CoA. This serves as a substrate for the glyoxylate cycle, which generates other cyclic intermediates and serves as a primary nutrient source prior to the production of sugars from photosynthesis.

Activity of this enzyme is measured by determining the glyoxylate produced from threo-Ds-isocitrate in 10 minutes at 30° C.

Materials

- Tris-HCl buffer – 0.1 M pH 7.7 (25° C), containing 3 mM magnesium chloride ($MgCl_2$).
- Glutathion (GSH) – 0.125 M (384 mg/10 ml) in Tris-HCl buffer pH 7.7 having $MgCl_2$.
- Tris sodium-DL-Isocitrate – 40 mM (103 mg/10 ml) in Tris-HCl buffer pH 7.7 having $MgCl_2$, store at 2° C.
- Tri chloro acetic acid (TCA) - 10% (w/v), Dissolve 10 g of trichloro acetic acid in distilled water and make the volume to 100 ml with distilled water.
- Mixture of 5 parts of oxaloacetic acid 10 mM and 1 part of 1% phenylhydrazine hydrochloride.
- Potassium ferricyanide ($K_3Fe(CN)_6$) – 5.0 g of potassium ferricyanide is dissolve in distilled water and volume is made to 100 ml - 5% (w/v).

Procedure

- 2.0 of fresh leaves is homogenized with chilled Tris-HCl buffer (pH 7.7) and centrifuged at 15,000 rpm at 1° C for 10 minutes. The supernatant is used as the enzyme source.
- To 1.5 ml of Tris-HCl buffer (pH 7.7) 0.2 ml each of glutathione solution and enzyme extract is added.
- The reagent mixture is incubated for 10 minutes at 30° C.
- The reaction is initiated by the addition of 0.2 ml of isocitrate solution followed by through mixing. Reaction is allowed to proceed for 10 minutes at 30° C.

- Reaction is stopped by the addition of 1.0 ml of TCA solution.
- From this reaction mixture, 1.0 ml is taken to 50 ml beaker and 6.0 ml of oxalic-phenyl hydrazine hydrochloride solution is added and heated to just boiling. Solution is first cooled to room temperature and then chilled for 2 minutes.
- 4.0 ml of concentrated HCl is added and then 1.0 ml of potassium ferricyanide. Colour is allowed to develop for 8 minutes and optical density is recorded at 520 nm against a reagent blank.
- At 520 nm and for a path length of 1.0 cm, the yield of glyoxylate in μmoles per reaction vessel *i.e.*, 2.0 ml of original incubation mixture is given by $(OD_{520} - 0.05) / 1.15$.
- *Units* – One unit of enzyme is that which catalyzes the disappearance of 1.0 micromole of threo-DL-isocitrate per minute at 30^0 C under conditions of assay. The amount of isocitrate that disappear is equivalent to the glyoxylate produced.

Reading

Mc Fadden (1969) In "*Method of Enzymology*", **13:** 163-170.

6.9. Estimation of Glycolate Content

Glycolic acid (or hydroxyacetic acid) is the smallest α-hydroxy acid (AHA). This colourless, odorless, and hygroscopic crystalline solid is highly soluble in water. It is used in various skin-care products. Glycolic acid is found in some sugar-crops.

O
HO OH

2-Hydroxyethanoic acid

Reagents

Napthalene-diol reagent: 0.01% of 2,7-naphthalene diol reagent is prepared by dissolving 10 mg of the chemical in 100 ml of concentrated H_2SO_4. Fresh solution is prepared before use.

Estimation

- 0.5 g of leaf material is grounded in a pre chilled mortar with acid washed sand and 4 ml of distilled water.

- The homogenate is centrifuged at 10,000 x g for 10 minutes and the supernatant is made up to 5 ml.
- 1.8 ml of 0.01 2,7-napthalenel diol solution is added to 0.2 ml of the supernatant. The mixture is heated for 20 minutes in boiling water bath and 2 ml of distilled water is added, mixed thoroughly and cooled.
- Optical density is recorded at 530 nm against reagent blank.
- A stamdard curve is plotted using several concentration, 0.005 to 0.07 mg, of sodium glycolate in a volume of 0.2 ml. To this 1.8 ml of 0.01% naphthalene-diol reagent is added.
- The mixture is heated on a boiling water bath for 20 minutes and then 2 ml of distilled water is added and cooled.
- Optical density is recorded at 530 nm and standard curve is plotted by plotting the ODs against their respective concentrations of sodium glycolate.

Reference

Smith, FD and Smell. CT (1961). Estimation of glycolate acid. In "*Colorimetric Methods of analysis*" **Vol III A**, pp. 315-316.

6.10. Estimtion of Starch Synthase (EC. 2.4.1.21) Activity

Starch synthase (EC 2.4.1.21) is an enzyme that catalyses the following chemical reaction

$$\text{ADP-glucose} + (\text{1,4-alpha-D-glucosyl})_n \rightarrow \text{ADP} + (\text{1,4-alpha-D-glucosyl})_{n+1}$$

Thus, the two substrates of this enzyme are ADP-glucose and a chain of D-glucose residues joined by 1,4-alpha-glycosidic bonds, whereas its two products are ADP and an elongated chain of glucose residues. Plants use these enzymes in the biosynthesis of starch.

This enzyme belongs to the family of hexosyltransferases, specifically the glycosyltrnsferases. The systematic name of this enzyme class is ADP-glucose:1,4-alpha-D-glucan 4-alpha-D-glucosyltransferase. Other names in common use include ADP-glucose-starch glucosyltrans-ferase, adenosine diphosphate glucose-starch glucosyltransferase, adenosine diphospho glucose starch glucosyltransferase, ADP-glucose starch synthase, ADP-glucose synthase, ADP-glucose transglucosylase, ADP-glucose-starch glucosyl-transferase, ADPG starch synthe-tase, and ADPG-starch glucosyltransferase

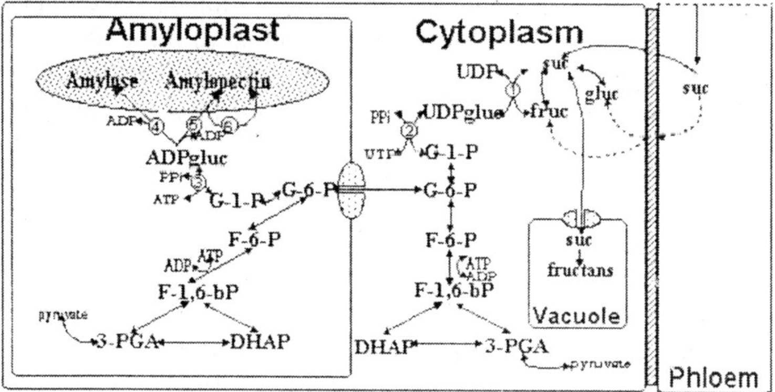

Five isoforms seems to be present. GBSS which is linked to amylose synthesis. The others are SS1, SS2, SS3 and SS4. These have different roles in amylopectin synthesis. New work implies that SS4 is important for granule initiation.

The starch synthases catalyze the transfer of the glucosyl moiety of ADP-glucose to the non-reducing end of an α- (1-4)-linked glucan primer in higher plants. Among the entire starch biosynthesis enzymes, SS has the highest number of isoforms. This group of enzymes is divided into two groups; first, the granule–bound starch synthases (GBSS) which are encoded by the *Waxy* (*Wx*) gene are involved in amylose biosynthesis. The second class of starch synthases consists of four major isoforms SSI, SSII, SSIII, and SSIV which are involved in amylopectin synthesis. Isoforms of the major classes of SSs are highly conserved in higher plants. A region of approximately 60kDa is highly conserved in C-terminus of all these enzymes in 14 higher plants and green algae, whereas this region is distributed across the protein sequence in prokaryotic glycogen synthases. The K–X–G–G–L motif is thought to be responsible for substrate (ADP-glucose) binding in prokaryotic glycogen synthase (GSs) and in higher plant SSs, and is also found only in the C-terminus of higher plants and green algal SSs (Nichols *et al.* 2000) where as the K-X-G-G-L domains are distributed across the GSs protein sequence in prokaryotes. The presence of lysine in the K–X–G–G–L domain determines glucan primer preference. Further, the glutamate and aspartate are found as important residues for catalytic activity and substrate binding in maize SSs. SSs show considerable variation within the N-terminus upstream of the catalytic core, and this region can vary greatly in length, from 2.2 kDa in granule-bound starch synthase I (GBSSI) to approximately 135 kDa in maize SSIII. The phylogenetic and sequence analysis of plants SS (*Arabidopsis thaliana,* wheat and rice) and algal SS and prokaryotic GS isoforms on the basis of predicted amino acid sequence suggests that SSIs, SSIIs and GBSSIs have distinct evolutionary origins as compared to SSIIIs and SSIVs. Especially, the valine residue within the highly conserved

K-X-G-G-L motif appears to have faced strong evolutionary selection in SSIII and SSIVs and it may affect primer/substrate binding of these SSs compared to SSIs, SSIIs and GBSSIs. The other prominent difference in SSIII and SSIV from other SSs is the highly conserved G-X-G motif near the nucleotide-binding cleft.

Procedure

- Frozen plant sample (approximately 1.0 – 2.0 g) is homogenized in pre chilled pastle and mortar with 10.0 ml of extraction medium having 50.0 mM HEPES-NaOH buffer (pH 7.5), 10.0 mM $MgCl_2$, 2.0 mM EDTA, 50.0 mM 2-mercaptoethanol, 12.5% (v/v) glycerol and 5% (w/v) insoluble polyvinyl pyrolidon-40 (PVP-40).
- 30 μl of the homogenate is added to 1.8 ml of buffer solution and centrifused at 2,000 x g at 4^0 C for 20 minutes.
- The sediment is then resuspended in 2.0 ml of the buffer solution for GBSS activity assay.
- Rest of the homogenate is centrifused at 10,000 x g for 10 minutes at 4^0C for assay of ATP content and activities of starch synthase, ADPGPPase, UDPGPPase and SSS.
- The assay of starch synthase is similar to that proposed by Nakamura *et al.* (1989).
- The reaction solution contain 100 μl of 14 mM ADPG and 70 μl of 50 mg/ml amylopectic.
- After incubation at 30^0 C for 5 minutes the reaction is started by the addition of 50 μl enzyme extract and reaction is stopped after 20 minutes by heating in boiling water.
- The ADP produced by starch synthase of GBSS is converted to ATP by the addition of 100.0 μl of 10.0 mM of PEP, 50 mM $MgCl_2$ and 1.0 IU pyruvate kinase (EC. 2.7.1.40) and then incubated at 30^0 C for 30 minutes.
- The resultant ATP is determined by adding 5.0 ml of luciferin-luciferase reagent (Li and Sun, 1980).
- Five ml of luciferin-luciferase reagent for measurement of ATP content is added to 50 μl of enzyme extract.

Reading

Jiang, D., Wei-Xing, CAO., Dai, Ting-Bo and Jing Qi. (2004). Diurnal changes in activities of related enzymes to starch synthesis in grains of winter wheat. *Acta Botanica* Sinica, **46 (1):** 51-57.

Jeng, TL., Tseng, TH., Wang, CS., Chen, CL. And Sung, JM. (2003). Starch biosynthesisizing enzymes in developing grains of rice cultivars Tainung 67 and its sodium azide-induced rice mutant. *Field Crop Res.*, **84:** 267-269.

Nakamura, Y., Kubo, A., Shimanune, T., Matsuda, T., Harada, K. and satoh, H. (1997). Correlation between activities of starch debranching enzyme and α-polyglucan structure in endosperm of sugary-1 mutants of rice. *Plant J.*, **12:** 143-153.

6.11. Estimation of Starch Phosphorylase (EC. 2.4.1.1) Activity

Starch phosphorylase, exists in both tetrameric and dimeric states and catalyses the reversible transfer of glucosyl units from glucose-1-phosphate (G-1-P) to the non-reducing end of α-1-4 linked glucan chains as shown in the following equation.

$$\text{G-1-P} + (\text{1,4-}\alpha\text{-D-Glucose})_n \rightleftharpoons (\text{1,4-}\alpha\text{-D-Glucose})_{n+1} + \text{Pi}$$

SP has often been regarded as a glucan degradative enzyme. The α-glucan phosphorylase (EC 2.4.1.1) found in animals, fungi, and prokaryotes plays a major role in glucan catabolism and the amino acid sequence of the enzyme is found to be highly conserved among prokaryotes and eukaryotes. Genetic analyses in *Chlamydomonas* showed that the mutation of plastidial SP affected starch accumulation. In addition, the mutation of plastidial α-glucan phosphorylase could not change the total accumulation of starch or the starch structure during the day or its remobilization at night when the phosphorylase gene activity was eliminated by T-DNA insertion in *Arabidopsis thaliana* leaves, where transient starch is synthesized. In contrast, research evidence demonstrated that the SP has a certain effect on the storage starch biosynthesis, that the development of plastidial SP activity coincides with starch accumulation in developing cereal endosperms; in rice, in wheat and in maize. Above evidence further suggests that the plastidial forms of SP are involved in starch synthesis rather than the degradation in higher plants. Starch phosphorylase plays an important role in starch metabolism in plants. It can be used for the production of glucose-1-phosphate, a cytostatic compound used in cardio-therapy. Starch phosphorylase may also be used to estimate inorganic phosphate in serum under the pathological conditions as well as to detect amount of inorganic phosphate pollution in the environment. Immobilized enzymes are in great demand in industries due to their reusability. In spite of its great importance, starch phosphorylase has been immobilized from few sources only.

Two major isoforms of SP are present in plants and differ in their intracellular localization, and are designated as plastidic (Pho1) and cytosolic (Pho2) isoforms. In developing rice endosperm, plastidial Pho1 accounts for about 96% of the total phosphorylase activity and it is restricted to the stroma. The predicted protein sequence alignment of Pho1 and Pho2 isoforms show a significant 50 amino acid extension in the N-terminus of Pho1, which represent the transit peptide. In this thesis the term SP is generally used for the plastidial form.

The plastidial form of SP (112 kDa in maize) is known to be the second most abundant protein in the maize amyloplast stroma next to SBEIIb. Peptide sequences of plastidial SP in maize showed higher identities to potato, sweet potato and spinach and the N-terminus sequence was unique in maize amyloplast; it can not be aligned with any other N-terminus sequences of Pho1 available in the gene bank. Excluding the N-terminus difference between Pho1 and Pho2, a unique 78-amino acid insertion in the middle of the Pho1 sequence is a prominent characteristic of the plastidial isoform in higher plants. In potato, Pho1 and Pho2 showed 81% - 84% amino acid sequence similarity over most part of the sequence with the exception of N-terminal transit peptide and the large L-78 Two major isoforms of SP are present in plants and differ in their intracellular localization, and are designated as plastidic (Pho1) and cytosolic (Pho2) isoforms. In developing rice endosperm, plastidial Pho1 accounts for about 96% of the total phosphorylase activity and it is restricted to the stroma. The predicted protein sequence alignment of Pho1 and Pho2 isoforms show a significant 50 amino acid extension in the N-terminus of Pho1, which represent the transit peptide.

Significant variation is found in the molecular mass of the Pho1 and Pho2 in wheat endosperm as 100 kDa and 90 kDa respectively. The peptide sequence ILDNADLPASVAELFVK is a common sequence fragment found in the L-78 region in maize and potato. In addition, the sequence comparison among SP from potato tuber, rabbit muscle, and *Escherichia coli* revealed the presence of the characteristic 78-residue insertion only in the middle of the polypeptide chain of the potato enzyme suggesting the L-78 region is specific to plants. The proposed function of the L-78 insertion is, thought to be the obstruction of the binding of Pho1 to large, highly branched polysaccharides. This idea was further confirmed by the observation that the L-78 insertion in sweet potato (*Ipomea batatas*) blocked the starch-binding site in Pho1 molecule showing low affinity towards starch. Several serine phosphorylation sites were also found in the L-78 insertion suggested that the regulation of Pho1 is phosphorylation dependent. This research group was able to purify a 338 kDa protein kinase activity from sweet potato roots using liquid chromatography methods and which actively phosphorylates the L-78 insertion. Interestingly, this phosphorylation modification was not found in Pho2 isoform or after L-78 insertion was proteolytically removed from Pho1.

According to their affinities for glucan substrates, SPs are further classified as low affinity (SP-L) and high affinity (SP-H) isoforms respectively in potato tuber and leaf. When the L-78 insertion in SP-L was replaced by high affinity SP-H sequence, the SP-L showed less affinity to glycogen compared to SP-H form (*Km*=10,400 and *Km*=10 μg/ml). The L-78 insertion-replaced chimeric enzyme was five times less active than the SP-L isoform but still showed low affinity to glycogen than in SP-L (*Km*= 24 μg/ml). However, when the glycogen was replaced by amylopectin and amylose (DP=30), the affinity increased in SP-L (*Km*= 82 and *Km*=76 μg/ml respectively), in SP-H form (*Km*=3.6 and *Km*=8.7 μg/ml respectively), and in chimeric form (*Km*=5.3 and *Km*=2 μg/ml respectively). Among all the isoforms, the SP-H form has the highest affinity to amylopectin, suggesting that the L-78 region has greater affinity towards low molecular weight substrates. In addition, two isoforms named Pho1a and Pho1b were identified in potato. The homodimeric form of Pho1a isoform was immunochemically detectable only in tuber extracts where both Pho1a and heterodimeric Pho1b were present in leaf extracts in potato. Wheat has three forms of SP (designated as P1, P2, P3) which are distinguished in non-denaturing separation gels containing glycogen. The activity form P3 is plastidic in where as P1 and P2 are cytosolic and found mainly in younger leaves. However, mature leaves only contain the plastidic form which was also strongly evident in the endosperm of the developing seeds. Cytosolic forms are more prominent in germinating seeds suggestive of the involvement of cytosolic SP forms in the utilization of α-glucans resulting from starch degradation.

The plastidial and cytosolic SP show different affinity towards high and low molecular glucan polymers in synthetic direction. Plastidial SP prefers amylopectin than the glycogen potato tuber, spinach leaf, and sweet corn and maize. In maize endosperm, the *Km* value for amylopectin in the synthetic direction of the SP reaction was 3.4-fold lower and the *Kd* value was 40-fold lower than of glycogen. The kinetic analysis indicated that the *Km* value for amylopectin was eight-fold lower than that of glycogen and the phosphorolytic reaction was favored over the synthetic reaction when malto-oligosaccharides (DP= 4 to 7 units) were used as substrates.

Procedure

- Fresh plant tissue (*e.g.*, cabbage) approximately 10 g is cut into small pieces and blended with 90 ml of isolation medium (buffer A) using a waring blender for 30 sec at low speed and 60 sec at high speed.
- The isolation medium is consisted of 0.01 M tris-HCl buffer, pH 7.5 containing 20 mM 2-mercaptoethanol and 0.05 M EDTA.
- The homogenate is made to volume 100 ml with the isolation medium and centrifuged at 15,000 x g for 30 minutes at 4°C.
- The supernatant containing the enzyme activity was taken as initial extract.
- To the initial extract, powdered ammonium sulfate is slowly added with constant stirring to get 0-30% saturation and the pH is maintained at 7.5 by the addition of dilute ammonia.
- After storage for 3 h, it is centrifuged at 15000 x g for 20 minutes at 4° C and the supernatant having most of the activity was brought to 60% saturation with powdered ammonium sulfate.
- After overnight incubation, the suspension was centrifuged at 15000 x g for 20 minutes at 2°C. The pellet is dissolved in buffer A, centrifuged and the supernatant is desalted using Sephadex-G-25 column chromatography. The desalted enzyme was used for further study.
- The enzyme assay for the soluble and the immobilized enzyme was carried out in the direction of polysaccharide synthesis as described by Kumar and Sanwal (1981) with some modifications.
- The enzyme assay system for the soluble enzyme was consisted of 0.2 ml of 0.2 M tris-maleate buffer, pH 6.0; 0.1 ml of sodium fluoride; 0.1 ml of 3% soluble starch and 0.5 ml of the enzyme preparation and water, pre-incubated at 37° C for 2 min.
- The reaction was started by the addition of 0.1 ml of 0.05 M glucose-1-phosphate.
- After 30 min, the reaction was stopped by the addition of 0.1 ml of 50% TCA and the tubes were put in an ice bath.
- The precipitate formed was removed by centrifugation in the cold. In the clear supernatant, inorganic phosphate formed was estimated using colorimetric method of Fiske and Subbarow (1925).

Reading

Fiske, C.H. and Subbarow, Y. (1925) The colorimetric determination of phosphorus. *J. Biol. Chem.*, **66:** 375-400.

Kumar, A. and Sanwal, G.G. (1981) Immobilization of starch phosphorylase from mature banana leaves. *Indian J. Biochem. Biophys.*, **18:** 114-119.

6.12. Estimation of Glucose Pyrophosphorylase (EC 2.4.2.13) activity

The accumulation of α-1,4-polyglucans is an important strategy to cope with transient starvation conditions in the environment. In bacteria and plants, the synthesis of glycogen and starch occurs by utilizing ADP-glucose as the glucosyl donor for elongation of the α-1,4-glucosidic chain. The main regulatory step takes place at the level of ADP-glucose synthesis, a reaction catalyzed by ADP-Glc pyrophosphorylase (PPase). Most of the ADP-Glc PPases are allosterically regulated by intermediates of the major carbon assimilatory pathway in the organism. Based on specificity for activator and inhibitor, classification of ADP-Glc PPases has been expanded into nine distinctive classes. According to predictions of the secondary structure of the ADP-Glc PPases, they seem to have a folding pattern common to other sugar nucleotide pyrophosphorylases. All the ADP-Glc PPases as well as other sugar nucleotide pyrophospho-rylases appear to have evolved from a common ancestor, and later, ADP-Glc PPases developed specific regulatory properties, probably by addition of extra domains. Studies of different domains by construction of chimeric ADP-Glc PPases support this hypothesis. In addition to previous chemical modification experiments, the latest random and site-directed mutagenesis experiments with conserved amino acids revealed residues important for catalysis and regulation.

Materials

- MOPS (pH 7.4) 50 μ Mole
- $MgSO_4.7H_2O$ 7.5 μ mole
- 3-PGA 3.0 μ mole
- $NADP^+$ 0.5 μ mole
- ADP-glucose 0.5 μ mole
- Phosphoglucomutase
- Glucose-6-phosphate dehydrogenase
- Sodium pyrophosphate 2.5 μ mole

Procedure

1. Set up a reaction mixture containing following reagents:

Reagents	Final concentration	Volume in 1 ml reaction
50 mM MOPS (pH 7.4)	1.06 g/30 ml	300 µl
7.5 ìmole $MgSO_4.7H_2O$	18.45 mg/ml	100 µl
Enzyme extract	————	100 µl
3 ìmole 3-PGA	6.912 mg/ml	100 µl
0.5 ìmole NADP+	3.827 mg/ml	100 µl
0.5 ìmole ADP-glucose	3.166 mg/ml	100 µl
Phosphoglucomutase	2 units	3 µl
Glucose-6-phosphate dehydrogenase	2 units	2 µl

Estimation of reducing sugar by Nelson-Somogyi method - Sugars with reducing property (because of the presence of a potential aldehyde or keto groups) are called reducing sugars. Some of the reducing sugars are glucose, galactose, lactose and maltose. The Nelson-Somogyi method is one of the classical and widely used methods for the quantitative determination of reducing sugars.

Principle

In the first step glucose (or a reducing sugar) is oxidised using a solution of Cu(II) ion which in the process is reduced to Cu(I). In the second step the Cu(I) ions are then oxidised back to Cu(II) using a colourless hetero-polymolybdate complex, which is, in the process, reduced to give the characteristic blue colour. Finally the absorption of the hetero-poly molybdenum blue is measured at 620nm using a colorimeter and compared to standards prepared from reacting sugar solutions of known concentration, to determine the amount of reducing-sugar present.

Materials

- Alkaline copper tartrate - Dissolve 2.54g anhydrous sodium carbonate, 2g sodium bicarbonate, 2.5g potassium sodium tartrate and 20g anhydrous sodium sulphate in 80ml water and make up to 100ml.
- Dissolve 15g copper sulphate in a small volume of distilled water. Add one drop of sulphuric acid and make up to 100ml. Mix 4ml of B and 96ml of solution A before use.
- Arsenomolybdate Reagent: Dissolve 2.5g ammonium molybdate in 45 ml water. Add 2.5ml sulphuric acid and mix well. Then add 0.3 g disodium hydrogen arsenate dissolved in 25 ml water. Mix well and incubate at 37°C for 24 to 48 hours.

- Standard Glucose Solution: Stock: 100mg in 100ml distilled water.
- Working Standard: 10ml of stock diluted to 100ml with distilled water [100µg/ml].

Procedure

- Take clean and dry test tube.
- Pipette out standard solution in the range of 0 to 2 ml.
- Make up the final volume in all the tubes to 2 ml with distilled water.
- Pipette 2 ml of distilled water in separate tube to set a blank.
- Add 1ml of alkaline copper tartrate reagent to each tube.
- Place the tubes in a boiling water for 10 minutes.
- Cool the tubes and add 1ml of arsenomolybdate reagent to all the tubes.
- Make up the volume to 10 ml with distilled water in all the tubes.
- Read the absorbance of blue colour at 620 nm after 10 minutes.
- From the graph calculate the amount of reducing sugar present in the sample.

Calculate

Absorbance corresponds to 0.1mg of test = 'x' mg of glucose

10 ml contains = ('x' / 0.1) x 10 mg glucose

6.13. Estimation of Malate Synthase (EC. 2.3.3.9) Activity

Chan, M and Sim, T.S. (1998). Malate synthase from Streptomyces clavuligerus NRRL3585 : cloning, molecular characterization and its control by acetate. *Microbiology*, **144:** 3229-3237.

Method 1 - Malate synthase (EC 2.3.3.9) is an enzyme that catalyses the following chemical reaction

$$\textbf{acetyl-CoA} + \textbf{H}_2\textbf{O} + \textbf{glyoxylate} \rightleftharpoons \textbf{(S)-malate} + \textbf{CoA}$$

The 3 substrates of this enzyme are acetyle-CoA, H_2O, and glyoxylate, whereas its two products are (S)-malate and CoA. Malate synthase, which catalyses the reaction

Acetyl-CoA + glyoxylate + H,O $\rightleftharpoons$ + malate + CoA

is a key component of the glyoxylate pathway, leading to the net synthesis of one molecule of malate from two molecules of acetate. This pathway is significant in acetate metabolism, as it allows the replenishment of three-carbon molecules such as phosphoenolpyruvate, which are normally depleted from the TCA cycle for gluconeogenesis. Thus, malate synthase activity is detectable in microrganisms grown in acetate, and bacteria which are defective in malate synthase have been

It is well known that genes encoding primary metabolic enzymes sometimes occur in multiple copies, which are regulated differently to respond to diverse physiological conditions. Consistent with this observation, Escherichia coli is known to possess two forms of malate synthase, *i.e.* A and G, encoded by the aceB and gfcB genes, respectively. The A form is predominant in cells growing on acetate, while the presence of the G form accounts for almost the entire malatesynthesizing activity of cells grown on glycolate. Similarly, *Saccharomyces cereuisiae* was also found to possess at least two genes encoding malate ynthase, namely MLSl and DAL7. MLSlp participates in carbon metabolism and is susceptible to carbon catabolite repression, while insensitive to nitrogen catabolite repression. Conversely, DAL7p, which participates in catabolism of the nitrogenous compound allantoin via

This enzyme belongs to the family of transferases, specifically those acetyltransferases that convert acyl groups into alkyl groups on transfer. The systematic name of this enzyme class is acetyl-CoA:glyoxylate C-acetyltransferase (thioester-hydrolysing, carboxymethyl-forming). Other names in common use include L-malate glyoxylate-lyase (CoA-acetylating), glyoxylate transacetylase, glyoxylate transacetase, glyoxylic transacetase, malate condensing enzyme, malate synthetase, malic synthetase, and malic-condensing enzyme. This enzyme participates in pyruvate metabolism and glyoxylate and dicarboxylate metabolism.

$$\begin{array}{ccc} CH_2\text{-}CO_2^- & & CH_2\text{-}CO_2^- & & CH_2\text{-}CO_2^- \\ | & & | & & | \\ CH-CO_2^- & \longrightarrow & H^+ \; CH-CO_2^- & \longrightarrow & CH_2\text{-}CO_2^- \\ | & & & & \text{succinate} \\ H-O-CH-CO_2^- & & O=C-CO_2^- & & \\ \text{isocitrate} & & H \quad \text{glyoxylate} & & \Delta G^{o\prime} = +8.7\ \text{kJ/mol} \end{array}$$

Estimation of malate synthase is based the consumption of acetyl-CoA at 232 nm.

Procedure

- *Preparation of crude extract:* - Break open 15-20 mg dry mass in 0.5 ml extraction buffer (Hepes 20 mM, + pH 7.1 Dithiothreitol (DTT) 1mM + potassium chloride (KCl) 100 mM + protease cocktail inhibitor)/1 g glass beads by vortexing 6 times 30 sec with 30 sec interval in ice.
- Centrifuge tubes at 3000g for 5 min at 4°C
- Transfer supernatant into eppendorf tube
- Centrifuge 15min in a microcentrifuge at full speed at 4°C

Enzymatic assay

- In 1 ml (spectrophotometer cuvettes) Mixture Reaction 0.5 ml (Kpi 0.1 M, pH 6.5; $MgCl_2$ 10 mM) + Acetyl-CoA 5 mM 0.02 ml + Crude extract 10 to 100 μl + Water to 0.98 ml
- Read optical density at 232nm until stabilization
- Add DL glyoxylate 100 mM (prepared in water) 0.02 ml
- Follow increase optical density at 232 nm for 5 -15 min

Note

To calculate the activity, consider that 1 unit at 232 = 222 nmols/ml of acetyl-CoA

MLS (in nmoles/min/ml) = optical density at 232(milliunits) – OD232 x 222 x 1000/ vol crude extract x 1/t1-2

Method 2 - based on reaction of CoASH with DTNB **MLS**

Acetyl-CoA + glyoxylate → Malate + CoASH

CoASH + DTNB CoA-TNB + TNB

- In 1 ml (spectrophotometer cuvettes) Mixture Reaction 0.5 ml (Kpi 0.1 M, pH 6.5; $MgCl_2$ 10 mM) + Acetyl-CoA 5 mM 0.02 ml + DTNB 10 mM 10 μl + Crude extract 10 to 100 μl + Water to 0.98 ml
- Read optical density at 412 nm until stabilization
- Add DL glyoxylate 100 mM (prepared in water) 0.02 ml
- Follow increase in optical density at 412 nm for 5 -15 min

6.14. Estimation of Sucrose Synthase (EC 2.4.1.13) Activity

In heterotrophic organs sucrose synthase (SuSy) is a major determinant of sink strength that highly controls the channeling of incoming sucrose into starch and cell wall polysaccharides. SuSy is a highly regulated enzyme that catalyzes the reversible conversion of sucrose and a nucleoside diphosphate into the corresponding nucle-oside diphosphate glucose and fructose. Although UDP is the preferred nucleoside diphosphate substrate of SuSy to produce UDPglucose (UDPG), ADP is also an effective acceptor molecule of this sucrolytic enzyme to produce ADP glucose (ADPG).

Surose synthase catalyses reversible conversion of sucrose and a nucleoside diphosphate into the corresponding nucleoside diphosphate-glucose and fructose.the chemical reaction

$$\textbf{UDP-glucose + D-fructose} \rightleftharpoons \textbf{UDP + sucrose}$$

Thus, the two substrates of this enzyme are NDP-glucose and D-fructose, whereas its two products are NDP and sucrose. This enzyme belongs to the family of glycosyltransferases, specifically the hexosyltransferases. The systematic name of this enzyme class is NDP-glucose:D-fructose 2-alpha-D-glucosyltransferase. Other names in common use include UDP glucose-fructose glucosyltransferase, sucrose synthetase, sucrose-UDP glucoseyltransferase, sucrose-uridine diphosphate glucosyltransferase, and uridine diphosphoglucose-fructose glucosyltransferase. This enzyme participates in starch and sucrose metabolism.

Sucrose synthase catalyses the cleavage of sucrose, the main transported form of assimilates in wheat plants to form UDP-glucose and fructose and is thought to be the first step in the sucrose-to-starch conversion. The cleavage activity of the enzyme is easily assayed by coupling the formation of UDP-glucose to the rediced NAD^+ in the presence of excess UDP-glucose dehydrogenase, and the change in absorbance at 340 nm followed.

Material

- HEOPES-KOH buffer – 0.1 M pH 7.5
- Sucrose 0.5 M – (117.11 g /100 ml)
- Uridine Diphosphate - 10.0 μ mole (4.0 mg/ml)
- NAD+ 0.015 M (9.95 mg.ml)
- UDP-glucose dehydrogenase (0.25 mg/ml)

Procedure

- Fresh tissue, approximately 10.0 g is homogenized in 20.0 ml of 10.0 mM potassium phosphate buffer (pH7.2) having 1 mM EDTA and 5 m M 2-mercaptoethanol. The homogenate is centrifused at 15,000 x g for 15 minutes at 4^0 C.
- Take 0.2 ml of HEPES-KOH buffer and add 0.2 ml each of sucrose solution, UDP solution, UDP-glucose DH solution, enzyme extract and 0.1 ml of NAD^+ solution. In control enzyme extract is replaced with 0.1 ml of buffer and NAD^+ solution is omitted.
- Set the spectrophotometer to zero absorbance at 340 nm without adding NAD^+ in the blank.
- Add NAD+ quickly to the sample mix well and initial absorbance is recorded.
- Decrease in absorbance at 340 nm in every minute is recorded till no further reduction is noticed.

Calculation

- The change in absorbance in 340 nm is 12.0 for each micromole of UDPG per milliliter. Express the emzyme activity as micromoles UDPG formed per mg of protein.

Note

- The enzyme may also be assayed by following the fructose released. In this case the assay mixture contain 20 mM HEPES-KOH, 100 mM sucrose solution and 2 mM UDP along with appropriate volume of enzyme extract. The reaction is stopped, after 30 minutes of incubation, by heating in a boiling water bath for 2 minutes. Fructose released is determined by reducing sugar method of Nelson-Somogyi method.

Reading

Strominger *et al.* (1957). In "*Methods of Enzymology*" **3:** pp. 974.

Morell, M. and Copeland, L. (1985). Sucrose synthase of soybean nodules. *Plant Physiol*, **78:** 149 - 154.

6.15. Estimaion of Invertase (EC 3.2.1.26) Activity

Invertase (EC 3.2.1.26), saccharase, glucosucrase, beta-h-fructosidase, beta-fructosidase, invertin, sucrase, maxinvert L 1000, fructosylinvertase, alkaline invertase, acid invertase, systematic name: beta-fructofuranosidase) is an enzyme that catalyses the hydrolysis (breakdown) of sucrose (table sugar). The resulting mixture of fructose and glucose is called inverted sugar syrup. Related to invertases are sucrases. Invertases and sucrases hydrolyze sucrose to give the same mixture of glucose and fructose. Invertases cleave the O-C (fructose) bond, whereas the sucrases cleave the O-C (glucose) bond. For industrial use, invertase is usually derived from yeast. It is also synthesized by bees, who use it to make honey from nectar. Optimum temperature at which the rate of reaction is at its greatest is 60 °C and an optimum pH of 4.5. Typically, sugar is inverted with sulfuric acid.

The enzyme invertase catalyzes the irreversible hydrolysis of sucrose to glucose and fructose. The invertases of higher plants are classified according to their solubility, localization and pH optima and include three types of enzymes: cytoplasmic, vacuolar and cell wall. The invertase activity is detected by the incubating reaction mixture containing substrate (sucrose), buffer,and enzyme at a particular temperature for specified incubation time. After incubation, glucose released as a product is estimated.

Material

- Potassium dihydrogen orthophosphate-dipotassium hudrogen phosphate buffer 0.1 M pH 7.5.
- Sucrose (25 μ mole)
- Copper reagent – Dissolve 15 g of copper sulphate in small volume of distilled water. Add one drop of sulphuric acid and makeup the volume to 100 ml with water.
- Arsenomolybdate – Dissolve 25 g of ammonium molybdate in 45 ml of water, add 2.5 ml of sulphuric acid. Then add 0.3 g of disodium hydrogen arsenate in 25 ml of water. Mix well and incubate at 37^0 C for 24 to 48 hours.

Procedure

- Prepare the reaction mixture (1 ml) containing: phosphate buffer (pH 7.5) 20 mM; sucrose, 25 μ mole and enzyme extract.
- Incubate the reaction mixture at 30^0C for 15 min.
- Stop the reaction by adding 1 ml of copper reagent and placing in a boiling water bath for 20 min.
- Cool and add 1 ml of arsenomolybdate reagent.

- Make the final volume to 25 ml and read the absorbance at 520 nm against reagent blank.
- Standard curve of glucose is prepared in range of 10 – 100 μ moles.

6.16. Estimation of Keto Organic Acid

Keto acids or ketoacids (also called oxo acids or oxoacids) are organic compounds that contain a carboxylic acid group and a ketone group. In several cases, the keto group is hydrated. The alpha-keto acids are especially important in biology as they are involved in the Kerbs citric acid cycle and in glycolysis.

Common types of keto acids include:

- Alpha-keto acids, or 2-oxoacids, such as pyruvic acid, have the keto group adjacent to the carboxylic acid. Another important member is oxaloacetic acid, a component of the Kerbs cycle. Alpha-ketoglutarate, (also known as 2-oxo glutarate) is a 5-carbon keto acid derived from glutamic acid. Alpha-ketoglutarate participates in cell signaling by functioning as a cofactor for a variety of iron-containing redox enzymes, for example the proline hydroxylases that modify the hypoxia inducible factor leading to the destruction of HIF except in hypoxia when this pathway is blocked.
- Beta-keto acids, or 3-oxoacids, such as acetoacetic acid, have the ketone group at the second carbon from the carboxylic acid
- Gamma-keto acids, or 4-oxoacids, such as levulinic acid, have the ketone group at the third carbon from the carboxylic acid.

When ingested sugars and carbohydrate levels are low, stored fats and proteins are the primary source of energy production. Glucogenic amino acids from proteins are converted to glucose and fats can be used to form ketone bodies. Ketonic amino acids can be deaminated to produce alpha keto acids and ketone bodies.

Alpha keto acids are used primarily as energy for liver cells and in fatty acid synthesis, also in the liver.

Materials

- 2,4-dinitrophenyl hydrazine solution – Dissolve 1.25 g of 2,4-dinitrophenyl hydrazine in 2.0 N HCl. Take 1.0 ml of this solution and make the final volume of 100 ml with 2.0 N HCl.
- NaOH 0.6 N – Dissolve 2.4 g of NaOH in 100 ml of distilled water.

- Pyruvic acid: Dissolve 880.6 mg of puruvic acid in 10 ml of distilled water to get a solution of 100 µM of puruvic acid/0.1 ml, and from this solution take a gradient from 0.1 ml to 0.9 ml in triplicate and make the final volume to 1.0 ml with distilled water.

Procedure

- Extract the macro molecules from plant tissue with 80% ethanol or 80% methanol after keeping in hot water for 10-15 minutes. Repeat the extraction process three times and pool the extracts.
- Using water bath, remove the alcohol from extracts and redissolve the residues in distilled water to known volume.
- To 1.0 ml of extract add 1.0 ml of 2,4-dinitrophenyl hydrazine solution and incubate for 10 minutes at 37^0 C.
- To the reaction mixture add 2.0 ml of distilled water and 5.0 ml of NaOH (0.6 N) and allow the colour to develop for 10-15 minutes.
- Optical density is recorded at 510 nm against reagent blank.
- Standard curve is obtained from the puruviv acid solution and following the above sated procedures.

CHAPTER 7

Carbohydrates

A carbohydrate is a large biological molecules, or macromolecules, consisting of carbon (C), hydrogen (H), and oxygen (O) atoms, usually with a hydrogen : oxygen atom ratio of 2:1 (as in water); in other words, with the empirical formula $C_m(H_2O)_n$ (where m could be different from n). Some exceptions exist; for example, deoxyribose, a sugar component of DNA, has the empirical formula $C_5H_{10}O_4$. Carbohydrates are technically hydrates of carbon; structurally it is more accurate to view them as polyhydroxy aldehydes and ketones.

The term is most common in biochemistry, where it is a synonym of saccharide. The carbohydrates (saccharides) are divided into four chemical groups: monosaccharides, disaccharides, oligosaccharides, and polysaccharides. In general, the monosaccharides and disaccharides, which are smaller (lower molecular weight) carbohydrates, are commonly referred to as sugars. While the scientific nomenclature of carbohydrates is complex, the names of the monosaccharides and disaccharides very often end in the suffix -ose. For example, grape sugar is the monosaccharide glucose, cane sugar is the disaccharide sucrose, and milk sugar is the disaccharide lactose.

Carbohydrates perform numerous roles in living organisms. Polysaccharides serve for the storage of energy (*e.g.*, starch and glycogen), and as structural components (*e.g.*, cellulose in plants and chitin in arthropods). The 5-carbon monosaccharide ribose is an important component of coenzymes (*e.g.*, ATP, FAD, and NAD) and the backbone of the genetic molecule known as RNA. The related deoxyribose is a component of DNA. Saccharides and their derivatives include many other important biomolecules that play key roles in the immune system, fertilization, preventing pathogenesis, blood clotting, and development. In food science and in many informal contexts, the term carbohydrate often means any food that is particularly rich in the complex carbohydrate starch (such as cereals, bread, and pasta) or simple carbohydrates, such as sugar (found in candy, jams, and desserts).

Formerly the name "carbohydrate" was used in chemistry for any compound with the formula $C_m (H_2O)_n$. Following this definition, some chemists considered formaldehyde (CH_2O) to be the simplest carbohydrate, while others claimed that title for glycolaldehyde. Today the term is generally understood in the biochemistry sense, which excludes compounds with only one or two carbons.

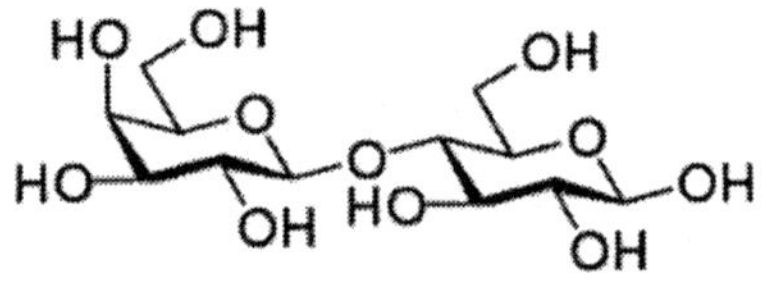

Lactose is a disaccharide found in milk. It consists of a molecule of D-galaxtose and a molecule of D-glucose bonded by beta-1-4 glycosidic linkage. It has a formula of $C_{12}H_{22}O_{11}$.

Natural saccharides are generally built of simple carbohydrates called monosaccharides with general formula $(CH_2O)_n$ where n is three or more. A typical monosaccharide has the structure H-$(CHOH)_x$(C=O)-$(CHOH)_y$-H, that is, an aldehyde or ketone with many hydroxyl groups added, usually one on each carbon atom that is not part of the aldehyde or ketone functional group. Examples of monosaccharides are glucose, fructose, and glyceraldehydes. However, some biological substances commonly called "monosaccharides" do not conform to this formula (*e.g.*, uronic acids and deoxy-sugars such as fucose), and there are many chemicals that do conform to this formula but are not considered to be monosaccharides (*e.g.*, formaldehyde CH_2O and inositol $(CH_2O)_6$). The open-chain form of a monosaccharide often coexists with a closed ring form where the aldehyde / ketone / cabonyl group carbon (C=O) and hydroxyl group (-OH) react forming a hemiacetol with a new C-O-C bridge. Monosaccharides can be linked together into what are called polysaccharides (or oligosaccharides) in a large variety of ways. Many carbohydrates contain one or more modified monosaccharide units that have had one or more groups replaced or removed. For example, deoxyribose, a component of DNA, is a modified version of ribose; chitin is composed of repeating units of N-acetyl glucosamine, a nitrogen-containing form of glucose.

Nutritionists often refer to carbohydrates as either simple or complex. However, the exact distinction between these groups can be ambiguous. However, the report put "fruit, vegetables and whole-grains" in the complex carbohydrate column, despite the fact that these may contain sugars as well as polysaccharides. This confusion persists as today some nutritionists use the term complex carbohydrate to refer to any sort of digestible saccharide present in a whole food, where fibre, vitamins and minerals are also found (as opposed to processed carbohydrates, which provide calories but few other nutrients). The standard usage, however, is to classify carbohydrates chemically: simple if they are sugars (monosaccharides and disaccharides) and complex if they are polysaccharides (or oligosaccharides).

In any case, the simple vs. complex chemical distinction has little value for determining the nutritional quality of carbohydrates. Some simple carbohydrates (*e.g.* fructose) raise blood glucose slowly, while some complex carbohydrates (starches), especially if processed, raise blood sugar rapidly. The speed of digestion is determined by a variety of factors including which other nutrients are consumed with the carbohydrate, how the food is prepared, individual differences in metabolism, and the chemistry of the carbohydrate.

The glycine index (GI) and glycemic load concepts have been developed to characterize food behavior during human digestion. They rank carbohydrate-rich foods based on the rapidity and magnitude of their effect on blood glucose levels. Glycemic index is a measure of how quickly food glucose is absorbed, while glycemic load is a measure of the total absorbable glucose in foods. The insulin index is a similar, more recent classification method that ranks foods based on their effects on blood insulin levels, which are caused by glucose (or starch) and some amino acids in food.

Carbohydrates are widely prevalent in the plant kingdom, comprising the mono-, di-, oligo-, and polysaccharides. The common monosaccharides are glucose, fructose, galactose, ribose etc. The disaccharides, *i.e.*, the combination of two monosaccharides include sucrose, lactose and maltose. Starch and cellulose are polysaccharides consisting of many monosaccharide residues. Cellulose is the most abundant organic compound on this planet since it forms part of the cell wall in plants.

7.1. Qualitative Reaction of Carbohydrates

Materials

- *Iodine solution*: Add a few crystals of iodine to 2% potassium iodide solution till the colour becomes deep yellow.
- *Fehling's reagent A:* Dissolve 34.65 g copper sulphate in distilled water and make up to 500 ml.
- *Fehling's reagent B:* Dissolve 125 g potassium hydroxide and 173 g Rochelle salt (potassium sodium tartrate) in distilled water and make up to 500 ml.

- *Benedict's qualitative reagent:* Dissolve 173 g sodium citrate and 100 g sodium carbonate in about 500 ml water. Heat to dissolve the salts and filter, if necessary. Dissolve 17.3 g copper sulphate in about 100 ml water and add it to the above solution with stirring and make up the volume to 1 L with water.
- *Barfoed's reagent:* Dissolve 24 g copper acetate in 450 ml boiling water. Immediately add 25 ml of 8.5% lactic acid to the hot solution. Mix well, Cool and dilute to 500 ml.
- *Seliwanoff's reagent:* Dissolve 0.05 g resorcinol in 100 ml dilute (1:2) hydrochloric acid.
- *Bial's reagent:* Dissolve 1.5 g orcinol in 500 ml of concentrated HCl and add 20 to 30 drops of 10% ferric chloride.

Molisch's test

Add Add two drops of Molisch's reagent (5% 1-naphthol in ethaol) to about 2 ml of test solution and mix well. Incline the tube and add about 1 ml of concentrated sulphuric acid along the sides of the tube. Observe the colour at the junction of the two liquids. A red cum violet ring appears at the junction of the two liquids. The red colour is due to the dehydration of the carbohydrate by sulfuric acid to produce an aldehyde, which condenses with two molecules of phenol (usually α-napthol, though other phenols (*e.g.* resorcinol, thymol) also give coloured products), resulting in a red- or purple-coloured compound. All carbohydrates react positively with this reagent.

Iodine test

Add a few drops of iodine solution to about 1 ml of the test solution. Appearance of deep-blue colour. This indicates the presence of starch in the solution. The blue colour is due to the formation of starch-iodine complex

Fehling's test

To 1 ml of Fehling's solution 'A', add 1 ml of Fehling's solution 'B' and a few drops of the test solution. Boil for a few minutes. Formation of yellow or brownish-red precipitate. The blue alkaline cupric hydroxide present in Fehling's solution, when heated in the presence of reducing sugars, gets reduced to yellow or red cuprous oxide and it gets precipitated. Hence, formation of the coloured precipitate indicates the presence of reducing sugars in the test solution.

Benedict's test

To 2 ml of Benedict's reagent add five drops of the test solution. Boil for five minutes in a water bath. Cool the solution. It gives red, yellow or green colour/ precipitate. As in Fehling's test, the reducing sugars because of having potentially free aldehyde or keto group reduce cupric hydroxide in alkaline solution to red coloured cuprous oxide. Depending on the sugar concentration yellow to green colour is developed.

Barfoed's test

To 1 ml of the test solution add about 2 ml of Barfoed's reagent. Boil it for one minute and allow to stand for a few minutes. It gives brick-red precipitate. Only monosaccharides answer this test. Since Barfoed's reagent is weakly acidic, it is reduced only by monosaccharides.

Selliwanoff's test

To 2 ml of Seliwanoff's reagent add two drops of test solution and heat the mixture to just boiling. Appearance of deep red colour. In concentrated HCl, ketoses undergo dehydration to yield furfural derivatives more rapidly than do aldoses. These derivatives form complexes with resorcinol to yield deep red colour. It is a timed colour reaction specific for ketoses.

Bial's test

To 5 ml of Bial's reagent add 2–3 ml of solution and warm gently. When bubbles rise to the surface cool under the tap. Appearance of green colour or precipitate. It is specific for pentoses. They get converted to furfural. In the presence of ferric ion orcinol and furfural condense to yield a coloured product.

Test for non-reducing sugars such as sucrose

a) Do Benedict's test with the test solution. No characteristic colour formation. This indicates the absence of reducing sugars in the given solution

b) Add 5 drops of concentrated HCl to 5 ml of test solution in another test tube. Heat for five minutes on a boiling water bath. Add 10% sodium hydroxide solution to give a slightly alkaline solution (test with red litmus paper). Now perform Benedict's test with this hydrolysed solution. It gives red or yellow colour. This indicates the formation of reducing sugars from non-reducing sugars after hydrolysis with acid.

Mucic acid test

Add a few drops of conc. HNO_3 to the concentrated test solution or substance directly and evaporate it over a boiling water bath till the acid fumes are expelled. Add a few drops of water and leave it overnight. It forms crystals. This indicate that the both end carbon groups are oxidized to carboxylic groups. The resultant saccharic acid of galactose is called mucic acid which is insoluble in water.

Osazone test

To 0.5 g of phenylhydrazine hydrochloride add 0.1 g of sodium acetate and 10 drops of glacial acetic acid. To this mixture add 5 ml of test solution and heat on a boiling water bath for about half an hour. Allow the tube to cool slowly and examine the crystals under a microscope. It give glucose, fructose and mannose produce needle-shaped yellow osazone crystals, whereas lactosazone is mushroomshaped. Different osazones show crystals of different shapes. Maltose produces flower-shaped crystals. This indicates that the ketoses and aldoses react with phenylhydrazine to produce a phenylhydrazone which in turn reacts with another two molecules of phenylhydrazine to form the osazone.

Notes

- For osazone test, the reaction mixture should be between pH 5 and 6. Fructose takes 2 min to form the osazone whereas for glucose it is 5 min. The disaccharides take a longer time to form osazones. Dissacharides form crystals only on cooling.
- When a mixture of carbohydrates is present in the test sample, chromatographic methods should be employed to identify the individual sugars.

7.2. Estimation of Reducing Sugars by Nelson-Somogyi Method

Sugars with reducing property (arising out of the presence of a potential aldehyde or keto group) are called reducing sugars. Some of the reducing sugars are glucose, galactose, lactose and maltose. The Nelson-Somogyi method is one of the classical and widely used methods for the quantitative determination of reducing sugars. The reducing sugars when heated with alkaline copper tartrate reduce the copper from the cupric to cuprous state and thus cuprous oxide is formed. When cuprous oxide is treated with arsenomolybdic acid, the reduction of molybdic acid to molybdenum blue takes place. The blue colour developed is compared with a set of standards in a colorimeter at 620 nm.

Materials

- *Alkaline Copper Tartrate* – (a) Dissolve 2.5 g anhydrous sodium carbonate, 2 g sodium bicarbonate, 2.5 g potassium sodium tartrate and 20 g anhydrous sodium sulphate in 80 ml water and make up to 100 ml. (b) Dissolve 15 g copper sulphate in a small volume of distilled water. Add one drop of sulphuric acid and make up to 100 ml. Mix 4 ml of B and 96 ml of solution A before use.
- *Arsenomolybdate reagent:* Dissolve 2.5 g ammonium molybdate in 45 ml water. Add 2.5 ml sulphuric acid and mix well. Then add 0.3 g disodium hydrogen arsenate dissolved in 25 ml water. Mix well and incubate at 37°C for 24–48 hours.
- *Standard glucose solution: Stock:* 100 mg in 100 ml distilled water.
- *Working standard:* 10 ml of stock diluted to 100 ml with distilled water [100 μg/ml].

Procedure

- Weigh 100 mg of the sample and extract the sugars with hot 80% ethanol twice (5 ml each time).
- Collect the supernatant and evaporate it by keeping it on a water bath at 80°C.
- Add 10 ml water and dissolve the sugars.
- Pipette out aliquots of 0.1 or 0.2 ml to separate test tubes.
- Pipette out 0.2, 0.4, 0.6, 0.8 and 1 ml of the working standard solution into a series of test tubes.
- Make up the volume in both sample and standard tubes to 2 ml with distilled water.
- Pipette out 2 ml distilled water in a separate tube to set a blank.
- Add 1 ml of alkaline copper tartrate reagent to each tube.
- Place the tubes in a boiling water for 10 minutes.
- Cool the tubes and add 1 ml of arsenomolybolic acid reagent to all the tubes.
- Make up the volume in each tube to 10 ml with water.
- Read the absorbance of blue colour at 620 nm after 10 min.

- From the graph drawn, calculate the amount of reducing sugars present in the sample.

Calculation

Absorbance corresponds to 0.1 ml of test = *x* mg of glucose

$$10 \text{ ml contains} = \frac{x}{0.1} \times 10 \text{ mg of glucose} = \% \text{ Reducing sugars}$$

7.3. Estimation of Reducing Sugar by Dinitrosasylic Acid Method

For sugar estimation an alternative to Nelson-Somogyi method is the dinitrosalicylic acid method. It is simple, sensitive and adoptable during handling of a large number of samples at a time. 3,5-Dinitrosalicylic acid (DNS or DNSA, IUPAC name 2-hydroxy-3,5-dinitrobenzoic acid) is an aromatic compounds that reacts with reducing sugars and other reducing molecules to form 3-amino-5-nitrosalicylic acid, which absorbs light strongly at 540 nm. It was first introduced as a method to detect reducing substances in urine and has since been widely used, for example, for quantifying carbohydrates levels in blood. It is mainly used in assay of alpha-amylase. However, enzymatic methods are usually preferred due to DNS lack of specificity.

This method tests for the presence of free carbonyl group (C=O), the so-called reducing sugars. This involves the oxidation of the aldehyde functional group present in, for example, glucose and the ketone functional group in fructose. Simultaneously, 3,5-dinitrosalicylic acid (DNS) is reduced to 3-amino, 5-nitrosalicylic acid under alkaline conditions:

$$\text{aldehyde group} \xrightarrow{\text{Oxidation}} \text{carboxyl group}$$

$$\text{3,5-dinitrosalicylic acid} \xrightarrow{\text{reduction}} \text{3-amino,5-nitrosalicylic acid}$$

Because dissolved oxygen can interfere with glucose oxidation, sulfite, which itself is not necessary for the colour reaction, is added in the reagent to absorb the dissolved oxygen.

The above reaction scheme shows that one mole of sugar will react with one mole of 3,5-dinitrosalicylic acid. However, it is suspected that there are many side reactions, and the actual reaction stoichiometry is more complicated than that previously described. The type of side reaction depends on the exact nature of the reducing sugars. Different reducing sugars generally yield different colour intensities; thus, it is necessary to calibrate for each sugar. In addition to the oxidation of the carbonyl groups in the sugar, other side reactions such as the decomposition of sugar also competes for the availability of 3,5-dinitrosalicylic acid. As a consequence, carboxymethyl cellulose can affect the calibration curve by enhancing the intensity of the developed colour.

Although this is a convenient and relatively inexpensive method, due to the relatively low specificity, one must run blanks diligently if the colorimetric results are to be interpreted correctly and accurately. One can determine the background absorption on the original cellulose substrate solution by adding cellulase, immediately stopping the reaction, and measuring the absorbance, *i.e.* following exactly the same procedures for the actual samples. When the effects of extraneous compounds are not known, one can effectively include a so-called internal standard by first fully developing the colour for the unknown sample; then, a known amount of sugar is added to this sample. The increase in the absorbance upon the second colour development is equivalent to the incremental amount of sugar added.

Materials

- *Dinitrosalicylic Acid Reagent* (*DNS Reagent*) - Dissolve by stirring 1 g dinitrosalicylic acid, 200 mg crystalline phenol and 50 mg sodium sulphite in 100 ml 1% NaOH. Store at 4°C. Since the reagent deteriorates due to sodium sulphite, if long storage is required, sodium sulphite may be added at the time of use.

Procedure

- Weigh 100 mg of the sample and extract the sugars with hot 80% ethanol twice (5 ml each time).
- Collect the supernatant and evaporate it by keeping it on a water bath at 80°C.
- Add 10 ml water and dissolve the sugars.
- Pipette out 0.5 to 3 ml of the extract in test tubes and equalize the volume to 3 ml with water in all the tubes.
- Add 3 ml of DNS reagent.
- Heat the contents in a boiling water bath for 5 min.
- When the contents of the tubes are still warm, add 1 ml of 40% Rochelle salt solution.

- Cool and read the intensity of dark red colour at 510 nm.
- Run a series of standards using glucose (0–500 μg) and plot a graph.

Calculation

Calculate the amount of reducing sugars present in the sample using the standard graph.

Reading

Miller, G.L. (1959). Use of dinitrosalicylic acid reagent for determination of reducing sugar, *Anal. Chem.*, **31:** 426.

7.4. Estimation of Total Carbohydrate by Anthrone Method

Carbohydrates are the important components of storage and structural materials in the plants. They exist as free sugars and polysaccharides. The basic units of carbohydrates are the monosaccharides which cannot be split by hydrolysis into simpler sugars. The carbohydrate content can be measured by hydrolysing the polysaccharides into simple sugars by acid hydrolysis and estimating the resultant monosaccharides. Carbohydrates are first hydrolysed into simple sugars using dilute hydrochloric acid. In hot acidic medium glucose is dehydrated to hydroxymethyl furfural. This compound forms with anthrone a green coloured product with an absorption maximum at 630 nm.

Materials

- HCl 2.5 N
- *Anthrone reagent:* Dissolve 200 mg anthrone in 100 ml of ice-cold 95% H_2SO_4. Prepare fresh before use.
- *Standard glucose:* Stock—Dissolve 100 mg in 100 ml water. Working standard—10 ml of stock diluted to 100 ml with distilled water. Store refrigerated after adding a fe4w drops of toluene.

Procedure

- Weigh 100 mg of the sample into a boiling tube.
- Hydrolyse by keeping it in a boiling water bath for three hours with 5 ml of 2.5 N HCl and cool to room temperature.
- Neutralise it with solid sodium carbonate until the effervescence ceases.
- Make up the volume to 100 ml and centrifuge.

- Collect the supernatant and take 0.5 and 1 ml aliquots for analysis.
- Prepare the standards by taking 0, 0.2, 0.4, 0.6, 0.8 and 1 ml of the working standard.'0' serves as blank.
- Make up the volume to 1 ml in all the tubes including the sample tubes by adding distilled water.
- Then add 4 ml of anthrone reagent.
- Heat for eight minutes in a boiling water bath.
- Cool rapidly and read the green to dark green colour at 630 nm.
- Draw a standard graph by plotting concentration of the standard on the X-axis *versus*
- absorbance on the Y-axis.
- From the graph calculate the amount of carbohydrate present in the sample tube.

Calculation

$$\text{Amount of carbohydrate present in 100 mg of the sample} = \frac{\text{mg of glucose 100}}{\text{Volume of test sample}} \times 100$$

Note

Cool the contents of all the tubes on ice before adding ice-cold anthrone reagent.

7.5. Estimation of Total Carbohydrate by Phenol Sulphuric Acid

The phenol sulphuric acid method to estimate total carbohydrates is described below. In hot acidic medium glucose is dehydrated to hydroxymethyl furfural. This forms a green coloured product with phenol and has absorption maximum at 490 nm.

Materials

- *Phenol 5%:* Redistilled (reagent grade) phenol (50 g) dissolved in water and diluted to one litre.
- Sulphuric acid 96% reagent grade.
- *Standard glucose:* Stock—100 mg in 100 ml of water. Working standard—10 ml of stock diluted to 100 ml with distilled water.

Procedure

- Weigh 100 mg of the sample into a boiling tube.
- Hydrolyse by keeping it in a boiling water bath for three hours with 5 ml of 2.5 N HCl and cool to room temperature.
- Neutralise it with solid sodium carbonate until the effervescence ceases.
- Make up the volume to 100 ml and centrifuge.
- Pipette out 0.2, 0.4, 0.6, 0.8 and 1 ml of the working standard into a series of test tubes.
- Pipette out 0.1 and 0.2 ml of the sample solution in two separate test tubes. Make up the volume in each tube to 1 ml with water.
- Set a blank with 1 ml of water.
- Add 1 ml of phenol solution to each tube.
- Add 5 ml of 96% sulphuric acid to each tube and shake well.
- After 10 min shake the contents in the tubes and place in a water bath at 25–30°C for 20 min.
- Read the colour at 490 nm.
- Calculate the amount of total carbohydrate present in the sample solution using the standard graph.

Calculation

Absorbance corresponds to 0.1 ml of the test = x mg of glucose

$$100 \text{ ml of the sample solution contains} = \frac{X}{0.1} \times 100 \text{ mg of Glucose}$$

= % of total carbohydrate present.

7.6. Estimation of Starch by Anthrone

Starch is an important polysaccharide. It is the storage form of carbohydrate in plants abundantly found in roots, tubers, stems, fruits and cereals. Starch, which is composed of several glucose molecules, is a mixture of two types of components namely amylose and amylopectin. Starch is hydrolysed into simple sugars by dilute acids and the quantity of simple sugars is measured colorimetrically. The sample is treated with 80% alcohol to remove sugars and then starch is extracted with perchloric acid. In hot acidic medium starch is hydrolysed to glucose and dehydrated to hydroxymethyl furfural. This compound forms a green coloured product with anthrone.

Materials

- *Anthrone:* Dissolve 200 mg anthrone in 100 ml of ice-cold 95% sulphuric acid.
- 80% ethanol.
- 52% perchloric acid.
- *Standard glucose:* Stock—100 mg in 100 ml water. Working standard—10 ml of stock diluted to 100 ml with water.

Procedure

- Homogenize 0.1–0.5 g of the sample in hot 80% ethanol to remove sugars. Centrifuge and retain the residue. Wash the residue repeatedly with hot 80% ethanol till the washings do not give colour with anthrone reagent. Dry the residue well over a water bath.
- To the residue add 5.0 ml of water and 6.5 ml of 52% perchloric acid.
- Extract at 0°C for 20 min. Centrifuge and save the supernatant.
- Repeat the extraction using fresh perchloric acid. Centrifuge and pool the supernatants and make up to 100 ml.
- Pipette out 0.1 or 0.2 ml of the supernatant and make up the volume to 1 ml with water.
- Prepare the standards by taking 0.2, 0.4, 0.6, 0.8 and 1 ml of the working standard and make up the volume to 1 ml in each tube with water.
- Add 4 ml of anthrone reagent to each tube.
- Heat for eight minutes in a boiling water bath.
- Cool rapidly and read the intensity of green to dark green colour at 630 nm.

Calculation

Find out the glucose content in the sample using the standard graph. Multiply the value by a factor 0.9 to arrive at the starch content.

7.7. Estimation of Amylose

Amylose is a spiral polymer made up of D-glucose units. This polysaccharide is one of the two components of starch, making up approximately 20-30% of the structure. The other component is amylopectin, which makes up 70-80%

of the structure. Because of its tightly packed structure, amylose is more resistant to digestion than other starch molecules and is therefore an important form of resistant starch, which has been found to be an effective prebiotic.

Amylose is made up of α (1 → 4) bound glucose molecules. The carbon atoms on glucose are numbered, starting at the aldehyde (C=O) carbon, so, in amylose, the 1-carbon on one glucose molecule is linked to the 4-carbon on the next glucose molecule (α 1→ 4) bonds). The number of repeated glucose subunits (n) is usually in the range of 300 to 3000, but can be many thousands. There are three main forms of amylose chains can take. It can exist in a disordered amorphous conformation or two different helical forms. It can bind with itself in a double helix (A or B form), or it can bind with another hydrophobic guest molecule such as iodine, a fatty acid, or an aromatic compound. This is known as the V form and is how amylopectin binds to amylose to form starch. Within this group, there are many different variations. Each is notated with V and then a subscript indicating the number of glucose units per turn. The most common is the V6 form, which has six glucose units a turn. V8 and possibly V7 forms exist as well. These provide an even larger space for the guest molecule to bind.

This linear structure can have some rotation around the phi and psi angles, but for the most part bound glucose ring oxygens lie on one side of the structure. The α(1→ 4) structure promotes the formation of a helix structure, making it possible for hydrogen bonds form between the oxygen atoms bound at 2-carbon of one glucose molecule and the 3-carbon of the next glucose molecule.

CH_2OH CH_2OH CH_2OH O O O OH OH OH OH O O OH OH OH OH 300–600

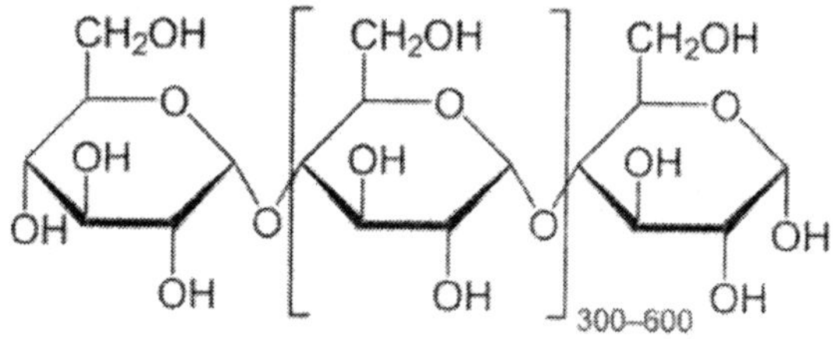

Amylose

Amylose is important in plant energy storage. It is less readily digested than amylopectin; however, because it is more linear than amylopectin, it takes up less space. As a result, it is the preferred starch for storage in plants. It makes up about 30% of the stored starch in plants, though the specific percentage varies by species. The digestive enzyme α-amylase is responsible for the breakdown of the starch molecule into maltotrios and maltose, which can be used as sources of energy. Amylose is also an important thickener, water binder, emulsion stabilizer, and gelling agent in both industrial and food-based contexts. Loose helical amylose chains have a hydrpphobic interior that can bind to hydrophobic molecules such as lipids and aromatic compounds. The one problem with this is that, when it crystallizes or associates, it can lose some stability, often releasing water in the process (syneresis). When amylose concentration is increased, gel stickiness decreases but gel firmness increases. When other things including amylopectin bind to amylose, the viscosity can be affected, but incorporating α-carrageenan, alginate, xanthun gum, or low-molecular-weight sugars can reduce the loss in stability. The ability to bind water can add substance to food, possibly serving as a fat replacement. For example, amylose is responsible for causing white sauce to thicken, but, upon cooling, some separation between the solid and the water will occur.

In a laboratory setting, it can act as a marker. Iodine molecules fit neatly inside the helical structure of amylose, binding with the starch polymer that absorbs certain known wavelength of light. Hence, a common test is the iodine test for starch. Mix starch with a small amount of yellow iodine solution. In the presence of amylose, a blue-black colour will be observed. The intensity of the colour can be tested with a colorimeter, using a red filter to discern the concentration of starch present in the solution. It is also possible to use starch as an indicator in titrations involving iodine reduction. It is also used in amylose magnetic beads and resin to separate maltose-binding protein.

Materials

- Distilled ethanol.
- NaOH, 1.0 N
- Phenolphthalein 0.1%
- *Iodine reagent:* Dissolve 1 g iodine and 10 g KI in water and make up to 500 ml.
- *Standard:* Dissolve 100 mg amylose in 10 ml 1 N NaOH; make up to 100 ml with water.

Procedure

- Weigh 100 mg of the powdered sample, and add 1 ml of distilled ethanol. Then add 10 ml of 1 N NaOH and leave it overnight.
- Make up the volume to 100 ml.
- Take 2.5 ml of the extract, add about 20 ml distilled water and then three drops of phenolphthalein.
- Add 0.1 N HCl drop by drop until the pink colour just disappears.

- Add 1 ml of iodine reagent and make up the volume to 50 ml and read the colour at 590 nm.
- Take 0.2, 0.4, 0.6, 0.8 and 1 ml of the standard amylose solution and develop the colour as in the case of sample.
- Calculate the amount of amylose present in the sample using the standard graph.
- Before use dilute 1 ml of iodine reagent to 50 ml with distilled water for a blank.

Calculation

Absorbance corresponds to 2.5 ml of the test solution

$= x$ mg amylose 100 ml contains

$$= \frac{X}{2.5} \times 100 \text{ mg amylose} = \% \text{ Amylose}$$

- The sample suspension may be heated for 10 min in a boiling water-bath instead of overnight dissolution.
- The amount of amylopectin is obtained by subtracting the amylose content from that of starch.

7.8. Estimation of Cellulose

Cellulose is an organic compound with the formula $(C_6H_{10}O_5)_n$, a polysaccharide consisting of a linear chain of several hundred to over ten thousand β (1→4) linked D-glucose units. Cellulose is an important structural component of the primary cell wall of green plants, many forms of algae and the oomycetes. Some species of bacteria secrete it to form biofilms. Cellulose is the most abundant organic polymer on Earth. The cellulose content of cotton fibre is 90%, that of wood is 40–50% and that of dried hemp is approximately 45%. Cellulose is mainly used to produce paperboard and paper. Smaller quantities are converted into a wide variety of derivative products such as cellophane and rayon. Conversion of cellulose from energy crops into biofuels such as cellulosic ethanol is under investigation as an alternative fuel source. Cellulose for industrial use is mainly obtained from wood pulp and cotton.

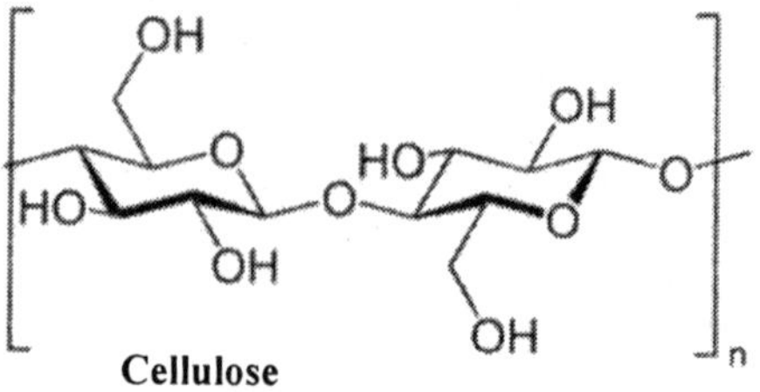

Cellulose

Some animals, particularly ruminants and termites, can digest cellulose with the help of symbiotic micro-organisms that live in their guts, such as *Trichonympha.* Humans can digest cellulose to some extent, however it mainly acts as a hydrophilic bulking agent for feces and is often referred to as α "dietary fibre". Cellulose is derived from D-glucose units, which condense through α (1→ 4)-glycosidic bonds. This linkage motif contrasts with that for a (1→4)-glycosidic bonds present in starch, glycogen, and other carbohydrates. Cellulose is a straight chain polymer: unlike starch, no coiling or branching occurs, and the molecule adopts an extended and rather stiff rod-like conformation, aided by the equatorial co nformation of the glucose residues. The multiple hydroxyl group on the glucose from one chain form hydrogen bonds with oxygen atoms on the same or on a neighbor chain, holding the chains firmly together side-by-side and forming *microfibrils* with high tensile strength. This confers tensile strength in cell walls, where cellulose microfibrils are meshed into a polysaccharide *matrix*. Compared to starch, cellulose is also much more crystalline. Whereas starch undergoes a crystalline to amorphous transition when heated beyond 60–70 °C in water (as in cooking), cellulose requires a temperature of 320 °C and pressure of 25 MPa to become amorphous in water.

Cellulose, a major structural polysaccharide in plants, is the most abundant organic compound in nature, and is composed of glucose units joined together in the form of the repeating units of the disaccharide cellobiose with numerous cross linkages. It is also a major component in many of the farm wastes. Cellulose undergoes acetolysis with acetic/nitric reagent forming acetylated cellodextrins which get dissolved and hydrolyzed to form glucose molecules on treatment with 67% H_2SO_4. This glucose molecule is dehydrated to form hydroxymethyl furfural which forms green coloured product with anthrone and the colour intensity is measured at 630 nm.

Materials

- *Acetic/Nitric reagent:* Mix 150 ml of 80% acetic acid and 15 ml of concentrated nitric acid.
- *Anthrone reagent:* Dissolve 200 mg anthrone in 100 ml concentrated sulphuric acid.
- Prepare fresh and chill for 2 h before use.
- Sulphuric acid 67%

Procedure

- Add 3 ml acetic/nitric reagent to a known amount (0.5 g or 1 g) of the sample in a test tube and mix in a vortex mixer.
- Place the tube in a water-bath at 100°C for 30 min.
- Cool and then centrifuge the contents for 15–20 min.
- Discard the supernatant.
- Wash the residue with distilled water.
- Add 10 ml of 67% sulphuric acid and allow it to stand for 1 h.
- Dilute 1 ml of the above solution to 100 ml.
- To 1 ml of this diluted solution, add 10 ml of anthrone reagent and mix well.
- Heat the tubes in a boiling water-bath for 10 min.
- Cool and measure the colour at 630 nm.
- Set a blank with anthrone reagent and distilled water.
- Take 100 mg cellulose in a test tube and proceed from Step No. 6 for standard. Instead of just taking 1 ml of the diluted solution take a series of volumes (say 0.4–2 ml corresponding to 40–200 ìg of cellulose) and develop the colour.

Calculation

Draw the standard graph and calculate the amount of cellulose in the sample.

7.9. Estimation of Hemicellulose

A hemicellulose (also known as polyose) is any of several heteropolymers (matrix polysaccharides), such as arabinoxylans, present along with cellulose in almost all plant cell walls. While cellulose is crystalline, strong, and resistant to hydrolysis, hemicellulose has a random, amorphous structure with little strength. It is easily hydrolyzed by dilute acid or base as well as myriad hemicellulase enzymes. Hemicelluloses are non-cellulosic, non-pectic cell wall polysaccharides. They are regarded as being composed of xylans, mannans, glucomannans, galactans and arabinogalactans. Hemicelluloses are categorized under 'unavailable carbohydrates' since they are not split by the digestive enzymes of the human system. Refluxing the sample material with neutral detergent solution removes the water-solubles and materials other than the fibrous component. The left out material is weighed after filtration and expressed as Neutral Detergent Fibre (NDF).

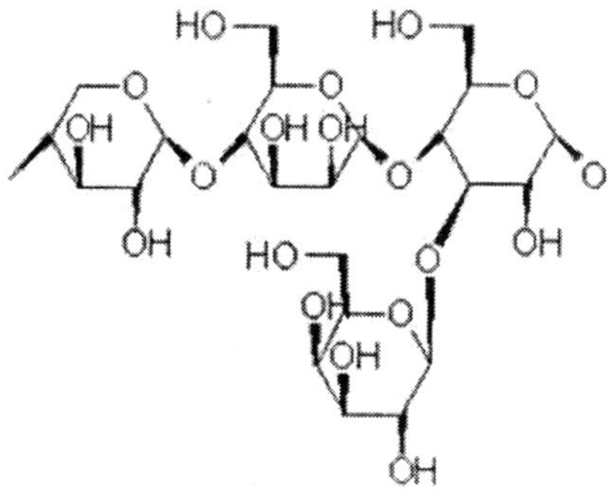

- Xylose - ß(1,4) - Mannose - ß(1,4) - Glucose -
- alpha(1,3) - Galactose
Hemicellulose

Hemicelluloses include xylan, gluconoxylan, arabinoxylan, glucomannan, and xyloglucan. These polysaccharides contain many different sugar monomers. In contrast, cellulose contains only anhydrous glucose. For instance, besides glucose, sugar monomers in hemicellulose can include xylose, mannose, galactose, rahmnose, and arabinose. Hemicelluloses contain most of the D-pentose sugars, and occasionally small amounts of L-sugars as well. Xylose is in most cases the sugar monomer present in the largest amount, although in softwoods mannose can be the most abundant sugar. Not only regular sugars can be found in hemicellulose, but also their acidified form, for instance glucorunic acid and galactorunic acid can be present.

Materials

- *Neutral Detergent Solution*
- Weigh 18.61 g disodium ethylenediamine tetraacetate and 6.81 g sodium borate decahydrate. Transfer to a beaker. Dissolve in about 200 ml of distilled water by heating and to this, add a solution (about 100–200 ml) containing 30 g of sodium lauryl sulphate and 10 ml of 2-ethoxy ethanol. To this add a solution (about 100 ml) containing 4.5 g of disodium hydrogen phosphate. Make up the volume to one litre and adjust the pH to 7.0.
- Decahydronaphthalene.
- Sodium sulphite.
- Acetone.

Procedure

- To 1 g of the powdered sample in a refluxing flask add 10 ml of cold neutral detergent solution.
- Add 2 ml of decahydronaphthalene and 0.5 g sodium sulphite.
- Heat to boiling and reflux for 60 min.
- Filter the contents through sintered glass crucible (G-2) by suction and wash with hot water.
- Finally give two washings with acetone.
- Transfer the residue to a crucible, dry at 100°C for 8 h.
- Cool the crucible in a desiccator and weigh.

Calculation

Hemicellulose = Neutral detergent fibre (NDF) – Acid detergent fibre (ADF)

7.10. Estimation of Fructose and Inulin

Inulins are a group of naturally occurring polysaccharides produced by many types of plants, industrially most often extracted from chicory. The inulins belong to a class of dietary fibres known as fructans. Inulin is used by some plants as a means of storing energy and is typically found in roots or rhizomes. Most plants that synthesize and store inulin do not store other forms of carbohydrate such as starch. Inulin is a heterogeneous collection of fructose polymers. It consists of glucosyl moiety and fructosyl moiety, which are linked by β (2,1) bonds. The degree of polymerization (DP) of standard inulin ranges from 2 to 60. After removing the fractions with DP lower than 10 during manufacturing process, the remaining product is high performance inulin. Some articles considered the fractions with DP lower than 10 as short-chained fructooligosaccharides, and only called the longer-chained molecules inulin.

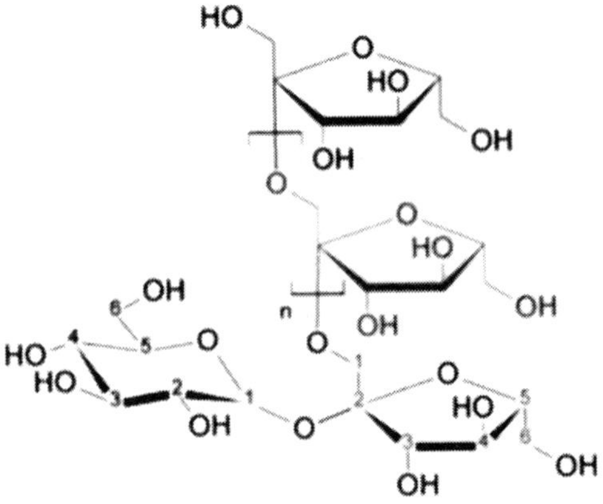

Inulin

Because of the β (2,1) linkages, inulin is not digested by enzymes in the human alimentary system, contributing to its functional properties: reduced calorie value, dietary fibre and prebiotic effects. Without colour and odor, Inulin has little impact on sensory characteristics of food products. Oligofructose has 35% of the sweetness of sucrose, and its sweetening profile is similar to sugar. Standard inulin is slightly sweet, while high performance inulin is not. Its solubility is higher than the classical fibres. When thoroughly mixed with liquid, inulin forms a gel and a white creamy structure, which is similar to fat. Tri-dimensional gel network, consisting of insoluble sub-micron crystalline inulin particles, immobilizes large amount of water, assuring its physical stability. It can also improve the stability of foams and emulsions.

Inulin is increasingly used in processed foods because it has unusually adaptable characteristics. Its flavour ranges from bland to subtly sweet (approx. 10% sweetness of sugar/sucrose). It can be used to replace sugar, fat, and flour. This is advantageous because inulin contains 25-35% of the food energy of carbohydrates (starch, sugar). In addition to being a versatile ingredient, inulin has many health benefits. Inulin increases calcium absorption and possibly mangansium absorption, while promoting the growth of beneficial intestinal bacteria. Chicory inulin is reported to increase absorption of calcium in girls with lower calcium absorption and in young men. In terms of nutrition, it is considered a form of soluble fibre and is sometimes categorized as a prebiotic. Conversely, it is also considered a FODMAP, a class of carbohydrates which are problematic for some individuals through causing overgrowth of intestinal methanogenic bacteria. The consumption of large quantities (in particular, by sensitive or unaccustomed individuals) can lead to gas and bloating, and products that contain inulin will sometimes include a warning to add it gradually to one's diet.

Due to the body's limited ability to process fructans, inulin has minimal increasing impact on blood sugar. It is considered suitable for diabetics and potentially helpful in managing blood sugar-related illnesses.

Fructose, or fruit sugar, is a simple monosachharide found in many plants, where it is often bonded to glucose to form the disaccharides sucrose. It is one of the three dietary monosaccharides, along with glucose and galactose, that are absorbed directly into the bloodstream during digestion. Pure, dry fructose is a very sweet, white, odorless, crystalline solid and is the most water-soluble of all the sugars. From plant sources, fructose is found in honey, tree and vine fruits, flowers, berries, and most root vegetables.

Commercially, fructose is frequently derived from sugar cane, sugar beets, and maize. Crystalline fructose is the monosaccharide, dried, ground, and of high purity. High-fructose corn syrup (HFCS) is a mixture of glucose and fructose as monosaccharides. Sucrose is a compound with one molecule of glucose covalently linked to one molecule of fructose. All forms of fructose, including fruits and juices, are commonly added to foods and drinks for palatability and taste enhancement, and for browning of some foods, such as baked goods.

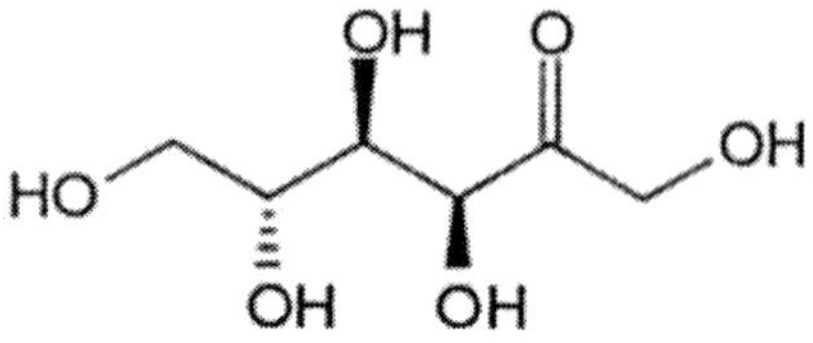

Fructose

Fructose is a 6-carbon polyhydroxyketone. It is an isomer of glucose; *i.e.*, both have the same molecular formula ($C_6H_{12}O_6$) but they differ structurally. Crystalline fructose adopts a cyclic six-membered structure owing to the stability of its hemiketal and internal hydrogen-bonding. This form is formally called D-fructopyranose. In solution, fructose exists as an equilibrium mixture of 70% fructopyranose and about 22% fructofuranose, as well as small amounts of three other forms, including the acyclic structure.

Natural sources of fructose include fruits, vegetables (including sugar cane), and honey. Fructose is often further concentrated from these sources. The highest dietary sources of fructose, besides pure crystalline fructose, are foods containing table sugar (sucrose), high fructose corn syrup, agave nectar, honey, molasses, maple syrup, and fruit juices, as these have the highest percentages of fructose (including fructose in sucrose) per serving compared to other common foods and ingredients. Fructose exists in foods either as a free monosaccharide or bound to glucose as sucrose, a disaccharide. Fructose, glucose, and sucrose may all be present in a food; however, different foods will have varying levels of each of these three sugars. In general, in foods that contain free fructose, the ratio of fructose to glucose is approximately 1:1; that is, foods with fructose usually contain about an equal amount of free glucose. A value that is above 1 indicates a higher proportion of fructose to glucose, and below 1 a lower proportion. Some fruits have larger proportions of fructose to glucose compared to others. For example, apples and pears contain more than twice as much free fructose as glucose, while for apricot the proportion is less than half as much fructose as glucose.

Apple and pear juices are of particular interest to paediatricians because the high concentrations of free fructose in these juices can cause diarrhoea in children. The cells (entrocytes) that line children's small intestines have less affinity for fructose absorption than for glucose and sucrose. Unabsorbed fructose creates higher osmolarity in the small intestine, which draws water into the gastrointestinal tract, resulting in osmotic diarrhoea. Sugarcane and sugar beet have a high concentration of sucrose, and are used for commercial preparation of pure sucrose. Extracted cane or beet juice is clarified, removing impurities; and concentrated by removing excess water. The end product is 99.9% pure sucrose. Sucrose-containing sugars include common table white granulated sugar and powdered sugar, as well as brown sugar. Fructose is usually accompanied by sucrose in fruits like apple. Honey is a rich source of fructose. The hydroxymethyl furfural formed from fructose in acid medium reacts with resorcinol to give a red colour product.

Materials

- *Resorcinol reagent:* Dissolve 1 g resorcinol and 0.25 g thiourea in 100 ml glacial acetic acid. This solution is indefinitely stable in the dark.
- *Dilute HCl:* Mix five parts of conc. HCl with one part of distilled water.
- *Standard fructose solution:* Dissolve 50 mg of fructose in 50 ml water. Dilute 5 ml of this stock to 50 ml for a working standard.

Procedure

- To 2 ml of the solution containing 20–80 μg of fructose add 1 ml of resorcinol reagent.
- Then add 7 ml of dilute hydrochloric acid.
- Pipette out 0.2, 0.4, 0.6, 0.8 and 1 ml of the working standard and make up the volume to 2 ml with water. Add 1 ml of resorcinol reagent and 7 ml of dilute HCl as above.
- Set a blank along with the working standard.
- Heat all the tubes in a water-bath at 80°C for exactly 10 min.
- Remove and cool the tubes by immersing in tap water for 5 min.
- Read the colour at 520 nm within 30 min.
- Draw the standard graph and calculate the amount of fructose present in the sample using the standard graph.

Inulin

Inulin is a polymer made of fructose units with β-2-1 linkage. It is found in onion, garlic and in many other plant parts.

Sample Extraction

Grind the sample and extract in 80% ethanol for six hours to remove free sugars. Dry the sample and take 500 mg in a 100 ml conical flask. Add 20 ml of water and heat it in a water bath at 90°C for 10 min. Collect the extract and then add 70 ml of water. Replace the flask for another 30 min with occasional shaking to dissolve the fructosan, then remove and cool it at room temperature. Combine the extracts and filter the solution if it is not clear and make up to 100 ml in a standard flask. To estimate the inulin content in the extract follow the procedure given for fructose estimation. The amount of inulin is expressed in terms of fructose concentration.

7.11. Estimation of Pectic Substances

Pectin is a structural heteropolysaccharide contained in the primary cell walls of terrestrial plants and is produced commercially as a white to light brown powder, mainly extracted from citrus fruits, and is used in food as a gelling agent particularly in jams and jellies. It is also used in fillings, medicines, sweets, as a stabilizer in fruit juices and milk drinks, and as a source of dietary fibre. In plant biology, pectin consists of a complex set of polysaccharides that are present in most primary cell walls and are particularly abundant in the non-woody parts of terrestrial plants. Pectin is a major component of the middle lamella, where it helps to bind cells together, but is also found in primary cell walls.

The amount, structure and chemical composition of pectin differs among plants, within a plant over time, and in various parts of a plant. Pectin is an important cell wall polysaccharide that allows primary cell wall extension and plant growth. During fruit ripening, pectin is broken down by the enzymes pectinase and pectinesesterase, in which process the fruit becomes softer as the middle lamellae break down and cells become separated from each other. A similar process of cell separation caused by the breakdown of pectin occurs in the abscission zone of the petioles of deciduous plants at leaf fall. Pectin is a natural part of the human diet, but does not contribute significantly to nutrition. The daily intake of pectin from fruits and vegetables can be estimated to be around 5 g (assuming consumption of approximately 500 g fruits and vegetables per day). In human digestion, pectin binds to cholesterol in the gastrointestinal tract and slows glucose absorption by trapping carbohydrates. Pectin is thus a soluble dietary fibre.

Consumption of pectin has been shown to reduce blood cholesterol levels. The mechanism appears to be an increase of viscosity in the intestinal tract, leading to a reduced absorption of cholesterol from bile or food. In the large intestine and colon, microorganisms degrade pectin and liberate short-chain fatty acids that have positive influence on health (prebiotic effect). Pectins, also known as pectic polysaccharides, are rich in galacturonic acid. Several distinct polysaccharides have been identified and characterised within the pectic group. Homogalac-turonans are linear chains of α-(1–4)-linked D-galacturonic acid. Substituted galacturonans are characterized by the presence of saccharide appendant residues (such as D-xylose or D-apiose in the respective cases of xylogalacturonan and apiogalacturonan) branching from a backbone of D-galacturonic acid residues. Rhamnogalacturonan I pectins (RG-I) contain a backbone of the repeating disaccharide: 4)-α-D-galacturonic acid-(1,2)-α-L-rhamnose-(1. From many of the rhamnose residues, sidechains of various neutral sugars branch off. The neutral sugars are mainly D-galactose, L-arabinise and D-xylose, with the types and proportions of neutral sugars varying with the origin of pectin.

Another structural type of pectin is rhamnogalacturonan II (RG-II), which is a less frequent complex, highly branched polysaccharide. Rhamnogalacturonan II is classified by some authors within the group of substituted galacturonans since the rhamnogalacturonan II backbone is made exclusively of D-galacturonic acid units. Isolated pectin has a molecular weight of typically 60–130,000 g/mol, varying with origin and extraction conditions. In nature, around 80 percent of carboxyl groups of galacturonic acid are esterified with methanol. This proportion is decreased to a varying degree during pectin extraction. The ratio of esterified to non-esterified galacturonic acid determines the behavior of pectin in food applications. This is why pectins are classified as high- vs. low-ester pectins (short HM vs. LM-pectins), with more or less than half of all the galacturonic acid esterified. The non-esterified galacturonic acid units can be either free acids (carboxyl groups) or salts with sodium, potassium, or calcium. The salts of partially esterified pectins are called pectinates, if the degree of esterification is below 5 percent the salts are called pectates, the insoluble acid form, pectic acid. Some plants such as sugar beet, potatoes and pears contain pectins with acetylated galacturonic acid in addition to methyl esters. Acetylation prevents gel-formation but increases the stabilising and emulsifying effects of pectin. Amidated pectin is a modified form of pectin. Here, some of the galacturonic acid is converted with ammonia to carboxylic acid amide. These pectins are more tolerant of varying calcium concentrations that occur in use.

To prepare a pectin-gel, the ingredients are heated, dissolving the pectin. Upon cooling below gelling temperature, a gel starts to form. If gel formation is too strong, syneresis or a granular texture are the result, whilst weak gelling leads to excessively soft gels. Pectins gel according to specific parameters, such as sugar, pH and bivalent salts (especially Ca^{2+}). In high-ester pectins at soluble solids content above 60% and a pH-value between 2.8 and 3.6, hydrogen bonds and hydrophobic interactions bind the individual pectin chains together. These bonds form as water is bound by sugar and forces pectin strands to stick together. These form a 3-dimensional molecular net that creates the macromolecular gel. The gelling-mechanism is called a low-water-activity gel or sugar-acid-pectin gel. In low-ester pectins, ionic bridges are formed between calcium ions and the ionised carboxyl groups of the galacturonic acid. This is idealised in the so-called "egg box-model". Low-ester pectins need calcium to form a gel, but can do so at lower soluble solids and higher pH-values than high-ester pectins. Normally low-ester pectins form gels with a range of pH from 2.6 to 7.0 and with a soluble solids content between 10 and 70%. Amidated pectins behave like low-ester pectins but need less calcium and are more tolerant of excess calcium. Also, gels from amidated pectin are thermo-reversible; they can be heated and after cooling solidify again, whereas conventional pectin-gels will afterwards remain liquid. High-ester pectins set at higher temperatures than low-ester pectins. However, gelling reactions with calcium increase as the degree of esterification falls. Similarly, lower pH-values or higher soluble solids (normally sugars) increase gelling speed. Suitable pectins can therefore be selected for jams and for jellies, or for higher sugar confectionery jellies.

Pectic acid is water soluble whereas pectin forms a colloidal solution. Protopectin is a larger molecule than pectic acid and pectin. During ripening of fruits, conversion of protopectin into pectic acid and pectin takes place. The pectins in fruits vary in their methoxyl content and in jellying power. Galacturonic acid is reacted with carbazole in the presence of H_2SO_4 and the colour developed is measured at 520 nm.

Materials

- 60% Ethyl alcohol (Mix 500 ml 95% alcohol and 300 ml water).
- 95% Ethyl alcohol.
- Purified ethyl alcohol (Reflux 1 L of 95% ethyl alcohol with 4 g zinc dust and 2 ml conc. H_2SO_4 for 15 h and distill in all glass distillation apparatus. Redistill with 4 g zinc dust and 4 g KOH).
- 1 N and 0.05 N Sodium hydroxide.
- H_2SO_4 (Analytical grade).
- *0.1% Carbazole reagent:* Weigh 100 mg recrystallized carbazole, dissolve and dilute to
- 100 ml with purified alcohol.

Procedure

- Weigh 100 mg pectin (see notes section for the preparation of pectin) and dissolve in 100 ml of 0.05 N NaOH.
- Allow it to stand for 30 min to deesterify the pectin.
- Take 2 ml of this solution and make up to 100 ml with water.
- Pipette out 2 ml of deesterified pectin solution and add 1 ml carbazole reagent. A white precipitate will be formed.
- Add 12 ml conc. H_2SO_4 with constant stirring.
- Close the tubes with rubber stopper and allow to stand for 10 min to develop the colour.
- To set a blank add 1 ml of purified ethyl alcohol in the place of carbazole reagent.
- Read the colour at 525 nm against blank, exactly 15 min after the addition of acid.

Standard

- Weigh 120.5 mg galacturonic acid monohydrate (from a sample vacuum dried for 5 h at 30°C) and transfer to a 1 L volumetric flask. Add 10 ml 0.05 N NaOH and dilute to volume with water. After mixing, allow it to stand overnight. Dilute 10, 20, 40, 50, 60 and 80 ml of this standard solution to 100 ml with water. Take 2 ml of these solutions for colour developing and proceed as in the case of the sample. Draw a standard curve—the absorbance *versus* concentration.

Calculation

Read the concentration of the anhydrogalacturonic acid corresponding to the reading of

the sample, and calculate as follows:

$$\%\ \text{anhydrogalacturonic acid} = \frac{\mu\text{g of anhydrogalacturonic acid in the aliquot} \times \text{Dilution} \times 100}{\text{ml taken for estimation} \times \text{Wt. of pectin sample} \times 1{,}000{,}000}$$

Notes

- Carbazole is recrystallized from toluene.
- An alternate procedure adopted for colour development is as follows:
- Take 12 ml of conc. H_2SO_4 in a test tube, cool in an ice-bath, and add 2 ml of the deesterified pectin solution and again cool. Heat the contents in a boiling water-bath for 10 min, cool to 20°C and add 1 ml of 0.15% carbazole reagent in purified ethyl alcohol. Allow it to stand for 25 ± 5 min at room temperature to develop the colour. Read the absorbance at 520 nm. Standards should also be treated similarly.

Extraction and Purification of Pectin

- Blend the fresh sample. If the material is dry grind.
- Transfer 100 g macerated sample (10 g dry tissue) to a pre-weighed 1 L beaker containing 400 ml water.
- Add 1.2 g freshly ground sodium hexametaphosphate and adjust to pH 4.5.
- Heat with stirring at 90–95°C for 1 h. Check the pH in every 15 min and maintain at pH with citric acid or NaOH. Replace water lost by evaporation at intervals. However, do not add water at the last 20 min.
- Add 4 g filter aid and 4 g ground paper pulp. Filter rapidly through a fast filter paper coated with 3 g moistened fast filter aid.
- Collect at least 200 ml of the filtrate in a preweighed container. Cool as rapidly as possible. Now, note the weight of the filtrate.
- If the filtrate contains less than 0.2% pectin, concentrate the filtrate under vacuum to attain this concentration.
- To three volumes of ethanol, isopropanol or acetone containing 0.5 N HCl, pour the cooled, weighed filtrate. The slurry should be at pH 0.7–1. Stir for 30 min.
- Centrifuge or filter. Wash the precipitate with the same solvent containing HCl. Then, wash repeatedly with 70% alcohol or acetone until the precipitate is essentially chloride free or the pH is above 4.
- Dehydrate the precipitate further in 400 ml acetone. Dry overnight *in vacuo* with a slow stream of dry air passing through the oven.
- Weigh the precipitate and use this pectin for analysis.

- The dried pectin should be free from ammonia for which a small sample of the pectin is heated with 1 ml of 0.1 N NaOH and ammoniacal odour can be noticed or tested with a moistened litmus paper. If ammonium ions are present wash with acidified 6% alcohol, followed by neutral alcohol to remove the acid and dry.

7.12. Estimation of Crude Fibre

Originally, fibre was defined to be the components of plants that resist human digestive enzymes, a definition that includes lignin and polysaccharides. The definition was later changed to also include resistant starch, along with inulin and other oligosaccharides. Dietary fibre, or sometimes roughage and ruffage is the indigestible portion of food derived from plants and waste of animals that eat dietary fibre.

There are two main components:

- Soluble fibre dissolves in water. It is readily fermented in the colon into gases and physiologically active byproducts, and can be prebiotic and/or viscous. Soluble fibres tend to slow the movement of food through the system.
- Insoluble fibre does not dissolve in water. It can be metabolically inert and provide bulking or prebiotic, metabolically fermenting in the large intestine. Bulking fibres absorb water as they move through the digestive system, easing defecation. Fermentable insoluble fibres mildly promote stool regularity, although not to the extent that bulking fibres do, but they can be readily fermented in the colon into gases and physiologically active byproducts. Insoluble fibres tend to accelerate the movement of food through the system.

Dietary fibres can act by changing the nature of the contents of the gastrointestinal tract and by changing how other nutrients and chemicals are absorbed. Some types of soluble fibre absorb water to become a gelatinous, viscous substance which is fermented by bacteria in the digestive tract. Some types of insoluble fibre have bulking action and are not fermented. Lignin, a major dietary insoluble fibre source, may alter the rate and metabolism of soluble fibres. Other types of insoluble fibre, notably resistant starch, are fully fermented.

Chemically, dietary fibre consists of non-starch polysaccharides such as arabinoxylans, cellulose, and many other plant components such as resistant starch, resistant dextrins, inulin, lignin, waxes, chitins, pectins, beta-glucans, and oligosaccharides. A novel position has been adopted by the US Department of Agriculture to include *functional fibres* as isolated fibre sources that may be included in the diet. The term "fibre" is something of a misnomer, since many types of so-called dietary fibre are not actually fibrous.

Food sources of dietary fibre are often divided according to whether they provide (predominantly) soluble or insoluble fibre. Plant foods contain both types of fibre in varying degrees, according to the plant's characteristics. Advantages of consuming fibre are the production of healthful compounds during the fermentation of soluble fibre, and insoluble fibre's ability (via its passive hygroscopic properties) to increase bulk, soften stool, and shorten transit time through the intestinal tract. Disadvantages of a diet high in fibre is the potential for significant intestinal gas production and bloating. Constipation can occur if insufficient fluid is consumed with a high-fibre diet.

Some plants contain significant amounts of soluble and insoluble fibre. For example plums and purnes have a thick skin covering a juicy pulp. The skin is a source of insoluble fibre, whereas soluble fibre is in the pulp. Grapes also contain a fair amount of fibre. The root of the konjac plant, or glucomannan, produces results similar to fibre and may also be used to relieve constipation. Glucomannan is sold in various forms, and while safe in some forms, it can be unsafe in others, possibly leading to throat or intestinal blockage.

Crude fibre consists largely of cellulose and lignin (97%) plus some mineral matter. It represents only 60–80% of the cellulose and 4–6% of the lignin. The crude fibre content is commonly used as a measure of the nutritive value of poultry and livestock feeds and also in the analysis of various foods and food products to detect adulteration, quality and quantity. During the acid and subsequent alkali treatment, oxidative hydrolytic degradation of the native cellulose and considerable degradation of lignin occur. The residue obtained after final filtration is weighed, incinerated, cooled and weighed again. The loss in weight gives the crude fibre content.

Materials

- *Sulphuric acid solution* (0.255 ± 0.005 N): 1.25 g concentrated sulphuric acid diluted to 100 ml (concentration must be checked by titration).
- *Sodium hydroxide solution* (0.313 ± 0.005 N): 1.25 g sodium hydroxide in 100 ml distilled water (concentration must be checked by titration with standard acid).

Procedure

- Extract 2 g of ground material with ether or petroleum ether to remove fat (Initial boiling temperature 35–38°C and final temperature 52°C). If fat content is below 1%, extraction may be omitted.
- After extraction with ether boil 2 g of dried material with 200 ml of sulphuric acid for 30 min with bumping chips.
- Filter through muslin and wash with boiling water until washings are no longer acidic.
- Boil with 200 ml of sodium hydroxide solution for 30 min.
- Filter through muslin cloth again and wash with 25 ml of boiling 1.25% H_2SO_4, three 50 ml portions of water and 25 ml alcohol.
- Remove the residue and transfer to ashing dish (preweighed dish W_1).
- Dry the residue for 2 h at 130 ± 2°C. Cool the dish in a desiccator and weigh (W_2).
- Ignite for 30 min at 600 ± 15°C.
- Cool in a desiccator and reweigh (W_3).

Calculation

$$\%\ \text{crude fibre in ground sample} = \frac{\text{Loss in weight on ignition } (W_2 - W_1) - (W_3 - W_1)}{\text{Weight of the sample}} \times 100$$

7.13. Estimation of Pyruvic Acid

Pyruvic acid ($CH_3COCOOH$) is an organic acid, a ketone, as well as the simplest of the alpha-keto acids. The carboxylate (COO^-) anion of pyruvic acid, its Bronsted-Lowry conjugate base, CH_3COCOO^-, is known as pyruvate, and is a key intersection in several metabolic pathways. Pyruvic acid can be made from glucose through glycosis, converted back to carbohydrates (such as glucose) via gluconeogenesis, or to fatty acids through acetyl-CoA. It can also be used to construct the amino acid alanine and be converted into ethanol. Pyruvic acid supplies energy to living cells through the citric acid cycle when oxygen is present (aerobic respiration), and alternatively ferments to produce lactic acid when oxygen is lacking (fermentation).

Pyruvate is an important chemical compound in biochemistry. It is the output of the anaerobic metabolism of glucose known as glycolysis. One molecule of glucose breaks down into two molecules of pyruvate, which are then used to provide further energy, in one of two ways. Pyruvate is converted into acetyl-coenzyme A, which is the main input for a series of reactions known as the Krebs cycle. Pyruvate is also converted to oxaloacetate by an anaplerotic reaction, which replenishes Krebs cycle intermediates; also, the oxaloacetate is used for gluconeogenesis. If insufficient oxygen is available, the acid is broken down anaerobically, creating lactate in animals and ethanol in plants and microorganisms. Pyruvate from glycolysis is converted by fermementation to lactate using the enzyme lactate dehydrogenase and the coenzyme NADH in lactate fermentation, or to acetaldehyde and then to ethanol in alcoholic fermentation. Pyruvate is a key intersection in the network of metabolic pathways. Pyruvate can be converted into carbohydrates via gluconeogenesis, to fatty acids or energy through acetyl-CoA, to the amino acid alanine, and to ethanol. Therefore, it unites several key metabolic processes.

It can be determined following the procedure given below: The DNPH (2,4-dinitrophenyl hydrazine) reacts with pyruvate after the addition NaOH giving a brown coloured hydrazone product which can be estimated colorimetrically at 510 nm.

Materials

- *Phosphate buffer pH 9.4 – (*A) 0.2 M solution of monobasic sodium phosphate $NaH_2PO_4\ H_2O$ (27.8 g in 1000 ml), (b) 0.2 M solution of dibasic sodium phosphate (53.65 g of $NaH_2PO_4\ 7H_2O$ in 1000 ml or 17.7 g of $NaH_2PO_4\ 12H_2O$ in 1000ml). 19 ml of A and 81 ml of B, diluted to a total of 200 ml. Store in refrigerator.
- *Pyruvate, Standard:* Dissolve 22 mg sodium pyruvate in 100 ml water in a standard flask.
- *2, 4-Dinitrophenyl hydrazine (DNPH)* - Dissolve 19.8 mg of DNPH in 10 ml of conc. HCl and make to 100 ml with water. Store it in an amber bottle at room temperature.
- *Sodium hydroxide 0.8 N* - Dissolve 16 g sodium hydroxide in one litre water.
- *Plant extract* - Grind 6 g of plant material in 15 ml of phosphate buffer. Centrifuge at 25,000 g for 15 min. Use the supernatant as plant extract.

Procedure

- Pipette out 50 µL, 75 µL, 100 µL, 150 µL, 200 µL of pyruvate standard solution and 0.5 ml, 1.0 ml, 1.5 ml, and 2.0 ml of sample extract into test tubes and make up the volume to 2.0 ml with phosphate buffer (pH 7.4).

- Set a blank with no pyruvate solution.
- Add 0.5 ml of DNPH solution to each tube.
- Incubate at 37°C for 20–30 min.
- Add 5 ml of NaOH solution to each tube, mix well and incubate for 10 min at room temperature.
- Record the absorbance at 610 nm.
- Draw the standard graph and calculate the amount of pyruvic acid present in the sample using the graph.

7.14. Estimation of α-Amylase Activity (EC. 3.2.1.1)

Amylase is an enzyme that catalyses the hydrolysis of starch into sugars. Amylase is present in the saliva of humans and some other mammals, where it begins the chemical process of digestion. Foods that contain large amounts of starch but little sugar, such as rice and potato, may acquire a slightly sweet taste as they are chewed because amylase degrades some of their starch into sugar. The pancreas and salivary gland make amylase (alpha amylase) to hydrolyse dietary starch into disaccharidess and trisaccharides which are converted by other enzymes to glucose to supply the body with energy. Plants and some bacteria also produce amylase. Specific amylase proteins are designated by different Greek letters. All amylases are glycoside hydrolases and act on α-1,4-glycosidic bonds.

α-Amylase

The α-amylases (EC3.2.1.1) (alternative names: 1,4-α-D-glucan glucanohydrolase; glycogenase) are calcium mretaloenzymes, completely unable to function in the absence of calcium. By acting at random locations along the starch chain, α-amylase breaks down long-chain carbohydrates, ultimately yielding maltotriose and maltose from amylose, or maltose, glucose and "limit dextrin" from amylopectin. Because it can act anywhere on the substrate, α-amylase tends to be faster-acting than β-amylase. In animals, it is a major digestive enzyme, and its optimum pH is 6.7–7.0.

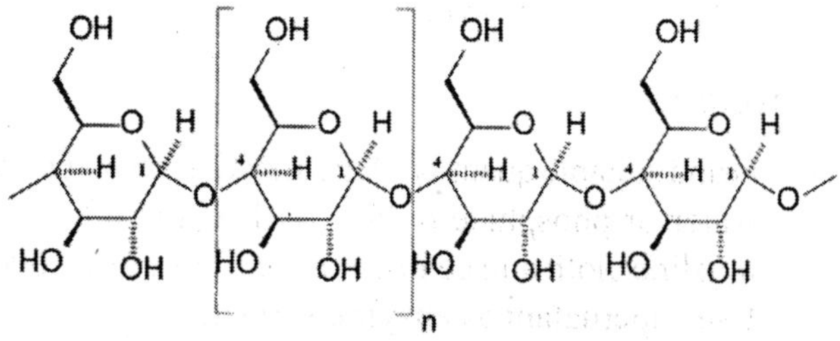

Amylose

This is a chain of amylose. The simple glucose molecules have a bond between the 1 carbon in the ring and the 4 caron in the ring. They are joined by an oxygen. The α-amylases form is also found in plants, fungi (scomycetes and basidiomycetes) and bacteria (*Bacillus*)

β-Amylase

Another form of amylase, β-amylase (EC 3.2.1.2) (alternative names: 1,4-α-D-glucan maltohydrolase; glycogenase; saccharogen amylase) is also synthesized by bacteria, fungi, and plants. Working from the non-reducing end, β-amylase catalyzes the hydrolysis of the second α-1,4 glycosidic bond, cleaving off two glucose units (maltose) at a time. During the ripening of fruit, β-amylase breaks starch into maltose, resulting in the sweet flavor of ripe fruit.

Both α-amylase and β-amylase are present in seeds; β-amylase is present in an inactive form prior to germination, whereas α-amylase and proteases appear once germination has begun. Cereal grain amylase is key to the production of malt. Many microbes also produce amylase to degrade extracellular starches. Animal tissues do not contain β-amylase, although it may be present in microorganisms contained within the digestive tract. The optimum pH for β-amylase is 4.0–5.0.

Materials

- Phosphate buffer (0.01M) pH 6.8 - 0.056g KOH (mol. Wt. 56.11 g) dissolved in 100 ml of water. 0.1361g of KH_2PO_4 (mol Wt. 136.09 g) dissolved in 100 ml of distilled water. Mix 50.0 ml of KH_2PO_4 solution of with 24.0 ml of KOH solution and final volume made to 100.0 ml with water.
- Soluble Starch Solution - 100 mg of soluble starch is added to the above buffer solution and heated to dissolve the starch.
- Iodine solution - Mix 20 mg of iodine crystal with 200 mg of potassium iodide, add 0.83 ml of concentrated HCl and final volume is made to 100.0 ml with water.

Procedure

- Grind suitable quantity of germinating seeds/plant material in cold distilled water or phosphate buffer (0.01 M pH 6.8). Pass homogenate through musline cloth and centrifuge the filtrate at 10,000 rpm for 10 min. at $4^{0}C$. Use supernatant as enzyme extract.

- Starch solution (0.1 ml) is mixed with 0.9 ml water and 0.1 ml enzyme extract (enzyme + water in total volume of 1 ml) and incubated for 5 min. at 30^0 C.
- Reaction is stopped by the addition of 1.0 ml iodine reagent. Thereafter 3.0 ml water is added and optical densityis recorded at 620 nm against a reagent blank.
- Use blank by taking 0.1 ml boiled enzyme extract.
- Calculate enzyme units as mg starch hydrolyzed per g of fresh weight (for total activity) or per mg protein per min. (for specific activity).

Reading

Shuster, L. and Gifford, R. H. (1962). Changes in 3' nucleotidase during the germination of wheat embryos. *Arch. Biochem. Biophys.* **96:** 534-540.

7.15. Estimation of β-Amylase Activity (EC. 3.2.1.2)

For quantitative estimation of β-amylase activity, total amylase activity is determined. This total amylase activity is the net total of α-amylase and β-amylase. After determining the total amylase activity, α-amylase is selectively deactivated by heating the enzyme extract to 70°C in presence of 3.0 mM $CaCl_2$. α-Amylase which is heat stable is then assayed.

Materials

- Sodium acetate buffer – (a) 0.1 M acetic acid is preapred by taking 5.88 ml of glacial acetic acid and volume is made to 1000 ml. with water, (b) 0.1 M of sodium acetate is prepared by dissolving 8.2 g of sodium acetate in 1000 ml of water, (c) 46.3 ml of 'a' is mixed with 3.7 ml of 'b' and volume is raised to 100 ml with water.
- Starch (1%) solution – Dissolve 1.0 g of soluble starch in 100.0 ml of acetate buffer.
- Dinitrosalicylic acid (DNS) reagent – Dissolve by stirring 1.0 g of 3,5-dinitrosalicylic acid in 20.0 ml of 2.0 N NaOH and 50.0 ml of water. To it 30.0 g of sodium potassium tartarate is added and volume is made to 100 ml. The solution is protected from CO_2.
- 40% Rochelle salt (Potassium sodium tartarate).

Procedure

Plant material (~10.0 g) is homogenised with 5.0 ml of 0.5 M Tris-HCl buffer pH 7.6 containing 5.0 mM 2-mercaptoethanol. The homogenates is centrifused at 20,000 x g for 10 minutes. Supernatants divided in two parts and one part is heated for 5 minutes at 70^0 C in the presence of 3.0 mM $CaCl_2$ and cooled to 0^0 C.

- Total amylotic as well as α-amylase activity is assayed in both the fraction.
- To 1.0 ml of starch solution 1.0 ml of water and 0.2 ml of enzyme extract is added
- Incubation is done for 10 minutes at 30^0 C.
- DNS reagent 2.0 ml is mixed and kept in boiling water.
- Maltose 100 mg is dissolved in 100 ml of water. This solution has 1000 μg of maltose/ml of solution. Standard curve is prepared by taking 0.2, 0.4, 0.6, 0.8 and 1.0 ml of this solution and volume of them is made to 2.0 ml by addition of respective volumes of water. Colour is developed as described above.

7.16. Estimation of Cellulase (E.C. 3.2.1.4) Activity

Cellulase (EC 3.2.1.4, *endo-1,4-beta-D-glucanase, beta-1,4-glucanase, beta-1,4-endoglucan hydrolase, celluase A, cellulosin AP, endoglucanase D, alkali cellulase, cellulase A 3, celludextrinase, 9.5 cellulase, avicelase, pancellase SS, 1,4-(1,3, 1,4)-beta-D-glucan 4-glucanohydrolase*) refers to a suite of enzymes produced chiefly by fungi, bacteria, and protozoans that catalyse cellulolysis (*i.e.* the hydrolysis of cellulose). However, there are also cellulases produced by a few other types of organisms, such as some termites and the microbial intestinal symbionts of other termites. Several different kinds of cellulases are known, which differ structurally and mechanistically. Reaction: Hydrolysis of 1,4-beta-D-glycosidic linkages in cellulose, lichenin and cereal beta-D-glucans.

Other names for 'endoglucanases' are: endo-1,4-beta-glucanase, carboxymethyl cellulase (CMCase), endo-1,4-beta-D-glucanase, beta-1,4-glucanase, beta-1,4-endoglucan hydrolase, and celludextrinase. The other types of cellulases are called exocellulases. The expression 'avicelase' refers almost exclusively to exo-cellulase activity as avicel is a highly micro-crystalline substrate. Cellulase action is considered to be synergistic as all three classes of cellulase can yield much more sugar than the addition of all three separately. Beta-glucosidases can also be considered as yet another group of cellulases.

Five general types of cellulases based on the type of reaction catalyzed:

- Endocellulase (EC 3.2.1.4) randomly cleaves internal bonds at amorphous sites that create new chain ends.
- Exocellulase (EC 3.2.1.91) cleaves two to four units from the ends of the exposed chains produced by endocellulase, resulting in the tetrasaccharides or disaccharides, such as cellobiose. There are two main types of exocellulases [or cellobiohydrolases (CBH)] - CBHI works processively from the reducing end, and CBHII works processively from the nonreducing end of cellulose.
- Cellobiase (EC 3.2.1.21) or beta-glucosidase hydrolyses the exocellulase product into individual monosaccharides.
- Oxidative cellulases depolymerize cellulose by radical reactions, as for instance cellobiose dehydrogenase (acceptor).

  Cellulose phosphorylases depolymerize cellulose using phosphates instead of water.

In the most familiar case of cellulase activity, the enzyme complex breaks down cellulose to beta-glucose. This type of cellulase is produced mainly by symbiotic bacteria in the ruminating chambers of herbivores. Aside from ruminants, most animals (including humans) do not produce cellulase in their bodies and can only partially break down cellulose through fermentation, limiting their ability to use energy in fibrous plant material. Enzymes that hydrolyze hemicellulose are usually referred to as hemicellulase and are usually classified under cellulase in general. Enzymes that cleave lignin are occasionally classified as cellulase, but this is usually considered erroneous.

Within the above types there are also progressive (also known as processive) and nonprogressive types. Progressive cellulase will continue to interact with a single polysaccha-ride strand, nonprogressive cellulase will interact once then disengage and engage another polysaccharide strand.

Most fungal cellulases have a two-domain structure, with one catalytic domain and one cellulose binding domain that are connected by a flexible linker. This structure is adapted for working on an insoluble substrate, and it allows the enzyme to diffuse two-dimensionally on a surface in a caterpillar-like fashion. However, there are also cellulases (mostly endoglucanases) that lack cellulose binding domains. These enzymes might have a swelling function.

In many bacteria, cellulases *in vivo* are complex enzyme structures organized in supramolecular complexes, the cellulosomes. They contain roughly five different enzymatic subunits represent-ting namely endocellulases, exocellulases, cellobiases, oxidative cellulases and cellulose phosphorylases wherein only endocellulases and cellobiases participate in the actual hydrolysis of the β(1→4) linkage. Recent work on the molecular biology of cellulosomes had led to the discovery of numerous cellulosome-related "signature" sequences known as dockerins and cohesins. Depending on their amino acid sequence and tertiary structures, cellulases are divided into clans and families.

Materials

- Sodium citrate buffer, 0.1 M (pH 4.8) – Dissolve 210.0 g citric acid monohydrate $C_6H_80_7$. H_20 in 750 ml of water and to it add NaOH (1.0 N) until pH equals 4.3 (50-60 g). Dilute to 1000 ml and check pH. If necessary add NaOH until pH =4.5. This is 1.0 M citrate buffer pH 4.5. When diluted to 0.05 M, pH should be 4.8.
- Carboxymethyl cellulose (CMC) – 1.0 g of carboxymethyl cellulose is dissolved in 100.0 ml of sodium citrate buffer, 0.1 M (pH 4.8).
- Dinitrosalicylic acid (DNS) reagent – 1.0 g of dinitrosalicylic acid, 200 mg crystalline phenol and 50 mg sodium sulphite is dissolved in 100 ml 1% NaOH. Store the reagent at 4°C. Since the reagent deteriorates due to sodium sulphite, if long storage is required, sodium sulphite may be added at the time of use. 40% Rochelle salt solution (Potassium sodium tartrate).
- Rochelle salt solution – dissolve 40.0 g of potassium sodium tartarate in in water and raise the volume to 100 ml.
- Glucose Standards: 0.2-5.0 mg of glucose per ml or per 0.5 ml as appropriate.

Procedure

- Homogenize 2.0 g of frsh plant material with 10.0 ml of prechilled sodium citrate buffer 0.1 M pH 5.0 and centrifuge at 15,000 x g for 15 minutes at 4^0 C. The supernatant is used as enzyme solution.
- Blanks of enzyme without substrate and substrate without enzyme are included with all enzyme assays and sample values are corrected for any blank value.
- Incubate the 0.45 ml of 1% CMC solution in waterbath and bting the temperature to 55^0 C and then 0.05 ml of enzyme extrct is added.
- Reaction mixture is incubated for 15 minutes at 55^0 C.

- After incubation period, the reaction is stopped by the addition of 0.5 ml of DNS reagent and the mixture is heated in a boiling waterbath for 5 minutes.
- To the hot reaction mixture, 1.0 ml of potassium sodium tartarate solution is added.
- Cool the reaction mixture to room temperature and total volume is made to 5.0 ml with water.
- Optical density is measured at 540 nm against a reagent blank.
- Standard curve is prepared with glucose in the concentration range of 50 to 1000 μg/ml.

Calculation

- The enzyme activity is expressed as the mg of glucose released per minute per mg protein.

Reading

Denison, DA and Koehn, RD. (1977). Cellulase activity of *Poronia Oedipus*. *Mycologia*, **69:** 592- 601.

7.17. Estimation of Phosphorylase (EC. 2.4.1.1) Activity

Phosphorylases are enzymes that catalyze the addition of a phosphate group from an inorganic phosphate (phosphate+hydrogen) to an acceptor.

$$A\text{-}B + P \rightarrow A + P\text{-}B$$

They include allosteric enzymes that catalyses the production of glucose-1-phosphate from a glucan such as glycogen, starch or maltodextrin. Phosphorylase is also a common name used for glycogen phosphorylase. Not to be confused with phosphotases. In more general terms, phosphorylases are enzymes that catalyze the addition of a phosphate group from an inorganic phosphate (phosphate+hydrogen) to an acceptor, not to be confused with a phosphatase (a hydrolase) or a kinase (a phosphotransferase). A phosphatase removes a phosphonate group from a donor using water, whereas a kinase transfers a phosphonate group from a donor (usually ATP) to an acceptor. The phosphorylases fall into the following categories:

- Glycosyltransferases (EC 2.4)
  - Enzymes that break down glucans by removing a glucose residue (break *O*-glycosidic bond)
  - Lycogen phosphorylase
  - Starch phosphorylase
  - Maltodextrin phosphorylase – Enzymes that break down nucleosides into their constituent bases and sugars (break *N*-glycosidic bond)
  - Purine nucleoside phophorylase (PNPase)
- Nucleotidyltransferases (EC 2.7.7)
  - Enzymes that have phosphorolytic 3' to 5' exoribonuclease activity (break phosphodiester bond)
    - RNAse PH
    - Polynucleotide pohosphorylase (PNPase)

All known phosphorylases share catalytic and structural properties.

The alpha-glucan phosphorylases of the glycosyltransferase family are important enzymes of carbohydrate metabolism in prokaryotes and eukaryotes. The plant alpha-glucan phosphorylase, commonly called starch phosphorylase (EC 2.4.1.1), is largely known for the phosphorolytic degradation of starch. Starch phosphorylase catalyzes the reversible transfer of glucosyl units from glucose-1-phosphate to the nonreducing end of alpha-1,4-D-glucan chains with the release of phosphate. Starch phosphorylase is industrially useful and a preferred enzyme among all glucan phosphorylases for phosphorolytic reactions for the production of glucose-1-phosphate and for the development of engineered varieties of glucans and starch.

The primer and glucose-1-phosphate are mixed with enzyme and orthophosphate set free during amylose synthesis is measured spectrophotometrically.

$$\text{Starch}_{(n)} + \text{G-1-P} \Longleftrightarrow \text{Starch}_{(n+1)} + \text{Pi}$$

Reagents

- Glucose-1-phosphate: Solution of G-1-P was prepared by dissolving 3.722 g of G-1-P in 100 ml of distilled water.
- Citrate buffer 0.05 M, pH 6.0: (a) 0.1 N citric acid is prepared by dissolving 2.101 g of citric acid in 100 ml of water. (b) 0.1 M sodium citrate is prepared by dissolving 2.941 g of sodium citrate in 100 ml of distilled

water. (c) 9.5 ml of 'A' is mixed with 41.55 ml of 'B' and made to 100 ml with distilled water and final pH is adjusted with pH meter.

- Trichloro acetic acid (5%): 20.0 ml of 70% TCA is diluted to 280.0 ml.
- Soluble starch solution: 5 g of soluble starch is dissolved in 100 ml of distilled water.

Procedure

The incubation mixture of 0.5 ml have the following reagents

- Citrate buffer 1.00 ml
- G-1-P 0.25 ml
- Soluble starch 0.25 ml
- Water 0.50 ml
- Enzyme extract 0.50 ml

A blank is taken with boiled enzyme. All tubes are kept on a water bath at 35^0 C. to equilibrate the temperature of the mixture, the reaction is initiated by the addition of 0.25 ml G-1-P and the reaction was stopped after 10 minutes by 5% TCA solution. The reaction mixture is centrifuged at 6,000 x g for 10 minutes. Phosphours released is estimated colorimetrically as described in **7.19.**

7.18. Estimation of Phosphatase (EC. 3.1.3.1, & EC. 3.1.3.2) Activity

A phosphatase is an enzyme that removes a phosphate group from its substrate by hydrolysis phosphoric acid monoesters into a phosphate ion and a molecule with a free hydroxyl group. This action is directly opposite to that of phosphorylases and kinases, which attach phosphate groups to their substrates by using energetic molecules like ATP. A common phosphatase in many organisms is alkaline phosphatase. Another large group of proteins present in *Arachea*, bacteria, and eukaryote exhibits deoxyribonucleotide and ribonucleotide phosphatase or pyrophosphatase activities that catalyse the decomposition of dNTP / NTP into dNDP / NDP and a free phosphate ion or dNMP / NMP and a free pyrophosphate ion.

Protein phosphorylation is the most common and important form of reversible protein posttranslational modification (PTM), with up to 30% of all proteins being phosphorylated at any given time. Protein kinases (PKs) are the effectors of phosphorylation and catalyse the transfer of a γ-phosphate from ATP to specific amino acids on proteins. Several hundred PKs exist in mammals and are classified into distinct super-families. Proteins are phosphorylated predominantly on Ser, Thr and Tyr residues, which account for 86, 12 and 2% respectively of the phosphoproteome, at least in mammals. In contrast, protein phosphatases (PPs) are the primary effectors of dephosphorylation and can be grouped into three main classes based on sequence, structure and catalytic function. The largest class of PPs is the phosphoprotein phosphatase (PPP) family comprising PP1, PP2A, PP2B, PP4, PP5, PP6 and PP7, and the protein phosphatase Mg^{2+}- or Mn^{2+}-dependent (PPM) family, composed primarily of PP2C. The protein Tyr phosphatase (PTP) super-family forms the second group, and the aspartate-based protein phosphatases the third.

Phosphatases act in opposition to kinases / phosphorylases, which add phosphate groups to proteins. The addition of a phosphate group may activate or de-activate an enzyme (*e.g.*, kinase signalling pathways) or enable a protein-protein interaction to occur (*e.g.*, SH2 domains); therefore phosphatases are integral to many signal transduction pathways. It should be noted that phosphate addition and removal do not necessarily correspond to enzyme activation or inhibition, and that several enzymes have separate phosphorylation sites for activating or inhibiting functional regulation. CDK, for example, can be either activated or deactivated depending on the specific amino acid residue being phosphorylated. Phosphatase are important in signal transduction because they regulate the proteins to which they are attached. To reverse the regulatory effect, the phosphate is removed. This occurs on its own by hydrolysis, or is mediated by protein phosphatases.

Protein phosphorylation plays a crucial role in biological functions and controls nearly every cellular process including metabolism, gene transcription and translation, cell-cycle progression, cytoskeletal rearrangement, protein-protein interactions, protein stability, cell movement, and apoptosis. These processes depend on the highly regulated and opposing actions of PKs and PPs, through changes in the phosphorylation of key proteins. Histone phosphorylation, along with methylation, ubiquitination, sumoylation and acetylation, also regulates access to DNA through chromatin reorganisation.

One of the major switches for neuronal activity is the activation of PKs and PPs by elevated intracellular calcium. The degree of activation of the various isoforms of PKs and PPs is controlled by their individual sensitivities to calcium. Furthermore, a wide range of specific inhibitors and targeting partners such as scaffolding, anchoring, and adaptor proteins also contribute to the control of PKs and PPs and recruit them into signalling complexes in neuronal cells. Such signalling complexes typically act to bring PKs and PPs in close proximity with target substrates and signalling molecules as well as enhance their selectivity by restricting accessibility to these substrate proteins. Phosphorylation events, therefore, are controlled not only by the balanced activity of PKs and PPs but also by their restricted localisation. Regulatory subunits and domains serve to restrict specific proteins to particular subcellular compartments and to modulate protein specificity. These are essential for maintaining the coordinated action of signalling cascades, which in neuronal cells include short-term (synaptic) and long-term (nuclear) signalling. These functions are, in part, controlled by allosteric modification by secondary messengers and reversible protein phosphorylation.

It is thought that around 30% of known PPs are present in all tissues, with the rest showing some level of tissue restriction. While protein phosphorylation is a cell-wide regulatory mechanism, recent quantitative proteomics studies have shown that phosphorylation preferentially targets nuclear proteins. Many PPs that regulate nuclear events, are often enriched or exclusively present in the nucleus. In neuronal cells, PPs are present in multiple cellular compartments and play a critical role at both pre- and post-synapses, in the cytoplasm and in the nucleus where they regulate gene expression.

Glucose 6-phosphatase (EC 3.1.3.9, G6Pase) is an enzyme that hydrolyzes glucose-6-phosphate, resulting in the creation of a phosphate group and free glucose. Glucose is then exported from the cell via glucose transporter membrane proteins. This catalysis completes the final step in gluconogenesis and glyconogenesis and therefore plays a key role in the homeostatic regulation of blood glucose levels. In humans, there are three isozymes, G6PC, G6PC2, and G6PC3.

The G6Pase family includes two functional phosphohydrolases; G6Pase-α and G6Pase-β, the former of which is the prototype. G6Pase-α and G6Pase-β have similar active site structure, topology, mechanism of action, and kinetic properties with respect to G6P hydrolysis.

Acid Phosphatase

Material

- Sodium hydroxide 0.085 N – 0.85 g of NaOH is dissolved in 250 ml of water.
- Substrate solution – Dissolve 1.49 g of EDTA, 0.84 g of citric acid and 0.03 g p-nitrophenyl phosphate in 100 ml of water and adjust the pH at 5.3.
- Standard – 69.75 mg of p-nitrophenol is dissolved in 5.0 ml of water (100 mM).

Procedure

- Homogenise 1.0 g of fresh leaves in 10.0 ml of chilled 50 mM citrate buffer (pH 5.3) and centrifuge for 10 minutes at 10,000 x g at 4^0 C. The supernatant is used as enzyme source.
- Incubate 3.0 ml of substrate solution at 37^0C for 5 minutes and then add 0.5 ml of enzyme extract and mix well.
- Remove immediately 0.5 ml and mix it with 9.5 ml of NaOH solution (0.085 M). This cooresponds to zero time assay.
- Incubate the remaining mixture for 15 minutes at 37^0 C.
- Draw 0.5 ml and mix it with 9.5 ml of sodium hydroxide solution (0.085 M).
- Optcial density for zerp time and after incubation period is recorded at 405 nm.
- A gradient of standard (0.2 – 1.0 ml) (4 – 20 mM) is diluted to 10 ml with NaOH solution and optical density os measured at 405 nm against reagent blank.
- Activity level of plant sample is calculated by referring to standard curve and the activity is expressed as m moles p-nitrophenol released per minutes per mg protein.

Notes

- For alkaline phosphatase enzyme is extracted in 50.0 mM glycine NaOH buffer pH 10.4.
- Alkaline phosphatase functions optimally at about pH 10.5. The assay procedure is similar to that of acid phosphatase, except for the substrate solution. The substrate solution is prepared as given below:
- Substrate solution – 375 mg of glycine is taken along with 10.0 mg of magnesium chloride ($MgCl_2$), 165 mg p-nitrophenyl phosphate and dissolved in 0.1 N NaOH solution and diluted to 100 ml. pH is adusted at 10.5.dissolved.

7.19. Assay of Phosphatase Using Glycerophosphate

Acid and alkaline phosphatase can also be assayed by following the amount of inorganic phosphorus released from sodium glycerophosphate. Phosphorus is estimated by the following method:

Materials

- Substrate – 0.1 M solution of β-glycerophosphate 3.153 g in 100.0 ml of water.
- Trichloro acetic acid – 15% (w/v) solution
- Ammonium molybdate (2.5%) – 25.0 g of ammonium molybdate is dissolved in 400.0 ml of water and 500.0 ml of 10 N H_2SO_4 is added and final volume is made to 1000.0 ml with water.
- 1-Amino 2-Napthol 4-Sulphonic acid (ANSA) reagent – Dissolve 30.0 g of sodium metabisulphite, 6.0 g of sodium sulphite and 500.0 mg of ANSA separately in small quantity of water. Combine all the solutions and make up to 250.0 ml with water. Allow to stand overnight and filter.
- Standard phosphate solution – 439.0 mg of potassium dihydrogen phosphate is dissolved in water and 10.0 ml of 10.0 N H_2SO_4 is added to it. Final volume is made to 1000.0 ml with water. This solution would give 100.0 µg phosphate/ml. Add 10.0 ml of chloroform as preservative.
- Dilute 10 ml of standard solution to 50.0 ml with water and it would give 20.0 µg phosphate/ml.
- Magnesium acetate solution – 0.2 M of magnesium acetate is prepared by dissolving 4.289 g in 100 ml of water.

Procedure

- Take 0.2 ml of respective buffer (acidic or alkaline) add 0.5 ml of magnesium acetate solution and 2.0 ml of β-glycerophosphate solution. Final volume is made up to 7.0 ml with buffer.
- Enzyme extract (0.1 ml) is added to the reagent mixture and incubated for 60 minutes at 37^0 C. After incubation period, reaction is stopped by the addition of 2.0 ml of 10% TCA solution. For the control boiled enzyme extract is added.
- Centrifuge at 10,000 x g for 10 minutes.
- A known amount of reaction mixture is atken and volume is made to 5.0 ml with water. To it 1.0 ml of ammonium molybdate is added and mixed thoroughly.
- To the above reaction mixture 0.4 ml of ANSA reagent is added and allow to stand for 10 minutes. Water is added to make the volume to 10.0 ml.

- Optical density is recorded at 660 nm against a reagenr blank.
- Standard curve is obtained by taking a gradient of phosphate standard solution (5 – 50 μg/ml) and colour is develop by the addition of ammonim molybdate and ANSA reagent as described above and optical density is recorded at 660 nm/ Standard curve is obtained by plotting the optical densities against respective concentrations of phosphate.
- Phosphatase activity is expressed as amount of phosphorus realsed per unit of time per unit of mg of protein.

Notes

- Any traces of phosphorus impurity in the distilled water interferes with the determination. This is easily noticed from the reagent blank which should be colourless otherwise.
- The blue colour so developed, intensify with time and it is preferable to read the optical density after allowing the colour to develop for 10 minutes.

7.20. Estimation of Malate Dehydrogenase (L-malate: NAD^+ Oxidore-ductase EC 1.1.1.37) Activity

Malate dehydrogenase (EC 1.1.1.37) (MDH) is an enzyme that reversibly catalyses the oxidation of malate to oxaloacetate using the reduction of NAD^+ to NADH. This reaction is part of many metabolic pathways, including the citric acid cycle. Other malate dehydrogenases, which have other EC numbers and catalyze other reactions oxidizing malate, have qualified names like malate dehydrogenase (NADP+). Malate dehydrogenase is also involved in gluconogenesis, the synthesis of glucose from smaller molecules. Pyruvate in the mitochondria is acted upon by pyruvate carboxylase to form oxaloacetate, a citric acid cycle intermediate. In order to get the oxaloacetate out of the mitochondria, malate dehydrogenase reduces it to malate, and it then traverses the inner mitochondrial membrane. Once in the cytosol, the malate is oxidized back to oxaloacetate by cytosolic malate dehydrogenase. Finally, phosphoenol-pyruvate carboxylase (PEPCK) converts oxaloacetate to phosphoenolpyruvate (PEP).

Several isozymes of malate dehydrogenase exist. There are two main isoforms in eukaryotic cells. One is found in the mitochondrial matrix, participating as a key enzyme in the citric acid cycle that catalyzes the oxidation of malate. The other is found in the cytoplasm, assisting the malate aspartate shuttle with exchanging reducing equivalents so that malate can pass through the mitochondrial membrane to be transformed into oxaloacetate for further cellular processes.

The active site of malate dehydrogenase is a hydrophobic cavity within the protein complex that has specific binding sites for the substrate and its coenzyme, NAD^+. In its active state, MDH undergoes a conformational change that encloses the substrate to minimize solvent exposure and to position key residues in closer proximity to the substrate. The three residues in particular that comprise a catalytic triad are histidine (His-195), aspartate (Asp-168), both of which work together as a proton transfer system, and arginines (Arg-102, Arg-109, Arg-171), which secure the substrate. Kinetic studies show that MDH enzymatic activity is ordered. NAD^+/NADH is bound before the substrate.

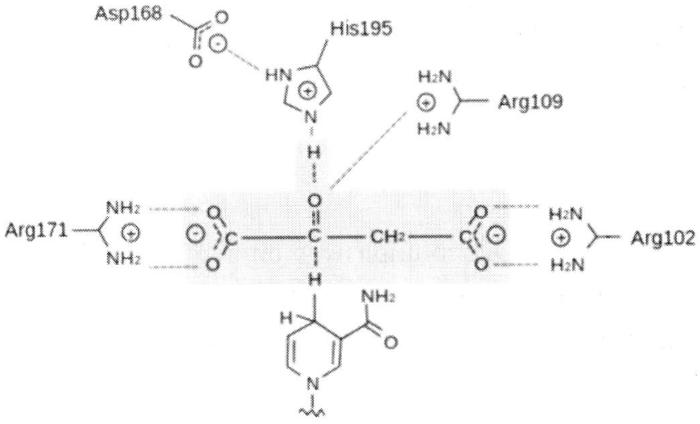

Active site of malate dehydrogenase

Malate dehydrogenas (MDH) is one of the enzymes involved in TCA cycle. It catalyses the reversible conversion of oxaloacetic acid to malic acid.

Oxaloacetic acid + NADH ↔ Malic acid + NAD^+

This enxyme is also involved in carbon ioxide assimilation in C_4 plants. It is coupled with PEPcase in CO_2 assimilation in the chloroplasts of mesophyll cells. Here it uses NADPH as the coenzymes. Since it is an oxidoreductase involving NAD the decrease in absorbance due to the oxidation of NADH is followed.

Materials

- Oxaloacetic acid (5 µ moles) 66 mg/50 ml of distilled water.
- Magnesium chloride (10 µ moles) 203 mg/50 ml of distilled water.
- Tris-HCl buffer (0.1 M) pH 7.8.
- NADH (0.4 µmoles) 5.32 mg/10 ml of distilled water.

Procedure

- *Crude enzyme extract* - Grind the tissue throughly with acid washed sand in a pre chilled pestle and mortar in grinding medium (1 ml/ 1 g tissue) containing 50 mM Tris-HCl (pH8.0), 50 mM $MgCl_2$, 5 mM 2-mercaptoethanol and 1 mM EDTA. Pass the homogenate through four layers of chese cloth and centrifuge the filterate at 5000 g for 10 minutes at 5^0 C. Use the supernatant as emzyme source.
- Take the following reaction mixture in triplicate
  - Oxaloacetic acid.......................................0.5 ml
  - Magnesium chloride.......................................0.5 ml
  - Tris-HCl buffer, pH 7.8.......................................1.3 ml
  - Enzyme extract.......................................0.2 ml
  - NADH Solution.......................................0.5 ml
- For blank, the NADH solution was omited and another 0.5 ml of Tris-HCl buffer was added.
- The reaction mixture were taken in the cuvetts of UV-Vis spectrophotometer.
- Zero absorbance was set at 340 nm against the blank in reference cuvette.
- Add NADH as quickly as possible, mix well and note the initial OD.
- Record the OD every 30 seconds for at least 3 minutes.

Calculation

Calculate the enzyme activity as follows with the decrease in absorbance for one minute.

μ mole of NADH oxidised per minute per 0.2 ml of enzyme extract

= Absorbance decrease / minute x 0.1613 x 3 (vloume of the reaction mixture in ml)

Determine the protein content of the emzyme extract by Lowery *et al* method. Compute the values for mg protein to calculate the specific activity.

7.21. Estimating Alcohol Dehydrogenase (EC. 1.1.1.1) Activity

Alcohol dehydrogenases (ADH) (EC 1.1.1.1) are a group of dehydrogenase enzymes that occur in many organisms and facilitate the interconversion between alcohols and aldehydes or ketones with the reduction of nicotinamide adenine dinucleotide (NAD^+ to NADH). They serve to break down alcohols that otherwise are toxic, and they also participate in generation of useful aldehyde, ketone, or alcohol groups during biosynthesis of various metabolites. In yeast, plants, and many bacteria, some alcohol dehydrogenases catalyses the opposite reaction as part of fermentation to ensure a constant supply of NAD^+. In plants, ADH catalyses the same reaction as in yeast and bacteria to ensure that there is a constant supply of NAD^+. maize has two versions of ADH - ADH1 and ADH2, *Arabidopsis thaliana* contains only one ADH gene. The structure of *Arabidopsis* ADH is 47%-conserved, relative to ADH from horse liver. Structurally and functionally important residues, such as the seven residues that provide ligands for the catalytic and noncatalytic zinc atoms, however, are conserved, suggesting that the enzymes have a similar structure. ADH is constitutively expressed at low levels in the roots of young plants grown on agar. If the roots lack oxygen, the expression of *ADH* increases significantly. Its expression is also increased in response to dehydration, to low temperatures, and to abscisic acid, and it plays an important role in fruit ripening, seedling development, and pollen development. Differences in the sequences of *ADH* in different species have been used to create phylogenesis showing how closely related different species of plants are. It is an ideal gene to use due to its convenient size (2–3 kb in length with a ~1000 nucleotide coding sequence) and low copy number.

Acetic acid bacteria oxidising primary and secondary alcohols and glycerol with a primary alcohol function to the corresponding carboxylic or hydrocarxylic acids. Secondary alcohols and glycerol with a secondary alcohols function are oxidised to the corresponding ketones. These oxidations are catalyzed by two different enzymes systems: one is soluble and other is particulate. These particulate enzymes are probably localised on the cell envelop considered to be cytoplasmic membranes.

1. Soluble NAD – linked primary and secondary alcohol dehydrogenasees:

$$R\text{-}CH_2OH + NAD+ \leftrightarrow R\text{-}CHO + NADH + H^+$$

$$R\text{-}CHOH\text{-}R' + NAD^+ \leftrightarrow R\text{-}CO\text{-}R' + NADH + H^+$$

Procedure for assay

- The assay is based on the rate of reduction of NAD in presence of alcohol and monitored at 340 nm.
- Fresh tissue (~ 2.0 g) is extracted with 10.0 ml of chilled 0.1 M phosphate buffer pH 6.5 in pre chilled mortar.

- The mixture is centrifuged at 20,000xg for 20 minutes at 2^0 C.
- The supernatant is collected and then fractionated by 25 – 55% with the use of ammonium sulphate.
- A 2.0 M solution is used for primary alcohol and glycols with a primary alcohol function. A 0.2 M solution is used for all other compound. The primary and secondary ADH are routinly tested with respectively n-propanol and meso-2,3-butanediol as substrate.
- The enzyme extract is diluted in 0.01 M phosphate buffer pH 7.0 to give a solution with an activity not greater than 0.3 μ ml/ml.
- To 0.6 ml of Tris-HCl buffer (0.1 M pH 8.8) 0.1 ml of $MgCl_2$ solution (0.05 M) and 0.1 ml of enzyme extract are mixed in a spectrophotometer cell with a 1.0 cm path length.
- The reaction is started with the addition of 0.1 ml of alcohol solution. The extintion at 340 nm is measured at one minute intervals.
- One unit of ADH is defined as that amount which causees the reduction of 1 μm of NAD/minute under the assay condition.

2. Particulate primary and secondary alcohol dehydrogenase

$$\text{R-CH}_2\text{OH} \leftrightarrow \text{R-CHO} + 2\ \text{H}^+ + 2\ \text{e}^-$$

$$\text{R-CHOH-R} \leftrightarrow \text{R-CO-R'} + 2\ \text{H}^+ + 2\ \text{e}^-$$

The particles contain the complete ETS. The oxidation of alcohol and glycols by the particulate enzyme can be followed by measuring the rate of O_2 uotake by Warburg respirometer.. Once solubilized the ETS to O_2 is broken. Therefore, the oxidation of primary and secondary alcohol is routinely determined spectrophotometrically by the reduction of 2,6-dichlorophenol-indophenol.

Materials

- Potassium dihydrogen orthophosphate (KH_2PO_4) - sodium hydrogen phosphate (Na_2HPO_4) buffer 0.03 M pH 5.8.
- 2,6-dichlorophenol indophenol solution 1.0 mM.
- Primary or secondary alcohol – 0.2 M ethanol and meso-2,3-butanediol.
- The enzyme solution is diluted in 0.03 M phosphate buffer pH 5.8 to give a solution with an activity not greater than 0.2 unit/ml.
- Suitably diluted enzyme solution (0.1 ml) is taken in a colorimetric cuvette added with 0.9 ml of phosphate buffer pH 5.8 and 0.3 ml of 2,6-dichlorophenol indophenol.

- Reaction is initiated bythe addition of 0.2 ml of alcohol solution.
- The rate of decolorization is followed during 3 minutes at 660 nm against reagent blank with bolied enzyme solution.

Enzymatic Assay of Alcohol Dehydrogenase (EC 1.1.1.1)

The continuous spectrophotometric rate determination (A_{340}, Light path = 1 cm) is based on the following reaction:

$$\text{Ethanol} + \beta\text{-NAD} \xrightarrow{\text{Alcohol Dehydrogenase}} \text{Acetaldehyde} + \beta\text{-NADH}$$

Unit Definition: One unit of Alcohol Dehydrogenase will convert 1.0 μmole of ethanol to acetaldehyde per minute at pH 8.8 at 25 °C.

Materials

- Sodium pyrophosphate, tetrabasic decarbohydrate.
- 95% (v/v) ethanol.
- β-NAD hydrate.
- Sodium phosphate, monobasic monohydrate.
- Sodium phosphate dibasic.

Preparation Instructions

- Use ultrapure water (>18 MΩ x cm resistivity at 25 °C) for the preparation of reagents.
- 50 mM Sodium Phosphate Buffer, pH 8.8 (25 °C) – Dissolve 2.23 g of sodium pytrophosphate, tetrabasic decarbohydrate, in 100 ml of ultrapure water. Adjust the pH to 8.8 with 8% (v/v) phosphoric acid.
- 95% (v/v) ethanol
- 15 mM β-NAD (ß-Nicotinamide Adenine Dinucleotide) Solution – Prepare 50 ml in ultrapure water using the corrected formula molecular weight of β-NAD hydrate. Prepare Fresh.
- 10 mM Sodium Phosphate Buffer, pH 7.5 (25 °C) – Dissolve 552 mg of sodium phosphate monobasic monohydrated in 400 ml of ultrapure water. Dissolve 710 mg of sodium phosphate dibasic in 500 ml of ultrapure water. Adjust the pH of the dibasic solution to 7.5 (25 °C) with the monobasic solution.

- Enzyme Diluent (10 mM Sodium Phosphate Buffer, pH 7.5 with 0.1% (w/v) Bovine Serum Albumin) – Dissolve 500 mg of bovine serum albumin in 500 ml of 10 mM Sodium Phosphate Buffer, pH 7.5. Adjust pH to 7.5 (25 °C) with 1 N HCl.
- Enzyme Solution – Alcohol Dehydrogenase is unstable in solution and should be assayed immediately following preparation of solutions. Prepare a 1 mg/ml solution of Alcohol Dehydrogenase in cold (2–8 °C) 10 mM Sodium Phosphate Buffer, pH 7.5. Then dilute 0.05 ml of the 1 mg/ml solution to 25.0 ml with cold Enzyme Diluent.
- Note: If this enzyme concentration yields a Δ A/minute >0.15, dilute 0.05 ml of the 1 mg/ml solution to 50.0 ml with cold Enzyme Diluent. *Do not deviate from this dilution scheme.*

Procedure

In a 3.00 ml reaction mix, the final concentrations are 22 mM sodium pyrophosphate, 3.2% (v/v) ethanol, 7.5 mM β-nicotinamide adenine dinucleotide, 0.3 mM sodium phosphate, 0.003% (w/v) bovine serum albumin, and 0.05–0.10 unit of alcohol dehydrogenase.

Into appropriate cuvettes, accurately pipette the following:

Reagent	Test (ml)	Blank (ml)
50 mM Sodium Phosphate Buffer, pH 8.8	1.30	1.30
95% (v/v) Ethanol	0.10	0.10
15 mM ß-NAD Solution	1.50	1.50

Mix by inversion and equilibrate to 25 °C. Monitor the A_{340} until constant, using a suitably thermostatted spectrophotometer. Then add:

Reagent	Test (ml)	Blank (ml)
Enzyme Solution	0.10	–
Enzyme Diluent	–	0.10

Immediately mix by inversion and record the increase in A_{340} for ~6 minutes. Obtain the A_{340nm}/minute using the one to six minute range for both the Test and Blank.

Calculations

Units of enzyme = (ΔA_{340}/min Test – ΔA_{340}/Min Blank) (3.0) (df))/((6.22)* (0.1))

- 3.0 = Total Volume (ml) of assay
- df = Dilution Factor
- 6.22 = Millimolar extinction coefficient of β-NADH at 340 nm
- 0.1 = Volume (ml) of enzyme solution used

Reference

Kagi, J.H.R., and Vallee, B.L., (1960). The Role of Zinc in Alcohol Dehydrogenase: V. The effect of metal-binding agents on the structure of the yeast alcohol dehydrogenase molecule. *Journal of Biological Chemistry*, **235:** 3188-3192.

7.22. Estimation of Lactate Dehydrogenase (E.C. 1.1.1.27) Activity

A lactate dehydrogenase (LDH or LD) is an enzyme found in animals, plants, and prokaryotes. A dehydrogenase is an enzyme that transfers a hydride from one molecule to another. Lactate dehydrogenase catalyzes the conversion of pyruvate to lactate and back, as it converts NADH to NAD^+ and back.

Lactate dehydrogenases exist in four distinct enzyme classes. Each one acts on either D-lactate (D-lactate dehydrogenase (cytochrome)) or L-lactate (L-lactate dehydrogenase (cytochrome)). Two are cyctochrome c-dependent enzymes. Two are NAD(P)-dependent enzymes. Lactate dehydrogenase catalyzes the interconversion of pyruvate and lactate with concomitant interconversion of NADH and NAD^+. It converts pyruvate, the final product of glycolysis, to lactate when oxygen is absent or in short supply, and it performs the reverse reaction during the Cori cycle in the liver. At high concentrations of lactate, the enzyme exhibits feedback inhibition, and the rate of conversion of pyruvate to lactate is decreased.

Lactate dehydrogenase is of medical significance because it is found extensively in body tissues, such as blood cells and heart muscle. Because it is released during tissue damage, it is a marker of common injuries and disease. It also catalyzes the dehydrogenation of 2-hydroxybutyrate, but it is a much poorer substrate than lactate. There is little to no activity with beta-hydroxybutyrate.

This enzyme reaction is assayed by a decrease in absorbance at 340 nm resulting from the oxidation of NADH. One unit causes the oxidation of one micromole of NADH per minute at 25°C and pH 7.3, under the specified conditions.

Materials

- 0.2 M Tris – HCl, pH 7.3
- 6.6 mM NADH in above 0.2 M Tris – HCl buffer, pH 7.3
- 30 mM Sodium pyruvate in above 0.2 Tris – HCl buffer, pH 7.3

Procedure

- During extraction for enzyme, 1% BSA or 10^{-3} M of β-mercaptoethanol is used for enzyme stability.
- Dissolve at 1 mg/ml in 0.2 M Tris – HCl buffer. Dilute enzyme prior to use to obtain a rate of 0.02-0.04 ΔA/min. in Tris buffer and keep cold.
- Set spectrophotometer at 340 nm and 25°C.
- Pipette into cuvette as follows:

  Tris – HCl, 0.2 M pH 7.3 2.8 ml

  6.6 mM NADH 0.1 ml

  30 mM Sodium pyruvate 0.1 ml
- Incubate in the spectrophotometer 4-5 minutes to achieve temperature equilibration and establish a blank rate, if any.
- Add 0.1 ml of appropriately diluted enzyme and record ΔA_{340}/min from initial linear portion.

Calculation

$$\text{Units/mg} = \frac{\text{D A 340/min}}{\text{6.22 x mg enzyme/ml reaction mixutre}}$$

Reading

Chenault, H.K. and Whitesides, G.M. (1989). Lactate dehydrogenase-catalyzed regeneration of NAD From NADH for use in enzyme-catalyzed synthesis. *Bioorganic Chemistry*, **17:** 400-409.

CHAPTER 8

Nitrogen Assimilation

Nitrogen assimilation is the formation of organic nitrogen compounds like amino acids from inorganic nitrogen compounds present in the environment. Organisms like plants, fungi and certain bacteria that cannot fix nitrogen gas (N_2) depend on the ability to assimilate nitrate or ammonia for their needs. Other organisms, like animals, depend entirely on organic nitrogen from their food.

Plants absorb nitrogen from the soil in the form of nitrate (NO_3^-) and ammonia (NH_4). In aerobic soils where nitrification can occur, nitrate is usually the predominant form of available nitrogen that is absorbed. However this need not always be the case as ammonia can predominate in grasslands and in flooded, anaerobic soils like rice paddies. Plant roots themselves can affect the abundance of various forms of nitrogen by changing the pH and secreting organic compounds or oxygen. This influences microbial activities like the inter-conversion of various nitrogen species, the release of ammonia from organic matter in the soil and the fixation of nitrogen by non-nodule-forming bacteria.

Ammonium ions are absorbed by the plant via ammonia transporters. Nitrate is taken up by several nitrate transporters that use a proton gradient to power the transport. Nitrogen is transported from the root to the shoot via the xylem in the form of nitrate, dissolved ammonia and amino acids. Usually (but not always) most of the nitrate reduction is carried out in the shoots while the roots reduce only a small fraction of the absorbed nitrate to ammonia. Ammonia (both absorbed and synthesized) is incorporated into amino acids via the glutamine–glutamate synthase (GS-GOGAT) pathway. While nearly all the ammonia in the root is usually incorporated into amino acids at the root itself, plants may transport significant amounts of ammonium ions in the xylem to be fixed in the shoots. This may help avoid the transport of organic compounds down to the roots just to carry the nitrogen back as amino acids.

Nitrate reduction is carried out in two steps. Nitrate is first reduced to nitrite (NO_2^-) in the cytosol by nitrate reductase using NADH or NADPH. Nitrite is then reduced to ammonia in the chloroplasts (plastids in roots) by a ferredoxin dependent nitrite reductase. In photosynthesizing tissues, it uses an isoform of ferredoxin (Fd1) that is reduced by PSI while in the root it uses a form of ferredoxin (Fd3) that has a less negative midpoint potential and can be reduced easily by NADPH. In non photosynthesizing tissues, NADPH is generated by glycolysis and the pentose phosphate pathway.

In the chloroplasts, glutamine synthetase incorporates this ammonia as the amide group of glutamine using glutamate as a substrate. Glutamate synthase (Fd-GOGAT and NADH-GOGAT) transfer the amide group onto an 2-oxoglutarate molecule producing two glutamates. Further transaminations are carried out make other amino acids (most commonly asparagine) from glutamine. While the enzyme glutamate dehydrogenase (GDH) does not play a direct role in the assimilation, it protects the mitochondrial functions during periods of high nitrogen metabolism and takes part in nitrogen remobilization.

pH and Ionic balance during nitrogen assimilation

Different plants use different pathways to different levels. Tomatoes take in a lot of K^+ and accumulate salts in their vacuoles, castor reduces nitrate in the roots to a large extent and excretes the resulting alkali. Soybean plants moves a large amount of malate to the roots where they convert it to alkali while the potassium is recirculated.

Every nitrate ion reduced to ammonia produces one OH^- ion. To maintain a pH balance, the plant must either excrete it into the surrounding medium or neutralize it with organic acids. This results in the medium around the plants roots becoming alkaline when they take up nitrate.

To maintain ionic balance, every NO_3^- taken into the root must be accompanied by either the uptake of a cation or the excretion of an anion. Plants like tomatoes take up metal ions like K^+, Na^+, Ca^{2+} and Mg^{2+} to exactly match every nitrate taken up and store these as the salts of organic acids like malate and oxalate. Other plants like the soybean balance most of their NO_3^- intake with the excretion of OH^- or HCO_3^-.

Plants that reduce nitrates in the shoots and excrete alkali from their roots need to transport the alkali in an inert form from the shoots to the roots. To achieve this they synthesize malic acid in the leaves from neutral precursors like carbohydrates. The potassium ions brought to the leaves along with the nitrate in the xylem are then sent along with the malate to the roots via the phloem. In the roots, the malate is consumed. When malate is converted back to malic acid prior to use, an OH^- is released and excreted. ($RCOO^- + H_2O \rightarrow RCOOH + OH^-$) The potassium ions are then recirculated up the xylem with fresh nitrate. Thus the plants avoid having to absorb and store excess salts and also transport the OH^-.

Plants like castor reduce a lot of nitrate in the root itself, and excrete the resulting base. Some of the base produced in the shoots is transported to the roots as salts of organic acids while a small amount of the carboxylates are just stored in the shoot itself.

The major source of inorganic nitrogen available for plants is a mixture of nitrate and ammonium, with nitrate being the predominant form in well-aerated soils because of nitrification. The nitrate present in the soils is then taken up by the roots of the plant and converted to ammonium by the sequential reductive action of the enzymes nitrate reductase (NR) and nitrite reductase (NiR). Some plants, most notably legumes, can also obtain nitrogen from atmospheric N_2. The symbiotic interaction of these plants with rhizobia, which are able in the nodules to reduce N_2 to NH_4^+ by the action of nitrogenase, makes it possible that ammonium is transferred from the microbe to the plant. Despite the ability of legumes to form nitrogen-fixing symbiosis with *Rhizobium* and *Bradyrhizobium* spp. there are many reasons why nodulation may not occur, reason why combined forms of inorganic nitrogen are important nitrogen sources for legumes as they are for non-legumes. Plants also produces significant amounts of ammonium from photorespiration, phenylpropanoid biosynthesis and amino acid catabolism. Consequently, the reduced form of inorganic nitrogen ultimately available for the plants is ammonium, irrespective of the primary nitrogen source utilized. Assimilation of ammonium produced from external nitrogen sources is commonly described as primary ammonium assimilation, to be distinguished from other forms of ammonium released endogenously in the plants, which is known as secondary ammonium assimilation. Glutamine synthetase (GS) and glutamate synthases (GOGAT) are the key enzymes responsible for all forms of ammonium assimilation that take place in higher plants. A set of different isoforms of these enzymes, having specific patterns of expression in several tissues, is a distinctive peculiarity in higher plants. Glutamine and glutamate are the first organonitrogen compunds produced as a result of the GS-GOGAT biosynthetic pathway. Nitrogen is then redistributed from these two compounds to the rest of N-containing metabolites and macromolecule.

The proportion of nitrate that can be assimilated in the shoot and in the root of different plant species varies considerably. Three main groups of plants exist in this respect: a) those that assimilate nitrate significantly both in shoots and roots (temperate annual non-legume species); b) those assimilate nitrate preferentially in leaves (tropical and subtropical legume and non-legume species); c) those assimilate nitrate preferentially in roots (temperate perennial and annual legume species). Many temperate legumes, growing in non-agricultural soils with low nitrate concentrations (1 mM) use their roots to assimilate most of the nitrate they absorb, although shoot assimilation becomes more important at higher external concen-trations. Other legume species, particularly amongst the tropical and subtropical legumes, have been found to assimilate most of their nitrate in the shoot, regardless of the external nitrate concentration.

8.1. Estimation of Nitrogenase (EC. 1.18.6.1 or 1.19.6.1) Activity

Nitrogenases (EC 1.18.6.1, EC 1.19.6.1) are enzymes used by leguminous plants to fix atmospheric nitrogen gas (N_2) with symbiotic association with *Rhizobium*. Nodules formed by *Rhizobium* are supplied necessary photosynthates by the host plant and in turn bacterium gives assimilated nitrogen. There is only one known family of enzymes that accomplishes this process. Dinitrogen is quite inert because of the strength of its N≡N triple bond. Whilst the equilibrium formation of ammonia from molecular hydrogen and nitrogen has an overall negative enthalpy of reaction ($\Delta H^0 = -45.2$ kJmol^{-1} NH_3), the energy barrier to activation is very high (EA = 420 kJmol^{-1}) without the assistance of catalysis.

In addition to reducing agents, such as dithionite *in vitro*, or ferrodoxin or flavodoxin *in vivo*, the enzymatic reduction of dinitrogen to ammonia therefore also requires an input of chemical energy, released from the hydrolysis of ATP, to overcome the activation energy barrier. The enzyme is composed of the heterotetrameric MoFe protein that is transiently associated with the homodimeric Fe protein. Electrons for the reduction of nitrogen are supplied to nitrogenase when it associates with the reduced, nucleotide-bound homodimeric Fe protein. The hetero-complex undergoes cycles of association and disassociation to transfer one electron, which is the rate-limiting step in nitrogen reduction. ATP supplies the energy to drive the transfer of electrons from the Fe protein to the MoFe protein. The reduction potential of each electron transferred to the MoFe protein is sufficient to break one of dinitrogen's chemical bonds, though it has not yet been shown that exactly three cycles are sufficient to convert one molecule of N_2 to ammonia. Nitrogenase ultimately bonds each atom of nitrogen to three hydrogen atoms to form ammonia (NH_3), which is in turn bonded to glutamate to form glutamine. The nitrogenase reaction additionally produces molecular hydrogen as a side product.

All nitrogenases have an iron- and sulfur-containing cofactor that includes a heterometal complex in the active site (*e.g.*, FeMoCo). In most, this heterometal complex has a central molybdenum atom, though in some species it is replaced by a vanadium or iron atom. Due to the oxidative properties of oxygen, most nitrogenases are irreversibly inhibited by dioxygen, which degradatively oxidizes the Fe-S cofactors. This requires mechanisms for nitrogen fixers to protect nitrogenase from oxygen *in vivo*. Despite this problem, many use oxygen as a terminal electron acceptor for respiration. One known exception is the nitrogenase of *Streptomyces thermoautotrophicus*, which is unaffected by the presence of oxygen. Although the ability of some nitrogen fixers such as Azobacteraceae to employ an oxygen-labile nitrogenase under aerobic conditions has been attributed to a high metabolic rate, allowing oxygen reduction at the cell membrane, the effectiveness of such a mechanism has been questioned at oxygen concentrations above 70 μM (ambient concentration is 230 μM O_2), as well as during additional nutrient limitations. The reaction that this enzyme performs is:

$$\mathbf{N_2 + 8\ H^+ + 8\ e^- + 16\ ATP + 16\ H_2O \rightarrow 2\ NH_3 + H_2 + 16\ ADP + 16\ P_i}$$

Nitrogenase catalyses the reduction of not only nitrogen but also a variety of other substances, many of which are characterised by a triple bond. The reduction of acetylene (C_2H_2) to ethylene (C_2H_4) is the best known example, because it is widely used as a method of measuring nitrogenase activity in natural samples, isolates and cell free extracts. Acetylene reduction assay (ARA) owes its popularity to its low cost. But ARA is indirect method, and its theoretical conversion factor to dinitrogen fixation of 4:1 is derived from the stoichiochemistry of the reactions:

$$\mathbf{N_2 + 8[H] \rightarrow 2NH_3 + H_2 \text{ and } C_2H_2 + 2[H] \rightarrow C_2H_4}$$

Acetylene is reduced to ethylene by nitrogenase. The ethylene produced is measured in a gas liquid chromatograph (GLC) and the activity is expressed as n mole ethylene produced per unit time per g dry weight of nodules.

Materials

- Gas chromatograph with Flame Ionization Detector (FID)
- Air tight syringes
- Conical flasks (100 ml) with small mouth to fit Serum Caps.
- Acetylene gas.

- Ethylene gas.
- GLC operating conditions:
- Carrier gas – nitrogen/Helium/Argon with a flow rate 30-45 ml.
- Gas for detection – Hydrogen and air.
- Cloum – Propak N, R, T and Q or Silica Gel.

	Poropak	Silica Gel
Oven / Colum Temperature	60^0 C	150^0 C
Injector Temperature	65^0 C	160^0 C
Detector temperature	85^0 C	175^0 C
Retention time for ethylene (minutes)	13	1.5
Retention time acetylene (minutes)	1.3	3.0

Procedure

- Root are collected without disruption of root nodules and roots with nodules are removed carefully.
- Roots are taken in 100 ml flask and flasks are caped with rubber septum.
- Using air tight syringe 10 ml of air is removed from the flask.
- Acetylene 10 ml is injected in the flask.
- Incubation is done for 45-60 minutes at room temperature.
- After incubation period, 0.5 – 1.0 ml of gas mixture from the flask is removed with air tight syringe and the same is injeced into a pre-conditioned GLC.
- Peak heights for acetylene and ethylene peakes are observed.
- After assay, roots are detached and their dry weight are measured after oven drying them at 80^0 C.
- For standard curve, 10 μg (Z) of pure ethylene is injected into a 100 ml sealed flasks. 0.5 – 1.0 ml of air removed from the flask and injected into GLC and height of ethylene is measured.

Calculation

- Standard amount of ethylene (E) in μ mole
  - = (0.446 x Z μl) / (Peak height in mm x Attenuation)

- Amount of ethylene produced in μ mole in the sample

  = E x Peak height of sample ethylene in mm x Attenuation

- Activity of nitrogenase = n mole or μ mole ethylene per unit sample per unit time

Notes

- Detaching nodule from root will give low nitrogenase activity. There sould be no moisture at the roots.
- The minize residual acetylene peak, 1 ml of air in the flask may be removed and 1 ml of acetylene injected instead of 10 ml.
- If any sub normal acetylene peak is noticed, discard the sample. This may be due to leakage in the septum of the flask.
- If no peak is seen, check the needle for blockage. Clogging of needles often happens when very sharp needles are used.
- The septum in the injecting point of GLC should be changed for every 25 injections.
- Pure acetylene should be injected into GLC as blank to check for any contamination with ethylene. If any ethylene peak is noticed in acetlene blank, this peak height multiplied by the attenuation should be deducted from the sample ethylene peak height and attenuation.

Colorimetric determination of nitrogenase activity

The acetylene reduction assay is now well established in the study of biological nitrogen fixation. Its utility is attested to by its reported use in over 200 publications. Until now gas chromatography has been the only means of measuring ethylene in the presence of acetylene. However, there are reports for colorimetric assay for ethylene sufficiently sensitive to estimate nitrogenase in nodulated legume roots. In designing the procedure, stress was placed on economy, safety, and simplicity. All reagents are inexpensive, none are particularly toxic or dangerous to the operator, and no heating steps are necessary.

Materials

- Sodium metaperiodate: 0.05 M $NaIO_4$, 16.5 g/l; store in a dark bottle for no more than 4 weeks. Potassium permanganate: 0.005 M KM_nO_4, 0.79 g/l; store in a dark, well stoppered bottle. A slight precipitate usually forms after 1 to 2 days; if there is much precipitate, use another lot of KM_nO_4.

- Sulphuric acid: 4 N.
- Sodium arsenite: 4 M, $NaAsO_2$, 52 g/ 100 ml.
- Oxidant solution: 80 ml of 0.05 M $NaIO_4$, 10 ml of 0.005 M $KMnO_4$, adjust pH to 7.5 with KOH, dilute to 100 ml.
- Acetyl acetone (2, 5-pentanedione): purify by distillation. Nash reagent (1, 6): 150 g of ammonium acetate, 3 ml of acetic acid, 2 ml of acetyl acetone diluted to 1 liter.

Procedure

- Oxidant solution (1.5 ml) is placed in a 10-ml conical flask which is then sealed with a rubber serum bottle cap.
- From 1 to 5 ml of gas containing up to 1 µmole of CH_4 are transferred by syringe from a chamber in which plant roots are incubated with C_2H_2. If necessary, some of the atmosphere in the conical flask is removed first with a syringe.
- The conical flask is agitated vigorously on a rotary shaker at 300 rpm for 90 min at room temperature (22^0 C).
- One-fourth ml of 4 M $NaAsO_2$ and one-fourth ml of 4 N H_2SO_4 are added, mixing to destroy excess oxidant.
- One ml of Nash reagent is added, and the absorbance at 412 nm is determined after 60 min. Standards containing known amounts of C_2H_4 are carried through the analysis at the same time as the samples.

8.2. Estimation of Nitrate Reductase (EC. 1.6.6.1-3.) Activity

Nitrate reductase (EC: 1.6.6.1-3) represents a short, soluble electron transport chain localized in the cytosol, catalyzes the NAD(P)H reduction of NO_3 to NO_2 and plays a key role in the regulation of NO_3 assimilation in lower and higher plants. Assimilatory nitrate reductase of higher plants is a most interesting enzyme, both from its central function in plant primary metabolism and control of catabolic activity. It has been shown that light and plant hormone availability is the major external triggers for the rapid and reversible modulation of nitrate reductase activity. As it is well known that products of light reaction reflects the need to co-ordinate carbon (C) and nitrogen (N) assimilation. The biochemistry of nitrate reductase has been elucidated to a great extent and the role that nitrate reductase plays in regulation of nitrate assimilation is becoming understood.

Materials

- Phosphate buffer: 0.1 M phosphate buffer pH 7.5 is prepared by dissolving 6.8045 g of potassium dihydrogen phosphate in 500 ml of distilled water and 8.709 g of potassium hydrogen phosphate in 500 ml of distilled water. The buffer of pH 7.5 is made by adding 16.ml of potassium dihydrogen phosphate and 84 ml of potassium hydrogen phosphate and final pH is adjusted by pH meter.
- Potassium nitrate: 0.1 M of potassium nitrate (KNO_3) is prepared by dissolving 1.011 g of KNO_3 in distilled water and the final volume is made to 100 ml with distilled water. 0.4 M solutuopn of potassium nitrate made by dissolving 4.044 g of KNO_3 in 100 ml water.
- Zinc acetate: 1.0 M solution is prepared by dissolving 21.94 g of zinc acetate in 100 ml of distilled water.
- Alcohol (90%): 10 ml of distilled water + 90 ml of alcohol
- NADH Solution: 4.6 mg of NADH is dissolved in 1.0 ml of 0.5% sodium bicarbonate solution to give 0.68 μ mole/0.1 ml. Fresh solution should be prepared before use.
- Sulphanilamide (1%): 1.0 g of sulphanilamide is dissolved in 100 ml of 1.0 N HCl.
- N-(1-Napthyl) ethylene diamine dihydroxide (NDD): 0.02% of NDD solution is repared by dissolving 20 mg of NDD in 100 ml of distilled water.

Reading

Radin, J.W. (1973). *In vivo* assay of nitrate reductase in cotton leaf discs. Effect of oxygen and ammonium. *Plant Physiol.*, **51(2)**: 332–336.

In Vitro Method

Grinding Media: The grinding media is prepared by adding the following in 20 ml of 0.1 M phosphate buffer pH 7.5

- 12 mg cysteine (4×10^{-3} M)
- 37.2 mg EDTA (5×10^{-3} M)
- 660 mg BSA (3%)

Extraction of enzyme: Leaf samples are collected in ice buckets at 0^0 C. The leaves are washed and wiped out by filter paper. A known quantity of leaf material is taken in pre chilled mortar and is grounded in presence of grinding media. The homogenates is centrifuged at 12,000 x g at 0^0 C for 30 minutes. The supernatant is used as the enzyme source.

Assay of activity

- The assay mixture has the following
    - Enzyme extract 0.5 ml
    - KNO_3 0.5 ml (50 μ moles)
    - NADH 0.1 ml (0.68 μ moles)
    - Phosphate buffer 1.9 ml (190 μ moles)
- A blank with boiled enzyme extract is used as control
- Incubation is done at 33^0 C (± 1^0 C) for 15 minutes and reaction is stopped by adding 0.1 ml of zinc acetate and 1.9 ml of 90% ethanol.
- The contents are centrifuged at 3,000 x g for 10 minutes to remove the precipitates.
- To the supernatant, 1.0 ml of sulphanilamide and 1.0 ml of NDD solutions are added to develop colour.
- After 15 minutes, optical density is recorded at 540 nm against blank.

In Vivo Method

- Leaf tissues are washed and wiped with filter paper to dry them out. Known amount of leaf (~0.5 mg) samples are taken 50 ml Erlenmeyer Flask having 2.5 ml of phosphate buffer (pH 7.5) and 2.5 ml of 0.4 M potassium nitrate solution.
- The flasks are placed in ice and the reaction mixture is infiltrated using vacuum pump for one minutes. The process of infiltration is repeated till the leaf tissues are visibly wet.
- Incubation is done at 35^0 C for 30 minutes after covering them with black cloth to prevent photo-oxidation of nitrites.
- The reaction is stopped by boiling the reaction mixture and the flasks are removed from hot plates to avoid chlorophyll coming out of the leaf tissues in the reaction mixture.
- A known volume (0.2 ml) of the reaction mixture is taken in test tubes.

- To the reaction mixture, 1.0 ml of sulphanilamide and 1.0 ml of NDD solutions are added and let it stand for 15 minutes.
- Final volume is made to 6.0 ml with distilled water and optical density is recorded at 540 nm against distilled water.

Preparation of standard curve

- 69.0 mg of sodium nitrite is dissolved in 10 ml of distilled water to have 10 μ moles per one ml. By successive dilution, solutions containing 5 to 50 n moles of nitrate are prepared.
- To these concentrations, one ml each of sulphanilamide and NDD solutions are added and colour is allowed to develop for 15 minutes.
- Final volume is made up to 6 ml with distilled water and optical densities are recorded at 540 n m.

Reading

Klepper, L., Flesher, D. and Hageman, R.H. (1971). Generation of reduced nicotinamide adenine dinucleotide for nitrate reduction in green leaves. *Pl. Physiol.*, **48:** 580-590.

8.3. Estimation of Nitrite Reductase (NADPH): Nitrite Oxidoreductase (EC. 1.6.6.4) Activity

Nitrite reductase refers to any of several classes of emzymes that catalyze the reduction of nitrite. There are two classes of NIR's. A multi haem enzyme reduces NO_2^- to a variety of products. Copper containing enzymes carry out a single electron transfer to produce nitric oxide. Assimilatory nitrate reductase is an enzyme of the assimilative metabolism involved in reduction of nitrate to nitrite. The nitrite is immediately reduced to ammonia (probably via hydroxylamine) by the activity of nitrite reductase. The term assimilatory refers to the fact that the product of the enzymatic activity remains in the organism. In this case, the product is ammonia which has an inhibitive effect on assimilatory nitrate reductase, thus ensuring that the organism produces the ammonia according to its requirements.

In enzymology, a ferredoxin—nitrite reductase (EC 1.7.7.1) is an enzyme that catalyses the following chemical reaction

NH_3 + 2 H_2O + 6 oxidized ferrodoxin $\rightleftharpoons$ nitrite + 6 reduced ferredoxin + 7 H^+

The 3 substrates of this enzyme are NH_3, H_2O, and oxidized ferrodoxin, whereas its 3 products are nitrite, reduced ferrodoxin, and H^+.

This enzyme belongs to the family of oxidoreductase, specifically those acting on other nitrogenous compounds as donors with an iron-sulfur protein as acceptor. The systematic name of this enzyme class is ammonia:ferredoxin oxidoreductase. This enzyme participates in nitrogen metabolism and nitrogen assimilation. It has 3 cofactor iron: iron, Siroheme, and iron-sulphur.

This enzyme can use many different isoforms of ferredoxin. In photosynthesizing tissues, it uses ferredoxin that is reduced by PSI and in the root it uses a form of ferredoxin (FdIII) that has a less negetive midpoint potential and can be reduced easily by NADPH.

Materials

- Tris-Hcl buffer 0.5 M pH 7.5
- Sodium nitrite solution – 43.2 mg of $NaNO_2$ is dissolved in 20 ml of distilled water.
- Methyl Viologen solution – 60.1 mg of methyl viologen is dissolved in 20 ml of distilled water.
- Sodium dithionite-bicarbonate solution – 250 mg each of $Na_2S_2O_4$ and $NaHCO_3$ in 10 ml of distilled water.

Procedure

- Homogenize fresh leaves (1.0 g/10 ml) with Tris-HCl buffer (pH7.5) having 1 mM EDTA and 10 mM mercaptoethanol., centrifuge at 8000 rpm for 15 minutes at 4^0 C. The supernatant was removed for analysis and kept on ice.
- A reaction mixture is prepared by mixing 6.25 ml of Tris-HCl buffer, 2 ml of sodium dinitrite solution, 2 ml of methyl viologen solution and 14.75 ml of water.
- To a portion (1.5 ml) of the above reaction mixture 0.3 ml of enzyme extract is added. In blank enzyme extract volume is replaced with Tris-HCl buffer (ph 7.5).
- The reaction was started by addition of 0.2 ml of freshly prepared sodium dithionite-sodium bicarbonate solution.
- The reaction mixture was incubated at 30° C for 15 minutes.
- Reaction is stopped by shaking the tube vigoursly until blue colour is disappeares.

- 0.5 ml of each reaction were removed and mixed with 0.5 ml of 1% (w/v) sulfanilamide in 1.0 N HC1 and 0.5 ml of 0.02% (w/v) n-napthylethylene-diamine dichydrochloride (N-NED) and left for 15 minutes to allow any colour to develop.
- The absorbance was read at 540 nm.
- Potassium nitrite was used to generate a standard curve.
- The enzyme activity is expressed as the amount nitrite reduced per minute per mg of protein.

Reading

R. W. Jones and R. W. Sheard. (1973). Application of automated dialysis and colorimetry to the assay of nitrate and nitrate reductase in plant extracts. *Can J Pl. Sci.*, **53:** 207-2013.

8.4. Estimation of Glutamate Dehydrogenase (L-glutamate: NADP Oxidoreductase (deaminase) EC. 1.4.1.4 and L-glutamate NAD Oxidoreductase (deaminating) EC 1.4.1.2) Activity

Glutamate dehydrogenase (GDH) occurs in almost all living organisms. In higher plants, GDH activity has been found in most species tested. The existance of two distinct GDH enzymes in higher plants is now well documented.

a. A mitochonrial enzyme which specifically requires conenzyme NAD.
b. A chloroplast enzyme which require specifically coenzyme NADP.

Both enzymes have dual coenzyme specificity to a certain extent and have different pH optima. The fungal and bacterial enzyme has a single coenzyme specificity. GDH activity has been found in both bacteriod and cytosol fractions of root nodules of a number of legumes. The enzyme has high Km for ammonia (10-80 mM). The reaction catalyzed by this enzyme is given below:

L-glutamate + H_2O + NAD^+ (P) $\leftrightarrow$ 2-Oxoglutarate + NH_4^+ +NAD(P)H + H^+

GDH is like other oxidoreducatses assayed by following the oxidation of the reduced conenzymes, NADH or NADPH. These reduced coenzymes absorb light at 340 nm. Thus the crude enzyme extracts the absorbance of NADH at 340 nm is easily detected. The molar extinction coefficient of NADH at 340 nm is 6.22 x 10^3. This indicates one micromole of NADH per ml wil have an absorbance of 6.22.

Materials

- Potassium phosphate buffer 1.0 M (pH 78 and pH 7.0)
- 2-oxo-glutarate 0.1 M, dissolve 14.6 g in one litre of distilled water.
- NH_4Cl (1.0 M). Dissolve 53.5 in one litre of distilled water
- NADH (10 mg / ml)
- NADPH (10 mg / ml)
- Enzyme extract:

Procedure

Take the following reaction mixture is taken in triplicate

- Potassium phosphate buffer pH 7.0..1.00 ml
- Potassium phosphate buffer pH 7.8..1.00 ml
- 2-Oxo-glutarate..0.30 ml
- NH_4Cl...0.50 ml
- NADH / NADPH Solution...0.12 ml
- Enzyme extract...0.20 ml
- Water...8.00 ml
- For blank, the 2-oxoglutarate solution was omited and another 0.3 ml of Tris-HCl buffer was added.
- Incubate the reaction mixture at 370 C for 15-30 minutes.
- Record the changes in OD at 340 nm.

Calculation

- The amount of NADH or NADPH oxidised is calculated from the molar extinction coefficient. Activities are expressed as n moles of NAD(P)H oxidised per minute per mg of protein.
- Nanomole of NAD(P)H oxidised/min/mg protein =

  (A340 x Vol. of assay solution x 100) / (6.22 x time of incubation x mg protein in enzyme extract).

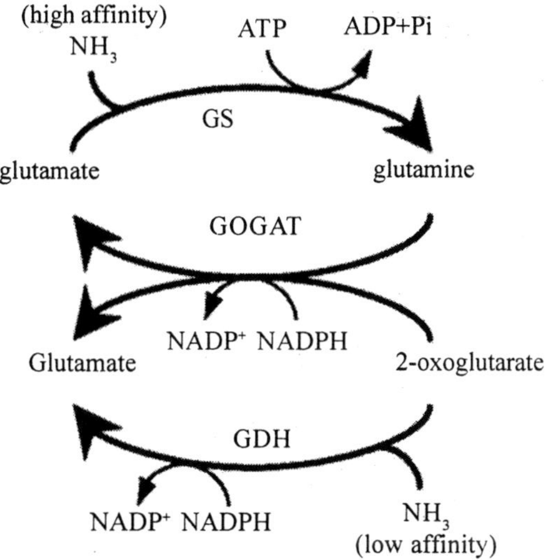

Reading

Delma Doherty (1970). In "*Methods of Enzymology*", Ed. H. Tabor and C.W. Tabor, **17-A:** pp.850

8.5. Estimation of Glutamine Synthatase (EC. 6.3.1.2) Activity

Glutamine synthetase (GS) (EC 6.3.1.2) is an enzyme that plays an essential role in the metabolism of nitrogen by catalyzing the condensation of glutamate and ammonia to form glutamine:

$$\textbf{Glutamate + ATP + NH}_3 \rightarrow \textbf{Glutamine + ADP + phosphate}$$

Glutamine synthetase uses ammonia produced by nitrate reduction, amino acid degradation, and photorespiration. The amide group of glutamate is a nitrogen source for the synthesis of glutamine pathway metabolites. Other reactions may take place via GS. Competition between ammonia ion and water, their binding affinities, and the concentration of ammonium ion, influences glutamine synthesis and glutamine hydrolysis. Glutamine is formed if an ammonium ion attacks the acyl-phosphate intermediate, while glutamate is remade if water attacks the intermediate. Ammonium ion binds more strongly than water to GS due to electrostatic forces between a cation and a negatively charged pocket. Another possible reaction is upon NH_2OH binding to GS, rather than NH_4^+, yields γ-glutamylhydroxamate.

The activity of ths enzyme is measured by estimating the production of inorganic phosphate, GS also catalyses the γ-glutamyl transfer reaction in presence of ADP, Mn^+ and arsenate

$$\textbf{Glutamate + Hydroxylamine} \rightarrow \textbf{Glutamyl hydroxamate}$$

When the activity is measured in presence of Mn^{+}, it represents total glutamyl synthatase activity (adenylated and unadenylated forms). The biologically active unadenylated form may be measured by inhibiting the adenylated form by the addition of 60 mM Mg^{++}.

Materials

- The following reagents are prepared in 20 mM Tris+HCl buffer (pH 8.0). The concentrations of the stock solutions are given in parenthesis.
- L-glutamine 0.2 M (700 mg/12 ml)
- Sodium arsenate 20 mM (500 mg of disodium hydrogen arsenate in 10 ml)
- Manganese chloride 3 mM (83 mg of $MnCl_2$/10 ml)
- Adenosine diphosphate 1 mM (40 mg/10 ml)
- Ferric chloride reagent – 10 g of trichloro acetic acid and 8 g ferric chloride in 250 ml of 0.5 N hydrochloric acid.

Procedure

- Enzyme extraction – green leaves 1.0 g is extracted in 5.0 ml of 50 mM imidazole-acetate buffer pH 7.8 containing 0.5 M EDTA, 1 mM dithiothereitol, 2 mM $MnCl_2$ and 20% glycerol at 4^0 C. Centrifuge at 10,000 x g for 30 minutes. If purification is required, precipitate the enzyme with $(NH_4)_2SO_4$ at 60% saturation. Resuspende the precipitate in extraction buffer. Desalt over sephadex G 25.
- 2.0 ml of glutamine solution is mixed with 0.5 ml of sodium arsenate, 0.3 ml of MnCl2, 0.5 ml of hydroxylamine, 0.5 ml of ADP and 0.2 ml of enzyme extract.
- The reaction mixture is incubated for 30 minutes at 37^0 C.
- After incubation, reaction is stopped by the addition of 1.0 ml of ferric chloride reagent.
- Optical density is measured at 540 nm against reagent blank.
- A blank is run separately where 2.0 ml of 20 mM Tris-Hcl buffer is taken instead of glutamine.
- A standard gradient of γ-glutamyl hydroxamate from 100 to 500 μg is prepared in 4 ml of buffer solution and colour is developed in accordance to above stated procedures.

- Standard curve is obtained by plotting the optical densities against their respective concentrations.

Calculation

- The amount of γ-glutamylhydroxamate formed in the reaction mixture is obtained from standard curve. Enzyme activity is expressed as nano moles γ-glutamyl hydroxamate formed per minute per mg protein.

Reading

Pateman, J.A. (1969). Regulation of synthesis of glutamate dehydrogenase and glutamine synthetase in micro-organisms. *Biochem J.*, **115:** 769-780

8.6. Estimation of Glutamate Synthase (EC. 1.4.7.1.) Activity

Glutamine oxoglutarate aminotransferase (also known as Glutamate synthase) is an enzyme and frequently abbreviated as GOGAT. This enzyme manufactures glutamate from glutamine and á-ketoglutarate, and thus along with glutamine synthesis (abbreviated GS) plays a central role in the regulation of nitrogen assimilation in photosynthetic eukaryotes and prokaryotes. This is of great importance as primary productivity in many marine environments is regulated by the availability of inorganic nitrogen. The primary sources of inorganic nitrogen used by marine algae are nitrate and ammonia. Both forms are ultimately incorporated into amino acids through the sequential reaction of glutamine synthetase (GS) and glutamate synthase (glutamine 2-oxyoglutarate aminotransferase; GOGAT). GOGAT isoenzymes catalyze the transfer of the amido nitrogen of glutamine to 2-oxoglutarate using pyridine nucleotides (NADH / NADH- / NADH- / NADPH-dependent) or ferrodoxin (ferredoxin dependant) as reductants. They are called NADH-GOGAT and Fd-GOGAT respectively. In photosynthetic eukaryotes, GS and GOGAT isoenzymes are localized in the cytosol and chloroplast.

Glutamine + 2-oxoglutarate + NADPH + $H^+ \rightarrow$ 2 Glutamate + $NADP^+$

Fd-GOGAT is found strictly in cyanobacteria and photosynthetic eukaryotes, and the gene is located in the chloroplast of rhodophytes and in the nucleus of vascular plants, but in both cases its product is active in the chloroplast. NADH-GOGAT is found in the nucleus of vascular plants, fungi, and diatoms, while NADPH-GOGAT is found in non-photosynthetic bacteria and archaea.

Glutamate synthase is assayed spectrophotometrically by measuring the rate of oxidation of NADPH or NADH as indicated by a change in optical density at 340 nm following the addition of enzyme extract.

Materials

- Tris-HCl buffer 50 mM pH 7.6
- Glutamine solution – 5 mM (36.5 mg/10 ml of Tris-HCl buffer 50 mM pH 7.6)
- 2-oxoglutarate solution- 5.0 mM (36.5 mg/10 ml of Tris-HCl buffer 50 mM pH 7.6)
- NADPH solution - 0.25 M (10 mg/10 ml of Tris-HCl buffer 50 mM pH 7.6).

Procedure

- Fresh leaves (~2.0 g) is extracted in 10 ml of 100 mM phosphate buffer pH 7.5 containing 1 mM disodium EDTA, 1 mM dithioerythritol and 1% polyvinyl pyrrolodine (PVP) and centrifuge at 10,000 g for 30 minutes at 4^0 C. Supernatant is used as the enzyme source.
- A reaction mixture having buffer 1.8 ml, glutamine solution 1.0 ml, 2-oxoglutarate solution 1.0 ml, NADPH solution 1.0 ml is prepared.
- To the above reaction mixture, 0.2 ml of enzyme extract is added.
- After addition of enzyme solution, optical density is recorded at 340 nm and it is initial NADPH concentration.
- In blank 2-oxoglutarate solution is replaced with 1.0 of buffer.
- Reaction mixture is incubated for 30 minutes at 37^0 C.
- After incubation period, change in optical density at 340 nm is observed. NADPH oxidized during incubation period is worked out by substracting from the initial value.
- Enzyme activity is expressed as n mole of NADPH oxidized per minute per mg of protein.

Reading

Tempest D.W., Meers, J.L. and Brown, C.M. (1970). Synthesis of glutamate in *Aerobacter aerogenes* by a hitherto unknown route. *Biochem. J.*, **117:** 405-407.

Van de Casteele, J.P., Lemal, J. and Coudest, M. (1975). Pathway and regulation of glutamate synthesis in a *Corynebacterium sp.* Overproducing glutamate. *J. Gen. Microbiol.* **90:** 178-180.

8.7. Estimation of Asparagine Synthetase (L-Aspartate; Ammonia Ligase EC. 6.3.1.4) Activity

Aspartate-ammonia ligase (EC 6.3.1.14) is an enzyme that catalyses the chemical rcactions involving the amidation of aspartate by either glutamine or ammonia.

ATP + L-aspartate + NH_3 AMP + PPi + H_2O + L-asparagine

Aspartate + Glutamine +ATP Asparagine + Glutamate + AMP + PPi

Asparagine synthetase is found in germinating cotyledons and legume plants.

The 3 substrates of this enzyme are ATP, L-aspartate, and NH_3^+, whereas its 3 products are AMP, diphosphate, and L-asparagine. This enzyme belongs to the family of ligases, specifically those forming carbon-nitrogen bonds as acid-D-ammonia (or amine) ligases (amide synthases). The systematic name of this enzyme class is L-aspartate:ammonia ligase (AMP-forming). Other names in common use include asparagine synthetase, and L-asparagine synthetase. This enzyme participates in 3 metabolic pathways: alanine and aspartate metabolism. cyanoamino acid metabolism, and nitrogen metabolism.

This enzyme is most easily measured by substituting the hydroxylamine for ammonia for above reaction

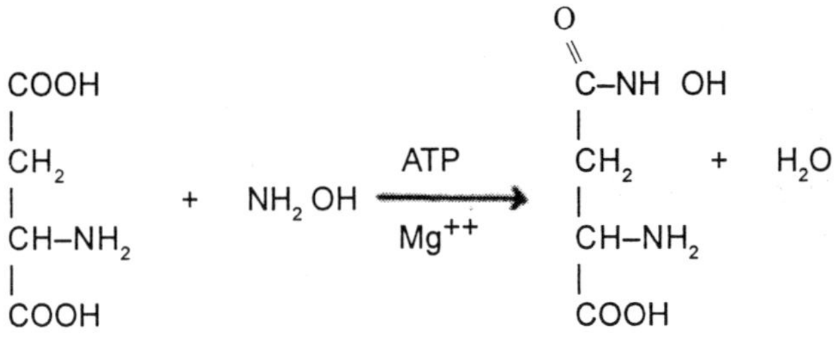

```
                                          O
                                          \\
COOH                                      C–NH OH
|                                         |
CH2                          ATP          CH2      +   H2O
|          +   NH2 OH   ----------->      |
CH–NH2                       Mg++         CH–NH2
|                                         |
COOH                                      COOH
```

Materials

- Tris-NH_2OH-$MnCl_2$ solution – Mix 2.42 g of Tris base (0.25 M), 11.1 g hydroxylamine hydrochloride (2 M) and 476 mg $MnCl_2.4H_2O$ and dissolve in about 40.0 ml of water. Adjust the pH to 6.4 by addition of 8.0 N KOH and final volume is made to 80.0 ml. This reagent is prepared fresh before use.
- Potassium or sodium ATP, 0.1 M - 551.0 mg of disodium salt of ATP is dissolved in 10.0 ml of water and neutralized. This solution is prepared fresh before use.

- L-aspartic acid, 0.1 M pH 6.4 – 133.0 mg of L-aspartic acid is dissolved in 10 ml of water and pH is adjusted to 6.4.
- Ferric chloride reagent – 125.0 ml of TCA (20% w/v) is mixed with 50.0 g of ferric chloride ($FeCl_3.6H_2O$) and 28.0 ml of concentrated HCl and volume is made to 500.0 ml with water. This reagent is stable reagent.

Procedure

- Chilled plant material (approximately 1.0 g) is homogenized with 10.0 ml of 100 mMTris-HCl buffer pH 8.5 having 15% (v/v) glycerol, 56 mM mercaptoethanol and 4mM potassium cyanide (KCN). Centrifuge the homogenate at 15,000 x g for 30 minutes at 4^0 C. Supernatant is usec as enzyme extract.
- To 0.4 ml of Tris-NH_2OH-$MnCl_2$ solution, 0.1 ml of ATP solution, 0.2 ml of enzyme extract, 0.2 ml of L-aspartic acid and 0.1 ml water is added.
- The reaction mixture is incubated at 37^0 C for 10 minutes.
- After incubation period, reaction is stopped by the addition of 3.0 ml of ferric chloride solution.
- Precipitated protein is removed through centrifugation and optical density of the supernatant is measured at 540 nm against a reagent blank (3.0 ml of ferric chloride solution + 1.0 ml of water).
- For blank, L-aspartic acid is replaced by water.
- Standard curve is prepared by using 0 – 2.5 μ moles of β-aspartyl hydroxamate.

Calculation

- Enzyme activity is expressed as micromoles of β-aspartyl hydroxamate formed per mg of protein per minute. The concentration of β-aspartyl hydroxamate can also be obtained by multiplication the assay absorbance value by a factor of 6.1, if the synthetic â-aspartyl hydroxamate is not available for standard curve.

Reading

Ravel, JM. (1970). In "*Methods of Enzymology*", (Ed. H. tabor and C. W. tabor, **XVII (A):** Academic Press, New York, pp. 722.

Rogenes, SE. (1975). Glutamine dependent asparagines synthetase from *Lupins luteus. Phytochem*, **14:** 1975.

8.8. Estimation of Glutathion Synthatase (E.C. 6.3.1.2) Activity

Glutamine synthetase (GS) (EC 6.3.1.2) is an enzyme that plays an essential role in the metabolism of nitrogen by catalyzing the condensation of glutamate and ammonia to form glutamine:

Glutamate + ATP + NH_3 → Glutamine + ADP + phosphate

Glutamine Synthetase uses ammonia produced by nitrate reduction, amino acid degradation, and photorespiration. The amide group of glutamate is a nitrogen source for the synthesis of glutamine pathway metabolites. Other reactions may take place via GS. Competition between ammonium ion and water, their binding affinities, and the concentration of ammonium ion, influences glutamine synthesis and glutamine hydrolysis. Glutamine is formed if an ammonium ion attacks the acyl-phosphate intermediate, while glutamate is remade if water attacks the intermediate. Ammonium ion binds more strongly than water to GS due to electrostatic forces between a cation and a negatively charged pocket. Another possible reaction is upon NH_2OH binding to GS, rather than NH_4^+, yields γ-glutamylhydroxamate.

GS catalyzes the ATP-dependent condensation of glutamate with ammonia to yield glutamine. The hydrolysis of ATP drives the first step of a two-part, concerted mechanism. ATP phosphorylates glutamate to form ADP and an acyl-phosphate intermediate, γ-glutamyl phosphate, which reacts with ammonia, forming glutamine and inorganic phosphate. ADP and P_i do not dissociate until ammonia binds and glutamine is released.

ATP binds first to the top of the active site near a cation binding site, while glutamate binds near the second cation binding site at the bottom of the active site. The presence of ADP causes a conformational shift in GS that stabilizes γ-glutamyl phosphate atom. Ammonium binds strongly to GS only if the acyl-phosphate intermediate is present. Ammonium, rather than ammonia, binds to GS because the binding site is polar and exposed to solvent. In the second step, deprotonation of ammonium allows ammonia to attack the intermediate from its nearby site to form glutamine. Phosphate leaves through the top of the active site, while glutamine leave through the bottom (between two rings).

The determination of GS activity is based on production of glutamyl hydroxamate.

Material

- *Extraction buffer* – 50 mM Tris-HCl buffer (pH7.5) containing 1% (W/v) polyvenyl pyrolidone.

- ATP – 0.05 M ATP at pH 7.5
- Sodium glutamate – 0.5 M of Na-glutamate
- Cystein – 0.1 M cystein at pH 7.5
- Magnesium sulphate – 1.0 M solution
- Hydroxylamine
- Ferric chloride reagent – equal volume of 10% ferric chloride ($FeCl_3.6H_2O$) in 0.2 N HCl, 24% (W/v) trichlloro acetic acid and 30% (V/v) HCl.

Procedure

- Approximately 2 g of fresh leaves are extracted with 10.0 ml of extraction buffer.
- The mixture is centrifuged for 15 minutes at 15,000xg at 2^0 C. The pallet part is discarded and the supernatant is then dialyzed for 24 hours against 0.01 M Tris-HCl buffer (pH 7.5).
- Assay mixture of 3.0 ml contain 33.33 mM Tris-HCl buffer (0.2 ml of buffer of 0.2 M at pH 7.5), 3.33 mM ATP, 3.33 mM cysteine , 33.33 mM of $MgSO_4$, 10.01 mM hydroxylamine at pH 7.5, 0.4 ml of dialysed enzyme extract and 0.9 ml of water.
- The reaction is initiated with the addition of 83.33 mM sodium glutamate
- The reaction mixture is incubated for 30 minutes at 30^0 C and the reaction is terminayed with the addition of 1.0 ml of ferric chloride reagent.
- Control was taken by omoting sodium glutamate.
- Precipitated protein, if any, from the reaction mixture is centrifuged and OD is recorded at 540 nm against reagent blank.
- Srtandard curve for sodium hydroxamate using a gradient from 0.5 to 10 µM is prepared using the above stated procedures.

8.9. Estimation of Nitrate Content

Materials

- Zinc dust powder: 100 g of barium sulphate, 75 g of citric acid. 10 g of manganese sulphate, 4 g of sulphanilamide, 2 g zinc powder and 2 g of N-(1 napthyl) ethylene diamine dihydrochloride were mixed thoroughly.
- Copper sulphate solution: 2 mg of copper sulphate ($CuSO_4$) is dissolved in 100 ml of distilled water.

- Acetic acid solution: 20% acetic acid solution having 0.2 ppm of copper is prepared by taking 1 ml of $CuSO_4$ solution and added to 20 ml of acetoc acid and the final volume is made to 100 ml by water.

Extraction of nitrate

- 100 mg of dried leaf powder is boiled with 10 ml of distilled water for few seconds and then filtered through Whatman's filter paper No. 1 at room temperature. The filtrate is used for nitrate estimation.

Estimation

- Take 2 ml of leaf extract and add 0.8 g of zinc powder and 8 ml of 20% acetic acid solution and shaken vigorously.
- Colour is allowed to develop for 30 minutes and then reagent mixture is centrifuged at 5000 x g for 5 minutes. Repeated centrifugation is done to get a clear solution.
- Optical density is recorded at 540 nm against reagent blank.

Preparation of standard curve

- Potassium nitrate solution (10 µ moles/ml) is prepared and from this different concentrations of 0.1 to 0.7 µ moles of nitrate is prepared in 2 ml of distilled water.
- 0.8 g of zinc dust powder and 8.0 ml of 20% acetic acid solution are added and colour is allowed to develop for 30 minutes.
- The contents are centrifuged at 5000 x g for 5 minutes.
- Optical density is recorded at 540 nm against reagent blank and standard curve is plotted by plotting the ODs against their respective concentrations of potassium nitrate.

Reference

Wooly, JT, Hicks, GP and Hageman, RH (1960). Plant analysis, rapid determination of nitrate and nitrite in plant materials. *J. Ag. Food Chem.*, **8:** 481-482.

Alternative method for nitrate estimation

This method Is based on measurement of UV (210 nm) absorption of nitrates.

- After removal of interfearence due to other ions mainly nitrites by trating the sample with perchloric acid and sulphamic (amido sulphuric) acid.

- Add 10.1 ml of 10% (w/v) suphamic acid to 1.5 ml of the sample containing nitrates (10 – 200 n moles), shake well using vortex mixture and let it stand for 2 minutes.
- Shake again and add 0.4 ml of 20% (v/v) perchloric acid.
- Shake again and measure the optical density at 210 nm.

$$E_{1cm}^{1mM} \text{ Nitrate (210 nm)} = 7.4$$

Reading

Cauire, P.A. (1967). The determination of nitrate in soil solution by ultra-violet spectroscopy. *Analyst*, **92:** 311-315.

8.10. Estimation of Protease (EC. 3.4.21.40) Activity

Proteolysis is the breakdown of proteins into smaller polypeptides or amino acids. In general, this occurs by the hydrolysis of the peptide bond, and is most commonly achieved by cellular enzymes called proteases, but may also occur by intramolecular digestion, as well as by non-enzymatic methods such as the action of mineral acids and heat. Proteolysis in organisms serves many purposes; for example, digestive enzymes break down proteins in food to provide amino acids for the organism, while proteolytic processing of polypeptide chain after its synthesis may be necessary for the production of an active protein. It is also important in the regulation of some physiological and cellular processes, as well as preventing the accumulation of unwanted or abnormal proteins in cells.

Some proteins and most eukaryotic polypeptide hormones are synthesized as a large precursor polypeptide known as polyprotein that require proteolytic cleavage into individual smaller polypeptide chains. The polyprotein pro-opiomelanocortin (POMC) contains many polypeptide hormones. The cleavage pattern of POMC, however, may vary between different tissues, yielding different sets of polypeptide hormones from the same polyprotein. Proteolytic cleavage breaks down proteins in food extracellularly into smaller peptides and amino acids so that they may be absorbed and used by an organism. Proteins in cells are also constantly being broken down into amino acids. This intracellular degradation of protein serves a number of functions: It removes damaged and abnormal protein and prevent their accumulation, and it also serves to regulate cellular processes by removing enzymes and regulatory proteins that are no longer needed. The amino acids may then be reused for protein synthesis. Many viruses also produce their proteins initially as a single polypeptide chain that were translated from a polycistronic mRNA. This polypeptide is subsequently cleaved into individual polypeptide chains.

A protease (also termed peptidase or proteinase) is any enzyme that performs proteolysis, that is, begins protein catabolism by hydrolysis of the peptide bonds that link amino acids together in the polypeptides chain forming the protein. Proteases have evolved multiple times, and different classes of protease can perform the same reaction by completely different catalytic mechanisms. Proteases can be found in animals, plants, bacteria, archea and viruses.

Proteases may be classified by the optimal pH in which they are active:

- *Acid proteases*
- *Neutral proteases* involved in type 1 hypersensitivity. Here, it is released by mast cells and causes activation of complement and kinins. This group includes the calpains.
- *Basic proteases* (or *alkaline proteases*)

Proteases occur in all organisms, from prokaryotes to eukaryotes to viruses. These enzymes are involved in a multitude of physiological reactions from simple digestion of food proteins to highly regulated cascades (*e.g.*, the blood clotting cascade, the complement system, apoptosis pathways, and the invertebrate prophenoloxidase-activating cascade). Proteases can either break specific peptide bonds (*limited proteolysis*), depending on the amino acid sequence of a protein, or break down a complete peptide to amino acids (*unlimited proteolysis*). The activity can be a destructive change (abolishing a protein's function or digesting it to its principal components), it can be an activation of a function, or it can be a signal in a signalling pathway.

Plant genomes encode hundreds of proteases, largely of unknown function. Those with known function are largely involved in developmental regulation. Plant proteases also play a role in regulation of photosynthesis.

Materials

- Phosphate buffer 0.2 M pH 7.5 - It is prepared by dissolving 35.598 g of dibasic sodium phosphate in 1000 ml of water and 27.218 g of potassium dihydrogen phosphate in 1000 ml of water. The desired pH is obtained by mixing 80.3 ml of sodium phosphate with 19.3 ml of potassium phosphate solution. Final pH is adjusted with pH meter and 1.0 HCl and/ or 1.0 NaOH.
- Phosphate buffer 0.2 M pH 7.0 - It is prepared by mixing 41.3 ml of sodium phosphate with 58.7 ml of potassium phosphate solution. Final pH is adjusted with pH meter and 1.0 HCl and/or 1.0 NaOH.

- Borate buffer 0.2 M pH 8.5 – is obtained by dissolving 1.237 g of boric acid with 243 mg of sodium hydroxide in 100 ml of water. Final pH is addusted with pH meter and 1.0 HCl and/or 1.0 NaOH.
- Bovine serum albumin (BSA) – 160 mg of BSA is dissolved in 10.0 ml of 0.2 phosphate buffer pH 8.5 to get a solution of 8.0 mg of BSA/0.5 ml.
- Trichloroacetic acid (TCA) – 20% (w/v) solution of TCA is obtained by dissolving 20 g of TCA in 80 ml of water.
- Cyanide acetate buffer – Cyanide acetate buffer pH 5.3 – 5.4 is obtained by dissolving 90.0 g of trihydrated sodium acetate in water and to it 16.75 ml of glacial acetic acid is added. Final volume is made to 250 ml with water. To this mixture 5.0 ml of 0.1 M sodium cyanide is added and mixed thoroughly.
- Ninhydrin – 3% ninhydrin solution is obtaine by dissolving 3.0 g of ninhydrin in 100.0 ml of mrthyl cellosolve (ethylene glycol monomethyl eather).
- Isopropyl alcohol – Isopropyl alcohol is diluted to 1:1 with water.

Procedure

- Fresh leaves approximately 2.0 g is extracted with 10.0 ml of prichilled phosphate buffer 0.2 M pH 7.5. Homogenates are centrifuged at 10,000 x g for 10 minutes at 2^0 C. Supernatant is saved and then dialized for 24 hours against 0.2 M phosphate buffer pH 7.0 with 4 – 5 changes of buffer at 4^0 C. Dialized estract is used as enzyme source.
- 2.5 ml of phosphate buffer 0.2 M pH 7.5 is taken and to it 0.5 ml of BSA solution and 0.2 ml of enzyme extract is mixed and incubated for 90 minutes at 40^0 C.
- Reaction is terminated by the addition of 0.8 ml of TCA solution. Precipitated protein is centrifuged at 5,000 rpm for 5 minutes.
- From 0.5 ml of assay mixture colour is developed with the addition of 0.5 ml of water, 0.5 ml of cyanide-acetate buffer and 0.5 ml of ninhydrin solution.
- This reaction mixture is kept in boiling water bath for colour to develop.
- 5.0 ml of iso propyl alcohol (diluted with water in 1:1 ratio) isadded and optical density is measured at 570 nm against a reagent blank.
- A standard curve is made from glycine (75 mg/100 ml of water) in range of 0.1 to 1.0 µmole and developing the coloure as described above.

- The protease activity is expressed as m mole of amino acid produced/ hour/g fresh wight.

Reading

Beevers, L. (1968). Protein degradation and proteolytic activity in the cotyledons of germinating pea seeds (*Pisum sativum* L.). *Phytochem.*, **7:** 1837-1844.

Rosen, H. (1957). A modified ninhydrin colorimetric analysis for amino acids. *Arch. Biochem. Biophys.*, **67:** 10-15.

8.11. Estimation of Aspartic Proteinase (EC. 3.4.23.23) Activity

Aspartic proteases are a family of protease enzymes that use an aspartate residue for catalysis of their peptide substrates. In general, they have two highly conserved aspartates in the active site and are optimally active at acidic pH. Nearly all known aspartyl proteases are inhibited by pepstatin. Aspartic endopeptidases EC 3.4.23.23 of vertebrate, fungal and retroviral origin have been characterised. More recently, aspartic endopeptidases associated with the processing of bacterial type 4 prepilin and archaean preflagellin have been described.

Eukaryotic aspartic proteases include pepsins, cathepsins, and renins. They have a two-domain structure, arising from ancestral duplication. Retroviral and retrotransposon proteases (Pfam PF00077) are much smaller and appear to be homologous to a single domain of the eukaryotic aspartyl proteases. Each domain contributes a catalytic Asp residue, with an extended active site cleft localized between the two lobes of the molecule. One lobe has probably evolved from the other through a gene duplication event in the distant past. In modern-day enzymes, although the three-dimensional structures are very similar, the amino acid sequences are more divergent, except for the catalytic site motif, which is very conserved. The presence and position of disulfide bridges are other conserved features of aspartic peptidases.

Aspartyl proteases are a highly specific family of proteases - they tend to cleave dipeptide bonds that have hydrophobic residues as well as a beta-methylene group. Unlike the closely related serine proteases these proteases do not form a covalent intermediate during cleavage.

While a number of different mechanisms for aspartyl proteases have been proposed, the most widely accepted is a general acid-base mechanism involving coordination of a water molecule between the two highly conserved aspartate residues. One aspartate activates the water by abstracting a proton, enabling the water to attack the carbonyl carbon of the substrate scissile bond, generating a tetrahedral oxyanion intermediate. Rearrangement of this intermediate leads to protonation of the scissile amides.

Aspartic proteinases constitute a relatively small group of homologous proteolytic enzymes involved in a wide range of cellular functions, including removal of singal peptides from nascent polypeptides, enzyme activation, protein degradation, and cellular reorganization. Notwithstan-ding, they have received much interest because several play significant roles in human diseases.These enzymes have been found to be widely distributed in fungi, higher plants and mammalian cells. They are active at low or neutral pH and are characterized by having two aspartic residues in their catalytic sites. Most of them have molecular mass of about 40 kDa and have homologous sequences varying between 323 and 340 amino acid residues in length. These hydrolases arise to be synthesized in the form of precursors and are found in the mature state predominantly in two chains form. As there is no information about the proteolytic maturation mechanism of the vacuolar enzymes of planaria and of the extensive protein degradation that takes place during the regeneration, it was of interest to purify proteinases of this animal in order to investigate its role in the course of the regenerative process.

Materials

- Sodium acetate buffer 0.2 M pH 4.1
- Alkaline copper solution – (A) 2.0 g of sodium carbonate is dissolved in approximately 75.0 ml of water and final volume is made to 100.0 ml. (B) 500 mg of copper sulphate ($CuSO_4.5H_2O$) is dissoved in water containing 1.0 g of potassium sodium tartarate and final volume is made to 100.0 ml with water. (C) 50.0 ml of solution A and 1.0 ml of solution B is mixed before use.
- Folin-Ciocalteau reagent – Dilute commercially available reagent to two folds volume with water.
- Trichloroacetic acid solution (TCA) – 10.0 g of TCA is dissolved in approximately 75.0 ml of water and final volume is made to 100.0 ml.
- Haemoglobin solution – 2.0 g of haemoglobin powder is mixed with 50.0 ml of water and stirred constantly for 30 minutes. Tritrate the haemoglobin solution with 0.3 N HCl to pH 1.7. After 10 minutes titrate with 0.5 M sodium acetate to pH 4.1. Final volume is made to 100.0 ml with water.
- Tyrosine standard – 50.0 mg of L-tyrosine is dissolved in water containing 0.1 NaOH and volume is made to 50.0 ml.
- Working standard – 10.0 ml of tyrosine stock solution is diluted to 50.0 ml with water. This standard solution has a strength of 200 μg/ml.

Procedure

- Haemoglobin solution (1.0 ml) is equilibrated for temperature by keeping at 37^0 C for 10 minutes. And 0.1 ml of enzyme extract solution is added.
- The reaction mixture is incubated for 30 minute at 37^0 C in a waterbath and reaction is terminated by the addition of 0.4 ml of 10% TCA solution and shake vigorously.
- Incubate at room temperature for 10 minutes. Centrifuge at 10,000 rpm for 10 minutes.
- To 0.5 ml of the clear supernatant 2.1 ml of alkaline copper reagent is added and reagent mixture is again incubated for 10 minutes at room temperature.
- To the above reaction mixture, 0.2 ml of Folin-Ciocalteu reagent is added and again incubate in dark for 10 minutes at room temperature.
- Optical density is measured at 660 nm against a reagent blank.
- For blank boiled enzyme extract is used.
- Standard curve is drawn using tyrosine standard (20 – 200 μg) and following the above steps.

Calculation

- One unit of acid proteinase is defined as the amount of enzyme required to liberate 1 μg of tyrosine per minute under assay conditions.

Reading

Anson, M. (1938). The estimation of papain and cathepsin with haemoglobin. *J. Gen. Physiol*, **22:** 79.

CHAPTER 9

Proteins and Enzymes

Proteins are large biological molecules, or macromolecules, consisting of one or more chains of amino acid residues. Proteins perform a vast array of functions within living organisms, including catalysing metabolic reactions, replicating DNA, responding to stimuli, and transporting molecules from one location to another. Proteins differ from one anothei primarily in their sequence of amino acids, which is dictated by the nucleotide sequence of their genes, and which usually results in folding of the protein into a specific three-dimensional structure that determines its activity. A polypeptide is a single linear polymer chain derived from the condensation of amino acids. The individual amino acid residues are bonded together by peptide bonds and adjacent amino acid residues. The sequence of amino acid residues in a protein is defined by the sequence of a gene, which is encoded in the genetic code. In general, the genetic code specifies 20 standard amino acids; however, in certain organisms the genetic code can include selenocysteine and, in certain archaea-pyrrolysine. Shortly after or even during synthesis, the residues in a protein are often chemically modified by post translational modification, which alters the physical and chemical properties, folding, stability, activity, and ultimately, the function of the proteins. Sometimes proteins have non-peptide groups attached, which can be called prosthetic groups or cofactors. Proteins can also work together to achieve a particular function, and they often associate to form stable protein complexes.

Like other biological macromolecules such as pollysaccharides and nucleic acids, proteins are essential parts of organisms and participate in virtually every process within cells. Many proteins are enzymes that catalyse biochemical reactions and are vital to metabolism. Proteins also have structural or mechanical functions, such as actin and myosin in muscle and the proteins in the cytoskeleton, which form a system of scaffolding that maintains cell shape. Other proteins are important in cell signalling, immune responses, cell adhesion, and the cell cycle. Proteins are also necessary in animals' diets, since animals cannot synthesize all the amino acids they need and must obtain essential amino acids from food. Through the process of digestion, animals break down ingested protein into free amino acids that are then used in metabolism.

Proteins may be purified from other cellular components using a variety of techniques such as ultracentrifugation, precipitation, electrophoresis, and chromatography; the advent of genetic engineering has made possible a number of methods to facilitate purification. Methods commonly used to study protein structure and function include immunohistochemistry, sit-directed mutagenesis, nuclear magnetic resonance and mass spectrometry.

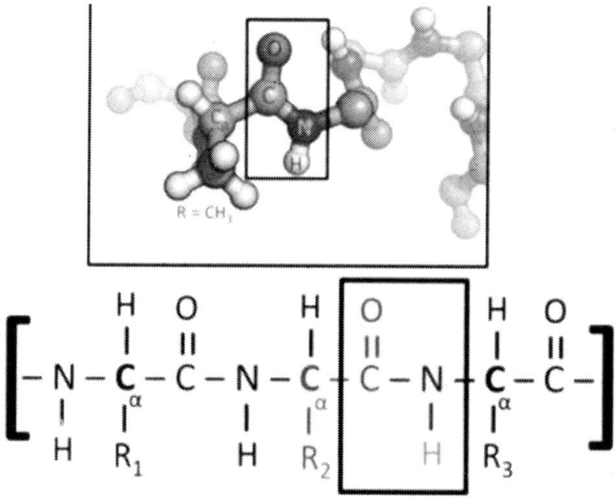

Chemical structure of the peptide bond (bottom) and the three-dimensional structure of a peptide bond between an alanine and an adjacent amino acid (top/inset)

Most proteins consist of linear polymers built from series of up to 20 different L-α-amino acids. All proteiogenic amino acids possess common structural features, including an α-carbon to which an amino group, a carboxyl group, and a variable side chain are bonded. Only proline differs from this basic structure as it contains an unusual ring to the N-end amine group, which forces the CO–NH amide moiety into a fixed conformation. The side chains of the standard amino acids, detailed in the list of standard amino acids, have a great variety of chemical structures and properties; it is the combined effect of all of the amino acid side chains in a protein that ultimately determines its three-dimensional structure and its chemical reactivity. The amino acids in a polypeptide chain are linked by peptide bonds. Once linked in the protein chain, an individual amino acid is called a *residue,* and the linked series of carbon, nitrogen, and oxygen atoms are known as the *main chain* or *protein backbone.*

The peptide bond has two resonance forms that contribute some double-bond character and inhibit rotation around its axis, so that the alpha carbons are roughly coplanar. The other two dihedral angles in the peptide bond determine the local shape assumed by the protein backbone. The end of the protein with a free carboxyl group is known as the C-terminus or carboxy terminus, whereas the end with a free amino group is known as the N-terminus or amino terminus. The words *protein*, *polypeptide,* and *peptides* are a little ambiguous and can overlap in meaning. *Protein* is generally used to refer to the complete biological molecule in a stable conformation, whereas *peptide* is generally reserved for a short amino acid oligomers often lacking a stable three-dimensional structure. However, the boundary between the two is not well defined and usually lies near 20–30 residues. *Polypeptide* can refer to any single linear chain of amino acids, usually regardless of length, but often implies an absence of a defined conformation.

Proteins are the chief actors within the cell, said to be carrying out the duties specified by the information encoded in genes. With the exception of certain types of RNA, most other biological molecules are relatively inert elements upon which proteins act. Proteins make up half the dry weight of an *Escherichia coli* cells, whereas other macromolecules such as DNA and RNA make up only 3% and 20%, respectively. The set of proteins expressed in a particular cell or cell type is known as its proteome. The chief characteristic of proteins that also allows their diverse set of functions is their ability to bind other molecules specifically and tightly. The region of the protein responsible for binding another molecule is known as the binding site and is often a depression or "pocket" on the molecular surface. This binding ability is mediated by the tertiary structure of the protein, which defines the binding site pocket, and by the chemical properties of the surrounding amino acids' side chains. Protein binding can be extraordinarily tight and specific; for example, the ribonuclease inhibitor protein binds to human angiogenin with a sub-femtomolar dissociation constant ($<10^{-15}$ M) but does not bind at all to its amphibian homolog onconase (>1 M). Extremely minor chemical changes such as the addition of a single methyl group to a binding partner can sometimes suffice to nearly eliminate binding; for example, the aminoacyl tRNA synthetase specific to the amino acid valine discriminates against the very similar side chain of the amino acid isoleucine.

Proteins can bind to other proteins as well as to small-molecule substrates. When proteins bind specifically to other copies of the same molecule, they can oligomerize to form fibrils; this process occurs often in structural proteins that consist of globular monomers that self-associate to form rigid fibres. Protein-protein interactions also regulate enzymatic activity, control progression through the cell cycle, and allow the assembly of large protein-complexes that carry out many closely related reactions with a common biological function. Proteins can also bind to, or even be integrated into, cell membranes. The ability of binding partners to induce conformational changes in proteins allows the construction of enormously complex signalling networks. Importantly, as interactions between proteins are reversible, and depend heavily on the availability of different groups of partner proteins to form aggregates that are capable to carry out discrete sets of function, study of the interactions between specific proteins is a key to understand important aspects of cellular function, and ultimately the properties that distinguish particular cell types.

The best-known role of proteins in the cell is as enzymes, which catalyse chemical reactions. Enzymes are usually highly specific and accelerate only one or a few chemical reactions. Enzymes carry out most of the reactions involved in metabolism, as well as manipulating DNA in processes such as DNA replication, DNA repair, and transcription. Some enzymes act on other proteins to add or remove chemical groups in a process known as post translational modification. About 4,000 reactions are known to be catalyzed by enzymes. The rate acceleration conferred by enzymatic catalysis is often enormous—as much as 10^{17}-fold increase in rate over the uncatalyzed reaction in the case of orotate decarboxylase (78 million years without the enzyme, 18 milliseconds with the enzyme).

The molecules bound and acted upon by enzymes are called substrates. Although enzymes can consist of hundreds of amino acids, it is usually only a small fraction of the residues that come in contact with the substrate, and an even smaller fraction—three to four residues on average—that are directly involved in catalysis. The region of the enzyme that binds the substrate and contains the catalytic residues is known as the active sites.

9.1. Protein Purification

Protein expression is regulated for normal functioning of a cell or organism. To understand protein structure and function in detail, they often need to be separated from other cellular components (lipids, nucleic acids, sugars, *etc.*) and isolated to homogeneity. After recovering a protein to near homogeneity, it should retain all its native biological characteristics of structure and activity. To achieve this objective, one needs to take into account the physical and chemical property of proteins (size, charge, solubility, hydrophobicity, precipitation, *etc.*). These common characteristics of the protein can be exploited to separate it from other components of the cell. With the introduction of recombinant DNA technology, protein purification technique has been enhanced and also simplified. Purification protocols vary, depending on the precise nature of the protein. General steps include (i) chromatography, (ii) precipitation and/ or (iii) extraction.

Ammonium sulphate fractionation of proteins

Ammonium sulphate precipitation is a method used to purify proteins by altering their solubility. It is a specific case of a more general technique known as salting out. Ammonium sulphate is commonly used as its solubility is so high that salt solutions with high ionic strength are allowed.

The solubility of proteins varies according to the ionic strength of the solution, and hence according to the salt concentration. Two distinct effects are observed: at low salt concentrations, the solubility of the protein increases with increasing salt concentration (*i.e.* increasing ionic strength), an effect termed salting in. As the salt concentration (ionic strength) is increased further, the solubility of the protein begins to decrease. At sufficiently high ionic strength, the protein will be almost completely precipitated from the solution (salting out). Since proteins differ markedly in their solubilities at high ionic strength, salting-out is a very useful procedure to assist in the purification of a given protein. The commonly used salt is ammonium sulphate, as it is very water soluble, forms two ions high in the Hofmeister series, and has no adverse effects upon enzyme activity. It is generally used as a saturated aqueous solution which is diluted to the required concentration, expressed as a percentage concentration of the saturated solution (a 100% solution).

In the preliminary test, the ammonium sulphate concentration is increased stepwise, and the precipitated protein is recovered at each stage. This is usually done by adding solid ammonium sulphate, but calculating how much ammonium sulphate to add to a solution at one concentration to achieve a desired higher concentration is tricky, since addition of ammonium sulphate significantly increases the volume of the solution. The amount to add can be determined either from published nomograms or by using an online calculator. Each protein precipitate is dissolved individually in fresh buffer and assayed for total protein content and amount of desired protein. The aim is to find the ammonium sulphate concentration which will precipitate the maximum proportion of undesired protein, whilst leaving most of the desired protein still in solution or vice versa.

The precipitated protein is then removed by centrifugation and then the ammonium sulphate concentration is increased to a value that will precipitate most of the protein of interest whilst leaving the maximum amount of protein contaminants still in solution. The precipitated protein of interest is recovered by centrifugation and dissolved in fresh buffer for the next stage of purification. This technique is useful to quickly remove large amounts of contaminant proteins, as a first step in many purification schemes. It is also often employed during the later stages of purification to concentrate protein from dilute solution following procedures such as gel filtration.

Ammonium sulphate is a particular useful salt for the fractional precipitation of proteins. It is available in highly purified form, has great solubility allowing for significant changes in the ionic strength and is inexpensive. Changes in the ammonium sulphate concentration of a solution can be brought about either by adding solid substance or by adding solution of known saturation, generally, a fully saturated (100%) solution.

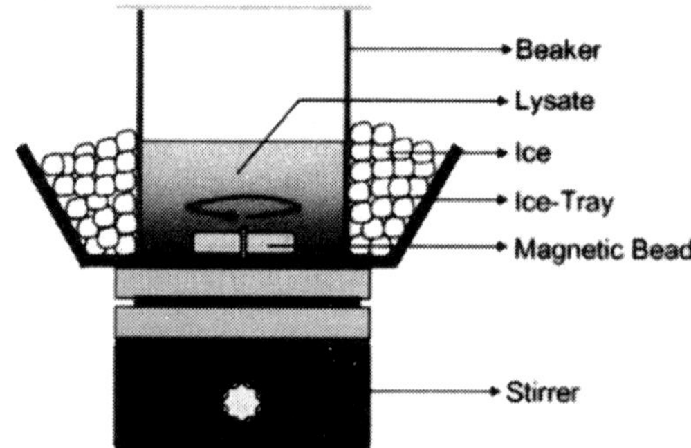

Protein Precipitation using ammonium sulphate

Amount of solid ammonium sulphate to be added to a solution to give the desired saturation at 0^0 C

	Final concentration of ammonium sulphate, % saturation at 0^0 C																
	10	20	25	30	33	35	40	45	50	55	60	65	70	75	80	90	100
	g solid ammonium sulphate to add to 100 ml of solution																
0	56	114	144	176	196	209	243	277	313	351	390	430	472	516	561	662	767
10		57	86	118	137	150	183	216	251	283	326	365	406	441	494	592	694
20			29	59	78	91	123	155	189	225	262	300	340	382	424	520	619
25				30	49	61	93	125	158	193	230	267	307	348	390	485	583
30					19	30	62	94	127	162	198	235	273	314	356	449	546
33						12	43	74	107	142	177	214	252	292	333	426	522
35							31	63	94	129	164	200	238	278	319	411	506
40								31	63	97	132	168	205	245	285	375	469
45									32	65	99	134	171	210	250	339	431
50										33	66	101	137	176	214	302	392
55											33	67	103	141	179	264	353
60												34	69	105	143	227	314
65													34	70	107	190	275
70														35	72	153	237
75															36	115	198
80																77	157
90																	79

Dialysis

Dialysis is a process of separation of alkaloids from crystaloids. It involves placing the protein solution in a semi-permeable membrane, and placing the membrane in a large container of buffer. Small molecules (such as salt ions) pass through the dialysis membrane (moving from high concentration to low concentration), while large molecules are unable to cross the membrane. Dialysis membranes come in a variety of pore sizes, and are therefore useful for removing a variety of different sized solutes. In principle, dialysis could allow separation of large proteins from small ones; in practice, however, the pores in the tubing are insufficiently uniform to allow this technique to be used effectively.

In practice, dialysis is often used to alter the concentration of salts and/or small molecules in protein solutions usually aimed at decreasing the concentration of these solutes. However, the composition of the solution can also be changed in additional ways. Dialysis is based on diffusion during which the mobility of solute particles between two liquid spaces is restricted, mostly according to their size. (In rarely used versions of dialysis, restriction of diffusion via polarity or charge is also possible.) Size restriction is achieved by using a porous material, usually a semi-permeable membrane called dialysis membrane. This membrane is permeable only for particles below a certain size. In the biochemical laboratory, this membrane is mostly a hose made from transparent material (also called dialysis bag) that can be tightly closed (tied) at its ends. The solution to be dialysed (with a volume V1) is loaded into the dialysis bag. The dialysis bag is then placed into a dialysis solution (with a volume V2) that is stirred slowly to aid the diffusion of the subset of solutes that can be released through the bag membrane, in order to achieve equilibrium between solute concentrations in the two liquid spaces. If the difference in volume between the two spaces is large (V2 >> V1, *e.g.* V2 = 10 L and V1 = 0.1 L, a 100-fold difference), the onset of the equilibrium will lead to a very significant dilution of the small solutes that were initially inside the bag (their concentration will change by a factor V1/(V1+V2), in this case << 1), with only a slight change in the concentration of small solutes in the outside solution (by a factor V2/(V1+V2) H" 1), whereas the concentration of the molecules inside the bag that cannot penetrate the membrane remains almost completely unchanged.

Practical aspects and applications of dialysis

The efficiency of dialysis, *i.e.* the extent to which the concentration and composition of the inside solution can be changed, is an important aspect. It follows from the above description of dialysis that the efficiency of dialysis largely depends on the difference between the volumes of the inside and outside liquid spaces. This is why we generally seek to use as large volume (V2) of the dialysing solution as possible. However, the efficiency of dialysis can be further increased by performing multi-step dialysis by exchanging the outer solution after the equilibrium has been reached. In this case, the attainable dilution of the inside solution will be $[V1/(V1+V2)]_n$ where n is the number of steps. It is easy to see that efficiency that can be achieved by applying a two-step dialysis at a 50-fold volume difference is much higher than the efficiency of a single-step dialysis at a 100-fold volume difference. The speed of dialysis can be increased not only by stirring the outside solution but also by increasing the surface/volume ratio of the inside solution, as the flux of diffusion is linearly proportional to the cross-section. It is, therefore, more practical to choose a narrower and longer tube than a wider and shorter one.

The semi-permeable membrane can be crossed not only by salts and small molecules but also by solvent particles (in most cases, water). The direction and extent of the net solvent flow is determined by the difference between the total concentration of solutes in the inside and outside solutions such that the solvent migrates from the less to the more concentrated solution (with regard to solutes). This way the equilibrium concentration of the solute(s) of the inside solution that cannot cross the membrane will be influenced also by the diffusion of the solvent. As the solute(s) that cannot cross the membrane also contribute to the total concentration of the inside solution, the net direction of solvent migration will almost always point towards the inside solution. Therefore, the volume of the inside solution will increase, thereby selectively decreasing the concentration of the membrane-impermeable solute(s)—but not that of the membrane-permeable ones, even if the relative increase in the volume is large. However, the relative increase in the volume is generally not large because (*i*) the concentration of the large impermeable solutes is low (much lower than that of the small permeable ones) (*ii*) the dialysis tube is largely unable to increase its volume. The occasional small (5-20 %) volume increase of the inside solution is associated with the compression of air above the liquid phase that was originally enclosed in the bag. Taken together, the decrease in the concentration of the large solutes (proteins) is usually negligibly small. The increase in the volume of the inside solution is remarkable from a technical point of view because it is accompanied by a (sometimes substantial) elevation of the pressure. Therefore, if there is a hidden "weakness" somewhere in the material of the membrane, the elevation of pressure may lead to bursting of the bag and, as a consequence, the complete loss of the dialysed material (*e.g.* protein preparation). To avoid this catastrophe, it is recommended to perform a pressure test on the bag in its water-filled state.

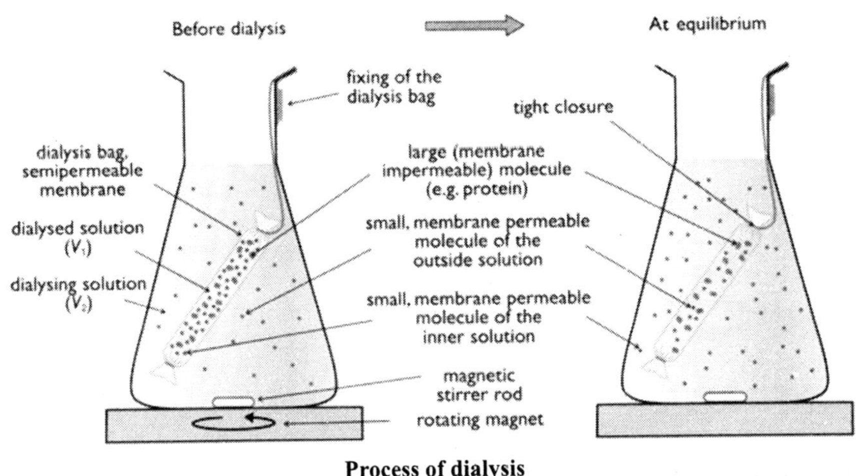

Process of dialysis

The other risk associated with pressure elevation occurs during the opening of the bag after completion of dialysis. In the absence of necessary care, the pressurised inside solution can sprinkle out, causing loss of material. In the biochemical laboratory practice, solutions of proteins are generally dialysed following fractioned ammonium sulphate precipitation as well as before or after ion exchange chromatography. A size selectivity (size exclusion or cut-off) specified as 4 or 11 kDa means that the pores of the dialysis membrane are impermeable for particles larger than 4 or 11 kDa, respectively. Besides the biochemical laboratory, dialysis is utilised in the field of life sciences also for therapeutic purposes during haemodialysis, *i.e.* in artificial kidneys. The principal difference between these two applications is that, in the artificial kidney, dialysis is executed under continuous counter-flow of the two solution spaces: both the inside solution (the blood of the patient) and the outside solution are pumped. Thus, in such a setting, also the inside liquid space is "open": it is not in a "bag" but flows inside a tube. Moreover, in order to increase the flux of diffusion, a large number of capillary tubes are employed in a bundle (which is actually the artificial kidney) by which the surface/volume ratio is increased enormously. The composition of the outside dialysing solution is very special as it must meet special requirements. In addition, the artificial kidney equipment is a very special apparatus because it must be able to ensure the appropriate pressure and temperature while the blood entering the body of the patient must be free of entrapped air bubbles that could lead to lethal consequences.

Column Chromatography

This method involves passing the protein through a column filled with resins of unique characteristics. Depending on the type of the resin or beads, purification can be achieved through (i) Ion Exchange, (ii) Size Exclusion or (iii) Affinity Chromatography.

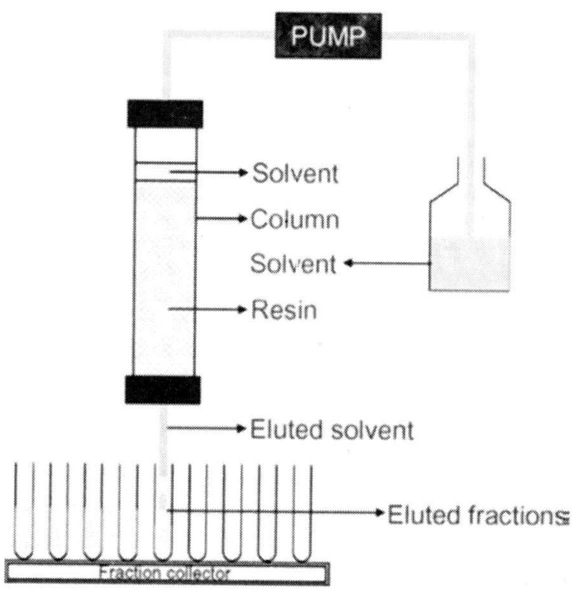

Flow Chart for column chromatography. The Central part is the column from the sample and/or solvent is loaded at a controlled flow rate with a pump. The eluates from the column are collected in tubes of a fraction collector.

Ionic Exchange Chromatography

This is one of the most useful methods of protein purification. Depending on the surface residues on the protein and the buffer conditions, the protein will have a net positive or negative charge. An ideal buffer should be in the physiological pH range of 6 to 8. At this pH range, most of the proteins have been observed to be negatively charged. Hence, proteins would bind to positively charged molecules of the resin. Change in the buffer pH condition could make the protein relatively positive, thereby allowing it to bind to a negatively charged resin material. Among the most commonly used charged molecules are DEAE and CM. These charged molecules are coupled to an inactive material, often nanoparticle beads, loaded into a column. The protein is loaded onto this packed column and is allowed to bind. The column is washed and the bound proteins are eluted depending on their tightness of binding, by subjecting them to either increasing concentrations of salt or changes in pH. Proteins with low charge will elute first.

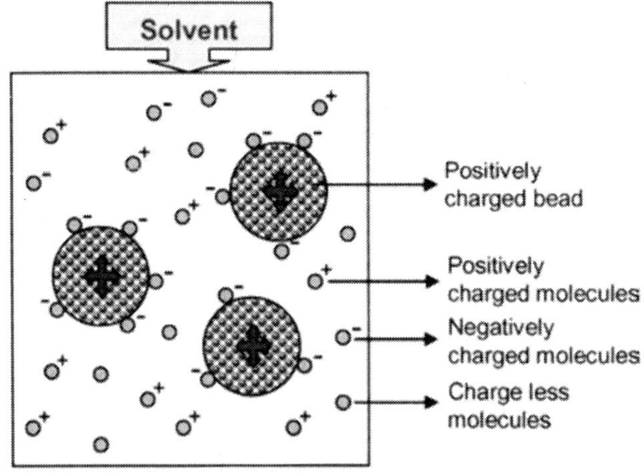

Ion Exchange Chromatography. The resins are charged and the protein molecules that bind are of opposite charge.

Size-Exclusion Chromatography

In this approach, the size of the protein is taken into consideration. The size of the protein depends on the number of amino acids it contains. This property can be used in protein purification. The column material consists of a porous matrix for proteins to diffuse into. The smaller proteins get entangled inside the porous material and hence their mobility is restricted. In contrast, the larger proteins do not get entangled and could just pass through. Hence, in the elution profile, the larger molecules would be the first ones to elute, while the smallest ones will be last to elute.

Affinity Chromatography

As the name suggests, the principle is the use of a moiety or molecule which has high affinity for the protein of interest. These molecules could either be co-factors, modified substrates, inhibitors or carbohydrates. This strategy of purification is used mostly in the later stages where the protein is relatively pure, and more specific approaches are required for additional purification. The affinity moiety or molecule is coupled to the matrix and used as a bait to fish the protein of interest. The protein could either be eluted with high salt in some cases or with increased amount of the affinity molecule itself.

Purification of Recombinant Proteins

This is the easiest method available for the purification of a protein, albeit it is a recombinantly expressed protein rather than an endogenous protein. The gene encoding a protein of interest is cloned into an expression vector (often with a tag such as GST or His) which is then introduced into the producer cell in order to express the protein as a fusion protein. The protein is then 'over-expressed' in higher than usual levels in a bacterial (*e.g.* BL21), yeast (*e.g. S. cerevisiae*), insect (*e.g.* sf9) or mammalian (*e.g.* CHO) cell system. The tag on the protein serves as a pull down, and thus separate and purify the protein from the cell lysate. The tag is usually a 6X His or Glutathione Transferase (GST). Thus, the column material is either Ni-NTA (Ninitrilotriacetic acid) which binds tightly to 6His, or Glutathione sepharose which binds to GST. Since these columns are very specific, the fusion protein is purified to near homogeneity. In order to attain complete purity, the protein then could be purified by other conventional chromatographic methods.

9.2. Determination of Protein Concentration

Determining the exact quantity of proteins in a solution is very often necessary in the biochemical practice. There are many ways to measure protein concentration. In chromogenic methods, the absorbance of a coloured product formed by the protein and an organic molecule is measured. Protein concentration can also be determined from the protein's own (intrinsic) UV absorbance. Note, however, that these methods may give different results for different proteins of the same concentration. Also, different methods can yield somewhat different results for the same protein. There is no absolute photometric protein concentration assay. All methods have advantages and disadvantages and we must choose among them by taking the following aspects into consideration: specificity, sensitivity, the measurable range of concentration, the accuracy, the nature of the protein to be examined, the presence of materials interfering with the measurement, and the time required for the measurement.

Biuret test

Molecules with two or more peptide bonds react with Cu^{2+} ions in alkaline solution and form a purple complex. Nitrogen atoms of the peptide bonds form a coordination bond with the metal ion. The quantity of the complexes formed is proportional to the number of peptide bonds. In practice, the determination of protein concentration is done using a calibration curve created using samples of known concentration. The protein treated with biuret reagent is measured at

540 nm after the purple product is formed. The advantages of the method include that only few materials (*e.g.* Tris and amino acid buffers) interfere with it, it can be done in a short time and does not depend on the amino acid composition of the protein. Its disadvantages are its low sensitivity and that it requires at least 1 mg of protein.

Lowry (Folin) protein assay

In this sensitive technique, a coloured product is formed similarly to the biuret reaction, but a second reagent (Folin-Ciocalteu reagent) is used in addition to strengthen the colour. The strong blue colour is created by two reactions: (1) formation of the coordination bond between peptide bond nitrogens and a copper ion and (2) reduction of the Folin-Ciocalteu reagent by tyrosine (phosphomolybdic and phosphotungstic acid of the reagent reacts with phenol). The measurement is carried out at 750 nm. As in the biuret reaction, a calibration curve is created (for example using BSA, bovine serum albumin), and the concentration of the unknown protein is determined from the curve. The advantages of the method include that it is quite sensitive and is able to detect even 1 μg of protein. Its disadvantages are that it takes rather long to carry out, is disturbed by various materials (including ammonium sulphate, glycine and mercaptans) and that the incubation time is critical. As different proteins contain different amounts of tyrosine, the amount of the coloured product will also be different. As a consequence, this method is more suited to compare the concentration of solutions of the same protein than to absolute measurement.

Bradford protein assay

Despite being relatively new, probably this is the most widely used protein assay. The method is based on the ability of the Coomassie Brilliant Blue dye to bind to proteins in acidic solution (via electrostatic and van der Waals bonds), resulting in a shift of the absorption maximum of the dye from 465 to 595 nm. The advantages of the method include that it is highly sensitive, is able to measure 1-20 μg of protein and is very fast. Only relatively few materials interfere with it (it works even in presence of urea or guanidine hydrochloride) but, importantly, detergents do. Even traces of detergent (*e.g.* cleaning products) can invalidate the results. Its disadvantages are that it depends strongly on amino acid composition and that it stains the cuvettes used.

Spectrophotometric protein assay

This method is based on the fact that two of the aromatic amino acids, tryptophan and tyrosine, show a peak in absorbance around 280 nm. It has the advantage of being quick and easy. Since it needs no chemical reaction to be performed, it is widely used for detection of proteins or peptides during their separation by chromatography. As proteins contain different ratios of aromatic amino acids, *per se* it is more suited to the comparison of solutions of the same protein and less to absolute measurement. The latter requires the knowledge of the molar extinction coefficients of proteins. For many proteins, these were determined and can be found in the literature. Moreover, if we know the number of tyrosine and tryptophan amino acids in the protein of interest, since their absorption values are additive, it is possible to calculate the molar extinction coefficient. This method has a moderate sensitivity with a material requirement of around 50 μg. It is disturbed by anything that has an absorbance at 280nm—most commonly, DNA. (Nucleic acids have an absorption maximum at 260 nm and their absorption at 280 nm is still considerable.) By using the correction introduced by Warburg and Christian, we can account for the error caused by nucleic acids. Absorption is measured at 260 and 280 nm and protein concentration can be calculated with the following equation: cprot (mg/ml) = 1.55 * A280nm % 0.76*A260nm

Absorption spectrum of proteins

Proteinogenic amino acids with an aromatic group in their side chain have a peak in absorption at 280 nm. As mentioned earlier, the absorption of amino acids is additive and thus the molar extinction coefficient of proteins can be calculated from their amino acid composition. Non-aromatic amino acids do not absorb at this wavelength, except for the low absorbance of cystine (a pair of cysteines forming a disulfide bond). Absorption values calculated based on amino acid composition generally fall close to values determined experimentally, making them suitable for the determination of protein concentration. The side chains of tryptophan and, to a lesser extent, that of tyrosine are also fluorescent. Tryptophan can be selectively excited at 295 nm, as it can be inferred from the shoulder in its absorption spectrum. In the spectrum of proteins, absorption at 220 nm is due to the peptide bonds and the peak at 280 nm is caused by the absorption of aromatic amino acids. The latter is better suited for concentration measurements because there are many other materials, including minor impurities in solvents, that exhibit absorption at 220 nm.

9.3. Cell Disruption, Cell Fractionation and Protein Isolation

How individual proteins can be isolated in homogeneous form from biological samples, especially from tissues of multicellular organisms. Naturally, the more information on protein to be investigated, the more straightforward it is to establish a well-suited and efficient isolation protocol. The first question to be addressed is the distribution of a given proteins among various tissues of an organism. Obviously, the tissue in which the given protein is the most abundant should be used as the starting point of the isolation protocol. The next question is whether the protein is intracellular, extracellular (secreted) or membrane-bound. If it is intracellular, the sub-cellular distribution of the protein should be considered. If the protein is associated to one of the many organelles of the eukaryotic cell, that organelle should be first isolated. The process through which individual organelles (plasma membrane, nucleus, mitochondria, *etc.*) are separated from one another is called cell fractionation. The process of cell fractionation starts with the disruption of the tissue and its cell constituents by a homogenisation procedure performed as gently as possible. Once the cells are opened up, individual organelle types can be separated from each other by various types of centrifugation techniques.

Cell disruption

The technical details of the cell disruption procedure largely depend on the type of tissue or cells to be homogenised. In the case of multi-cellular organisms, the first aim is to disintegrate the tissue into individual cells by abolishing the connections that organise the cells into the given tissue. Then the plasma membrane and, in case of plants, fungi and bacteria, the cell wall need to be ruptured. The harshness of the treatment can greatly vary depending on the tissue and cell type. For example, in the case of blood cells that do not need to be disintegrated from a solid tissue, even a mild osmotic shock using a hypotonic solution can lead to the rupture of the cell membrane. In the case of cells having a cell wall or tissues stabilised by a strong extracellular matrix, simple osmosis-based treatments are inefficient. In such cases various mechanical methods applying shearing force on the cells can be used. The two most frequently applied tools are high-speed laboratory blenders and ultrasonic cell disruptors. While blenders can disrupt even highly structured strong tissues, ultrasonic cell disruptors applying ultrasound (~20–50 kHz) to the sample (sonication) are used mostly in the case of cell suspensions. The ultrasonic cell disruptor generates the high-frequency waves electronically. These shock waves are transmitted to the cell suspension via an oscillating metal probe. The oscillation causes large localised pressure inhomogeneity resulting in cavitation eventually disrupting the cells. There are several other procedures that also use shearing force to open up cells. Some of these apply high pressure to pump the cell suspension through a very narrow channel or orifice into a low pressure container. Due to the sudden drop of pressure the cells explode in the container. Shearing force can be also generated by a pair consisting of a carefully designed glass tube and a tightly fitting glass pestle, called the Potter-Elvehjem homogeniser, or a potter in short. The diameter of the tube is just a little larger than that of the pestle. The sample is pushed into the very narrow space between the sides of the tube and the pestle. The shearing force is generated as the cell suspension squeezes up and past the pestle. This method is applied on the suspension of individual cells (already dissociated tissues, blood cells *etc.*). Plant cells protected by cell wall are most often disrupted by various grinding methods. Manual grinding is the most common method. The tissue is usually frozen in liquid nitrogen and then crushed using a mortar and pestle. Optimal cell disruption methods open up a high percentage of the cells in the sample while preserving the organelles or molecules to be investigated in their native state. This is not a trivial task. In order to preserve the native state of most organelles and molecules, the procedure should be quick and the heat generated by the disruption method should be dissipated by intensive cooling. This helps to avoid heat denaturation of proteins and also lowers the rate of unwanted chemical reactions such as oxidation or proteolytic cleavage of proteins. To further suppress these chemical reactions, oxidation can be prevented by the addition of reducing agents and proteolysis can be controlled by the addition of a mixture of protease inhibitors, often referred to as a protease inhibitor cocktail. In order to extract the content of the cell into a native-like solution, the buffers used for cell disruption often mimic the cytosol in terms of pH and ionic strength. The sample might contain trace amounts of heavy metal ions. Such ions can form complexes with various amino acid residues of proteins. To prevent such complex formation, a chelating agent, most often ethylene-diamine-tetraacetic acid (EDTA), is added to the homogenisation buffer to sequester the heavy metal ion components.

Cell fractionation

The major goal of cell fractionation is to separate the various types of cell organelles from each other and from the cytosol. As mentioned, cell disruption should be intensive enough to open up a large fraction of the cells but still gentle enough to preserve the native state of the organelles and the soluble components of the cytosol. Upon cell disruption, the plasma membrane and the endoplasmic reticulum become disintegrated into small membrane vesicles. These vesicles along with the native organelles of the cell and soluble molecules of the cytosol compose the mixture from which the components need to be separated into different fractions. The most important fractionation technique applied to this mixture is centrifugation using specialised laboratory equipment. The following section reviews the principles and major types of laboratory centrifugation.

Centrifugation

When an object attached to a rope is whirled around, one can feel that the rope must be pulled inward towards the centre of the rotation in order to keep the object on the orbit. This force prevents the object from getting away and move with a constant speed along a straight tangential line. The inward force with which one has to pull the rope is called the centripetal force. One can also define the outward force, the centrifugal force, by which the object pulls the rope. This force is equal in magnitude to the centripetal force but has the opposite direction. The centrifugal force (*Fc*) is a virtual, so-called fictional force emerging due to the inertia of the object. Yet, because it leads to a simpler mathematical formalism, equations describing the processes when solutions are centrifuged use the *Fc* force.

Differential centrifugation: cell fractionation based primarily on particle size

The density of the various organelles differs on a smaller scale than their size. Therefore, while both size and density affect sedimentation velocity, their size difference dominates when organelles are separated by centrifugation. In the procedure of differential centrifugation, cell constituents are separated from each other by their Svedberg value. Several consecutive centrifugation steps are applied in the order of increasing accelerating potential. Each individual centrifugation step relies on the different sedimentation speed of the different cell constituents at the given acceleration potential. At a properly chosen acceleration potential, almost 100 % of the largest component will sediment in the time span of the centrifugation. The sedimented organelles form a pellet at the bottom of the centrifuge tube. The potential should be set so that in the same period of time only a small portion of all smaller constituents latch on to the pellet.

The disrupted cell homogenate is centrifuged first at a relatively low accelerating potential of 500 g for 10 minutes. Under these conditions, only particles having the highest Svedberg value, intact cells and nuclei will form the pellet. All other cell constituents will sediment at a much lower rate and remain in the homogenate. The supernatant of the first centrifugation is transferred into an empty centrifugation tube and is subjected to another centrifugation step, now at a significantly higher accelerating potential of 10,000 g and for 20 minutes.

These conditions favour sedimentation of mitochondria, lysosomes and peroxisomes having lower Svedberg values than nuclei. Many cell constituents still remain in the supernatant, which is again transferred into an empty tube. This tube is placed into an ultracentrifuge and, with an accelerating potential of 100 000 g in one hour, the so-called microsomal fraction sediments. This fraction contains mostly artificial vesicles with a diameter of 50-150 nm that originate mostly from the endoplasmic reticulum and are generated by the cell disruption procedure. Other natural cell constituents of the same size range will also contribute to this fraction. After this third centrifugation step, the supernatant contains mostly macromolecules and supramolecular complexes such as ribosomes. By applying an accelerating potential as high as several hundred thousand *g*, ribosomes and large proteins can also be sedimented.

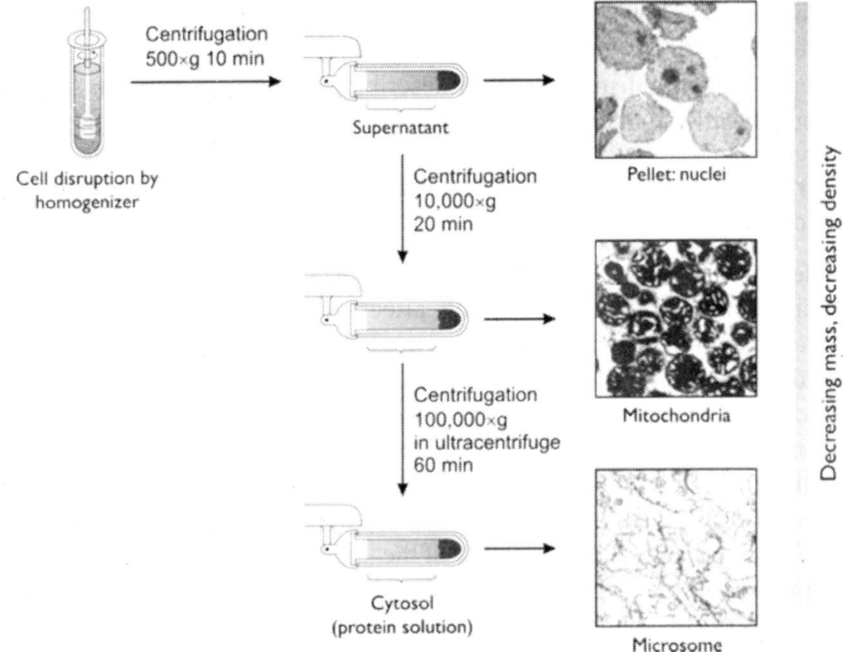

Differential centrifugation. In the course of differential centrifugation, consecutive centrifugation steps are applied. The consecutive centrifugation steps follow each other in the order of increasing centrifugal acceleration potential. During the first centrifugation, only the largest and/or heaviest cell constituents sediment in the time frame of the centrifugation. Typically, only nuclei and undisrupted whole cells form the pellet. The supernatant of the first centrifugation step is further centrifuged in the consecutive step at higher acceleration potential and typically for a longer period of time. Following this scheme, ever smaller and/or lower-density cell constituents can be sedimented.

Equilibrium density-gradient centrifugation: fractionation based on density

In the previous section we introduced the method of differential centrifugation. For simplicity, we stated that the constituents of the sample were separated in a medium of homogeneous density. This first approximation has didactical advantages as it makes the basic principle of differential centrifugation easier to comprehend. Nevertheless, it is sometimes advantageous to use a very shallow density gradient in the medium during differential centrifugation. This is done only to suppress convectional flows in the medium that could unsettle and mix layers of already separated cell constituents. The essence of equilibrium density gradient centrifugation is principally different. In this case, a rather steep density gradient is created in the medium in such a manner that the density of the medium gradually increases towards the bottom of the centrifuge tube. This is achieved by using a very high-density additive, for example caesium chloride (CsCl). The density gradient is created as follows. When the centrifuge tube is filled with the medium, a high concentration CsCl solution is added first. Subsequently, in the process of filling the tube, the concentration of CsCl is gradually decreased resulting in a CsCl gradient and, as a consequence, a density gradient in the tube. The sample is layered on the top of this special medium.

In the course of centrifugation, particles start to sediment moving towards the bottom of the centrifuge tube. By doing so, they travel through an increasing density medium. Each particle sediments to a section of the medium where its own density equals the density of the medium. At this section, the buoyancy factor becomes zero and, as a consequence, the accelerating force acting on the particle also becomes zero. The particle stops sedimenting. If it moved further towards the bottom of the tube, it would meet a higher density medium and a force opposing to its moving direction would be exerted on it, turning the particle back. If, by travelling backwards, it would meet a density lower than its own density, it would sediment again. As a consequence, this method separates particles exclusively based on their density. It is an equilibrium method in which, by the end of the separation, the system reaches a constant state. (In this aspect, this method shows an interesting analogy to the isoelectric focusing IEF). The two methods separate particles by entirely different characteristics (density versus isoelectric point), but in both cases, the separation leads to an equilibrium state. Both methods apply a gradient, but in the case of IEF a pH gradient is created.) Note that the two centrifugation approaches introduced above separate particles by partially different characteristics. Consecutive combination of the two methods can lead to a more efficient separation than achieved by any of the methods alone. Therefore, to increase separation efficiency, fractions generated by differential centrifugation can be subjected to a subsequent density-gradient centrifugation step to further separate individual components.

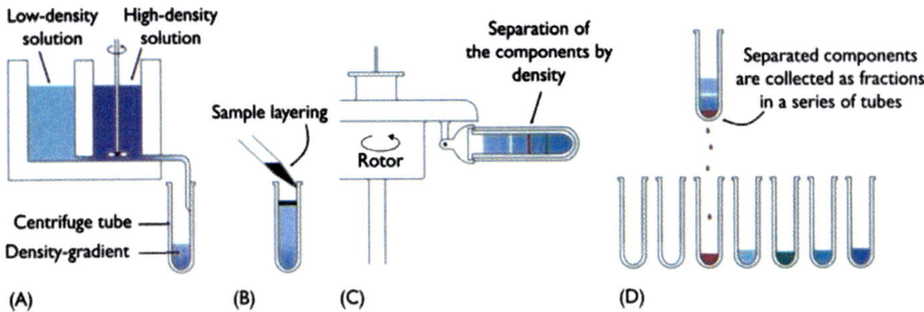

Equilibrium density-gradient centrifugation. In the course of equilibrium density-gradient centrifugation, a concentration gradient of a high density compound such as caesium chloride is generated. (The compound should not react with the biological sample.) The concentration gradient of this special additive creates a density gradient in the centrifuge tube. The density gradually increases toward the bottom of the centrifuge tube. The sample is layered on the low-density top of this gradient. As the centrifugation begins, each compound of the sample starts to sediment. By doing so, the compounds travel through layers of increasing density. As soon as a compound reaches the layer where the density equals its own density, the compound stops sedimenting. At this layer, no resultant force is exerted on the particle and thus it will float. As a result, equilibrium density-gradient centrifugation separates compounds from each other independently of their size, solely by their density, in a single run.

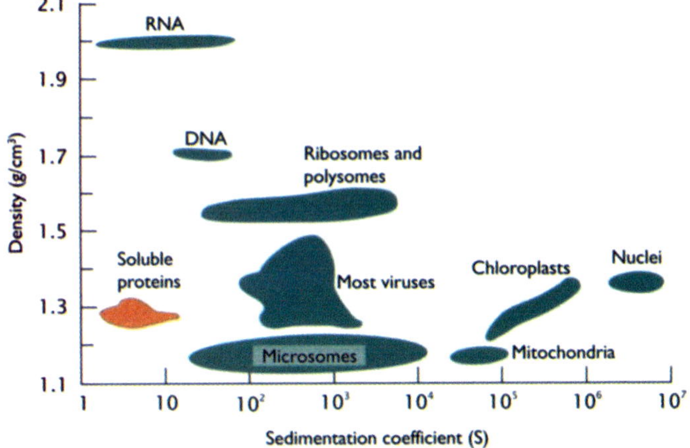

Combination of differential centrifugation and density-gradient centrifugation. Differential centrifugation separates compounds primarily based on their size, while density-gradient centrifugation separates compounds exclusively based on their density. Compounds that have different density but sediment in the same fraction during differential centrifugation can be separated by a subsequent step of density-gradient centrifugation. Two such consecutive steps of the two centrifugation methods can provide significantly higher separation efficiency than either procedure alone.

Fractionation methods based on solubility

The solubility of a protein is a measure of the maximal quantity of the given protein that a unit volume of solvent can keep in solution. Solubility is often expressed in units of mg/ml. Solubility depends largely on the composition of the solvent. Before going into details about the influence of solvent composition, it is important to clarify whether solubility refers to native or denatured proteins. A popular misconception based on oversimplification is that native proteins are always soluble while denatured proteins always precipitate from the solution. The following section focuses on the solubility of native proteins.

pH-dependence of solubility

The solubility of native proteins depends primarily on the pH and ionic strength of the solvent. It is widely known that proteins have the lowest solubility when the pH equals their pI value. At such pH value, the number of positive charges on the protein equals that of negative charges.

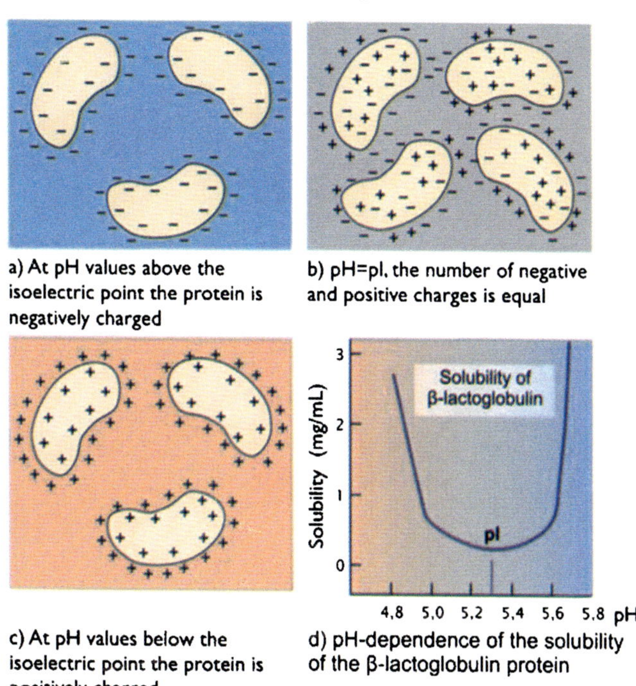

a) At pH values above the isoelectric point the protein is negatively charged

b) pH=pI, the number of negative and positive charges is equal

c) At pH values below the isoelectric point the protein is positively charged

d) pH-dependence of the solubility of the β-lactoglobulin protein

pH dependence of protein solubility. The solubility of proteins is at its minimum when the pH of the solution equals the pI value of the protein. At this pH, the number of the positive and negative charges of the protein is identical. As a consequence, the number of potential attracting intermolecular ionic interactions is at a maximum, which facilitates aggregation of the protein molecules. At pH values below or above the pI, the protein molecules have a net charge and so they tend to repel each other.

Consequently, in this state the level of ionic repulsion between the protein molecules is minimal, while the number of possible intermolecular attracting ionic interactions is at its maximum. Naturally, a minimum in solubility does not mean that all molecules of a given protein precipitate when the pH equals their pI value. When the concentration of the given protein is below the solubility value, all of its molecules still remain in solution. When the protein concentration exceeds the solubility value, a dynamic equilibrium exists between the precipitated and the soluble set of molecules. Note that both the soluble as well as the insoluble subset contains native proteins precipitation is not linked to denaturation. If one aims to precipitate native proteins, it is advisable to perform the experiment at a pH that equals the pI of the protein, since proteins are the most stable at pH values equal to their pI‘s.

Dependence of protein solubility on ionic strength

Solubility of proteins in distilled (deionised) water is significantly lower than in solutions containing ions. Accordingly, when salt is added to an ion-free protein sample in which the protein is dissolved above its solubility, the originally precipitated insoluble fraction becomes soluble. This phenomenon is called ‘salting-in‘. Apparently, ions added at a small concentration increase the solubility of proteins. As it has been discussed in the previous paragraph on pH-dependence, the solubility of proteins is lowered by attracting intermolecular ionic interactions between opposite surface charges of the protein molecules. Ions incorporated by the solvent act as counter ions and efficiently shield the intermolecular attraction between proteins. This effect readily explains the salting-in phenomenon.

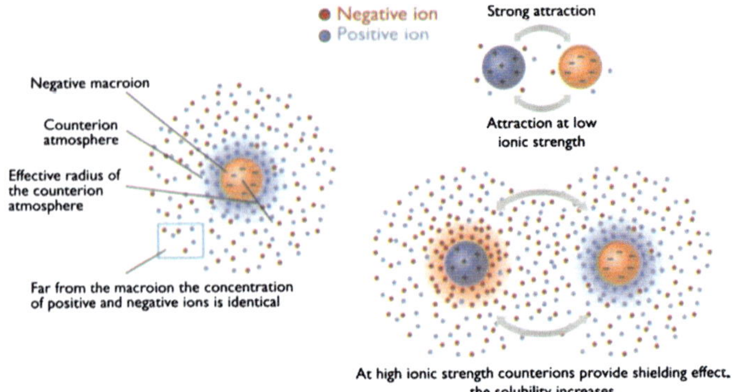

The salting-in‘ effect. The solubility of proteins is usually low in deionised water, but it increases with increasing ionic strength of the solution, up to a certain point. As the figure illustrates, dissolved salt (ions) facilitates solvation of proteins by shielding the exposed charges of protein molecules. This shielding effect hinders intermolecular attracting interactions between opposite charges of individual protein molecules and, by doing so, it also hinders protein aggregation.

When the concentration of ions is increased above a critical level, the solubility of proteins gradually decreases. Most importantly, the critical ion concentration level is characteristic of the identity of the protein. Accordingly, at high protein concentration, if the ionic strength of the solvent is increased, the excess of the proteins above the solubility level will precipitate while preserving their native state. This phenomenon is called salting-out. The salting-out effect can be explained as follows. When ions (salt) are added to aqueous solutions, ions become solubilised such that a multilayer water boundary, a hydration shell is formed around them. In chemically pure water, the concentration of water molecules is ~ 55.6 M. When the concentration of ions exceeds several mol/litre, a very high percentage of the originally free water molecules will become engaged in the hydration shell of ions. Formation of more and more new hydration shells around ions competes for water molecules hydrating proteins. As the hydration shells around proteins decrease intermolecular attracting forces, mostly apolar interactions between protein molecules take over leading to precipitation of native protein molecules. Generally, various salts of univalent positive ions are used for this purpose, with the most common salt being ammonium sulphate. As the solubility of different proteins depends differently on salt concentration, the salting-out phenomenon can be applied as a protein fractionation method. Let us suppose that we have a protein mixture and the protein to be isolated has a medium sensitivity to ammonium sulphate precipitation. When very slowly and gradually, *e.g.* by adding with mild stirring a saturated ammonium sulphate solution ammonium sulphate is introduced in the solution, many proteins precipitate. The ammonium sulphate concentration can be increased up to a level where the target protein remains in the solution while many other, more sensitive proteins precipitate. At this point, the precipitated proteins can be removed simply by centrifugation. (Clumps of precipitated proteins represent an enormous particle size and a very high corresponding Svedberg value. Sedimentation, therefore, at a very high rate). The supernatant will contain our target protein and many other proteins. The ammonium sulphate concentration can then be further increased to a level where the target protein and some other proteins precipitate, while many other proteins still remain in solution. After a second centrifugation step the supernatant can be disposed of and the pellet containing the target protein can be collected. The solution of the resolubilised pellet will contain a protein mixture in which the target protein will be enriched compared to the starting sample. Then the high concentration of ammonium sulphate can be decreased by dialysis or size exclusion chromatography.

Irreversible precipitation methods

As already mentioned, the relationship between solubility and native/denatured state of proteins is much more complex than usually considered. Native globular proteins have most of their apolar residues buried in the hydrophobic core of the molecule. Nevertheless, the dominantly polar surface of the protein still contains hydrophobic patches of apolar residues. The size and distribution of these hydrophobic spots widely varies with individual proteins. Upon unfolding, all apolar residues become solvent-accessible. Water molecules form a highly structured, low-entropy clathrate structure around apolar residues. The free enthalpy level of the thermodynamic system can decrease if the amount of the clathrate structure decreases. This indeed happens, because this way the entropy of the system increases. The amount of the clathrate structures can decrease if the apolar residues form short-range, weak, apolar-apolar interactions with each other. These interactions are weaker than the apolar-polar interactions between the apolar residues and the water molecules, but the newly formed polar-polar interactions between water molecules partially balance this loss of interaction energy. As already mentioned, the breakdown of the clathrate structures has a major contribution to lowering the free energy of the system. As a result, the apolar residues are expelled from water and interact with each other rather than with water molecules. This phenomenon is denoted as the hydrophobic effect. Due to the hydrophobic effect, denatured globular proteins tend to precipitate from solution. However, it might be confusing that many popular denaturing agents such as urea or sodium dodecyl sulphate (SDS) are highly effective solubilisers of denatured proteins. When urea or SDS is used as denaturing agents, denaturation is not accompanied by precipitation. Denaturation can be triggered by increasing the temperature or by applying extreme pH values. If the target protein is unusually resistant against certain denaturing effects (*e.g.* it is thermostable, acid-stable *etc.*), this property can be a good starting point for establishing an effective fractionation protocol. After high-temperature or extreme-pH treatment, many proteins unfold and precipitate, while the target protein (and, usually, several other resistant proteins) remains folded and stays in the solution. After a centrifugation step, the target protein will be in the supernatant and can be easily separated from the pellet of denatured "contaminating" proteins.

9.4. Protein Fractionation Based on Particle Size

Several low-resolution fractionation methods separate molecules from each other based on their size. The two methods reviewed below apply semipermeable membranes to separate the sample in two fractions by size. Semipermeable membranes are special "sieves" having pores with diameters in the size range of molecules. Molecules that are smaller than the pores of the membrane can permeate the membranes while molecules larger than the pores are retained. In the course of *dialysis*, the sample is poured in a tube made of a semipermeable membrane. The two ends of the tube are then sealed, and the resulting sachet is immersed into a container containing an appropriate buffer. All molecules that are smaller than the pores of the membrane start to diffuse across the membrane in both directions. This process leads to an equilibrium. At equilibrium, the concentration of the small molecules on the two sides of the membrane becomes identical, while large molecules remain in the dialysis sachet. The time required for reaching the equilibrium state can be shortened by stirring the buffer in the container. Dialysis is usually applied in order to remove small molecules from the sample and/or to replace the buffer around the retained large molecules (usually proteins) in the sample. The equilibrium concentration of the small molecules can be calculated based on their original concentrations inside and outside the sachet and the ratio of the volumes of the sachet and the container. The larger the container, the lower the concentration of the small molecules to be removed will be in the sachet. However, instead of increasing the volume of the container to an impractical volume, a better strategy is to apply a relatively small container and change the buffer in the container several times. This way, the concentration of the small molecule in the sample will decrease exponentially. At the end of the previously mentioned salting-out experiment, the protein sample will contain high concentration of ammonium and sulphate ions. These ions are typically removed by dialysis.

Semipermeable membranes are applied in the course of *ultrafiltration*, too. However, in this case the sample solution is forced to flow through the membrane. This comes with two advantages. First, the process is faster than dialysis. The second, more important difference is that by applying an external driving force, the method does not rely on equilibrium. This way, the concentration of the molecules that are retained by the membrane can be increased tremendously. This method is often applied to concentrate protein samples. Furthermore, the solvent of the large molecules can also be easily changed by this method by adding the appropriate buffer to the sample between several consecutive ultrafiltration steps. The external force can be generated either by centrifugation of the sample in Centricon vials, or by a high-pressure inert gas (usually nitrogen) in Amicon systems.

Lyophilisation (freeze-drying)

Protein (or any other non-volatile molecule) samples can be concentrated by evaporating water and other volatile compounds from the sample. In principle, this could be achieved by simply heating the sample. However, most proteins would unfold during such a simple evaporation process. In order to prevent the denaturation of proteins, the sample is transferred into a glass container and is frozen as quickly as possible, usually by immersing the outside of the container into liquid nitrogen. Moreover, the container is rotated in order to spread and freeze the sample on a large surface area. The glass container with the sample is then placed into an extremely low-pressure space (vacuum) that contains a cooling coil as well. The cooling coil acts as a condenser. The temperature of the coil is usually lower than -50°C. Volatile compounds of the frozen sample will evaporate (sublimate) in the vacuum. The process of evaporation (in this case, sublimation) absorbs heat. This effect keeps the sample frozen. Evaporated molecules are captured from the gas phase by the cooling coil, forming a frozen layer on it. At the end of the process, proteins and other non-volatile compounds of the sample remain in the container in a solid form. This process does not cause irreversible denaturation of proteins. Thus, it is a method frequently used not only to concentrate but also to preserve proteins or other sensitive biomolecules for long-term storage. Such samples can usually be stored for years without a significant deterioration of quality. However, before lyophilisation it is very important to carefully consider all non-volatile compounds of the initial sample as these will concentrate along with the proteins. Non-volatile acids or bases can cause extreme pH, and the presence of salts can result in very high ionic strength when the sample is resolubilised.

Chromatographic methods

Chromatography is the collective term for a set of separation techniques that operate based on the differential partitioning of mixture components between a mobile and a stationary phase. The mobile phase (a liquid or a gas) travels through the stationary phase (a liquid or a solid) in a defined direction. The distribution of components between the two phases depends on adsorption, ionic interactions, diffusion, solubility or, in the case of affinity chromatography, specific interactions. Depending on the experimental design, the separation in a liquid mobile phase may be carried out via column or planar chromatography, on analytical or preparative scales. Chromatographic methods are important in the analytical and preparative separation of biological samples. Gel filtration chromatography (size exclusion chromatography) is often the method of choice to purify macromolecules, taking advantage of their different sizes and shapes. Ion exchange chromatography is also useful for the separation of macromolecules, operating based on the various net charges on their surface, which can be tuned via the pH of the medium. Biological specificity in enzyme-substrate, enzyme-inhibitor, receptor-ligand, antigen-antibody (and other) interactions is utilised in affinity chromatography. In this method, one interaction partner is immobilised on a solid surface (stationary phase) and can selectively bind its interacting partner from a mixture in the mobile phase. The other components of the mixture can then be removed by replacing the mobile phase (washing). The pure material is then eluted by applying a mobile phase that disrupts the specific interaction. Of the quasi-infinite possibilities of analytical applications of liquid chromatography, only a few relevant ones will be mentioned in this chapter. The analysis of amino acid mixtures and that of the products of Edman degradation in the process of amino acid sequencing are both carried out using chromatography. Chromatographic methods coupled to on-line mass spectrometry are instrumental in current proteomics research.

Quantification of separation

(1) Resolution (RS) Resolution is a number describing the separation of chromatographic peaks. By definition, resolution is the distance between peak maxima (elution volumes) divided by the average of peak widths. The elution volumes and the peak widths are to be measured in identical units. RS is therefore a dimensionless number. In the case of a constant flow rate, the quantification of volumes can be replaced with the more convenient measurement of time. The RS value defines the extent of separation. The larger the RS between two peaks, the more ideal the separation.

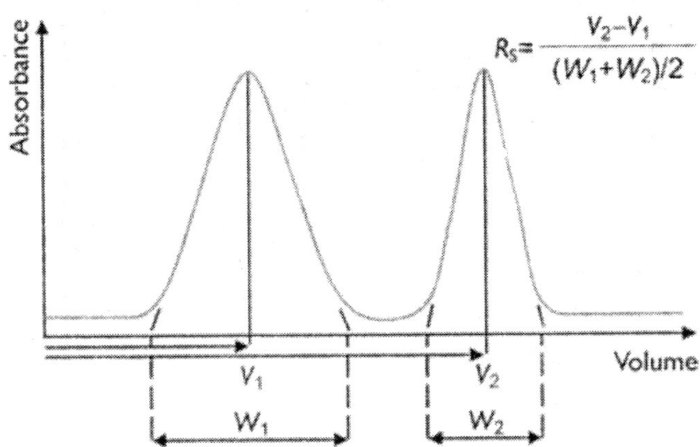

Determination of the resolution. RS = resolution; V1 = elution volume at the maximum of the first peak; W1 = base width of the first peak; V2 = elution volume at the maximum of the second peak; W2 = base width of the second peak. In the case of a constant flow rate, it is more convenient to measure time instead of elution volumes. (In this case, time will be directly proportional to the elution volume.) Thus, RS = 2(t2 – t1 / W2 + W1), where t2 and t1 are elution times corresponding to V2 and V1, respectively.

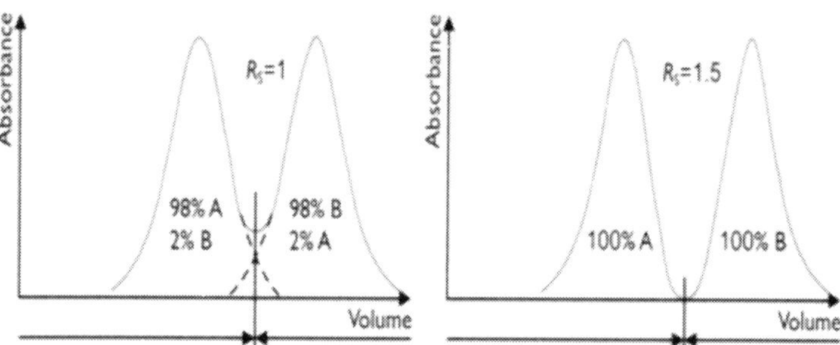

Relation between resolution and separation. It can be proven by calculation that, if RS = 1, and both peaks have ideal shapes (*i.e.* Gauss curves) and identical sizes, then the two components can be isolated at 98 % purity. Perfect, so-called base-line separation can be achieved in cases where RS > 1.5.

Gel filtration chromatography

The chromatographic medium for gel filtration is a hydrophilic gel made up from porous, fine-grain spheres of 10-300 ìm diameter. This type of medium defines two solution compartments within the column: one is the freely moving mobile phase outside the gel particles, while the other is the restricted liquid compartment inside the porous particles.

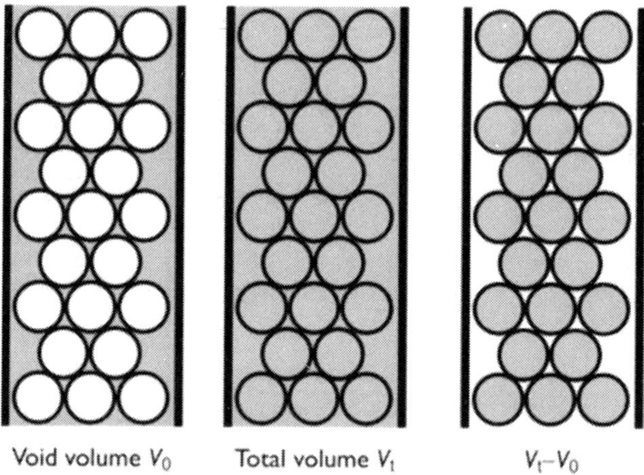

Liquid compartments inside a column packed with a porous gel. The individual liquid compartments are depicted as dark areas. V0 = exclusion volume (void volume); Vt = total volume of the column; Vt – V0 = combined volume of the liquid inside the gel particles and the material of the gel.

When a solution is moving through the gel filtration column, the movement of the solutes depends on two factors: the flow rate of the mobile phase and diffusion. Diffusion enables the molecules to explore the inside of the gel particles if their size so permits. The separation of a molecular mixture is based on the phenomenon that some molecules are excluded from the inside of the gel particles due to their size. These molecules travel quickly in the mobile phase of the column, which is the only compartment available to them. Smaller molecules, on the other hand, spend various amounts of time inside the particles (stationary phase) and flow through the column slower.

Travel of variably-sized molecules through a porous gel. During gel filtration (size exclusion) chromatography, molecules of different sizes will explore the available liquid spaces via diffusion. The largest molecules, due to their size, cannot enter the pores of the matrix at all. The movement of molecules with medium sizes is confined to the larger pores. The smallest molecules can enter the gel particles through all pores. Therefore, the largest molecules will advance most rapidly through the column, whereas the smaller molecules will be retarded.

The result of a gel filtration experiment is usually depicted as an elution diagram. In this diagram, the concentration of the eluted compound is plotted against the volume of the eluent. The appearance of a given compound occurs at its elution volume (Ve). As in other distribution chromatographic methods, the elution of a compound is best characterised by its distribution coefficient (Kd):

$$Kd = (Ve - Vo) / Vs$$

where Vo equals the exclusion volume, *i.e.* the elution volume of a molecule that is larger than the largest pore size of the separating gel. Such a molecule therefore explores only the mobile phase, and is entirely excluded from the gel. Vs equals the volume of the stationary phase, *i.e.* the volume of the liquid inside the gel particles that is fully accessible only to molecules small enough to travel smoothly even through the smallest pores of the gel. Vs itself is difficult to determine. Therefore, in practice, it is replaced by the Vt-Vo term, also accounting for the non- negligible volume of the gel itself. As a result, a constant pertinent to an apparent volume (Kav) is used instead of Kd (the latter would be valid only for real liquid volumes):

$$Kav = (Ve - Vo) / (Vt - Vo)$$

where Kav represents the portion of the gel volume that is accessible to a molecule of a given size. For a totally excluded macromolecule, Kav = 0; whereas, for small molecules diffusing freely in the entire volume of the gel, Kav = 1.

Ion exchange chromatography

Ion exchange chromatography is one of the most efficient methods for the separation of charged particles. Ion exchange chromatography is most often performed in the form of column chromatography. However, there are also thin-layer chromatographic methods that work basically based on the principle of ion exchange. Column materials used for ion exchange chromatography contain charged groups covalently linked to the surface of an insoluble matrix. When suspended in an aqueous solution, the charged groups of the matrix will be surrounded by ions of the opposite charge. In this "ion cloud", ions can be reversibly exchanged without changing the nature and the properties of the matrix. The charged groups of the matrix can be positively or negatively charged. A positively charged matrix will bind negatively charged ions from the solution. Therefore, it is called an anion exchanger. Cation exchanger matrices have negative charges. Based on the structure of the ion exchange matrix, we distinguish ion exchangers with hydrophobic and hydrophilic matrices. Ion exchangers with a hydrophobic matrix are most often highly substituted polystyrene resins. These are suitable for the binding of inorganic ions, *e.g.* in water softening applications. However, they tend to denature proteins due to the high hydrophobicity of their matrix and their high surface charge density. Ion exchangers with hydrophilic matrices were first produced from modified cellulose. However, cellulose has disadvantageous mechanical properties: cellulose fibres are prone to break, making it difficult to create a well-utilisable column. This disadvantage has been partially remedied in Sephadex (dextran-based) ion exchange matrices. In recent years, regular spherical and monodisperse matrices have been produced from synthetic hydrophilic polymers. The best known of such resins is the MonoBead-based ion exchange matrix. Table summarises the charged groups linked to ion exchange matrices.

Anion Exchanger	**Functional Group**
diethyl-aminoethyl (DEAE)	$-OCH_2CH_2N^+H(CH_2CH_3)^2$
quaternary aminoethyl (QAE)	$OCH_2CH_2N^+(C_2H_5)^2CH_2CH(OH)CH_3$
quaternary ammonium (Q)	$-CH_2N^+(CH_3)^3$
Cation exchangers	**Functional Group**
carboxymethyl (CM)	$-OCH_2COO^-$
sulfopropyl (SP)	$-CH_2CH_2CH_2SO_3^-$
methylsulfonate (S)	$-CH_2SO_3^-$

Ion exchangers containing sulfonyl and quaternary ammonium groups are called strong ion exchangers. These are practically completely charged between pH 3.0 and 11.0. The degree of dissociation of DEAE and CM groups—and thus their ion exchange capacity—depends on the pH of the medium. Most ion exchange experiments comprise five different phases.

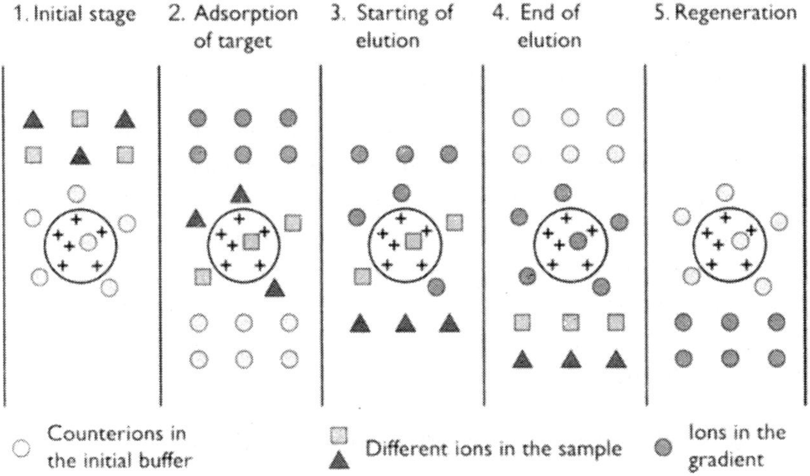

Phases of ion exchange chromatography (salt gradient elution). In the figure, a positively charged anion exchanger particle is shown, with counterions on its surface in the starting state (first phase). In the second phase, the binding of the ions to be separated takes place. At the start of the elution (third phase), weaker-binding ions are desorbed. At the end of the desorption, the stronger-binding ions are also desorbed (fourth phase). During regeneration (fifth phase), the starting state can be reconstituted via washing the column with the starting buffer.

The first phase is the equilibration of the ion exchange column with a so-called starting buffer, setting the conditions of the experiment (pH and ionic strength). In this phase, the charged groups of the ion exchanger will bind easily replaceable simple ions (*e.g.* chloride or sodium). The second phase is the loading of the sample and its reversible binding to the column. If some of the contaminating materials do not bind to the column, these can be removed via washing the column with the starting buffer. The third and fourth phases comprise the elution, *i.e.* the desorption of the bound molecules, which can be achieved via changing the composition of the elution buffer. The simplest form of elution is achieved via an increase in ionic strength, *i.e.* in the concentration of the counterions present. Another means of desorption is the change of the pH of the medium. The most effective method is the continuous change of the ionic strength or the pH, *i.e.* the application of a gradient elution. During gradient elution, molecules with smaller net charges (*i.e.* the weaker-binding ones) will be the first to leave the column. An important property of an ion exchange column medium is its *ion exchange capacity*. This parameter reflects the amount of counterions that can be bound to the column. Three types of capacity can be distinguished:

Total capacity: The number of charged groups per gram dry weight of the ion exchanger or per millilitre of swollen gel. This can be determined by titration with a strong acid or base.

Free capacity: Due to steric reasons, only a part of the full capacity is accessible for macromolecules (proteins, nucleic acids). This is the free capacity.

Dynamic capacity: The so-called dynamic capacity is determined when the binding of the given macromolecule to the column is measured during buffer flow. The free and the dynamic capacity values are dependent on the properties of the material to be separated, the properties of the ion exchanger, and the applied experimental conditions. With regard to separation, the important properties of the material to be separated are the size of the molecules and the pH dependence of their charge. This implies that the capacity of ion exchangers will de different for different proteins.

Frequently used Sephadex-based ion exchangers

Via the modification of the Sephadex G-25 and G-50 gel filtration matrices with four different groups, eight different ion exchangers were created). Matrices derived from G-25 (A-25 and C-25) contain more crosslinks. Therefore, they are more rigid and swell to a smaller extent than the ones derived from G-50. With the use of specially-treated polysaccharide and synthetic polymer matrices that are more pressure-resistant than Sephadex, it was possible to develop fine-grained regular spherical polymer beads for ion exchangers with significantly increased efficiency. Such matrices include monodisperse MonoBeads with a bead size of 10 nm, or MiniBeads with a 3-μm bead size. Such ion exchangers are used in FPLC techniques.

9.5. Important Parameters for Ion Exchange-based Separation

(1) *The charge of components present in the sample to be separated:* In the case of proteins, this will depend on the isoelectric point. Below the isoelectric point, proteins are cations; above the isoelectric point, they are anions. The applied chromatographic buffer should be chosen in a way that the proteins to be separated should bind to the given anion or cation exchange column, from which they can be eluted after the washout of the non-binding components. More rarely, it can occur that the unwanted contaminants are bound to the column, whereas the component to be isolated will freely flow through.

(2) *The amount of the sample:* The size of the column should be chosen so that the dynamic capacity of the medium should somewhat exceed the amount of the sample. If the column is too small, the sample will saturate it, and part of the sample will not bind to the column. If the column is too large, a significant loss can occur during elution.

(3) *The molecular mass of the protein to be isolated:* The pore size of the ion exchange matrix should be chosen in a way that gel filtration effects—*i.e.* size-based separation—do not occur during ion exchange.

Hydrophobic interaction chromatography

The method of hydrophobic interaction chromatography (HIC) is based on the observation that protein molecules can interact with fully hydrophobic adsorbents, and this interaction is dependent on the salt concentration of the solution. Similarly to the salting-out of proteins where the increasing salt concentration will lead to the aggregation and precipitation of protein molecules via the rearrangement of their hydrate shell, during HIC chromatography a high salt concentration (1-1.5 M neutral salt) facilitates the interaction between the hydrophobic chromatographic medium and the hydrophobic patches present on protein molecules. During separation, the decrease in salt concentration will lead to the elution of bound molecules. Based on these observations, bead polymers were created that are suitable for HIC chromatographic purposes. The surface of the beads was modified by hydrophobic alkyl or aryl groups. Such media include the polysaccharide-based Butyl-Sepharose, Octyl-Sepharose and Phenyl-Sepharose materials that are derived from polymers that had proven to be suitable for chromatographic separation of proteins.

Here we draw attention to the fact that the so-called reverse-phase chromatography is in principle very similar to HIC chromatography. In both cases, the separation is based on the strength of interaction forming between hydrophobic surfaces. The main difference is that, in the case of media used in HIC chromatography, the concentration of the hydrophobic ligand bound to the solid matrix is 10-20 mol/ml of column volume, whereas in RPC, the hydrophobic ligand concentration on the matrix surface is several 100 μmol/ml. Therefore, in the case of reverse-phase chromatography, the binding between the adsorbent and the molecules to be separated is very strong. Thus, for RPC elution, it is necessary to use solvents that are less polar than water (methanol, acetonitrile, *etc.*). Therefore, the isolation of native proteins is not always possible with RPC, the proteins can denature on the column as the amino acids forming their hydrophobic core can also strongly bind to the hydrophobic matrix. However, due to its high resolution, the RPC technique is very advantageous for the qualitative analysis of complex protein or peptide mixtures. During HIC, we can work in an aqueous, non-denaturing medium throughout the chromatographic procedure, from loading to elution.

Factors affecting HIC chromatography

(1) *The type and density of the hydrophobic ligand on the matrix surface:* The immobilised hydrophobic ligands will primarily determine the selectivity of the HIC adsorbent. Alkyl chains show purely hydrophobic interactions. In the case of aromatic ligands, beyond the hydrophobic effect, specific interactions between aromatic groups are also present. The choice between alkyl-containing versus aromatic-liganded matrices is mainly empirical and can be made based on preliminary binding experiments. In the case of alkyl matrices, the protein-binding capacity of the HIC adsorbent will increase with the length of alkyl chains. In addition to alkyl chain length, the binding capacity will, of course, also depend on the concentration of these chains on the matrix surface. In the case of HIC, the optimal hydrophobic ligand concentration is around 20 mol/ml of medium. The behaviour of the medium is, to some extent, also affected by the hydrophobicity of the polymer matrix. For instance, even in the case of the same hydrophobic ligands, the selectivity will differ in the case of agarose and synthetic polymer matrices.

(2) *The quality of the salt used and its concentration:* The effect of salts used in HIC is similar to their salting-out efficiency. Both effects are associated with the effect of the given salt on the surface tension of water, which can be arranged in the following order:

$$\mathbf{LiCl < NaCl < Na_2HPO_3 < (NH_4)_2SO_4 < Na_2SO_4}$$

Most often, $(NH_4)_2SO_4$ is used, as the salting-out efficiency of this salt is about four times that of NaCl. The initial salt concentration should be selected so that the protein to be isolated should bind to the column with sufficient efficiency. The determination of the salt concentration is empirical—a good approach may be testing the use of a 1M $(NH_4)_2SO_4$ solution.

One must also ensure that, at the applied initial salt concentration, protein precipitation does not occur due to salting-out.

(3) *The effect of pH:* The pH of the medium changes the charge of ionisable groups of protein molecules. This effect will obviously affect the separation based on hydrophobic interactions. It is found that, in general, the retention of protein samples changes dramatically below pH 5 and above pH 8.5. Thus, the states close to the zwitterionic state are advantageous for HIC. Therefore, similarly to the salt concentration, the pH of the solvent should also be optimised.

(4) *The effect of temperature:* The hydrophobic interaction is temperature dependent. Increasing temperature will increase the strength of the interaction in most cases, but this phenomenon is complex—the opposite effect has also been observed. In practical terms, it must be taken into account that the outcome of a method developed at room temperature may not be reproducible in the cold room (at 4°C).

(5) *The effect of additives:* Water-miscible organic solvents (alcohols, acetonitrile, dimethyl formamide, *etc.*) or added detergents can reduce the binding of proteins to be separated, even when present at low concentrations. These compounds "compete" with the protein for the adsorption sites on the matrix surface. Therefore, when added in low concentrations, these substances can increase the efficiency of the elution. They can be thus applied *e.g.* when the decrease in salt concentration alone does not lead to satisfactory results. One must ensure, however, that the additives used do not denature the protein to be isolated. Additives can be used even at high concentrations for the purification or regeneration of HIC columns.

Affinity chromatography

By affinity chromatography, high-selectivity separation of biomolecules can be achieved through their specific interactions. This separation technique is special because it is based on the biological function or the unique chemical structure of a given biomolecule. During affininty chromatography, the interacting partner of the biomolecule is immobilised on a chromatographic resin. This ligand, fixed to the stationary phase, reversibly binds the desired biomolecule present in the multi-component mobile phase. The materials can be eluted from the column by changing the composition of the mobile phase.

The technique provides high selectivity, high resolution and generally high capacity for the desired protein. The degree of purification can be thousands of times, and the achieved yield can also be usually very good. Affinity chromatography, as already mentioned, is unique in the sense that it is based on the specific biological function of the biomolecule of interest. This feature also makes affinity chromatography suitable for the selective separation of active biomolecules, and their isolation from the inactive or denatured forms. Another significant advantage of the method is that, in many cases, it allows for single-step isolation of the desired biomolecule. However, it is required that the sample to be separated should be a clear solution free of large particles. It is often advisable to prepare the sample for affinity chromatography via an initial partial separation. For instance, in the case of affinity isolation of very scarce components of the blood serum, it is advisable to perform an initial separation to eliminate serum albumin (which makes up more than 50 % of the serum protein content). Affinity chromatographic purification is frequently of great importance in the case of recombinant proteins. Recombinant proteins are often produced in a way that they contain a fused label at their N- or C- terminus, resulting from genetic engineering. This way, if the label endows the protein to enter into affinity binding, the recombinant protein can be simply fished out of the cell extract via affinity chromatography One of the most widely used of such labels fused to protein termini is the oligo-histidine tag (His-tag), which binds reversibly to metal chelates (*e.g.* Ni chelate immobilised on the stationary phase). Another frequently applied tag is glutathione S-transferase (GST), a fusion protein that can be used to isolate the protein of interest using a glutathione-conjugate matrix. These specific affinity matrices are commercially available as pre-packed columns. In other cases, the specific ligand is to be linked by the user to the chromatographic matrix. Various activated reactive chromatographic matrices are available for this purpose. Some of the commonly used interactions in affinity chromatography are listed in Table below:

Enzyme	**Substrate analogue or inhibitor**
Antibody	Antigen (virus, cells)
Nuclic acid	Complementary nuclic acid
Nuclic acid	Histone or other nuclic acid binding protein
Hormone	Hormone receptor
Glutathion	Glutathion S-transferase (GST) fusion protein
Metal chelate	His-tag fusion protein

The phases of affinity chromatography are shown in figure.

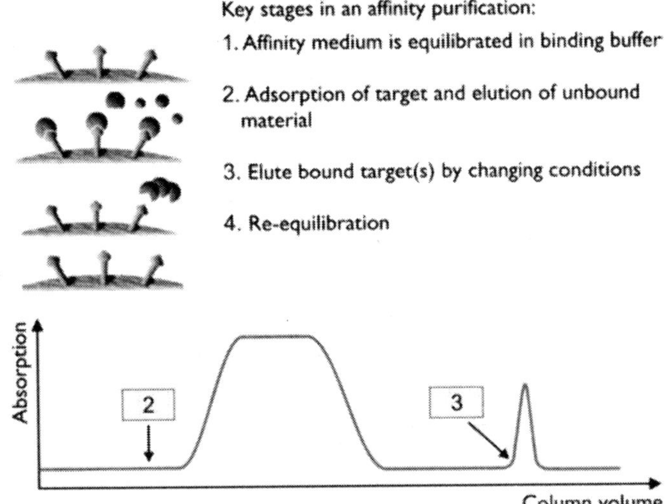

The phases of affinity chromatography. In the upper left side of the figure, an illustrative representation of affinity chromatography is shown. Of the molecules present in the sample, only the ones having a matching binding site can bind to the matrix-conjugated specific ligand molecules. Other molecules can be readily washed off. By changing the composition of the mobile phase, the molecules of interest can be isolated in a pure form. The lower part of the figure shows a typical affinity chromatographic elution profile.

Sample preparation: The sample must be a clear solution free from solid particles. This can be achieved by centrifugation or filtration. Protein solutions should be centrifuged at at least 10000 g. Cell lysates should be centrifuged at 40-50000 g. A 0.45-μm pore size filter can be used for filtration. (Such preparation of samples is also necessary in FPLC and HPLC methods.) One must also consider how the solubility and stability of the sample or the desired protein can be influenced by the pH, the salt concentration, or the presence of any organic solvent. The factors affecting the interactions between the desired target protein and the matrix-bound ligand (pH, salt concentration, temperature) should also be determined. The composition of the initial binding buffer must be adjusted accordingly. Sample components interfering with the target protein and/or the ligand (*e.g.* metabolites in cell lysates) should be removed before loading onto the column.

Equilibration with a buffer facilitating the specific interaction: The chromatographic column is washed with 3-4 column volumes of the starting (binding) buffer. The sample must also be equilibrated with this starting binding buffer (if necessary, via solvent exchange or dialysis).

Binding of the molecule of interest and wash-out of the unbound material During sample loading, consider the strength of the interaction. In case of high-affinity samples, a high flow rate may be applied. In case of a weak interaction and/or a slow equilibration process, reduce the rate of sample loading. After sample application, the column should be further washed with binding buffer until all unbound components are removed.

Elution of the molecules of interest by changing the composition of the mobile phase Elution via pH and/or ionic strength changes: One possible and simple means of elution is achieved through decreasing the interaction strength between the ligand and the target protein. Changes in the pH will change the ionisation state of charged groups of the ligand and/or the target protein, thereby changing the strength of the interaction. Similarly, increasing the ionic strength (usually by raising the NaCl concentration) will generally reduce the interaction strength. In either case, the solubility and stability of the target protein should be considered.

Competitive elution: For competitive elution, materials are applied that react with the target protein or the ligand, competing for the pre-existing interaction. For instance, His-Tag fusion proteins can be readily displaced from the metal chelate matrix by imidazole buffer. GST-tagged target proteins will detach from their column-conjugated glutathione ligand upon mixing excess glutathione into the elution buffer.

In all cases, the flow rate of the buffer should be reduced during elution, thereby avoiding excessive dilution of the target protein. In cases when the target-ligand interaction is very strong, the above elution methods may turn out insufficient for eluting the protein of interest. In these cases, chaotropic agents (urea, guanidine) can be used to wash off the target protein from the column. This naturally will involve the denaturation of the protein, which can then be renatured in some (lucky) cases under suitable conditions—in the case uf urea or guanidine there is a good chance for this.

Regeneration After successful completion of the elution, the column can be washed with several column volumes of binding buffer, and it can then be reused. For long-term storage, one must ensure that the column is not exposed to bacterial or fungal infection. The toxic compound sodium azide can be used to prevent such infections.

Specific protein detection methods: Western blot

If, instead of detecting all proteins, the goal is the highly selective detection of a certain protein, two different methods are available. One method called Western blot can be used if a highly selective antibody is available against the protein of interest. The principle of the procedure is as follows. In the first step, proteins are separated based on their size by using SDS-PAGE.

Subsequently, a nitrocellulose membrane is laid tightly on one side of the gel. This sandwich is placed into a special electrophoresis apparatus. By performing electrophoresis in a direction perpendicular to the plane of the gel, the proteins will be transferred from the gel onto the membrane. This step preserves the relative spatial arrangement of individual protein bands. The membrane carrying the tightly bound proteins is then incubated in a solution containing the specific antibody against the target protein, usually called the primary antibody. The primary antibody will bind only to the target protein. To visualise the sites where the antibody has bound to the membrane, a secondary antibody is added that had been conjugated (covalently bound) to an enzyme. The secondary antibody recognises the constant region of the primary antibody. Therefore, a given secondary antibody conjugate can be used with many primary antibodies. After this step, the membrane is soaked in a solution containing a substrate of the conjugated enzyme. Catalytic conversion of the substrate yields a colourful insoluble product that precipitates at the site where it was generated and, ultimately, labels the site of the target protein.

Specific protein detection methods: In-gel method based on enzyme activity

When electrophoresis is performed in native conditions, many enzymes retain their catalytic activity in the gel. There are robust and highly selective in-gel activity detection methods for several types of enzymes (dehydrogenases, ATPases, proteases *etc.*). For such specific detection, the gel is soaked after electrophoresis in the solution of the proper substrate. Usually, synthetic substrates are used that have a common built-in property: upon being converted by the enzyme, they generate a colourful and insoluble product. As a consequence, the end product stays in the gel very near to the enzyme and colours the spot where the enzyme is located.

Zymography, a special variant of activity detection methods, has been introduced for the detection of proteinases. While in the above example a native gel was run, in the case of zymography usually a modified SDS-PAGE is performed. When the resolving gel is prepared, some kind of large molecular weight protein (casein, denatured collagen *etc.*) is mixed into the buffer. This protein is so large that it practically does not migrate in the gel. The sample to be loaded onto the gel is SDS-treated, but no reducing agent is added and the sample is not treated at high temperature. Unlike in the case of normal SDS-PAGE, the goal here is to denature the proteins reversibly. After the proteins are separated by gel electrophoresis, the gel is soaked in a non-ionic detergent solution in order to completely remove SDS. As the SDS is removed, at least some percentage of the proteases (and other proteins) can regain their native conformation. The refolded proteases start to digest the protein substrate in the gel. The products of the proteolytic degradation are small peptides that diffuse out of the gel. After an appropriate time period, the gel is stained by one of the general methods. The gel will be intensely stained in most parts except where it contained proteinases. Around the proteinase molecules the staining will be lighter. This is a kind of negative staining approach in which the specific signal is actually the lack of signal.

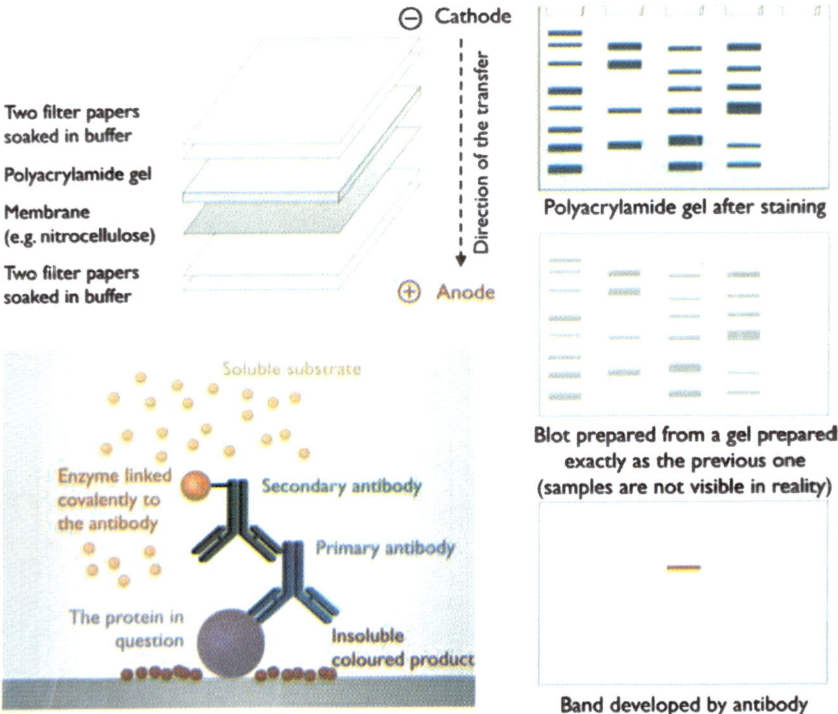

The scheme of Western blot. After (usually SDS-) PAGE separation, proteins are transferred to a membrane (usually made of nitrocellulose) by a second electrophoresis performed in a direction perpendicular to the plane of the gel. The membrane tightly binds the proteins. The target protein can be specifically detected among hundreds or thousands of other indifferent proteins by a method analogous to the sandwich ELISA assay. A critical reagent for this detection method is a highly selective antibody raised against the target protein. This antibody finds and binds specifically to the target protein even in the presence of a large excess of unrelated proteins. Visualisation of the location of the bound antibodies is usually achieved via enzymatic methods. In principle, the enzyme could be covalently linked (conjugated) to the target-recognising, so-called primary antibody. However, in this case, such antibody-enzyme conjugates would need to be prepared for each new target protein. A more straightforward approach has been the use of a secondary antibody that recognises the constant region of many different primary antibodies. In the course of the enzyme reaction, a soluble substrate is enzymatically converted into a coloured insoluble product that precipitates around the target protein.

9.6. Polyacrylamide-sodium dodecyl Sulphate Electrophoresis (SDS-PAGE) of Proteins

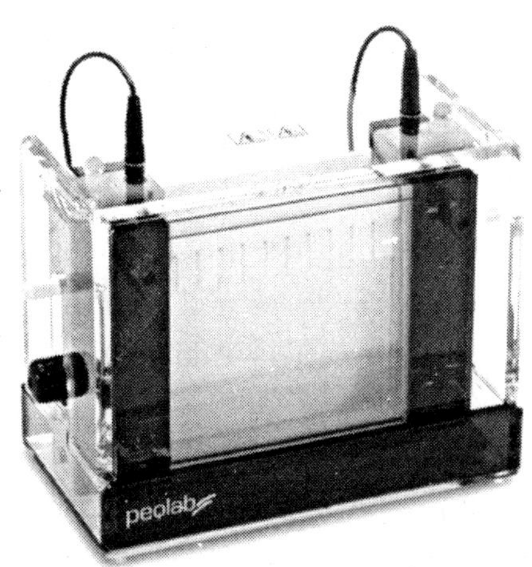

Electrophoresis is widely used to separate and characteize protcins by applying electric current. Electrophoresis procedures are rapid and relatively sensetive requiring only micro-weight of proteins. Electrophoresis in the polyacrylamide gel is more convenient than in any other medium such as paper and starh gel. Electrophoresis of proteins in polyacrylamide gel is acrried out in buffer gels (non-denaturing) as well as in sodium dodecyl sulphate (SDS) containing (denaturing) gels. Seperation in buffer gels relies on both the charges and size of the proteins whereas it depends only upon the size in SDS-gels. Analysis and comparison of proteins in a large number of samples is easily made on polyacrylamide gel slabs.

Polyacrylamide gels are formed by polymerising acralymide with a cross-linking agents (bisacrylamide) in the presence of a catalyst (persulphate ions) and chain initiator (TEMED; N,N,N',N',- tetramethylethylene diamines). Solutions are normally degassed by evacuation prior to polymerisation since oxygen inhibits polymerisation. The porosity of the gel is determined by the relative proportion of acrylamide monomer to bisacrylamide. Gels are usually referred to in terms of the total percentage of acrylamide and bis pesent, and most proteins separations are performed using gels in the range 7 – 15%. A low percentage gel (with large pore size) is used to separate high molecular weight proteins and *vice-versa.* At high concentration of persulphate and TEMED the rat of polymerisation is also high. Among a number of methods commonly used, the sodium dodecyl sulphate-polyacrylamide gel electrophoresis(SDS-PAGE) in slabs (facilitating characterization of polypeptides abd detrmination of their molecular weight by co-electrophoresis) is a common practice. It is carried as described belo:

SDS is an anioninc detergent which binds strongly to, and denature proteins. The number of SDS molecules bound to a polypeptide chain is approximately half the number of amino acid residues in that chain. The protein-SDS complex carries net negative charges, hence move towards anode and the separation is based on the size of the protein.

Materials

- Stock acrylamide solution – Acrylamide 30 g, bisacrylamide (0.8%) 0.8 g and water 100 ml.
- Separating gel buffer – 1.875 M Tris-HCl 22.7 g and water 100.0 ml (pH 8.8) pH 8.8.
- Stacking Gel buffer – 0.6 M Tris-HCl (7.26 g in 100 ml of water) pH 6.8.
- Polymerising agents – a). Ammonium persulhate 5% (0.5 g/10 ml) prepare freshly before use. b). TEMED fresh from refrigerator
- Electrode buffer – Add 12.0 g of Tris (0.05 M) to 28.8 g of glycine (0.192 M) and 2 g of SDS (1%) to 2000 ml of water, pH 8.2 - 8.4.
- Sample buffer – to 5000 ml of water add 5.0 ml of Tris-HCl buffer (pH 6.8), 0.5 g of SDS, 5.0 g of sucrose, 0.25 ml of mercaptoethnol, 1.0 ml of 0.5% (W/V) bromophenol blue solution. Make the final volume to 10 L with water.
- Sodium dodecyl sulphate 10% solution.
- Standard marker protein – Dissolve the following proteins in single strength sample buffer at a concentration each of 1 mg/ml. Load the well with 25-50μ.

Protein	*MW Dalton*
Á-Lactalbumin	14,200
Trypsin inhibitor soybean	20,100
Trypsinogen	24,000
Carbonic anhydrase	29,000
Gluceraldehyde-3-phosphare dehydrogenase, dabbit	36,000
Albumin, egg	45,000
Albumin bovine	66,000

- Protein stain solution – 0.1 g of Coomassic brilliant blue (R 250) in a mixture of 40.0 ml methanol + 10.0 ml of acetic acid + 50.0 ml of water.

Procedure

- With a thoroughly clean and dry glass plate and spacer all assembling are done and is secured by the clips. White petroleum jelly or 2% agar (melted in a boiling water bath) is then applied around the edges of the spacer.
- Sufficient quantity of separating gel mixture (30 ml for a chamber of 18 x 9 x 0.1 cm) by mixing the following:

	For 15% gel	For 10% gel
Stock acrylamide solution	20.0 ml	13.3 ml
Tris-HCl buffer pH 8.8	8.0 ml	8.0 ml
Water	11.4 ml	18.1 ml
Degas on a water pump for 3 – 5 minutes and then add		
Ammonium persulphate solution	0.2 ml	0.2 ml
10% SDS	0.4 ml	0.4 ml
TEMED	20 µl	20 µl

- Mix gently and carefully, pour the gel solution in the chamber between glass plates, and leave to set for 30-60 minutes.
- Prepare stacking gel (4%) by mixing the following solutions in a total volume of 10 ml
    - Stock acrylamide solution 1.35 ml
    - Tris-HCl buffer (pH6.8) 1.00 ml
    - Water 7.50 ml
    - Degas on a water pump for 3 – 5 minutes and then add
    - Ammonium persulphate solution (5%) 50.0 µl
    - 10% SDS 0.1 ml
    - TEMED 10.0 µl
- Remove the water from top of the gel and wash with little stacking gel solution. Pour stacking gel mixture, place the comb in the stacking gel and allow the gel to set for 30 – 60 minutes.
- Remove the comb after the stacking gel has set and carefully install the gel after removing the clips, agar *etc.* in the main apparatus. Fill the electrode buffer and remove any air bubble by gentle tapping. Connect the cathode at the top and turn on the DC-power supply to check electric circuit.
- Prepare sample for electrophoresis, following suitable extraction procedure. Adjust the protein concentration in each sample, using buffer and water, so that same amount of protein is present per unit of volume (50 – 200 µg) in a volume of (20 – 50 µl).
- Cool the sample solution and load appropriate volume of the sample in sample well through sample buffer using micro pipette.

- Mark the position of sample with a marker pen.
- Similarly load a few well with standard marker protein in the sample buffer.
- Turn the current to 10 – 15 mA for initial 10- 15 minutes then continue to run at 30 mA until the bromophenol blue reached at the bottom of the gel (about three hours).
- After run is complete, carefully remove the gel from between the plates and immerse in staining solution for at least 3 hours or over night with uniform shaking.
- Transfer the gel to a suitable container with at least 200 – 300 ml de-staining solution and shake gently and continuously. Dye which is bound to protein is thus removed. Change the de-strainer frequently, particularly during initial period.
- The protein fractioned into bands are seen as blue coloured bands. Gels can now be photographed or stored in polyethylene bags or dried using vacuum pump for a permanent record.

9.7. Estimation of Crude Protein Content

For many years, the protein content of foods has been determined on the basis of total nitrogen content, while the Kjeldahl (or similar) method has been almost universally applied to determine nitrogen content. Nitrogen content is then multiplied by a factor to arrive at protein content. This approach is based on two assumptions: that dietary carbohydrates and fats do not contain nitrogen, and that nearly all of the nitrogen in the diet is present as amino acids in proteins. On the basis of early determinations, the average nitrogen (N) content of proteins was found to be about 16 percent, which led to use of the calculation N x 6.25 (1/0.16 = 6.25) to convert nitrogen content into protein content.

This use of a single factor, 6.25, is confounded by two considerations. First, not all nitrogen in foods is found in proteins: it is also contained in variable quantities of other compounds, such as free amino acids, nucleotides, creatine and choline, where it is referred to as non-protein nitrogen (NPN). Only a small part of NPN is available for the synthesis of (non-essential) amino acids. Second, the nitrogen content of specific amino acids (as a percentage of weight) varies according to the molecular weight of the amino acid and the number of nitrogen atoms it contains (from one to four, depending on the amino acid in question). Based on these facts, and the different amino acid compositions of various proteins, the nitrogen content of proteins actually varies from about 13 to 19 percent. This would equate to nitrogen conversion factors ranging from 5.26 (1/0.19) to 7.69 (1/0.13).

In response to these considerations it was suggested that N x 6.25 be abandoned and replaced by N x a factor specific for the food in question. These specific factors, have been widely adopted. Factors for the most commonly eaten foods range from 5.18 (nuts, seeds) to 6.38 (milk). It turns out, however, that most foods with a high proportion of nitrogen as NPN contain relatively small amounts of total N. As a result, the range of Jones factors for major sources of protein in the diet is narrower. Jones factors for animal proteins such as meat, milk and eggs are between 6.25 and 6.38; those for the vegetable proteins that supply substantial quantities of protein in cereal/legume-based diets are generally in the range of 5.7 to 6.25. Use of the high-end factor (6.38) relative to 6.25 increases apparent protein content by 2 percent. Use of a specific factor of 5.7 rather than the general factor of 6.25 decreases the apparent protein content by 9 percent for specific foods. In practical terms, the range of differences between the general factor of 6.25 and Jones factors is narrower than it at first appears (about 1 percent), especially for mixed diets. Table gives examples of the Jones factors for a selection of foods.

Table: Specific factors for the conversion of nitrogen content to protein content (selected foods)

Food	Factor	Food	Factor	Food	Factor
Animal Origin					
Eggs	6.25	Meat	6.25	Milk	6.38
Vegetable Origine					
Barley	5.83	Corn	6.25	Millets	5.83
Oats	5.83	Rice	5.95	Rye	5.83
Sorghum	6.25	Wheat whole kernel	5.83	Bran	6.31
Endosperm	5.70	Beans: Castor	5.30	Jack, Lima, Navy, Mung	6.25
Soybean	5.71	Velvet bean	6.25	Peanuts	5.46

Because proteins are made up of chains of amino acids joined by peptide bonds, they can be hydrolysed to their component amino acids, which can then be measured by ion-exchange, gas-liquid or high-performance liquid chromatography. The sum of the amino acids then represents the protein content (by weight) of the food. This is sometimes referred to as a "true protein". The advantage of this approach is that it requires no assumptions about, or knowledge of, either the NPN content of the food or the relative proportions of specific amino acids - thus removing the two problems with the use of total N x a conversion factor. Its disadvantage is that it requires more sophisticated equipment than the Kjeldahl method, and thus may be beyond the capacity of many laboratories, especially those that carry out only intermittent analyses. In addition, experience with the method is important; some amino acids (*e.g.* the sulphur-containing amino acids and tryptophan) are more difficult to determine than others. Despite the complexities of amino acid analysis, in general there has been reasonably good agreement among laboratories and methods.

Recommendations

- It is recommended that protein in foods be measured as the sum of individual amino acid residues (the molecular weight of each amino acid less the molecular weight of water) plus free amino acids, whenever possible. This recommendation is made with the knowledge that there is no official Association of Analytical Communities (AOAC) method for amino acid determination in foods. Clearly, a standardized method, support for collaborative research and scientific consensus are needed in order to bring this about.
- Related to the previous recommendation, food composition tables should reflect protein by sum of amino acids, whenever possible. Increasingly, amino acid determinations can be expected to become more widely available owing to greater capabilities within government laboratories and larger businesses in developed countries, and to the availability of external contract laboratories that are able to carry out amino acid analysis of foods at a reasonable cost for developing countries and smaller businesses.
- To facilitate the broader use of amino acid-based values for protein by developing countries and small businesses that may lack resources, FAO and other agencies are urged to support food analysis and to disseminate updated food tables whose values for protein are based on amino acid analyses.
- When data on amino acids analyses are not available, determination of protein based on total N content by Kjeldahl or similar method x a factor is considered acceptable.
- A specific Jones factor for nitrogen content of the food being analysed should be used to convert nitrogen to protein when the specific factor is known. When the specific factor is not known, N x the general factor 6.25 should be used. Use of the general factor for individual foods that are major sources of protein in the diet introduces an error in protein content that is relative to the specific factors and ranges from -2 percent to +9 percent. Because protein contributes an average of about 15 percent of energy in most diets, the use of N x 6.25 should introduce errors of no more than about 1 percent in estimations of energy content from protein in most diets ([-2 to +9 percent] x 15).

Micro-Kjeldahl method

Total nitrogen in the samples may be estimated by conventional Micro-Kjeldahl's method. Multiply nitrogen content with suitable factor to calculate crude protein content. Generally 16% nitrogen is present in proteins. So, 100/16 = 6.25 factor may be used to convert nitrogen content to protein. If nitrogen content in protein of particular crop is more, value of factor will be low. So, factor for conversion of nitrogen to crude protein varies from crop to crop. Nitrogen present in proteins or other organic compounds reacts with sulphuric acid during digestion and form ammonium sulphate. Ammonium sulphate reacts with NaOH during distillation and releases ammonia, which may be trapped in boric acid or dilute sulphuric acid. Thus nitrogen content may be quantified by titrating it with standard acid or alkali.

Materials

- Concentrated sulphuric acid
- Mixed indicator
  - a. Bromocresol green (0.1% in 95% alcohol)
  - b. Methyl red (0.1% in 95% alcohol)
  - c. Mix 10 ml of (a) with 2 ml of (b)
- Boric acid solution (2%)
- Sodium hydroxide solution (40%)
- Digestion mixture - Mix thoroughly 2.5 g of powdered selenium dioxide (SeO_2), 100 g potassium sulphate (K_2SO_4) and 20g of copper sulphate ($CuSO_4.5H_2O$) or mix $CuSO_4$ and K_2SO_4 in 1:10 ratio. Copper sulphate act as catalyst and potassium sulphate increases boiling point of sulphuric acid.
- Standard hydrochloric acid solution (0.01N)

Procedure

Digestion

- Weigh exactly 0.1 to 1.0 g of dry homogenized sample into clean and dry digestion flask taking care to see that no sample/particles adhere to the sides of the flask.
- Add a pinch (approximately 0.5g) of digestion mixture, a dumping chip and 10.0 ml of concentrated sulphuric acid.

- Heat the flask on a burner/suitable electric heater in a fume hood. Initially the sample should be heated on a low flame/heat to prevent foaming/frothing. Gradually increase the heat to keep the contents boiling. Continue heating until the solution is colorless/very light green.
- Cool and add distilled water, shake carefully since the reaction is exothermic and after attaining room temperature carefully transfer the contents into 50 to 100 ml volumetric flask. Wash the digestion flask several times with small portion of distilled water and transfer the washings into the same volumetric flask. Shake carefully make up the volume and shake again for uniform concentration.

Distillation and titration

- Set up the distillation apparatus.
- Take 10 ml of 2% boric acid in a 100 ml conical flask. Add 2-3 drops of mixed indicator (colour should be pink) and place the flask under the condenser taking care to see the end of the condenser is dipping in boric acid solution.
- Pipette out 5.0 to 10 ml of digested material into the distilling portion (Vacuum jacket) of the set followed by 10.0 ml of distilled water and 10.0 ml 40% NaOH. Immediately stop the mouth of apparatus.
- Steam distills the mixture by opening the steam inlet and keeping the other outlets closed except the receiving end of the condensor. The liberated ammonia is collected in the boric acid which results in the colour change of the solution from pink to blue. Distil for 5-6 min or until the volume in the conical flask increases by 2 to 2.5 folds (30-40 ml).
- Close the steam inlet, wash down the tip with a few ml of distilled water, remove the conical flask containing the distillate and titrate against standard 0.01N HCl until the colour changes to light pink. Note the titre value.
- Run a blank preparation which has been identically prepared except that it does not contain the sample.

Calculation

1000 ml of 1N HCl = 14g N

Or 1 ml of 0.01N HCl = 0.00014 mg N

Volume of 0.01N HCl used for blank = v ml

Volume of 0.01N HCl used for sample = y ml

Titre volume of sample = (y-v) ml

Volume of sample taken for distillation = 5 ml

Total volume of the digested sample = 50 ml

N present in 5 ml of digested sample = (y-v) x 0.00014 g

N present in 50 ml of digested sample = (y-v) x 0.00014 x 50 / 5 g

N present in 1 g sample = (y-v) x 0.00014 x 50 / 5 g

N present in 100 g sample / per cent N = (y-v) x 0.00014 x 50 x 100 / 5 g

Calculate crude protein content by multiplying the nitrogen content with a suitable factor as mentioned below.

Crop	Factor
Wheat (whole), Rye, Barley, Oats	5.83
Wheat flour	5.70
Wheat bran	6.31
Rice	5.95
Groundnut	5.46
Soybean	5.71
Sesame, Safflower and Sunflower	5.30
Milk and Cheese	6.38
Other foods	6.25

Colorimetric estimation of total nitrogen

Material

- Digestion mixture - 6 g of salicylic acid + 82 ml of conc. H_2SO_4 + 18 ml of water.
- Hydrogen peroxide 30% (v/v).
- Reagent A – 100 ml of freshly prepared alkaline phenolate (22 g of phenol dissolved in 25 ml of 1.0 N NaOH and volume is made to 100 ml with water) is mixed with 200 ml of freshly prepared sodium nitropruside (0.05%) and 10 ml of 4% EDTA solution. The reagent is prepared immediately before use.
- Reagent B – 400 ml of sodium phosphate buffer (5.34 g of $Na_2HPO_4.2H_2O$ dissolved in 20 ml of 1.0 N MaOH and volume made to 400 ml) is mixed with 100 ml of (1:3 diluted) sodium hypochlorite.

Procedure

- Approximately 100 mg of sample is digested with 3.3 ml of acid mixture and with the help of H_2O_2.
- After digestion, the volume is raised to 100 ml with distilled water. Volume was further diluted by 10 times and is used as sample solution.
- To 1.0 ml of sample solution 3.0 ml of reagent A and 5.0 ml of reagent B is mixed and colour is allowed to develop for 90 minutes.
- Optical density is measured at 630 nm against reagent blank.
- A standard curve for nitrogen is made with $(NH_4)_2SO_4$ (ammonium sulphate) dissolved in 0.7 M H_2SO_4.

Reading

Novozamosky, T., van Eck, J.R., van Schouwenhuve, J. Ch and Wallingan, I. (1974). Total nitrogen determination in plant materials by means of indophenols blue method. *Neth. J. Agric. Sci.*, **22:** 3-5.

Protein Estimation by Lowry's Method

The Lowry protein assay is a biochemical assay for determining the total level of protein in a solution. The total protein concentration is exhibited by a colour change of the sample solution in proportion to protein concentration, which can then be measured using colorimetric techniques. It is named for the biochemist Oliver H. Lowry who developed the reagent in the 1940s. His 1951 paper describing the technique is the most-highly cited paper ever in the scientific literature, cited over 200,000 times.

The method combines the reactions of copper ions with the peptide bonds under alkaline conditions (the Biuret test) with the oxidation of aromatic protein residues. The Lowry method is best used with protein concentrations of 0.01–1.0 mg/ml and is based on the reaction of Cu^+, produced by the oxidation of peptide bonds, with Folin-Ciocalteu reagent (a mixture of phosphotungstic acid and phosphomolybdic acid in the Folin–Ciocalteu reaction). The reaction mechanism is not well understood, but involves reduction of the Folin reagent and oxidation of aromatic residues (mainly tryptophan, also tyrosine). Experiments have shown that cysteine is also reactive to the reagent. Therefore, cysteine residues in protein probably also contribute to the absorbance seen in the Lowry Assay. The concentration of the reduced Folin reagent is measured by absorbance at 750 nm. As a result, the total concentration of protein in the sample can be deduced from the concentration of Trp and Tyr residues that reduce the Folin reagent.

Materials

- Alkaline sodium carbonate solution – It is prepared by diisolving 2.0 g of anhydrous sodium carbonate in 100 ml of 0.1 N NaOH solution.
- Alkaline copper reagent – (a) it is prepared by dissolving 0.5 g of pentahydrate copper sulphate in 1.0% solution of sodium potassium tartarate. Mixing of the two solution is done fresh before use. (b) 1.0 ml of copper reagent is mixed with 50.0 ml of alkaline sodium carbonate solution.
- Follin-Ciocalteau reagent – This is obtained commercially and it contain sodium tungstate, sodium molybdate in phosphoric and hydrochloric acid. This is diluted to 1:2 with distilled water.
- Protein solution (stock solution) – 50.0 mg of bovine serum albumin (Fraction V) is dissolved in distilled and volume is made to 50.0 ml.
- Working standard – 10.0 ml of stock solution of protein in diluted to 50.0 ml with water and this solution contain 200 μg/ml.

Procedure

- Extraction of soluble proteins from green leaves or germinating seeds is usally carried out with 0.2 M phosphate buffer pH 7.5, centrifuged and supernatant is used as protein source.
- To 0.1 ml of protein extract, 1.0 alkaline copper reagent is mixed and kept in room temperature for 10 minutes and thereafter 0.1 ml of Folin-Ciocalteau reagent is added and colour is allowed to develop for 30 minutes.
- After 30 minutes, the final volume is made to 5.0 ml with water and optical density is recorded at 750 nm against reagent blank where 0.1 ml of protein extract is replaced with water.
- A gradientof soluble protein (20 – 200 μg/ml.) is taken from the working stock solution for protein and colour is developed in accordance to the methods described above. Standard graph is obtained byplotting the optical densities against respective protein concentrations.
- Soluble protein content of the sample is calculated after referring the standard curve and is expressed as mg of prtein/g fresh weight.

Notes

- For complete enzyme extraction, sometime the chemicals like ethylenediamine tetraacetic acid (EDTA), magnesium salts and mercaptoethanol are included. This method of protein estimation should not be followed if the extractant contains K^+, Mg^{++}, Tris, EDTA and thiol (mercaptoethanol) compounds as they interfere with this procedure. When these chemicals are present in the extract, protein is precipitated by the addition of 10% TCA and the precipateted protein is centrifuged out and redissolved in 2.0 N NaOH for protein estimation.
- If the protein concentration is very high (above 500 μg/ml) optical density is measured ar 550 nm.
- Follin-Ciocalteau reagent can be prepared in the laboratory and for this reflux gently for 10 hours a mixture having 100 g sodium tungstate ($Na_2WoO_4.2H_2O$), 25 g of sodium molybdate ($Na_2MoO_4.2H_2O$), 700 ml water, 50.0 ml of 85% phosphoric acid and 100.0 ml of concentrated hydrochloric acid. Lithium sulphide 150.0 g, water 50.0 ml and few drops of bromine water is added after reflux. Boil the mixture for 15 minutes without condenser to remove excess of bromine. Cool, dilute to 1000.0 ml and filter. The reagent should have no green tint.
- If protein estimation is desired in a sample with phenolic or pigment content, extract should be prepared with a reducing agent preferably cystein and NaCl. Precipitate the protein with 10% TCA, separate the protein and redissolve in 2.0 N NaOH.

Protein estimation with Bradford method

The Bradford assay, a colorimetric protein assay, is based on an absorbance shift of the dye Coomassoe Brilliant Blue G-250 in which under acidic conditions the red form of the dye is converted into its bluer form to bind to the protein being assayed. During the formation of this complex, two types of bond interaction take place: the red form of Coomassie dye first donates its free electron to the ionizable groups on the protein, which causes a disruption of the protein's native state, consequently exposing its hydrophobic pockets. These pockets in the protein's tertiary structure bind non-covalently to the non-polar region of the dye via van der Walls forces, positioning the positive amine groups in proximity with the negative charge of the dye. The bond is further strengthened by the ionic interaction between the two. The binding of the protein stabilizes the blue form of the Coomassie dye; thus the amount of the complex present in solution is a measure for the protein concentration, and can be estimated by use of an absorbance reading.

The (bound) form of the dye has an absorption spectrum maximum historically held to be at 595 nm. The cationic (unbound) forms are green or red. The binding of the dye to the protein stabilizes the blue anionic form. The increase of absorbance at 595 nm is proportional to the amount of bound dye, and thus to the amount (concentration) of protein present in the sample.

Unlike other protein assays, the Bradford protein assay is less susceptible to interference by various chemicals that may be present in protein samples. An exception of note is elevated concentrations of detergent. Sodium dodecyl sulphate (SDS), a common detergent, may be found in protein extracts because it is used to lyse cells by disrupting the membrane lipid bilayer. While other detergents interfere with the assay at high concentration, the interference caused by SDS is of two different modes, and each occurs at a different concentration. When SDS concentrations are below critical micelle concentration (known as CMC, 0.00333% w/v to 0.0667%) in a Coomassie dye solution, the detergent tends to bind strongly with the protein, inhibiting the protein binding sites for the dye reagent. This can cause underestimations of protein concentration in solution. When SDS concentrations are above CMC, the detergent associates strongly with the green form of the Coomassie dye, causing the equilibrium to shift, thereby producing more of the blue form. This causes an increase in the absorbance at 595 nm independent of protein presence.

Other interference may come from the buffer used when preparing the protein sample. A high concentration of buffer will cause an overestimated protein concentration due to depletion of free protons from the solution by conjugate base from the buffer. This will not be a problem if a low concentration of protein (subsequently the buffer) is used.

Materials

- Dye concentrate – 100 g of Coomassie brilliant blue G 250 is dissolved in 95% ethanol. Then 100 ml of concentrated (ortho) phosphoric acid is mixed and final volume is made to 200 ml with water. Store in refrigerator in amber coloured bottle.
- Mix water with dye (1:4) before use.
- Phosphate buffered saline.

Procedure (Standard Assay, 20-150 µg protein; 200-1500 µg/ml)

- Prepare a series of protein standards diluted with 0.15 M NaCl to final concentrations of 0 (blank = NaCl only), 250, 500, 750 and 1500 µg/ml. Also prepare serial dilutions of the unknown sample to be measured.

- Add 100 μl of each of the above to a separate test tube (or spectrophotometer tube if using a Spec 20).
- Add 5.0 ml of Coomassie Blue to each tube and mix by vortex, or inversion.
- Adjust the spectrophotometer to a wavelength of 595 nm, and blank using the tube which contains no protein.
- Wait 5 minutes and read each of the standards and each of the samples at 595 nm wavelength.
- Plot the absorbance of the standards vs. their concentration. Compute the extinction coefficient and calculate the concentrations of the unknown samples.

Procedure (Micro Assay, 1-10 μg protein/ml)

- Prepare standard concentrations of protein of 1, 5, 7.5 and 10 μg/ml. Prepare a blank of NaCl only. Prepare a series of sample dilutions.
- Add 100 μL of each of the above to separate tubes (use microcentrifuge tubes) and add 1.0 ml of Coomassie Blue to each tube.
- Turn on and adjust a spectrophotometer to a wavelength of 595 nm, and blank the spectrophotometer using 1.5 ml cuvettes.
- Wait 2 minutes and read the absorbance of each standard and sample at 595 nm.
- Plot the absorbance of the standards vs. their concentration. Compute the extinction coefficient and calculate the concentrations of the unknown samples.

9.8. Enzymes

Enzymes are large biological molecules responsible for the thousands of metabolic processes that sustain life. They are highly selective catalyst, greatly accelerating both the rate and specificity of metabolic reactions, from the digestion of food to the synthesis of DNA. Most enzymes are proteins, although some catalytic RNA molecules have been identified. Enzymes adopt a specific three-dimensional structure, and may employ organic (*e.g.* biotin) and inorganic (*e.g.* magnesium ion) cofactors to assist in catalysis. In enzymatic reactions, the molecules at the beginning of the process, called substrates, are converted into different molecules, called products. Almost all chemical reactions in a biological cell need enzymes in order to occur at rates sufficient for life. Since enzymes are selective for their substrates and speed up only a few reactions from among many possibilities, the set of enzymes made in a cell determines which metabolic pathways occur in that cell.

Like all catalysts, enzymes work by lowering the activation energy ($E_a^{\ddagger}$) for a reaction, thus dramatically increasing the rate of the reaction. As a result, products are formed faster and reactions reach their equilibrium state more rapidly. Most enzyme reaction rates are millions of times faster than those of comparable un-catalyzed reactions. As with all catalysts, enzymes are not consumed by the reactions they catalyze, nor do they alter the equilibrium of these reactions. However, enzymes do differ from most other catalysts in that they are highly specific for their substrates. Enzymes are known to catalyze about 4,000 biochemical reactions. A few RNA molecules called ribozymes also catalyze reactions, with an important example being some parts of the ribosome. Synthetic molecules called artificial enzymes also display enzyme-like catalysis. Enzyme activity can be affected by other molecules. Inhibitors are molecules that decrease enzyme activity; activators are molecules that increase activity. Many drugs and poisons are enzyme inhibitors. Activity is also affected by temperature, pressure, chemical environment (*e.g.*, pH), and the concentration of substrate. Some enzymes are used commercially, for example, in the synthesis of antibiotics. In addition, some household products use enzymes to speed up biochemical reactions (*e.g.*, enzymes in biological washing powders break down protein or fat stains on clothes; enzymes in meat tenderizers break down proteins into smaller molecules, making the meat easier to chew).

Some enzymes do not need any additional components to show full activity. However, others require non-protein molecules called cofactors to be bound for activity. Cofactors can be either inorganic (*e.g.*, metal ions and iron-sulphur clusters) or organic compounds (*e.g.*, flavin and heme). Organic cofactors can be either prosthetic groups, which are tightly bound to an enzyme, or coenzymes, which are released from the enzyme's active site during the reaction. Coenzymes include NADH, NADPH and adenosine triphosphate. These molecules transfer chemical groups between enzymes. An example of an enzyme that contains a cofactor is carbonic anhydrase. These tightly bound molecules are usually found in the active site and are involved in catalysis. For example, flavin and heme cofactors are often involved in redox reactions.

Enzymes that require a cofactor but do not have one bound are called *apoenzymes* or *apoproteins*. An apoenzyme together with its cofactor(s) is called a *holoenzyme* (this is the active form). Most cofactors are not covalently attached to an enzyme, but are very tightly bound. However, organic prosthetic groups can be covalently bound (*e.g.*, biotin in the enzyme pyruvate carboxylase).

The term "holoenzyme" can also be applied to enzymes that contain multiple protein subunits, such as the DNA polymerases; here the holoenzyme is the complete complex containing all the subunits needed for activity.

Coenzymes are small organic molecules that can be loosely or tightly bound to an enzyme. Tightly bound coenzymes can be called prosthetic groups. Coenzymes transport chemical groups from one enzyme to another. Some of these chemicals such as riboflavin, thiamin and folic acid are vitamins (compounds that cannot be synthesized by the body and must be acquired from the diet). The chemical groups carried include the hydried ion (H^-) carried by NAD or $NADP^+$, the phosphate group carried by adenosine triphosphate, the acetyl group carried by coenzyme A, formyl, methenyl or methyl groups carried by folic acid and the methyl group carried by S-adenosyl methionine.

Since coenzymes are chemically changed as a consequence of enzyme action, it is useful to consider coenzymes to be a special class of substrates, or second substrates, which are common to many different enzymes. For example, about 700 enzymes are known to use the coenzyme NADH. Coenzymes are usually continuously regenerated and their concentrations maintained at a steady level inside the cell: for example, NADPH is regenerated through the pentose phosphate pathway and *S*-adenosylmethionine by methionine adenosyl transferase. This continuous regeneration means that even small amounts of coenzymes are used very intensively. For example, the human body turns over its own weight in ATP each day. Enzyme reaction rates can be decreased by various types of enzyme inhibitors.

Competitive inhibition: In competitive inhibition, the inhibitor and substrate compete for the enzyme (*i.e.*, they can not bind at the same time). Often competitive inhibitors strongly resemble the real substrate of the enzyme. For example, methotrexate is a competitive inhibitor of the enzyme dihydrofolaye reductase, which catalyzes the reduction of dihydrofolae to tetrahydro-folate. In some cases, the inhibitor can bind to a site other than the binding-site of the usual substrate and exert an allosteric effect to change the shape of the usual binding-site. For example, strychnine acts as an allosteric inhibitor of the glycine receptor in the mammalian spinal cord and brain stem. Glycine is a major post-synaptic inhibitory neurotransmitter with a specific receptor site. Strychnine binds to an alternate site that reduces the affinity of the glycine receptor for glycine, resulting in convulsions due to lessened inhibition by the glycine. In competitive inhibition the maximal rate of the reaction is not changed, but higher substrate concentrations are required to reach a given maximum rate, increasing the apparent K_m.

Uncompetitive inhibition: In uncompetitive inhibition, the inhibitor cannot bind to the free enzyme, only to the ES-complex. The EIS-complex thus formed is enzymatically inactive. This type of inhibition is rare, but may occur in multimeric enzymes.

Non-competitive inhibition: Non-competitive inhibitors can bind to the enzyme at the binding site at the same time as the substrate, but not to the active site. Both the EI and EIS complexes are enzymatically inactive. Because the inhibitor can not be driven from the enzyme by higher substrate concentration (in contrast to competitive inhibition), the apparent V_{max} changes. But because the substrate can still bind to the enzyme, the K_m stays the same.

Mixed inhibition: This type of inhibition resembles the non-competitive, except that the EIS-complex has residual enzymatic activity. This type of inhibitor does not follow Michaelis-Menten equation.

In many organisms, inhibitors may act as part of a feedback mechanism. If an enzyme produces too much of one substance in the organism, that substance may act as an inhibitor for the enzyme at the beginning of the pathway that produces it, causing production of the substance to slow down or stop when there is sufficient amount. This is a form of negative feedback. Enzymes that are subject to this form of regulation are often multimeric and have allosteric binding sites for regulatory substances. Their substrate/velocity plots are not hyperbolar, but sigmoidal (S-shaped). Irreversible inhibitors react with the enzyme and form a covalent adduct with the protein. The inactivation is irreversible. These compounds include eflornithine a drug used to treat the parasitic disease sleeping sickness. Penicillin and Aspirin also act in this manner. With these drugs, the compound is bound in the active site and the enzyme then converts the inhibitor into an activated form that reacts irreversibly with one or more amino acid residues.

9.9. Enzyme Commission Nomenclature

The Enzyme Commission number (EC number) is a numerical classification scheme for enzymes, based on the chemical reactions they catalyse. As a system of enzyme nomenclature, every EC number is associated with a recommended name for the respective enzyme. Strictly speaking, EC numbers do not specify enzymes, but enzyme-catalyzed reactions. If different enzymes (for instance from different organisms) catalyze the same reaction, then they receive the same EC number. Furthermore, through convergent evolution, completely different protein folds can catalyze an identical reaction and therefore would be assigned an identical EC number (these are called non-homologous isofunctional enzymes, or NISE). By contrast, UniProt identifiers uniquely specify a protein by its amino acid sequence.

Group	Reaction catalysed	Typical reaction	Enzyme example with trivial name
EC 1 Oxido-reductase	To catalyze oxidation/reduction reactions; transfer of H and O atoms or electrons from one substance to another	$AH + B \rightarrow A + BH$ reduced) $A + O \rightarrow AO$ (oxidized)	Dehydrogenase, oxidase
EC 2 Transferase	Transfer of a functional group from one substance to another. The group may be methyl-, acyl-, amino- or phosphate group	$AB + C \rightarrow A + BC$	Transaminase, kinase
EC 3 Hydrolase	Formation of two products from a substrate by hydrolysis	$AB + H_2O \rightarrow AOH + BH$	Lipase, amylase, peptidases
EC 4 Lyases	Non-hydrolytic addition or removal of groups from substrates. C-C, C-N, C-O or C-S bonds may be cleaved	$RCOCOOH \rightarrow RCOH + CO_2$ or $[X\text{-}A\text{-}B\text{-}Y] \rightarrow [A\text{-}B + X\text{-}Y]$	Decarboxylase
EC 5 Isomerase	Intramolecule rearrangement, i.e. isomerization changes within a single molecule	$AB \rightarrow BA$	Isomerase, mutase
EC 6 Ligases	Join together two molecules by synthesis of new C-O, C-S, C-N or C-C bonds with simultaneous breakdown of ATP	$X + Y + ATP \rightarrow XY + ADP + Pi$	Synthetase

Format of number

Every enzyme code consists of the letters "EC" followed by four numbers separated by periods. Those numbers represent a progressively finer classification of the enzyme.

For example, the tripeptide aminopeptidases have the code "EC 3.4.11.4", whose components indicate the following groups of enzymes:

- *EC 3* enzymes are hydrolases (enzymes that use water to break up some other molecule)
- *EC 3.4* are hydrolases that act on peptide bonds
- *EC 3.4.11* are those hydrolases that cleave off the amino-terminal amino acid from a polypeptide.
- *EC 3.4.11.4* are those that cleave off the amino-terminal end from a tripeptide.

9.10. Enzyme Kinetics

One of the common fundamental characteristics of all living organisms is that thousands of chemical reactions occur in them at relatively low temperature. Moreover, these reactions happen at very high rates and in a highly regulated manner. High rates and regulation have a common origin: these reactions are catalysed by enzymes, which are mostly proteins and, in some important cases, RNA. Enzymes, either protein- or RNA-based, are all macromolecules. These macromolecules bind to the interacting chemical compounds in a highly specific manner using large complex binding sites. Regulation of enzyme-catalysed reactions is primarily achieved by direct regulation (activation or inactivation) of the enzyme. The discipline of enzyme kinetics studies—among many things—how the rate of enzyme-catalysed reactions depends on the concentration of the compounds directly interacting with the enzyme and what is the highest rate achievable by the enzyme. Further questions of interest include how the rate of the catalysed reaction depends on the temperature, pH, ionic strength *etc.* of the medium. All related studies provide pieces of information that serve as input data to establish a mechanistic model of the enzymatic reaction. Enzyme kinetics, as its name implies, studies primarily the rates of reactions and all factors that affect these rates. However, before introducing fundamental rate equations of enzyme kinetics, we need to review how the rate-increasing capacity of enzymes can be interpreted in the framework of thermodynamics.

Thermodynamic interpretation of enzyme catalysis

Let us first examine what determines the rate of a *non-catalysed* chemical reaction. The reaction scheme described in Equation 9.1 corresponds to a chemical reaction leading to equilibrium. The reacting compounds are denoted as A and B, while P is the product of the reaction.

$$A + B \Leftrightarrow P \qquad (9.1)$$

In the above reaction, as long as only A and B are present (no product has yet been generated), the reaction rate can be described as shown in Equation 9.2. "V" is the rate (velocity) of the reaction, square brackets indicate molar concentrations, whereas "t" denotes time:

$$V = \frac{d[P]}{dt} = [A][B] \qquad (9.2)$$

In order to react, the two molecules need to encounter each other and collide. It is therefore intuitively comprehensible that the rate of the reaction will be linearly proportional to the concentration of both molecules. The factor of proportionality, "k" is denoted as the reaction rate constant. This constant is a central parameter of the reaction. What is the deeper meaning of "k" and what determines its value? The simplest mechanistic model that successfully interpreted the meaning of "k" is illustrated in Equation 9.3:

$$A + b \Leftrightarrow X^* \Rightarrow P \qquad (9.3)$$

The interpretation of Equation 9.3 is as follows. Let us suppose that the reaction between A and B is instantaneous and results in a compound X* called the transition state. (Note that, while the unusual name describes a "state", X* is in fact an actual albeit very short-lived compound.) While compounds A and B are in ground state, X* is in an activated state, i.e. it is at a higher free enthalpy level. Let us also suppose that this first reaction step leads to equilibrium (or, rather, quasi-equilibrium) between the reactants and the transition state. In this model, the transition state will convert into the product in a separate step. The equilibrium constant (importantly, this is *not* a rate constant!) is denoted as K*, while the *rate constant* of the X*'!P reaction step is defined as k'. The reaction scheme of Equation 8.3 will thus lead to rate Equation 9.4:

$$V = \frac{d[P]}{dt} = k[A][B] = k'[X^*] \qquad (9.4)$$

Ultimately, the rate of product formation is the product of the concentration of X* and the newly introduced rate constant, k'. Introduction of a new rate constant introduces additional complexity but, as we will see, it will eventually simplify the case. Let us now examine what determines the concentration of X*. This will lead us to the topic of thermodynamics. As shown in Equation 9.5, the equilibrium between the reactants and the transition state can be defined by the K* equilibrium constant:

$$K^* = \frac{[X^*]}{[A][B]} \tag{9.5}$$

In thermodynamics, the equilibrium constant can be interchangeably defined by the standard free enthalpy change of the reaction leading to the given equilibrium. For the given case, it is shown in Equation 9.6:

$$-RT\ \mathrm{In}K^* = \Delta G^* \tag{9.6}$$

In Equation 9.6, ΔG* defines the free enthalpy difference between the activated state of compound X* and the ground state of the reactants. A more familiar term for the ΔG* parameter is as follows: it is the activation free enthalpy of the chemical reaction. As we will see, it is important that ΔG* corresponds to a free enthalpy change. Yet, for the sake of brevity, it is often referred to—rather incorrectly—as "activation energy".

From Equation 9.6, K* can be expressed by two simple algebraic transformations according to Equations 8.7 and 8.8:

$$\mathrm{In}\, K^* = \frac{-\Delta G^*}{RT} \tag{9.7}$$

$$K^* = e\frac{-\Delta G^*}{RT} \tag{9.8}$$

Equations 9.8 and 9.5 lead to Equation 9.9 that ultimately results in Equation 9.10:

$$V = k'[X^*] = k'[A][B] = k'e^{\frac{-\Delta G^*}{RT}}[A][B] \tag{9.9}$$

$$V = k'e^{\frac{-\Delta G^*}{RT}}[A][B] \tag{9.10}$$

The rate of the reaction is proportional to the concentration of the reactants and also to the k' rate constant. It is also a function of the ΔG* activation free enthalpy. As ΔG* represents an activated state, its value cannot be negative. If ΔG* is 0 (no activation barrier exists), the value of e- ΔG*/RT will be 1. The larger positive number the activation free enthalpy has, the lower positive value the $e^{-\Delta G^*/RT}$ factor will have, the latter approximating zero. The value of the e- ÄG*/RT factor can therefore be between 0 and 1. This can be interpreted by several similar phrasings of the physical meaning of this factor. The $e^{-\Delta G^*/RT}$ factor in the $e^{-\Delta G^*/RT}$ [A][B] product provides a measure of the fraction of A-B collisions that actually lead to the generation of X*. If this factor is 1, all collisions will generate a transition state. If this factor is 0, none of the collisions will generate X*. In the first case, the reaction rate is maximal, while in the second case it is, naturally, zero. In general, the higher the reaction-specific activation free enthalpy, the lower the rate of the reaction. Moreover, the above model provides a comprehensible explanation of the physical meaning of the k‘ rate constant as follows. In order for the transition state to convert into product, at least one chemical bond must be broken. The k‘ rate constant describes the rate, i.e. frequency, of the breakage of the chemical bond. This has been put in mathematical terms in Equation 9.11 in which “V” (Greek letter nu) denotes the vibration frequency of the chemical bond to be broken and “κ” (Greek letter kappa) provides a measure of the proportion of X* decaying forward (generating product) versus decaying backwards (regenerating the reactants). If the value of kappa is 0.5, half of the decays will generate product, whereas the other half will generate reactants. If the value of kappa is 1, X* will decay exclusively towards the product.

$$k' = kv \tag{9.11}$$

For simplicity, let us consider the case when the value of kappa is 1. The energy of chemical bond vibration is described in Equation 9.12, according to Planck‘s law:

$$V = \frac{E}{h} \tag{9.12}$$

In the above equation, E is the energy of the vibrating bond, while h is the Planck constant. The same energy can be expressed using Equation 9.13 in which kb is the Boltzmann constant and T is the temperature on the Kelvin scale. This describes the average kinetic energy of molecule X* at the given temperature:

$$E = k_b T \tag{9.13}$$

By combining Equations 9.11-9.13 and keeping the value of kappa at 1, Equation 9.14 is obtained:

$$k' = \frac{k_b T}{h} \tag{9.14}$$

Combining Equations 9.4, 9.10 and 9.14 leads to Equations 9.15 and 9.16. The latter equations ultimately explain the dependence of the rate of non-catalysed chemical reactions on all already mentioned factors:

$$k = \frac{kBT}{h} e^{\frac{-\Delta G^{*}}{RT}} \tag{9.15}$$

$$V = \frac{kBT}{h} e^{\frac{-\Delta G^{*}}{RT}} [A][B] \tag{9.16}$$

As evident, the rate constant k of the reaction is inversely proportional to the reaction-specific activation free enthalpy and is a complex function of the temperature. Temperature appears twice in the equation: it is present in the pre-exponential term as a multiplying factor and it is also present in the denominator of the exponential term. Both terms indicate that raising the temperature will increase the rate of the reaction. In the pre-exponential term, which corresponds to the k' rate constant, a higher temperature results in a higher vibration frequency of the chemical bond to be broken. In the exponential term, a higher temperature corresponds to a larger proportion of transition state-generating reactant collisions.

Based on Equation 9.4, the rate of the chemical reaction is a linear function of the rate constant and the concentration of the reactants. The thermodynamic model of the chemical reaction rate is illustrated below.

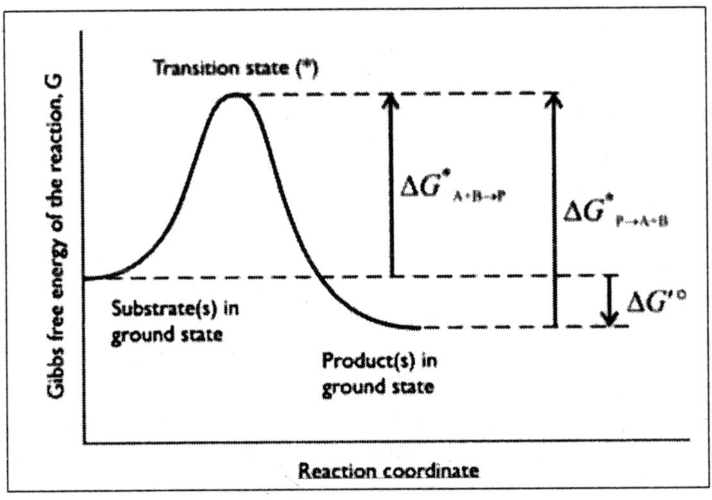

Thermodynamic description of the rate of the chemical reactions

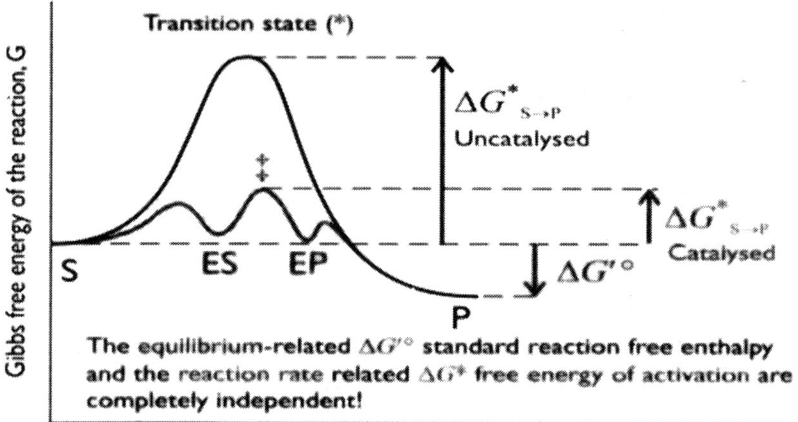

Enzymes increase reaction rates by decreasing the activation free enthalpy of a chemical reaction S → P (S, substrate; P, Product; ES and EP are their complexes with the enzyme, respectively).

Above figure provides a clear graphical illustration of the centrally important fact that the equilibrium proportion of the reactants described by the standard free enthalpy change (ΔG'°) of the chemical reaction is completely independent of the rate of the reaction expressed by the activation free enthalpy (ΔG*) term. The thermodynamic description of the rate of chemical reactions unequivocally points out how enzymes (and all catalysts) can increase the rate of chemical reactions: they do so by *decreasing the activation free enthalpy of the chemical reaction.*

This statement does not tell much about the molecular mechanism of action of enzymes. The question is how, in general, the ΔG* term can be decreased. To answer this, let us recall that free enthalpy can be decomposed to two different terms: an enthalpy-related one and an entropy-related one, as shown in Equation 9.17:

$$\Delta G^* = \Delta H^* - T\Delta S^* \qquad (9.17)$$

An unfavourably high ΔH* decreases the rate of the reaction, as it corresponds to a high-energy transition state. This can be due, for example, to a highly unfavourable electron distribution in the transition-state compound, which represents a high potential energy state. As a consequence of this, the proportion of the transition-state reactants to the ground-state ones will be low. An enzyme can decrease ΔH* by binding to—and thus stabilising—the transition state, providing attracting interactions through a proper binding site. Accordingly, the enzyme must possess a binding pocket that is spherically and electrostatically complementary to the transition state. A negative ΔS^* term is also unfavourable and thus decreases the reaction rate. Such a negative term corresponds to a more ordered transition state than ground state. This means that, by moving from the ground state towards the activated state, the system must get more ordered, which comes with an entropy-related energetic price, sometimes also called penalty, expressed in the $-T\Delta S^*$ term. Enzymes can increase the rate of chemical reactions by positioning and orienting the reactants into an arrangement that is optimal for the reaction.

9.11. Isoenzymes

Isoenzymes (also known as isoenzymes or more generally as multiple forms of enzymes) are enzymes that differ in amino acid sequence but catalyze the same chemical reaction. These enzymes usually display different kinetic parameters (*e.g.* different K_m values), or different regulatory properties. The existence of isozymes permits the fine-tuning of metabolism to meet the particular needs of a given tissue or developmental stage (for example lactate dehydrogenase (LDH). In biochemistry, isozymes (or isoenzymes) are isoforms (closely related variants) of enzymes. In many cases, they are coded for by homologous genes that have diverged over time. Although, strictly speaking, alloenzymes represent enzymes from different alleles of the same gene, and isozymes represent enzymes from different genes that process or catalyse the same reaction, the two words are usually used interchangeably.

Isozymes are usually the result of gene duplication, but can also arise from polyploidisation or nucleic acid hybridisation. Over evolutionary time, if the function of the new variant remains *identical* to the original, then it is likely that one or the other will be lost as mutation accumulate, resulting in a pseudogene. However, if the mutations do not immediately prevent the enzyme from functioning, but instead modify either its function, or its pattern of gene expression, then the two variants may both be favoured by natural selection and become specialised to different functions. For example, they may be expressed at different stages of development or in different tissues.

Allozymes may result from point mutations or from insertion-deletion (*indel*) events that affect the DNA coding sequence of the gene. As with any other new mutations, there are three things that may happen to a new allozyme:

1. It is most likely that the new allele will be non-functional, in which case it will probably result in low fitness and be removed from the population by natural selection.

2. Alternatively, if the amino acid residue that is changed is in a relatively unimportant part of the enzyme (*e.g.*, a long way from the active site), then the mutation may be selective neutral and subject to genetic drift.
3. In rare cases, the mutation may result in an enzyme that is more efficient, or one that can catalyse a slightly different chemical reaction, in which case the mutation may cause an increase in fitness, and be favoured by natural selection.

Isozymes (and allozymes) are variants of the same enzyme. Unless they are identical in terms of their biochemical properties, for example their substrates and enzyme kinetics, they may be distinguished by a biochemical assay. However, such differences are usually subtle (particularly between *allozymes* which are often neutral variants). This subtlety is to be expected, because two enzymes that differ significantly in their function are unlikely to have been identified as *isozymes*.

Whilst isozymes may be almost identical in function, they may differ in other ways. In particular, amino acid substitutions that change the electric charge of the enzyme (such as replacing aspartic acid with glutamic acid) are simple to identify by gel electrophoresis, and this forms the basis for the use of isozymes as molecular markers. To identify isozymes, a crude protein extract is made by grinding animal or plant tissue with an extraction buffer, and the components of extract are separated according to their charge by gel electrophoresis. Historically, this has usually been done using gels made from potato starch, but acrylamide gels provide better resolution.

All the proteins from the tissue are present in the gel, so that individual enzymes must be identified using an assay that links their function to a staining reaction. For example, detection can be based on the localised precipitation of soluble indicator dyes such as tetrazolium salts which become insoluble when they are reduced by cofactors such as NAD or NADP, which generated in zones of enzyme activity. This assay method requires that the enzymes are still functional after separation (negative gel electrophoresis), and provides the greatest challenge to using isozymes as a laboratory technique. Isoenzymes differ in kinetics (they have different Km and V_{max} values.

9.12. Isoenzyme Analysis

Isoenzyme analysis is based on the existence of enzymes with similar or identical specificity, but different molecular structure isoenzymes). Isoenzyme analysis is used to study the patterns of migration of isoenzymes present in cell lysates following electrophoresis using agarose gels. The patterns obtained are species specific and therefore are used as quality control and authentication procedures to confirm species of origin of material.

Technically the sample extract is elrctrophoresed in starch or polyacrylamide buffered slab gels at a low temperature (4–8°C). Each lane should be loaded with equal amount of protein ater normalizing the protein content in extract in as small volume as possible (25–50 µl). After electrophoresis thc gel is incubated in a solution containing all the necessary components for enzyme reaction. The coloured reaction products stain the gel where the enzymes are located.

Staining procedure

- Conduct electrophoresis in starch or polyacrylamide disc/slabs buffer gels at low temperature.
- Immediately after electrophoresis, incubate the gel in the substrate solution(s). The zones where the enzymes are located in the gel are visualized due to the appereance of colour reaction products. After syfficient incubation period, stop the reaction by adding the appropriate reaction stopping solution and photograph the zymogram. Otherwise the relative position of each visualized band in the gel may be drawn schematically for easy reference.

9.13. Native PAGE Separation and Detection of Lactate Dehydrogenase Isoenzymes

Enzymes that catalyse the same reaction in the same organism but have different chemical structure are called isoenzymes. The very first enzyme shown to be present in animal tissues in the form of isoenzymes was lactate dehydrogenase (LDH). It turned out that most tissues contain five different isoenzyme forms. LDH is a tetrameric protein. It has two types of subunits called M and H. The M subunit is characteristic mostly of skeletal muscle, while the H subunit is characteristic of cardiac muscle.

The two different subunits are encoded by two different genes. In the quaternary structure of the enzyme, the subunits are arranged as the four vertices of a tetrahedron. As the two subunits can combine randomly, the tetrameric enzyme can be of five isoenzyme forms: M4, M3H, M2H2, MH3 and H4. Due to random combination of the four subunits, the proportion of the five isoenzymes in different tissues will depend on the tissue-specific relative level of expression of the two encoding LDH genes. The shape and molecular mass (34,000 Da) of the two subunit types is identical. Therefore, the shape and molecular mass (136,000 Da) of the five different tetrameric isoenzymes will also be identical. However, the electric charge of the two subunits is different. Therefore, the charge of the five isoenzymes will also be different. H-type subunits carry more negative charges than M-type ones.

When LDH is subjected to native PAGE, the subunits remain associated and tetramers migrate in the gel. As the five different tetramers have identical shape and size but distinct charge, the native PAGE in this case will separate the isoforms based exclusively on charge differences. The larger number of H subunits present in the isoenzyme, the more negative charges it will carry. Accordingly, the H4 isoenzyme will have the highest and the M4 the lowest electropho-retic mobility. Based on the principles of electrophoresis, the relative mobility of the isoenzymes can be easily deduced.

The magnitude of the electric force (*Fe*) exerted on a given isoenzyme is linearly proportional to the number of H subunits it contains. On the other hand, due to their identical size and shape, the frictional coefficient (*f*) is identical for all isoenzymes. As a consequence of this, the electrophoretic mobility (*i*) of the isoenzymes will be a linear function of the number of their H subunits. Let us consider that, at a given electric field, the velocity of the M3H isoenzyme exceeds that of the M4 type by a value of X cm/h. This velocity change is due to the replacement of a single M subunit with a single H subunit. The same increase in velocity should apply to all additional M-to-H replacements. As a result, the electrophoretic mobility will be linearly proportional to the number of H subunits. As a remarkable consequence of this simple relationship, at the end of the separation, the distances between neighbouring isoenzyme banas will be identical.

The fact that the proportion of the various isoenzymes is tissue-specific provides a great diagnostic value. Thus, components of the cytoplasm of the damaged cells including the LDH isoenzymes enter the blood stream. The normal LDH isoenzyme distribution of the serum will therefore be shifted towards the distribution characteristic of the damaged tissue. Serum analysis of the altered LDH isoenzyme distribution would be a very complex and laborious procedure if the isoenzymes had to be purified to homogeneity. This is circumvented by applying native PAGE combined with a subsequent in-gel enzymatic assay. Although hundreds or thousands of different proteins are run in the gel, the specific enzymatic reaction will detect only the LDH isoforms. LDH catalyses the reversible chemical reaction shown in below. LDH can be detected by a special chromogenic reaction. The assay is based on the reducing power of NADH. After electrophoresis, the gel is incubated in a buffer solution containing lactate, NAD^+ coenzyme, phenazine methosulphate (PMS, an electron acceptor) and nitro-blue-tetrazolium-chloride (NBTC, a redox dye). In the multi-step reaction, lactate is oxidised to pyruvate, while NAD^+ is reduced to NADH. NADH will reduce PMS, which then will reduce the soluble, yellowish oxidised NBTC. The reduced form of NBTC is blue and has a very low solubility. The position of LDH enzymes in the gel will be detected through the formation of a dark blue precipitate in the gel.

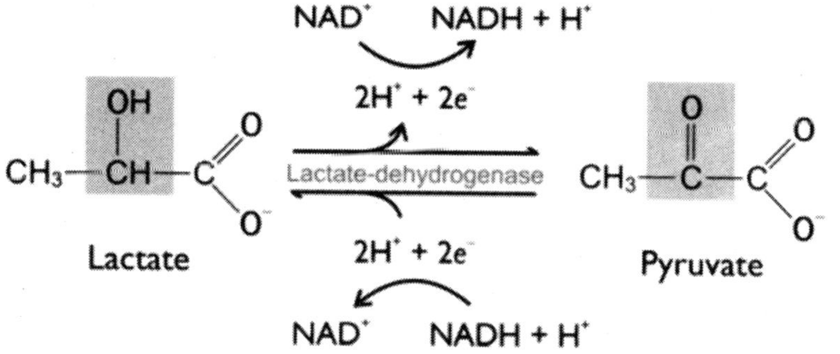

NAD^+ $NADH + H^+$

$2H^+ + 2e^-$

Lactate-dehydrogenase

$2H^+ + 2e^-$

NAD^+ $NADH + H^+$

Lactate

Pyruvate

The chemical reaction catalysed by lactate dehydrogenase. In the reversible reaction, depending on the actual direction of the reaction, either the oxidised or reduced coenzyme (NAD^+ or NADH, respectively) is produced. In the assay the gel was soaked in a buffer containing lactate and oxidised NAD^+. Two other consecutive chemical steps utilise the reduced NADH state to produce a coloured insoluble product that precipitates around the LDH enzymes (see main text for details).

CHAPTER 10

Amino Acid

Amino acids are biologically important organic compounds composed of amine ($-NH_2$) and carboxylic acid (-COOH) functional groups, along with a side-chain specific to each amino acid. The key elements of an amino acid are carbon, hydrogen, oxygen, and nitrogen, though other elements are found in the side-chains of certain amino acids. About 500 amino acids are known and can be classified in many ways. They can be classified according to the core structural functional groups' locations as alpha- (α-), beta- (β), gamma- (γ), or delta- (δ-) amino acids; other categories relate to polarity, pH level, and side-chain group type (aliphatic, acrylic, aromatic, containing hydroxyl or sulphur, *etc*.). In the form of proteins, amino acids comprise the second-largest component (water is the largest) of human muscles, cells and other tissues. Outside proteins, amino acids perform critical roles in processes such as neurotransmitter transport and biosynthesis. Amino acids having both the amine and the carboxylic acid groups attached to the first (alpha-) carbon atom have particular importance in biochemistry. They are known as 2-, alpha-, or α-amino acids (generic formula $H_2NCHRCOOH$ in most cases where R is an organic substituent known as a "side-chain"); often the term "amino acid" is used to refer specifically to these. They include the 22 proteinogenic ("protein-building") amino acids, which combine into peptide chains ("polypeptides") to form the building-blocks of a vast array of proteins. These are all L-stereoisomers ("left-handed" isomers), although a few D-amino acids ("right-handed") occur in bacterial envelopes and some antibiotics. Twenty of the proteinogenic amino acids are encoded directly by triplet codons in the genetic code and are known as "standard" amino acids. The other two ("non-standard" or "non-canonical") are pyrrolysine (found in methanogenic organisms and other eukaryotes) and selenocysteine (present in many noneukaryotes as well as most eukaryotes). For example, 25 human proteins include selenocysteine (Sec) in their primary structure, and the structurally characterized enzymes (selenoenzymes) employ Sec as the catalytic moiety in their active sites. Pyrrolysine and selenocysteine are encoded via variant codons; for example, selenocysteine is encoded by stop codon and SECIS element. Codon–tRNA combinations not found in nature can also be used to "expand" the genetic code and create novel proteins known as alloproteins incorporating non-proteinogenic amino acids.

Nine of the 22 standard amino acids are called "essential" for humans because they cannot be created from other compounds by the human body and, so, must be taken in as food. Others may be conditionally essential for certain ages or medical conditions. Essential amino acids may also differ between species. Because of their biological significance, amino acids are important in nutrition and are commonly used in nutritional supplements, fertilizers, and food technology. Industrial uses include the production of drugs, biodegradable plastics, and chiral catalysts.

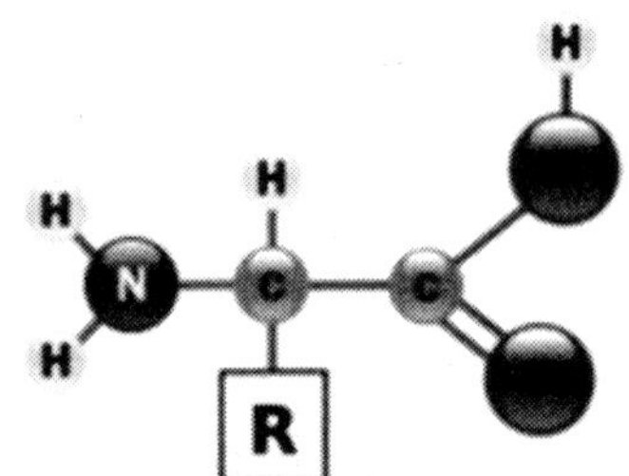

The generic structure of an alpha amino acid in its un-ionized form

In this structure R represents a side-chain specific to each amino acid. The carbon atom next to the carboxyl group is called the α-carbon and amino acids with a side-chain bonded to this carbon are referred to as *alpha amino acids*. These are the most common form found in nature. In the alpha amino acids, the α–carbon is a chiral carbon atom, with the exception of glycine. In amino acids that have a carbon chain attached to the α–carbon (such as lysine) the carbons are labelled in order as α, β, γ, δ, and so on. In some amino acids, the amine group is attached to the β or γ-carbon, and these are therefore referred to as *beta* or *gamma amino acids*.

Amino acids are usually classified by the properties of their side-chain into four groups. The side-chain can make an amino acid a weak-acid or a weak-base, and a hydrophile if the side-chain is polar or a hydrophobe if it is nonpolar. The phrase "branched-chain amino acids" or BCAA refers to the amino acids having aliphatic side-chains that are non-linear; these are leucine, isoleucine, and valine. Proline is the only proteinogenic amino acid whose side-group links to the α-amino group and, thus, is also the only proteinogenic amino acid containing a secondary amine at this position. In chemical terms, proline is, therefore, an imino acid, since it lacks a primary amino group, although it is still classed as an amino acid in the current biochemical nomenclature, and may also be called an "N-alkylated alpha-amino acid".

Zwitterions

The amine and carboxylic acid functional groups found in amino acids allow them to have amphiprotic properties. Carboxylic acid groups ($^-CO_2H$) can be deprotonated to become negative carboxylates ($^-CO_2^-$), and α-amino groups (NH_2^-) can be protonated to become positive α-ammonium groups ($^+NH_3^-$). At pH values greater than the pKa of the carboxylic acid group (mean for the 20 common amino acids is about 2.2), the negative carboxylate ion predominates. At pH values lower than the pKa of the α-ammonium group (mean for the 20 common α-amino acids is about 9.4), the nitrogen is predominantly protonated as a positively charged α-ammonium group. Thus, at pH between 2.2 and 9.4, the predominant form adopted by α-amino acids contains a negative carboxylate and a positive α-ammonium group, so has net zero charge. This molecular state is known as a zwitterion, from the German Zwitter meaning *hermaphrodite* or *hybrid*. Below pH 2.2, the predominant form will have a neutral carboxylic acid group and a positive α-ammonium ion (net charge +1), and above pH 9.4, a negative carboxylate and neutral α-amino group (net charge "1). The fully neutral form is a very minor species in aqueous solution throughout the pH range (less than 1 part in 10^7). Amino acids exist as zwitterions also in the solid phase, and crystallize with salt-like properties unlike typical organic acids or amines.

Isoelectric point

At pH values between the two pKa values, the zwitterion predominates, but coexists in dynamic equilibrium with small amounts of net negative and net positive ions. At the exact midpoint between the two pKa values, the trace amount of net negative and trace of net positive ions exactly balance, so that average net charge of all forms present is zero. This pH is known as the isoelectric point pI, so pI = ½(pKa_1 + pKa_2). The individual amino acids all have slightly different pKa values, so have different isoelectric points. For amino acids with charged side-chains, the pKa of the side-chain is involved. Thus for Asp, Glu with negative side-chains, pI = ½(pKa_1 + pKa_R), where pKa_R is the side-chain pKa. Cysteine also has potentially negative side-chain with pKa_R = 8.14, so pI should be calculated as for Asp and Glu, even though the side-chain is not significantly charged at neutral pH. For His, Lys, and Arg with positive side-chains, pI = ½(pKa_R + pKa_2). Amino acids have zero mobility in electrophoresis at their isoelectric point, although this behaviour is more usually exploited for peptides and proteins than single amino acids. Zwitterions have minimum solubility at their isolectric point and some amino acids (in particular, with non-polar side-chains) can be isolated by precipitation from water by adjusting the pH to the required isoelectric point.

10.1. Essential Amino Acids

An essential amino acid or indispensable amino acid is an amino acid that cannot be synthesized *de novo (from scratch)* by the organism being considered, and therefore must be supplied in its diet. The amino acids regarded as essential for humans are methionine, arginine, tryptophan, valine, isoleucine, leucine, phenylalanine, and lysine. Additionally, cysteine (or sulphur-containing amino acids), tyrosine (or aromatic amino acids) are required by infants and growing children. Essential amino acids are "essential" not because they are more important for protein synthesis or general homeostasis than the others, but because the body does not synthesize them. If they are not uptaken through diet, they will not be available for protein synthesis. In addition, the amino acids arginine, cysteine, glycine, glutamine, proline, serine and tyrosine are considered conditionally essential, meaning they are not normally required in the diet, but must be supplied exogenously to specific populations that do not synthesize them in adequate amounts. An example would be with the disease phenylketonuria (PKU). Individuals living with PKU must keep their intake of phenylalanine extremely low to prevent mental retardation and other metabolic complications. However, they cannot synthesize tyrosine from phenylalanine, so tyrosine becomes essential in the diet of PKU patients.

The distinction between essential and non-essential amino acids is somewhat unclear, as some amino acids can be produced from others. The sulphur-containing amino acids, methionine and homocysteins, can be converted into each other but neither can be synthesized *de novo* in humans. Likewise, cysteine can be made from homocysteine but cannot be synthesized on its own. So, for convenience, sulfur-containing amino acids are sometimes considered a single pool of nutritionally equivalent amino acids as are the aromatic amino acid pair, phenylalanine and tyrosine. Likewise arginine, ornithine, and citrulline, which are interconvertible by the urea cycle, are considered a single group.

Valine (abbreviated as Val or V) is an α-amino acid with the chemical formula $HO_2CCH(NH_2)CH(CH_3)_2$. L-Valine is one of 20 proteionogenic amino acids. Its codons are GUU, GUC, GUA, and GUG. This essential amino acid is classified as nonpolar. Human dietary sources are any proteinaceous foods such as meats, dairy products, soy products, beans and legumes. Along with leucine and isoleucine, valine is a branched chain amino acid. It is named after the plant valerian. In sickle-cell disease, valine substitutes for the hydrophilic amino acid, glutamic acid in haemoglobin. Because valine is hydrophobic, the haemoglobin is prone to abnormal aggregation.

Valine is an essential amino acid, hence it must be ingested, usually as a component of proteins. It is synthesized in plants via several steps starting from pyruvic acid. The initial part of the pathway also leads to leucine. The intermediate α-ketoisovalerate undergoes reductive amination with glutamate. Enzymes involved in this biosynthesis include:

Isoleucine (abbreviated as Ile or I) is an α-amino acid with the chemical formula $HO_2CCH(NH_2)CH(CH_3)CH_2CH_3$. It is an essential amino acid, which means that humans cannot synthesize it, so it must be ingested. Its codons are AUU, AUC and AUA. With a hydrocarbon side chain, isoleucine is classified as a hydrophobic amino acid. Together with threonine, isoleucine is one of two common amino acids that have a chiral side chain. Four stereoisomer of isoleucine are possible, including two possible diastereomers of L-isoleucine. However, isoleucine present in nature exists in one enantiomeric form, (2*S*,3*S*)-2-amino-3-methylpentanoic acid.

Leucine (abbreviated as Leu or L) is a branched-chain α-amino acid with the chemical formula $HO_2CCH(NH_2)CH_2CH(CH_3)_2$. Leucine is classified as a hydrophobic amino acid due to its aliphatic isobutyl side chain. It is encoded by six codons (UUA, UUG, CUU, CUC, CUA, and CUG) and is a major component of the subunits in ferritin, astacin and other 'buffer' proteins. Leucine is an essential amino acid, meaning that the human body cannot synthesize it, and it therefore must be ingested. Leucine is utilized in the liver, adipose tissue, and muscle tissue. In adipose and muscle tissue, leucine is used in the formation of sterols, and the combined usage of leucine in these two tissues is seven times greater than its use in the liver.

Leucine is the only dietary amino acid that has the capacity to stimulate muscle protein synthesis. As a dietary supplement, leucine has been found to slow the degradation of muscle tissue by increasing the synthesis of muscle proteins in aged rats. However, results of comparative studies are conflicted. Long-term leucine supplementation does not increase muscle mass or strength in healthy elderly men. More studies are needed, preferably those which utilize an objective, random sample of society. Factors such as lifestyle choices, age, gender, diet, exercise, *etc.* must be factored into the analyses in order to isolate the effects of supplemental leucine as a standalone, or if taken with other BCAA's Branched Chain Amino Acids. Until then, dietary supplemental leucine cannot be associated as the prime reason for muscular growth or optimal maintenance for the entire population. While once seen as an important part of the three branch chained amino acids in sports supplements, leucine has since earned more attention on its own as a catalyst for muscle growth and muscular insurance. Supplement companies once marketed the "ideal" 2:1:1 ratio of leucine, iso-leucine and valine; but with furthered evidence that leucine is the most important amino acid for muscle building, it has become much more popular as the primary ingredient in dietary supplements with a 4:1:1 ratio.

Phenylalanine abbreviated as Phe or F is an α-amino acid with the formula $C_6H_5CH_2CH(NH_2)COOH$. This essential amino acid is classified as nonpolar because of the hydrophobic nature of the benzyl side chain. L-Phenylalanine (LPA) is an electrically neutral amino acid, one of the twenty common amino acids used to biochemically form proteins, coded for by DNA. The codons for L-phenylalanine are UUU and UUC. Phenylalanine is a precursor for tyrosine, the monoamine signalling molecules dopamine, norepinephrine (noradrenaline), and epinephrine (adrenaline), and the skin pigment melanin. Phenylalanine is found naturally in the breast milk of mammals. It is used in the manufacture of food and drink products and sold as a nutritional supplement for its reputed analgesic and antidepressant effects. It is a direct precursor to the neuromodulator phenylethylamine, a commonly used dietary supplement.

L-Phenylalanine is biologically converted into L-tyrosine, another one of the DNA-encoded amino acids. L-tyrosine in turn is converted into L-DOPA, which is further converted into dopamine, norepinephrine (noradrenaline), and epinephrine (adrenaline). The latter three are known as the catecholamines. Phenylalanine uses the same active transport channel as tryptophan to cross the blood-brain barrier, and, in large quantities, interferes with the production of serotonin. Phenylalanine is the starting compound used in the flavpnoid biosynthesis. Lignan is derived from phenylalanine and from tyrosine. Phenylalanine is converted to cinnamic acid by the enzyme phenylalanine ammonia-lyase.

10.2. Estimation of Glycine, Alanine and Isoleucine

Materials

- sodium bicarbonate
- dichlone (2,3-dichloro-1,4-naphthoquinone)
- dimethyl sulphoxide
- hydrochloric acid
- Standards of glycine, alanine, and isoleucine.

Procedure

- Powder of glycine, alanine or isoleucine (125 mg) is weighed accurately and transferred to 25 ml volumetric flask.
- Sodium bicarbonate - 125 mg of Na_2CO_3 is added to each flask. The contents of flask are dissolved in water to obtain final concentration 5 mg/ml for each amino acid.
- Dichlone (62.5 mg) is weighed, dissolved and diluted with dimethyl sulfoxide to obtain final vconcentration of 2.5 mg/ml.
- Sodium bicarbonate - 125 mg of Na_2HCO_3 is dissolved in water to obtain final concentration of 5 mg/ml.
- Hydrochloric acid - 42.5 ml of concentrated HCl is diluted to 1000 ml with water to yield 0.5 M solution of acid.
- Amino acid solution (0.4 ml) is taken in 100 ml volumetric flask and 0.6 ml of sodium bicarbonate and 5.0 ml of dichlone are added.
- Mixture was heated on boiling water bath for 15 min, cooled to room temperature and volume was made up to 100 ml with 0.5 M HCl.
- Blank was prepared in same manner omitting amino acid solution.
- Optical density is recorded at 470 nm.
- Amino acid (glycine, alanine or isoleucine) stock solution (0.1, 0.2, 0.3, 0.4 and 0.5 ml) was transferred to a series of 100 ml volumetric flask, and colour is developed by following above stated procedures.
- Calibration curve was prepared by plotting absorbance versus its corresponding concentration.

10.3. Estimation of Lysine

Lysine (abbreviated as Lys or K) is an α-amino acid with the chemical formula $HO_2CCH(NH_2)(CH_2)_4NH_2$. It is an essential amino acid for humans. Lysine's codons are AAA and AAG. Lysine is a base, as are arginine and histidine. The ε-amino group often participates in hydrogen bonding and as a general base in catalysis. (The ε-amino group (NH_3^+) is attached to the fifth carbon beginning from the α-carbon, which is attached to the carboxyl (C=OOH) group. Common posttransitional modifications include methylation of the ε-amino group, giving methyl-, dimethyl-, and trimethyllysine. The latter occurs in calmodulin. Other posttransitional modifications at lysine residues include acettylation and ubiquitination. Collagen contains hydroxylasine, which is derived from lysine bylysyl hydroxylase. *O*-Glycosylation of hydroxyl-lysine residues in the endoplasmic reticulam or Golgi apparatus is used to mark certain proteins for secretion from the cell.

Lysine is a limiting amino acid in the cereal grains. Assessment of lysine in cereal grains for nutritional quality is hence one of the procedure adopted for screening varieties. Though an ideal simple method for direct estimation of lysine is yet to be found, the given procedure is sufficient for routine screening. The protein in the grain sample is hydrolysed with a proteolytic enzyme, papain. The alpha- amino groups of the derived amino acids are made to form a complex with copper. The ε amino group of lysine which does not couple with copper is made to form ε dinitropyridyl derivative of lysine with 2-chloro-3,5-dinitropyridine. The excess pyridine is removed with ethyl acetate and the colour of ε dintropyridyl derivative is read at 390 nm.

Materials

- Solution a: 2.89 $CuCl_2$, $2H_2O$ in 100 ml water
- Solution B: 13.6g Na_3PO_4,12 H_2O in 200ml water
- Sodium Borate Buffer 0.05 M pH 9.0
- *Copper phosphate Reagent*: Pour solution A (100ml) to B (200 ml) with swirling, centrifuge and discard the supernatant. Resuspend the pellet three times in 15 ml. Borate and centrifuge after suspension. After third washing resuspend the pellet in 80ml of borate buffer. The reagent must be prepared fresh for every seven days.
- 3% solution of 2-Chloro-3,5-Dinitropyridine in methanol. Prepare fresh just prior to use.
- *0*.05M Sodium Carbonate Buffer *pH* 9.0
- *Amino Acid Mixture*: Grind in a mortar 30mg alanine, 50mg glutamic acid, 60mg aspartic acid, 20 mg cystein, 300mg glutamic acid, 40mg glycine, 30mg histidine, 30 mg isoleucine, 80mg leucine, 30mg methionine, 40mg phenylalanine, 80mg proline, 50mg serine, 30mg theonine, 30mg tyrosine and 40mg valine. Dissolve 100mg of this mixture in 10 ml of sodium carbonate buffer (0.05M, pH 9.0).
- *Dissolve 400mg technical grade papain (Sigma Co*., USA) in 100ml 0.1 M sodium acetate buffer (pH 7.0)
- 1.2N HCL
- *Ethyl acetate*

Procedure

- To 100 mg of defatted grain sample add 5ml of pappain solution and incubate overnight at 65°C. Cool to room temperature, centrifuge and decant the clear digest.
- To one ml digest taken in a centrifuge tube add 0.5 ml carbonate and 0.5ml copper phosphate suspension.
- Shake the mixture for 5min in a vortex mix and centrifuge.
- To one ml supernatant add 0.1ml pyridine reagent, mix well for 2 h
- Add 5ml of 1.2 N HCL and mix
- Extract three times with 5ml ethyl acetate and discard the ethyl acetate (top) layer.
- Read the absorbance of aqueous layer at 390 nm
- Prepare a blank with 5 ml papain alone repeating above steps.
- Dissolve 62.5 mg lysine monohydrochloride in 50 ml carbonate buffer (1mg lysine/ml). Pipette out 0.2, 0.4, 0.6, 0.8 and 1ml and make up to one ml with carbonate buffer. Add 4ml papain to each tube and mix. Pipette out one ml from each and add 0.5ml of aminoacid mixture and 0.ml of copper phosphate suspension. Carry out steps 3 to 7. The standard curve represents absorbance values for 40, 80,120,160 and 200 µg lysine.

Calculation

Prepare a standard curve from the readings of the standard lysine. Substrate absorbance of the blank from that of the sample and calculate the content in the aliquot from the graph.

Lysine content of the sample =

Lysine value from graph in µg lysine x 0.16 / Percent nitrogen in the sample

= g per 16g N.

Readings

- Mertz, ET, Jambunathan,R and Mishra,PS (1975) In: Protein Quality, Agric Experiment Stn Bull No 70 Purdue Univ USA p 11.
- Theymoli Balasubramanian and Sadasivam S (1987) *Plant Foods Hum Nutr* **37:** 41.

10.4. Estimation of Methionine

Methionine is an α-amino acid with the chemical formula $HO_2CCH(NH_2)CH_2CH_2SCH_3$. This is an essential amino acid and is classified as non-polar. This amino-acid is coded by the initiation codon AUG which indicates mRNA's coding region where translation into protein begins. Together with cysteine, methionine is one of two sulphur-containing proteinogenic amino acids. Its derivative S-adenosyl methionine (SAM) serves as a methyl donor. Methionine is an intermediate in the biosynthesis of cysteine, carnitine, taurine, lacithin, phosphatidylcholine, and other phospholipids. Improper conversion of methionine can lead to atherosclerosis. This amino acid is also used by plants for synthesis of ethylene. The process is known as the Yang Cycle or the methionine cycle.

Methionine is one of only two amino acids encoded by a single codon (AUG) in the standard genetic code (tryptophan, encoded by UGG, is the other). The codon AUG is also the most common eukaryote "Start" message for a ribosome that signals the initiation of protein translation from mRNA when the AUG codon is in a Kozak consensus sequence. As a consequence, methionine is often incorporated into the N-terminal position of proteins in eukaryotes and archaea during translation, although it can be removed by post transitional modification. In bacteria, the derivative N-formylmethionine is used as the initial amino acid. Methionine estimation is a routine in nutritional screening programmes of grain legumes. The protein in the grain is first hydrolised under mild acidic condition. The liberated methionine gives an yellow colour with nitroprusside solution under alkaline condition and turns red on acidification. Glycine is added to the reaction mixture in order to inhibit colour formation with other amino acids.

Materials

- 2 N hydrochloric Acid
- 10 N NaOH (40%)
- 10% NaOH
- 10% Sodium Nitroprusside
- 3% Glycine
- Orthophosphoric Acid (Sp.gr.1.75)
- Standard Methionine: Dissolve 100mg of DL-Methionine in 4ml of 20% HCL and dilute with water to 100 ml.

Procedure

- Weigh 0.5g of defatted sample into a 50 ml conical flask. Add 6ml of 2 N HCL and autoclave at 15 psi pressure for one hour.
- Add a pinch of activated charcoal to the hydrolysate (autoclaved sample) and heat to boil. Filter when hot and wash the charcoal with hot water.
- Neutralise the filtrate with 10N NaOH to ph 6.5 Make up the volume to 50ml with water after cooling to ambient temperature.
- Transfer 25.0 ml of the made up solution into a 100 ml conical flsk.
- Add 3ml of 10 % NaOH followed by 0.15 ml solution sodium nitrprusside.
- After 10 min add 1ml of glycine solution.
- After another 10 min add 2 ml orthophosphoric acid and shake vigorously.
- Read the intensity of red colour after 10 min at 520 nm against a blank prepared in the same way but without nitroprusside.
- Standard Curve: Pipette out 0, 1, 2, 3, 4 and 5ml of standard methionine sol ution and make up to 25ml with water. Follow steps 5 to 8 to develop the colour in the standards. The 0 level serves as the blank.

Calculation

Draw a standard curve and calculate the methionine content from the graph methionine content in the sample =

(Methionine content from the graph x 4) mg per g

Methionine is usually expressed as percentage of protein or g per 16 g N.

Methionine content of the sample =

Methionine content from the graph x 6.4

Percentage of N in the sample

=g per 16 g N

Reading

Horn, MJ, Jones, DB and Blum, A E (1946). Colorimetric determination of methionine in proteins and foods. *J Biol Chem.* **166 (1):** 313 - 320.

Rapid colorimetric procedure of methionine estimation

Acid hydrolysis of sample

Defated sample (~1.0 g) is taken in hydrolysis tube and 10.0 ml of 40% hydroclric acid is added and partial evacuation the tubes are sealed and hydrolysed at 120^0 C for 12 hours. The samples are cooled and treated with about 5.0 g of carcoal. The contents are filtered and residue is washed with 25% ethyl alcohol. Final volume is made to 25 ml with ethanol.

Procedure

- To a suitable aliquots (1.0 ml) of acid hydrolysate, 1.0 ml of 13 N NaOH and 1.0 ml of 5 N NaOH are added, followed by 1.0 ml of 1% glycine. Mix thoroughly and add 1.0 ml of 1% solution of sodium nitropruside. Incubate for 10 minutes at 40^0 C. Cool the reaction mixture and 5.0 ml of 85% phosphoric acid is added. Optical density is measured at 520 nm against a reagent blank.

Reading

Lunder, T.J. (1973). An improved method for the colorimetric determination of methionine in acid hydrolysates of biological products. *Id. Aliment*, **12:** 94-98.

Mc Carthy, R.E. and Sullivan, M.X. (1941). A new and highly specific colorimetric test for methionine. *J. Biol. Chem.*, **141:** 871-876.

10.5. Estimation of Arginine

Arginine (abbreviated as Arg or R) is an essential α-amino acid. It was first isolated in 1886. The L-form is one of the 20 most common natural amino acids. At the level of molecular genetics, in the structure of the messenger ribonucleic acid mRNA, CGU, CGC, CGA, CGG, AGA, and AGG, are the triplets of nucleotide bases or codons that code for arginine during protein synthesis. In mammals, arginine is classified as a semi-essential or conditionally essential amino acid, depending on the developmental stage and health status of the individual. There are some conditions that put an increased demand on the body for the synthesis of L-arginine, including surgical or other trauma, sepsis and burns. Arginine was first isolated from a lupin seedling extract in 1886 by the Swiss chemist Ernst Schultze. In general, most people do not need to take arginine supplements because the body usually produces significant amounts in healthy individuals.

The calorimetric estimation of arginine is based on the unstable red colour produced by the addition of α-naphthol and 0.3 N sodium hypochlorite to an alkaline solution of the amino acid.

Materials

- Sodium hypochlorite (0.06 N) solution is prepared and necessary dilution of the stock product was ascertained iodometrically as taking 1.0 ml of sodium hypochlorite solution add 25.0 ml of chlorine free water having 1.0 g of potassium iodide. The mixture is titrated with 0.1 N sodium thiosulfate, using 1.0 ml of starch as indicator. The stock solution is stable for 3 to 4 months if kept in the refrigerator in a brown bottle.
- Sodium hydroxide - 10% solution.
- Urea - 20% solution.
- Permutit. A 60 mesh product was used and was activated.
- ac-Naphthol. 100 mg. of the resublimed product are dissolved in 100 ml of 95% ethanol. The solution is kept in a brown bottle and stored in the refrigerator.
- Arginine standard. 12.05 mg. of arginine hydrochloride, Merck (26.6 per cent N found), are weighed accurately and dissolved in 100 ml of water. 1.0 ml of this solution is equivalent to 100 y of the free base. This solution is stored in the refrigerator.

Procedure

- Hydrolysates were prepared by refluxing 5.0 gm. samples with 25 ml of constant boiling (6 N) hydrochloric acid for 24 hours. The total nitrogen content of the hydrolysate is determined directly by micro-Kjeldahl analysis; the excess of acid is then removed by concentration in vacua and the humin separated by filtration. (Sulfuric acid digests are prepared by refluxing 5.0 gm. of sample with 25 ml of 25% (by weight) sulfuric acid (6 N) for 24 hours and removing the acid as calcium sulfate. These amounts of protein are of course far in excess of the 0.5 to 1.0 mg. of hydrolysate nitrogen actually required for the arginine determination.
- Aliquots of these hydrolysates (5 ml) are taken in 10 ml graduated photoelectric calorimeter tubes. If necessary, the volume of the sample is adjusted to 5 ml; 1 ml of 10% NaOH and 1 ml of L naphthol reagent are then added with mixing and allowed to stand for 5 minutes.
- After 5 minutes, 1 ml of sodium hypochlorite is added, followed exactly 1 minute later by the addition of 2 ml of 20% urea solution.
- The resulting solutions are thoroughly mixed calorimeter with Filter S-54. The colour intensity of the reaction mixture has been found to remain constant for more than 15 minutes, so that there is no need to hasten the colour measurement.

- A parallel detêrmination is also done on an aliquot of the standard containing an amount of arginine comparable to that of the unknown.

Calculations

A = calorimetric reading of protein sample

u= 'I calorimetric reading of 5 ml of sample

p= (('I " 5 C' permutit filtrate

B= " 6' 'I arginine standard

c= " " " reagent blank

A' = A - C, corrected calorimetric reading of protein sample

TJ' = u - c, (1 " 'L " sample

p' = p - c, (6 " " " permutit filtrate

B' = B - C, '(" I' " arginine standard

Then for proteins, mg. of arginine in sample = (A'/B') X mg. of arginine in standard; for urine, mg. of arginine in 24 hour specimen = ((U' - P')/ B') X mg. of arginine in standard (total volume of 24 hour specimen)/5.

10.6. Estimation on Tryptophan

Tryptophan is one of the 22 standard amino acids and an essential amino acid in the human diet, as demonstrated by its growth effects on rats. It is encoded in the standard genetic code as the codon *UGG*. Only the L-stereoisomer of tryptophan is used in structural or enzyme proteins, but the R -stereoisomer is occasionally found in naturally produced peptides (for example, the marine venom peptide contryphan). The distinguishing structural characteristic of tryptophan is that it contains an indole functional group. Plants and microorganisms commonly synthesize tryptophan from shikimic acid or anthranilate. The latter condenses with phosphoribosyl-pyrophosphate (PRPP), generating pyrophosphate as a by-product. After ring opening of the ribose moiety and following reductive decarboxylation, indole-3-glycerinephosphate is produced, which in turn is transformed into indole. In the last step, tryptophan synthase catalyzes the formation of tryptophan from indole and the amino acid serine. Cereal like maize and sorghum are deficient in tryptophan in addition to lysine. Methods to determine this essential amino acid is useful to identify nutritionally superior types.

Tryptophan or (2S)-2-amino-3-(1H-indol-3-yl)propanoic acid

The indole ring of tryptophan gives an orange-red colour with ferric chloride under strongly acidic condition. The colour intensity is measured at 545 nm.

Materials

- Papain Solution: Dissolve 400 mg technical grade papain in 100 ml 0.1N sodium acetate buffer pH 7.0 Prepare the solution fresh every day
- Reagent A: Dissolve 135 mg $FeCl_3.6H_2O$ in 0.25 ml water and dilute to 500 ml with glacial acetic acid containing 2% acetic anhydride.
- Reagent B: 30 N H_2SO_4
- Reagent C: Mix equal volumes of reagent A and B about one hour before use
- Standard Tryptophan: Dissolve 5mg trptophan in 100ml water (50μg/ml)

Procedure

- Weigh 100mg of air-dried, Powdered and defatted grain sample into a small test tube.
- Add 5.0 ml papain solution, shake well and close the tube.
- Incubate at 65°C overnight.
- Cool and digest to room temperature, centrifuge and collect the clear supernatant.
- To one ml supernatant add 4ml reagent C.
- Mix in a votex mixer and incubate at 65°C for 15 min.
- Cool to room temperature and read the orange-red colour at 545nm.
- Set a blank with 5ml papain alone and repeat the above stated steps.
- Pipette out 0,0.2,0.4,0.6,0.8 and 1ml standard tryptophan and up to 1ml with water. Develop colour following the above mentioned steps.
- Standard curve. Subtract the absorbance value of the blank from that sample and calculate tryptophan content from the graph.

Reading

Mertz, E.T., Jambunathana, R. And Misra, P.S. (1975). In "*Protein Quality*", Agricultural Research, Stn Bul No. 7, Purdue University, USA, pp. 9.

Alternative method for tryptophane estimation

This method is bsed on the principal of Hopkins-Kole Reaction, wherein iron in the presence of sulphuric acid converts acetic acid to glyoxylic acid that in turn reacts with biological materials.

Materials

- Acetic acid-ferric chloride solution – Dissolve 0.54 g of ferric chloride in 10.0 ml of water containing few drops of acetic acid to prevent the formation of insoluble ferrous hydroxide.
- To 0.5 ml of above olution, glacial acetic acid containg 2.0% acetic anhydride is added to a final volume of one liter.
- Sulphuric acid (25.8 N)
- Sodium hydroxide (0.075 N).

Procedure

- Defated plant sample (100 – 150 mg) is taken in test tube. To the tube 10.0 ml of 0.075 N NaOH solution is added and mixed well using a vortex mixture.
- The tube is then shaken with mechanical device for another 60 minutes and after that the content is centrifuged at 12,000 x g for 15 minutes.
- Protein in the supernatant is estimated by micro-kjeldhal method.
- To 1.0 ml of protein extract 3.0 ml of glacial acetic acid-ferric chloride solution is added followed by 2.0 ml of sulphuric acid is also added and mixed well.
- Incubate the solution for 45 minutes at 60^0 C. After ibcubation period, sample is cooled to room temperature using ice water bath.
- A colour is developed and its optical density is measured at 545 nm against reagent blank.
- For standard curve preparation, different concentration of tryptophane solution(0 – 40 µg/ml) are prepared by appropriate dilution and are treated as stated above.

Reading

Friedman, M and Finley, J.W. (1971). Methods for tryptophane analysis. *J. Agril. Food Chem.*, **19:** 626-631.

Another alternative method for tryptophane estimation

This method for tryptophane estimation is based on the estimation of tryptophane by its reaction with p-dimethylaminobenzaldehyde (DAB) and subsequent development of a blue colour by oxidation with sodium nitrate.

Materials

- p-Dimethylaminobenzaldehyde solution – take 300 mg of p-dimethylaminobenzaldehyde and dissolve in 100 ml of 19.0 N H_2SO_4.
- Sulphuric acid – 19.0 N
- Sodium nitrate – 0.045%.

Procedure

- Suitable amount of plant sample (20 – 25 mg) is taken and to it 10.0 ml p-dimethylamino-benzaldehyde s added.
- The sample is incubated in dark at room temperature (25 ± 2° C) for 18 hours.
- After incubation period, 0.1 ml of sodium nitrate solution is added and mixed well with vortex mixture.
- The mixture is again incubated at room temperature for 30 minutes.
- After second incubation, optical density is measured at 590 nm against reagent blank.

Reading

Spies, J.R. and Chambers, D.C. (1949). Chemical determination of tryptophane in protein. *Analytical Chem.*, **21:** 1249-1266.

10.7. Estimation of Cystine

Cystine is the amino acid formed by the oxidation of two cysteine molecules that covalently link via a disulfide bond. This organosulphur compound has the formula $(SCH_2CH(NH_2)CO_2H)_2$. It is a white solid that is slightly soluble in water. Human hair and skin contain approximately 10-14% cystine by mass. Cystine serves as a substrate for the cystine-glutamate antiporter. This transport system,

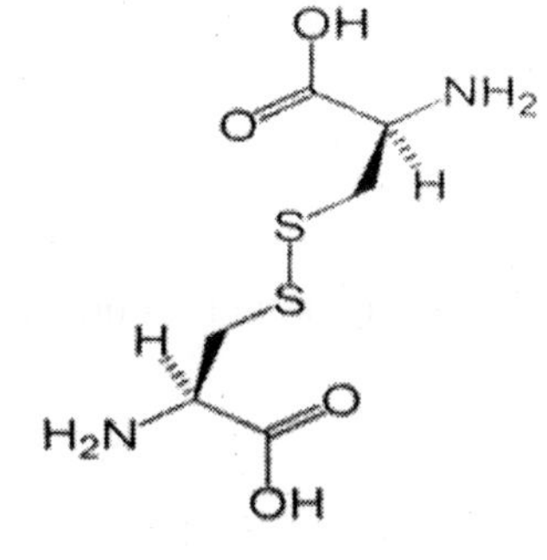

Cystine molecule

which is highly specific for cystine and glutamate, is used to increase the concentration of cystine inside the cell. In this system, the anionic form of cystine is transported in exchange for glutamate. Cystine is quickly reduced to cysteine. Cysteine prodrugs, *e.g.* acetylcysteine, increase glutamate release into the extracellular space.

Cystine (abbreviated as Cys or C) is an α-amino acid with the chemical formula $HO_2CCH(NH_2)CH_2SH$. It is a semi-essential amino acid, which means that it can be biosynthesized in humans. The thiol side chain in cysteine often participates in enzymatic reactions, serving as a nucleophile. The thiol is susceptible to oxidization to give the disulphide derivative cystine, which serves an important structural role in many proteins.

The procedure for quantitive determination of cystine is based on the reduction of cystine or cysteine sulphur by hydrazine hdrate or hydrogen sulphide which is determined colorimetrically.

$$\text{Cystine} + \text{Hydrazine} \rightarrow 2H_2S + NH_3 + \text{Unknown substance}$$

Materials

- Hydrazine hydrate : 99 – 100%.
- Bismuth nitrate solution – 2.2 g of bismuth nitrate pentahydrae is dissolved in 250 ml of 3.2% solution of manitol. To this mixture, 80 ml of glycerol and 360 ml of 2.5% gum Arabica solution in water is added. Final volume is made to 1 liter with 0.2 N acetate buffer (pH 4.4) and filter. Self life of this solution is 3 weeks ubder ambient condition.
- Sulphuric acid – 6.0 N.

Procedure

- Defated plant sample (40 – 50 mg) is taken in screw capped tubes (15 x 100 mm) and to it 1.0 ml of water and 2.0 ml of hydrazine hydrate (99 – 100%) is added.
- The tubes are closed and heated for 18 hours at 120^0 C.
- Tubes are cooled and 1.0 ml of water is added.
- The tubes are connected to glass manifold. The manifold has arrangement for continuous flow of nitrogen gas, hydrogen sulphide evolved on addition of of H_2SO_4 could be tapped into bismuth nitrate solution.
- After passing the gas stream from bismuth nitrate solution for 15 minutes, the optical density of the solution is recorded at 400 nm using bismuth nitrate as a blank.

Reading

Goa, J. (1961). Micro determination of cysteine and cystine in proteins. *Acta Chemica. Scaninavia*, **15:** 853-855.

10.8. Estimation of Proline

Proline (abbreviated as Pro or P) is an á-amino acid, one of the twenty DNA-encoded amino acids. Its codons are CCU, CCC, CCA, and CCG. It is not an essential amino acid, which means that the human body can synthesize it. It is unique among the 20 protein-forming amino acids in that the amine nitrogen is bound to not one but two alkyl groups, thus making it a secondary amine. The more common L form has *S* stereochemistry.

The distinctive cyclic structure of proline's side chain gives proline an exceptional conformational rigidity compared to other amino acids. It also affects the rate of peptide bond formation between proline and other amino acids. When proline is bound as an amide in a peptide bond, its nitrogen is not bound to any hydrogen, meaning it cannot act as a hydrogen donor, but can be a hydrogen bond acceptor.

Peptide bond formation with incoming Pro-tRNA is considerably slower than with any other tRNAs, which is a general feature of N-alkylamino acids. Peptide bond formation is also slow between an incoming tRNA and a chain ending in proline; with the creation of proline-proline bonds slowest of all.

The exceptional conformational rigidity of proline affects the secondary structure of proteins near a proline residue and may account for proline's higher prevalence in the proteins of thermophilic organisms. Protein secondary structure can be described in terms of the dihedral angles φ, Ψ, and ω of the protein backbone. The cyclic structure of proline's side chain locks the angle φ at approximately 60°.

Proline acts as a structural disruptor in the middle of regular secondary structure elements such as alpha helices and beta sheets; however, proline is commonly found as the first residue of an alpha helix and also in the edge strands of beta sheets. Proline is also commonly found in turns (another kind of secondary structure), and aids in the formation of beta turns. This may account for the curious fact that proline is usually solvent-exposed, despite having a completely aliphatic side chain.

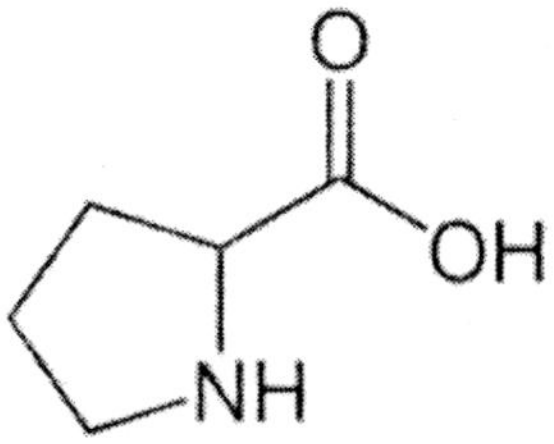

O
OH
NH

Proline molecule
(Pyrrolidine-2-carboxylic acid)

Multiple prolines and/or hydroprolines in a row can create a polyproline helix, the predominant secondary structure in collagen. The hydroxylation of proline by prolyl hydroxylase (or other additions of electron-withdrawing substituents such as fluorine) increases the conformational stability of collagen significantly. Hence, the hydroxylation of proline is a critical biochemical process for maintaining the connective tissue of higher organisms. Severe diseases such as scurvy can result from defects in this hydroxylation, *e.g.*, mutations in the enzyme prolyl hydroxylase or lack of the necessary ascorbate (vitamin C) cofactor.

Proline is a basic amino acid found in high percentage in basic proteins. Free proline is said to play a role in plants under stress conditions. Though the molecular mechanism has not yet been established for the increased levels of proline, one of the hupothesis refers to breakdown of proteins into amino acids and conversion to proline for storage. Many workers have reported a several fold increase in proline content in plants under physiological and pathological stress conditions.

During selective extraction with aqueous suphosalicylic acid, proteins are precipitated as a complex. Other interfering materials are also presumably removed by absorption to the protein-suphosalicylic acid complex. Extract of proline is made to react with ninhydrin in acidic conditions (pH 1.0) to form the chromophore (red colour) and read at 520 nm.

Materials

- Acid ninhydrin – Warm 1.25 g ninhydrin in 30 ml glacial acetic acid and 20.0 ml of 6M phosphoric acid, with agitation until dissolved. Store at 4^0 C and use it within 24 hours.
- Aqueous sulphosalicylic acid
- Glacial acetic acid
- Tolune
- Proline

Procedure

- Extract 0.5 g of plant material by homogenizing in 10.0 ml of 3% aqueous sulphosalicylic acid.
- Filter the homogenate through Whatman No. 2 filter paper.
- Take 2.0 ml of filterate in a test tube and add 2.0ml of glacial acetic acid and 2.0 ml acid-ninhydrin.

- Heat in a boiling water bath gor 60 minutes.
- Terminate the reaction by placing the tube in ice bath.
- Add 4.0 ml toluene to the reaction mixture and stir well for 20 – 30 seconds.
- Separate the toluene layer and warm to room temperature.
- Measure the red colour intensity at 520 nm. (*The colour is stable for at least 60 minutes.*)
- Run a series of standard with pure proline in a similar way and prepare a standard curve. (*The relationship between the amino acid concentration and absorbance is linear in the range of 0.02 to 0.1 μM per ml of roliene.*)
- Find the amoint of proline in the test samples from the standard curve.

Calculation

Express the proline content on fresh weight basis as follows:

μ moles per g tissue = (μg proline/ml x ml toluene/115.5) x (5/g sample)

where 115.5 is the molecular weight of proliene.

10.9. Estimation of Total Amino Acid using Cyanide-acetate Buffer

Materials

- Cyanide-acetate buffer (pH 5.3 to 5.4) – Take 90 g of trihydrated sodium acetate and dissolve in water and it add 16.75 ml of glacial acetic acid and the final volme is made to 250 ml with water. To this solution, 5.0 ml of 0.01 M sodium cyanide is added and mixed throughly.
- Ninhydrin solution – Three per cent ninhydrin solution is made by dissolving 3.0 g of ninhydrin in 100 ml of methyl cellosolve (ethylen glycol monomethyl eather).
- Isopropyl alcohol – Isopropyl alcohol is diluted 1:1 with water.

Procedure

- Extraction – 1.5 -2.0 g of dried and powdered plant sample is extracted with 80% ethanol by dipping the sample in alcohol and keeping the tubes in boiling water for 10 minutes. The material is centrifused at 10,000 rpm for 10 minutes and supernatant is saved. The extraction procedure is repeated two more times using the pellets and the supernatants are pooled.

Pooled supernatants are dried to near zero volume by keeping them over a boiling waterbath. The residue is redissolved and made to 50 ml with water and is used as source of amino acid..

- To 0.5 ml of amino acid extract add 0.5 ml of water, 0.5 ml of cyanide acetate buffer and 0.5 ml of ninhydrin solutions. The tubes are incubated at 90^0 C for 15 minutes and then cool to room temperature..
- To the above mixture add 5.0 ml of diluted (1:1) isopropyl alcohol and optical density is measured at 570 nm against reagent blank where plant extract is omitted by 0.5 ml of water.
- Amount of amino acid in extract is calculated by comparing the ODs of standard amino acid (glycine 0.1 to 1.0 μ mol/ml).

Reading

Rosen, H. (1957). A modified ninhydrin colorimetric analysis for amino acids. *Arch. Biochem. Biophys.*, **67:** 10-15.

Alternate colorimetric method for estimation free amino acids

Materials

- Citrate buffer 0.5 m pH 5.5.
- Ninhydrin (tri-ketohydrindene hydrate) solution – 1% solution in 0.5 M citrate buffer (pH 5.5).
- Glycerol 55% (v/v) in distilled water.

Procedure

- Dried and grounded 1.0 g of plant sample is extract 2 – 3 times with 10 ml of 80% (v/v) ethanol / methnol). Extracts are centrifuged and pooled and and dried to near zero volume (do not over dry or char the sample).
- The near zero volume of extract is redissolved in distilled water and volume is made to 50.0 ml.
- Plant sample (0.1 ml) is taken in test tube and is mixed with 0.5 ml of 1% ninhydrin solution, 0.2 ml of citrate buffer (pH 5.5), 1.2 ml of 55% (v/v) glycerol and the tubes are heated in boiling water bath for 20 minutes. And then cooled under running water.
- Volume of the reaction mixture is made to 5.0 ml with distilled water and optical density is measured at 570 nm against blank, where plant sample is replaced with 0.1 ml of distilled water.

- The free amino acid is expressed as μ mol/g dry weight.

10.10. Estimation of Free Amino Acid through Paper Chromatography

The Rf values of the common naturally occurring amino acids on filter paper has been determined using 55 different one-phase solvent mixture. Selection by a novel method, of some promising combinations of these mixtures have led to the development of techniques whereby 16 different amino acids may be identified rapidly on a single, two dimensional chromatogram. Treatment of the chromatograms with cyclohexamine or dicyclohexalmine prior to development of colour with ninhydrin brings about a striking variation in the usual colours and facilitates identification of individual amino acids while closely related amides and peptides have been found to give colours which differ from those obtained with the plant amino acids.

Effect of cyclohexamine and dicyclohexalamine on the ninhydrin reaction on filter paper

Amino Acids	Colour with cyclohexylamine	Colour with dicyclohexylamine
Aspartic acid	Royal blue	Tarquoise
Cystine	Orange	Carmine
Alanine	Purple	Blue purple
Histidine	Grey green	Grey
Serine	Purple	Grey purple
Phenylalanine	Blue grey	Grey brown
Proline	Yellow	Yellow
Hydroxyproline	Carmine	Yellow
Threonine	Grey purple	Grey
Tyrosine	Slate grey	Grey brown
Glycine	Red brown	Wine red
Other amino acid	Purple	Purple

Composition of solvent mixture

Code No.	10 volume	10 volume	5 volume	2 volume
A	Methanol	1-Butanol	Water	———
B	Ethanol	1-Butanol	Water	———
C	1-Butanol	Acetone	Water	———
D	1-Butanol	Butanol	Water	———
E	Ethanol	1-Butanol	Water	Propionic acid
F	1-Propanol	1-Butanol	Water	Cyclohexylamine
G	1-Butanol	Acetone	Water	Cyclohexylamine
H	1-Butanol	Butanol	Water	Cyclohexylamine
I	Ethanol	1-Butanol	Water	Diethylamine
J	1-Propanol	1-Butanol	Water	Diethylamine

Code No.	10 volume	10 volume	5 volume	2 volume
K	1-Propanol	Butanol	Water	Diethylamine
L	1-Butanol	Acetone	Water	Diethylamine
M	Ethanol	1-Butanol	Water	Dicyclohexylamine
N	1-Propanol	Butanone	Water	Dicyclohexylamine
O	1-Butanol	Acetone	Water	Dicyclohexylamine
P	1-Butanol	Butanone	Water	Dicyclohexylamine
Q	Ethanol	1-Butanol	Water	Piperidine

10.11. Estimation of Aspartate Aminotransferase (Glutamate: Oxaloacetate Aminotrasferase EC. 2.6.1.1) Activity

Aspartate transaminase (AST) or aspartate aminotransferase, also known as AspAT/ASAT/AAT or glutamic oxaloacetic transaminase (GOT), is a pyridoxal phosphate (PLP)-dependent transaminase enzyme (EC 2.6.1.1). AST catalyzes the reversible transfer of an α-amino group between aspartate and glutamate and, as such, is an important enzyme in amino acid metabolism. Aspartate transaminase catalyzes the interconversion of aspartate and α-ketoglutarate to oxaloacetate and glutamate.

Aspartate (Asp) + α-ketoglutarate → oxaloacetate + glutamate (Glu)

As a prototypical transaminase, AST relies on PLP as a cofactor to transfer the amino group from aspartate or glutamate to the corresponding keto acid. In the process, the cofactor shuttles between PLP and the pyridoxamine phosphate (PMP) form. The amino group transfer catalyzed by this enzyme is crucial in both amino acid degradation and biosynthesis. In amino acid degradation, following the conversion of α-ketoglutarate to glutamate, glutamate subsequently undergoes oxidative deamination to form ammonium ions, which are excreted as urea. In the reverse reaction, aspartate may be synthesized from oxaloacetate, which is a key intermediate in the citric acid cycle.

Aspartate transaminase, as with all transaminases, operates via dual substrate recognition; that is, it is able to recognize and selectively bind two amino acids (Asp and Glu) with different side-chains. In either case, the transaminase reaction consists of two similar half-reactions that constitute what is referred to as a ping-pong mechanism. In the first half-reaction, amino acid 1 (*e.g.*, L-Asp) reacts with the enzyme-PLP complex to generate ketoacid 1 (oxaloacetate) and the modified enzyme-PMP. In the second half-reaction, ketoacid 2 (α-ketoglutarate) reacts with enzyme-PMP to produce amino acid 2 (L-Glu), regenerating the original enzyme-PLP in the process. Formation of a racemic product (D-Glu) is very rare.

Rf values of 22 amino acids in solvent mixture A – Q

AA	A	B	C	D	E	F	G	H	I	J	K	L	M	N	O	P	Q
1	22	14	9	3	20	8	11	5	14	4	9	9	10	10	11	1	13
2	22	12	9	2	16	13	28	8	23	10	17	20	18	16	20	4	18
3	18	4	14	3	15	13	18	10	15	4	13	13	26	30	30	6	15
4	26	12	18	5	26	17	20	9	15	4	12	12	33	35	37	9	15
5	11	3	14	0	5	29	23	30	27	6	28	24	—	66	61	41	12
6	40	27	21	8	31	32	32	23	36	24	30	27	43	40	35	16	36
7	19	14	32	5	15	27	36	21	34	23	32	32	41	43	48	23	26
8	29	20	31	6	20	26	40	27	29	16	33	35	42	51	49	27	27
9	55	44	39	18	50	55	53	43	57	43	47	45	64	57	53	38	54
10	59	51	59	29	54	65	67	59	63	59	63	56	75	68	69	60	64
11	45	35	23	10	36	43	36	26	45	29	33	31	41	35	32	14	43
12	35	24	20	7	25	28	28	20	29	14	21	21	39	38	36	12	29
13	43	37	46	23	48	55	55	47	54	42	50	48	67	61	60	46	51
14	36	24	56	7	28	34	58	55	47	31	57	50	53	66	67	55	36
15	44	35	41	20	38	50	52	43	43	30	41	37	60	61	60	44	40
16	29	18	17	6	20	26	29	17	28	15	22	23	37	38	33	13	26
17	64	58	52	31	61	66	63	57	58	57	60	58	70	67	63	52	65
18	64	55	47	27	60	63	62	54	63	56	57	56	70	67	63	51	64
19	54	43	46	27	42	59	63	57	50	50	60	57	63	65	65	51	55
20	66	59	—	—	63	72	65	59	67	60	62	60	70	69	67	53	64
21	17	9	23	—	6	24	31	20	16	8	21	19	7	57	49	22	21
22	—	33	15	—	32	18	23	14	22	6	14	14	43	49	45	16	14

The specific steps for the half-reaction of Enzyme-PLP + aspartate $\rightleftharpoons$ Enzyme-PMP + oxaloacetate proceeds in the reverse manner, with α-ketoglutarate as the substrate.

Materials

- Phosphate buffer pH 7.4 – Dissolve 113.0 g of anhydrous disodium hydrogen phosphate and 2.7 g of anhydrous potassium dihydrogen phosphate and make the final volume to 1000.0 ml with water and check the pH.
- Substrate solution – Dissolve 13.3 g DL-aspartic acid in minimum amount of 1.0 N NaOH and prepare a solution with phosphate buffer pH 7.4. Add 0.146 g 2-oxoglutarate and dissolve it by adding a little more sodium hydroxide solution. Adjust the pH at 7.4 and make the final volume to 500.0 ml with phosphate buffer.
- Pyruvate standard – Dissolve 22.0 mg of sodium pyruvate in 100.0 ml of water.
- 2,4-Dinitrophenyl hydrazine (DNPH): Dissolve 19.8 mg of dinitrophenyl hydrazine in 10.0 ml of concentrated hydrochloric acid and make the volume to 100.0 ml with water.
- Sodium hydroxide 0.4 N – Dissolve 1.0 g of NaOH in 1000.0 ml of water.

Procedure

- Homogenate ~5.0 g of fresh plant tissue in pre chilled 0.2 M potassium phosphate buffer pH 7.5. Centrifuge at 25,000 x g for 15 minutes at 2^0 C. Supernatant is used as enzyme extract.
- Warm 0.5 ml of substrate solution in a warm water bath at 37^0 C for 5 minutes. To this solution add 0.2 ml of enzyme extract.
- Incubate for 60 minutes at 37^0 C. and terminate the reaction with the addition of 0.5 ml dinitrophenyl hydrazine solution. Mix well.
- Allow the reaction mixture to stand for 20 minutes at room temperature.
- To the above reaction mixture add 5.0 ml of 0.4 N NaOH, mix well and again allow to stand for 10 minutes.
- Optical density is recorded at 510 nm against reagent blank (0.5 ml each of substrate solution and DNPH and 0.2 ml of boiled enzyme extrzct).
- For preparing the standard curve a gradient of pyruvate solution, from 0.05 to 0.2 ml, is taken and volume is made to 0.2 ml with water. To this,

0.5 ml of substrate solution and 0.5 ml DNPH solution are added and colour is developed as described above.

Calculation

- The pyruvate formed by the enzymic reaction is esponsible for the differences in optical densities with sapmle and control. The pyruvate in standard produces the differences between standard and blank. Express the enzyme activity as µ mole of pyruvate formed per minute per mg of protein.

10.12. Estimation of Alanine Aminotrasferase (Glutamate Pyruvate Aminotransferase EC. 2.6.1.2) Activity

Alanine Transaminase (ALT) is a transaminase enzyme (EC. 2.6.1.2) and is known by multiple names including: alanine aminotransferase, glutamate-pyruvate transaminase (GPT), and serum glutamate-pyruvate transaminase (SGPT). As a member of the aminotransferase family, ALT catalyzes the reversible transfer of the amino group from glutamate to pyruvate while replacing the amino group of glutamate with a carbonyl group:

HO OH NH_2 + H_3C HO → HO OH + H_3C NH_2 HO

Glutamate Pyruvate α-Ketoglutarate Alanine

Alanine aminotransferase (AlaAT; E.C. 2.6.1.2) converts pyruvate and glutamate to alanine and 2-oxoglutarate. In doing so, it is involved in both carbon and nitrogen metabolism in plants. Arabidopsis contains four AlaAT homologues. Two of these homologues encode GGAT1 and GGAT2, which have both AlaAT activity and glutamate:glyoxylate aminotransferase (GGAT) activity. The presence of a peroxisomal targeting signal in both GGAT1 and GGAT2 suggests that they are localized in the peroxisomes, and subcellular fractionation analysis has found GGAT activity to be localized exclusively within Arabidopsis peroxisomes. GGATs appear to play important roles in photorespiration and amino acid metabolism. The other two genes, AlaAT1 and AlaAT2, have not been characterized in detail. In silico predictions suggest that AlaAT2 is mitochondrially localized while AlaAT1 is cytosolic. Cytosolic localization of a GFP–AlaAT1 fusion protein was observed in *Nicotiana tabacum* BY-2 cells.

ALT (and all transaminases) require the coenzyme pyridoxal phosphate, which is converted into pyridoxamine in the first phase of the reaction, when an amino acid is converted into a keto acid. The reaction catalyzed by ALT requires the cofactor pyridoxal-5'-phosphate (P5P), the active form of vitamin B6. P5P binds covalently but reversibly to an active-site lysine. This helps catalyze the reaction by accepting the amino group from the amino acid then transferring it to the carbonyl compound. Any sample, whether serum or purified extract, to which P5P is added may show a marked increase in ALT activity if some or all of the enzyme does not already have the cofactor bound to it.

Measuring ALT is usually done by an indirect enzymatic method. There are no useful spectrophotometric changes that occur during the transamination reaction. However a very useful and versatile technique is to couple a product of a reaction to another enzymatic reaction that either reduces or oxidizes NAD+/NADH respectively. There are many such reactions. Reduction of NAD+ results in an increase in absorption at 340 nm. Oxidation of NADH results in the opposite effect. In the case of ALT, the product pyruvate can be very conveniently coupled to the lactate dehydrogenase reaction. This results in the oxidation of NADH and the ALT-LDH coupled reaction is followed by monitoring the decrease in absorbance at 340 nm.

Materials

- Phosphate buffer pH 7.4 – Dissolve 113.0 g of anhydrous disodium hydrogen phosphate and 2.7 g of anhydrous potassium dihydrogen phosphate and make the final volume to 1000.0 ml with water and check the pH.
- Substrate solution – Dissolve 9.0 g alanine in 90.0 ml of water with addition of about 2.5 ml of 1.0 N sodium hydroxide. Adjust pH to 7.4. Then add 0.146 g of 2-oxoglutarate and dissolve it by addition of a little NaOH and adjust pH to 7.4. Make the final volume to 500.0 ml with phosphate buffer pH 7.4.
- Pyruvate standard – Dissolve 22.0 mg of sodium pyruvate in 100.0 ml of water.
- 2,4-Dinitrophenyl hydrazine (DNPH): Dissolve 19.8 mg of dinitrophenyl hydrazine in 10.0 ml of concentrated hydrochloric acid and make the volume to 100.0 ml with water.
- Sodium hydroxide 0.4 N – Dissolve 1.0 g of NaOH in 1000.0 ml of water.

Procedure

- Homogenate ~5.0 g of fresh plant tissue in pre chilled 0.2 M potassium phosphate buffer pH 7.5. Centrifuge at 25,000 x g for 15 minutes at 2^{0}C. Supernatant is used as enzyme extract.
- Warm 0.5 ml of substrate solution in a warm water bath at 37^{0}C for 5 minutes. To this solution add 0.2 ml of enzyme extract.
- Incubate for 60 minutes at 37^{0} C. and terminate the reaction with the addition of 0.5 ml dinitrophenyl hydrazine solution. Mix well.
- Allow the reaction mixture to stand for 20 minutes at room temperature.
- To the above reaction mixture add 5.0 ml of 0.4 N NaOH, mix well and again allow to stand for 10 minutes.
- Optical density is recorded at 510 nm against reagent blank (0.5 ml each of substrate solution and DNPH and 0.2 ml of boiled enzyme extrzct).
- For preparing the standard curve a gradient of pyruvate solution, from 0.05 to 0.2 ml, is taken and volume is made to 0.2 ml with water. To this, 0.5 ml of substrate solution and 0.5 ml DNPH solution are added and colour is developed as described above.

Calculation

The pyruvate formed by the enzymic reaction is esponsible for the differences in optical densities with sapmle and control. The pyruvate in standard produces the differences between standard and blank. Express the enzyme activity as μ mole of pyruvate formed per minute per mg of protein.

CHAPTER 11

Lipids

Lipids are a group of naturally occurring molecules that include fats, waxes, sterols, fat-soluble vitamins (such as vitamins A, D, E, and K), monoglycerides, diglycerides, triglycerides, phospholipids, and others. The main biological functions of lipids include storing energy, signalling, and acting as structural components of cell membrane. Lipids have applications in the cosmetic and food industries as well as in nanotechnology.

Lipids may be broadly defined as hydrophobic or amphiphilic small molecules; the amphiphilic nature of some lipids allows them to form structures such as vesicles, liposomes, or membranes in an aqueous environment. Biological lipids originate entirely or in part from two distinct types of biochemical subunits or "building-blocks": ketoacyl and isoperen groups. Using this approach, lipids may be divided into eight categories:fatty acids, glycerolipids, glycerophospholipids, sphingolipids, saccharolipids and polyketides (derived from condensation of ketoacyl subunits); and sterol lipids and prenol lipids (derived from condensation of isoprene subunits).

Although the term *lipid* is sometimes used as a synonym for fats, fats are a subgroup of lipids called triglycerides. Lipids also encompass molecules such as fatty acids and their derivatives (including tri-, di-, monoglycerides and phospholipids), as well as other sterol containing metabolites such as chloestrol. Although humans and other mammals use various biosynthetic pathways to both break down and synthesize lipids, some essential lipids cannot be made this way and must be obtained from the diet.

11.1. Categories of Lipids

Fatty acids

Fatty acids, or fatty acid residues when they form part of a lipid, are a diverse group of molecules synthesized by chain-elongation of an acetyl-CoA primer with malonyl-CoA or methylmlonyl-CoA groups in a process called fatty acid synthesis. They are made of a hydrocarbon chain that terminates with a carboxylic acid group; this arrangement confers the molecule with a polar, hydrophilic end, and a nonpolar, hydrophobic end that is insoluble in water. The fatty acid structure is one of the most fundamental categories of biological lipids, and is commonly used as a building-block of more structurally complex lipids. The carbon chain, typically between four and 24 carbons long, may be saturated or unsaturated, and may be attached to functional groups containing oxygen, halogens, nitrogen, and sulphur. If a fatty acid contains a double bond, there is the possibility of either a cis or trans geometric isomerism, which significantly affects the molecule's configuration. Cis-double bonds cause the fatty acid chain to bend, an effect that is compounded with more double bonds in the chain. Three double bonds in 18-carbon linolenic acid, the most abundant fatty-acyl chains of plant thylakoid membranes, render these membranes highly fluid despite environmental low-temperatures, and also makes linolenic acid give dominating sharp peaks in high resolution 13-C NMR spectra of chlorplasts. This in turn plays an important role in the structure and function of cell membranes. Most naturally occurring fatty acids are of the cis configuration, although the trans form does exist in some natural and partially hydrogenated fats and oils.

Examples of biologically important fatty acids include the eicosanoids, derived primarily from arachidonic and eicosapentaenoic acid, that include prostaglandins, leukotrienes, and thromboxanes. Docosahexaenoic acid is also important in biological systems, particularly with respect to sight. Other major lipid classes in the fatty acid category are the fatty esters and fatty amides.

Glycerolipids

Glycerolipids are composed of mono-, di-, and tri-substituted glycerols, the most well-known being the fatty acid triesters of glycerol, called triglycerides. The word "triacylglycerol" is sometimes used synonymously with "triglyceride". In these compounds, the three hydroxyl groups of glycerol are each esterified, typically by different fatty acids. Because they function as an energy store, these lipids comprise the bulk of storage fat in animal tissues. The hydrolysis of the ester bonds of triglycerides and the release of glycerol and fatty acids from adipose tissue are the initial steps in metabolising fat.

Additional subclasses of glycerolipids are represented by glycosylglycerols, which are characterized by the presence of one or more sugar residues attached to glycerol via a glycosidic linkage. Examples of structures in this category are the digalactosyldiacylglycerols found in plant membranes and seminolipid from mammalian sperm cells.

Glycerophospholipids

Glycerophospholipids, usually referred to as phospholipids, are ubiquitous in nature and are key components of the lipid bilayer of cells, as well as being involved in metabolism and cell signalling. Neural tissue (including the brain) contains relatively high amounts of glycerophospholipids, and alterations in their composition have been implicated in various neurological disorders. Glycerophospholipids may be subdivided into distinct classes, based on the nature of the polar headgroup at the *sn-3* position of the glycerol backbone in eukaryotes and eubacteria, or the sn-1 position in the case of archaebacteria.

Examples of glycerophospholipids found in biological membranes are phosphatidylcholine (also known as PC, GPCho orlectin), phosphotidy-lethanolamine (PE or GPEtn) and phosphati-dylserine (PS or GPSer). In addition to serving as a primary component of cellular membranes and binding sites for intra- and intercellular proteins, some glycerophospholipids in eukaryotic cells, such as phosphotidylinositols and phosphatidic acids are either precursors of or, themselves, membrane-derived second messenger. Typically, one or both of these hydroxyl groups are acylated with long-chain fatty acids, but there are also alkyl-linked and 1Z-alkenyl-linked (plasmalogen) glycerophospholipids, as well as dialkylether variants in archaebacteria.

Sphingolipids

Sphingolipids are a complicated family of compounds that share a common structural feature, a sphingoid base backbone that is synthesized *de novo* from the amino acid serine and a long-chain fatty acyl CoA, then converted into ceramides, phosphosphingolipids, glycosphingolipids and other compounds. The major sphingoid base of mammals is commonly referred to as sphingosine. Ceramides (N-acyl-sphingoid bases) are a major subclass of sphingoid base derivatives with an amide-linked fatty acid. The fatty acids are typically saturated or mono-unsaturated with chain lengths from 16 to 26 carbon atoms.

The major phosphosphingolipids of mammals are sphingomyelines (ceramide phosphor-cholines), whereas insects contain mainly ceramide phosphoe-thanolamines and fungi have phytoceramide phosphoinositols and mannose-containing headgroups. The glycophingo-lipids are a diverse family of molecules composed of one or more sugar residues linked via a glycosidic bond to the sphingoid base. Examples of these are the simple and complex glycosphingolipids such as cerebrosides and gangliosides.

Sterol lipids

Sterol lipids, such as cholesterol and its derivatives, are an important component of membrane lipids, along with the glycerophospholipids and sphingomyelins. The steroids, all derived from the same fused four-ring core structure, have different biological roles as hormones and signalling molecules. The eighteen-carbon (C18) steroids include the estrogen family whereas the C19 steroids comprise the androgens such as testosterone and androsterone. The C21 subclass includes the progestogens as well as the glucocorticoids and mineralo-corticoids. The secosteroids, comprising various forms of vitamin D, are characterized by cleavage of the B ring of the core structure. Other examples of steroids are the bile acids and their conjugates, which in mammals are oxidized derivatives of cholesterol and are synthesized in the liver. The plant equivalents are the phytosterols, such as α-sitosterol, stigmasterol, and brassicasterol; the latter compound is also used as a biomarker for algal growth. The predominant sterol in fungal cell membranes is ergosterol.

Prenol lipids

Prenol lipids are synthesized from the five-carbon-unit precursors isopentenyl diphosphate and dimethylallyl diphosphate that are produced mainly via the mevalonic (MVA) pathway. The simple isoprenoids (linear alcohols, diphosphates, *etc.*) are formed by the successive addition of C5 units, and are classified according to number of these terpene units. Structures containing greater than 40 carbons are known as polyterpenes. carotenoids are important simple isoprenoids that function as antioxidants and as precursors of vitamin A. Another biologically important class of molecules is exemplified by the quinines and hydroquinones, which contain an isoprenoid tail attached to a quinonoid core of non-isoprenoid origin. Vitamin E and vitamin K, as well as the ubiquinones, are examples of this class. Prokaryotes synthesize polyprenols (called bactoprenols) in which the terminal isoprenoid unit attached to oxygen remains unsaturated, whereas in animal polyprenols (dolichols) the terminal isoprenoid is reduced.

Saccharolipids

Saccharolipids describe compounds in which fatty acids are linked directly to a sugar backbone, forming structures that are compatible with membrane bilayers. In the saccharolipids, a monosaccharide substitutes for the glycerol backbone present in glycerolipids and glycerophospholipids. The most familiar saccharolipids are the acylated glucosamine precursors of the Lipd A component of the lipopolysaccharides in Gram-negative bacteria. Typical lipid A molecules are disaccharides of glucosamine, which are derivatized with as many as seven fatty-acyl chains. The minimal lipopolysaccharide required for growth in *E. coli* is Kdo_2-Lipid A, a hexa-acylated disaccharide of glucosamine that is glycosylated with two 3-deoxy-D-manno-octulosonic acid (Kdo) residues.

Polyketides

Polyketides are synthesized by polymerization of acetyl and propionyl subunits by classic enzymes as well as iterative and multimodular enzymes that share mechanistic features with the fatty acid synthases. They comprise a large number of secondary metabolites and natural products from animal, plant, bacterial, fungal and marine sources, and have great structural diversity. Many polyketides are cyclic molecules whose backbones are often further modified by glycosylation, methylation, hydroxylation, oxidation, and/ or other processes. Many commonly used anti-microbial, anti-parasitic, and anti-cancer agents are polyketides or polyketide derivatives, such as erythromycins, tetracyclines, avermectins, and antitumor epothilones.

11.2. Biological Functions of Lipids

Membranes

Eukaryotic cells are compartmentalized into membrane-bound organells that carry out different biological functions. The glycerophospholipids are the main structural component of biological membranes, such as the cellular plasma membrane and the intracellular membranes of organelles; in animal cells the plasma membrane physically separates the interceluular components from the extracellular environment. The glycerophospholipids are amphipathic molecules (containing both hydrophobic and hydrophilic regions) that contain a glycerol core linked to two fatty acid-derived "tails" by ester linkages and to one "head" group by a phosphate ester linkage. While glycerophospholipids are the major component of biological membranes, other non-glyceride lipid components such as sphingomyelin and sterols (mainly cholestrol in animal cell membranes) are also found in biological membranes. In plants and algae, the galactosyldiacylglycerols, and sulfoquinovosyldiacylglycerol, which lack a phosphate group, are important components of membranes of chloroplasts and related organelles and are the most abundant lipids in photosynthetic tissues, including those of higher plants, algae and certain bacteria.

Plant thylakoid membranes have the largest lipid component of a non-bilayer forming monogalactosyl diglyceride (MGDG), and little phospholipids; despite this unique lipid composition, chloroplast thylakoid membranes have been shown to contain a dynamic lipid-bilayer matrix as revealed by magnetic resonance and electron microscope studies.

Bilayers have been found to exhibit high levels of birefringence, which can be used to probe the degree of order (or disruption) within the bilayer using techniques such as dual polarisation interferometry and Circular dichroism.

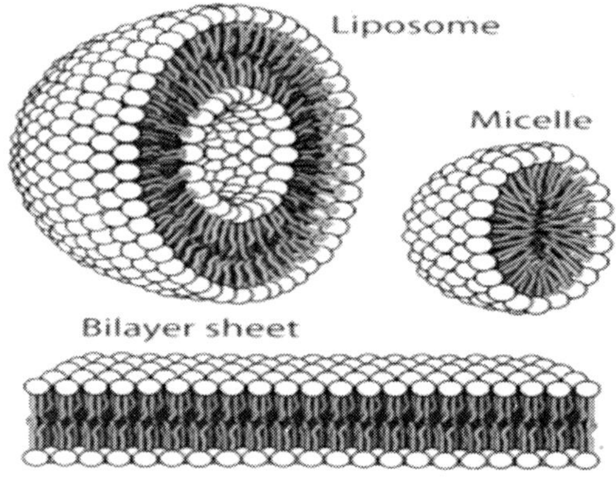

Self-organization of phospholipids: a spherical liposome, a micelle, and a lipid bilayer.

A biological membrane is a form of lipid bilayer. The formation of lipid bilayers is an energetically preferred process when the glycerophospholipids described above are in an aqueous environment. This is known as the hydrophobic effect. In an aqueous system, the polar heads of lipids align towards the polar, aqueous environment, while the hydrophobic tails minimize their contact with water and tend to cluster together, forming a vesicle; depending on the concentration of the lipid, this biophysical interaction may result in the formation of micelles, liposomes, or lipid bilayers. Other aggregations are also observed and form part of the polymorphism of amphiphile (lipid) behavior. Phase behavior is an area of study within biophysics and is the subject of current academic research. Micelles and bilayers form in the polar medium by a process known as the hydrophobic effect. When dissolving a lipophilic or amphiphilic substance in a polar environment, the polar molecules (*i.e.*, water in an aqueous solution) become more ordered around the dissolved lipophilic substance, since the polar molecules cannot form hydrogen bonds to the lipophilic areas of the amphiphile. So in an aqueous environment, the water molecules form an ordered "clathrate" cage around the dissolved lipophilic molecule.

Energy storage

Triglycerides, stored in adipose tissue, are a major form of energy storage both in animals and plants. The adipocyte, or fat cell, is designed for continuous synthesis and breakdown of triglycerides in animals, with breakdown controlled mainly by the activation of hormone-sensitive enzyme lipase. The complete oxidation of fatty acids provides high caloric content, about 9 kcl/g, compared with 4 kcal/g for the breakdown of carbohydrates and proteins. Migratory birds that must fly long distances without eating use stored energy of triglycerides to fuel their flights.

Signalling

In recent years, evidence has emerged showing that lipid signalling is a vital part of the cell signalling. Lipid signalling may occur via activation of G protein-coupled or nuclear receptors, and members of several different lipid categories have been identified as signalling molecules and cellular messengers. These include sphingosine-1-phosphate, a sphingolipid derived from ceramide that is a potent messenger molecule involved in regulating calcium mobilization, cell growth, and apoptosis; diacylglycerol (DAG) and the phosphatidyl-inositol phosphates (PIPs), involved in calcium-mediated activation of protein kinase C; the prostaglandins, which are one type of fatty-acid derived eicosanoid involved in inflammation and immunity; the steroid hormones such as estrogen, testosterone and cortisol, which modulate a host of functions such as reproduction, metabolism and blood pressure; and the oxysterols such as 25-hydroxy-cholesterol that are liver X receptior agonists. Phosphatidyl-serine lipids are known to be involved in signaling for the phagocytosis of apoptotic cells and/or pieces of cells. They accomplish this by being exposed to the extracellular face of the cell membrane after the inactivation of flippases which place them exclusively on the cytosolic side and the activation of scramblases, which scramble the orientation of the phospholipids. After this occurs, other cells recognize the phosphatidylserines and phagocytosize the cells or cell fragments exposing them.

Other functions

The "fat-soluble" vitamins (A, D, E and K), which are isoprene-based lipids are essential nutrients stored in the liver and fatty tissues, with a diverse range of functions. Acetyl-carnitines are involved in the transport and metabolism of fatty acids in and out of mitochondria, where they undergo beta-oxidation. Polyprenols and their phosphorylated derivatives also play important transport roles, in this case the transport of oligosaccharides across membranes. Polyprenol phosphate sugars and polyprenol diphosphate sugars function in extra-cytoplasmic glycosylation reactions, in extracellular polysaccharide biosynthesis (for instance, peptidoglycan polymerization in bacteria), and in eukaryotic protein N-glycosylation. Cardiolipins are a subclass of glycerophospholipids containing four acyl chains and three glycerol groups that are particularly abundant in the inner mitochondrial membrane. They are believed to activate enzymes involved with oxidative phosphorylation. Lipids also form the basis of steroid hormones.

11.3. Estimation of Oil in Oilseeds

Fats are fatty acid esters of glycerol. Fat as liquis is called oil. Seeds like gingerly, groundnut, castor, sunflower *etc.* contain oil as reserve material for the embryo. Oil from a known quantity of the seeds is extracted with petroleum eather. It is then distilled off completely, dried, the oil weighd and the per cent oil is calculated.

Materials

- Petroleum eather (40 – 60^0 C)
- Whatman No. 2 filter paper
- Absorbant cotton
- Soxhlet apparatus

Procedure

- Fold a pirce of filter paper in such a way to hold the seed material. Wrap around a second filter paper which is left open at the top like a thimble. A piece of cotton wool is placed at the top to evenly distribute the solvant as it drop on the sample during extraction.
- Place the sample packet in the butt tubes of the soxhlet apparatus.
- Extract with petroleum eather (150 drop / min.) for 60 hours without interruption by gentle heating.
- Allow to cool and dismantle extaction flask. Evaporate the eather on a stream or water bath until no odour of eather remains. Cool at room temperature.
- Carefully emove the dirt or moisture from outside the flask and weigh the flask.. Repeat the heating until constant weight is recorded.

Calculation

Oil in ground sample % = (Weight of the oil (g)) / (Weight of sample (g)) X 100

Oil to dry weight basis = (% oil in grounded sample) / (100 % moisture in whole seed)

11.4. Estimation of Free Fatty Acids

Volumetric method

A small quantity of free fatty acids is usuallly present in oils along with the triglycerides. The free fatty acid content is known as acid number / acid value. It increases during storage. The keeping quality of oil therefore relies upon the free fatty acid content. A known quantity of oil is refluxed with an excess amount of alcoholic KOH. After saponification, theremaining KOH is estimated against a standard acid.

Materials

- Hydrochloric acid 0.5 N, accurately standardised.
- Alcoholic KOH – Dissolve 40 g KOH in one litre of distilled alcohol keeping the temperature below 15.5^0 C while the alkali is being dissolved. This solution remain clear.
- Phenolphthalein indicator – 1% in 95% alcohol.
- Air condenser.

Procedure

- Melt the sample, if it is not already liquid, and filter through filter paper to remove any impurities and the traces of moisture. The sample must be completely moisture free.
- Weigh 4 – 5 g sample into flask. Add 50.0 ml of alcoholic KOH from burette by allowing it to drain for a definite period of time.
- Prepare a blank also by taking only 50.0 ml of alcoholic KOH allowing it to drain at the same duration of time.
- Connect the air condenser to the flask and boil them gently for about 60 minutes.
- After the flask and condenser get cooled, rinse down the inside of the condenser with a little distilled water and then remove the condenser.

- Add about 1.0 ml of indicatorand titrate against 0.5 N HCl untill the pink colour just disappear.

Calculation

Saponification value

= (28.05 X (titre value of blank – titre value of sample)) / (Weight of sample (g))

Colorimetric method

The samples is shaken with chloroform from buffered copper nitrate solution when the free fatty acid are removed from their complex with albumin and form copper fatty acid sops, which are extracted into the organic phase. The amount of copper in the chloroform is equivalent to the amount of the fee fatty acids in the sample and this is assayed by sodium diethyl-dithiocarbamate, a specific and very sensitive colorimetric reagent for estimation of copper.

Materials

- Copper reagent: 270 ml of 2 m mol/l triethanolamine, 30 ml of 1 mol/l acetic acid, and 30 ml of 100 g/l copper sulphate
- Sodium diethyl-dithiocarbamte: 1 g/l in n-butanol prepared fresh before use.
- Palmitic acid: 5 m mol/l in chloroform

Procedure

- In glass stoppered tube take 0.5 ml of sample + 2.5 ml of the copper reagent and shake vigoursky with voetex mixture.
- Add 5.0 ml of chloroform and again shake vigoursly for 1 minute. Three distict layers appear, an upper aquous layer , an interphase containing denatured proteins, and a layer of chloroform containing the salt of free fatty acids.
- Remove the upper aquous layer with Pasture pipette and discard.
- Break the interphase layer with fresh pipette and carefuly transfer the chloroform to a test tube. It is very important at this stage to avoid contaminating chloroform layer with the other two phases which contain high concentration of copper.

- Take 3 ml of chloroform, add 0.5 ml of sodium diethyl-dithiocarbamate reagent and observe the ooptical density at 440 nm.
- Prepae a range of concentrations of the standard solution of palmitic acid to cover the range 0 – 5.0 m mol.
- Take 0.2 ml of each standard concentration and evaporate to dryness on a boiling water bath.
- Add 0.5 ml of water and develop the colour as stated above.
- Plot the ODs against concentrations and plot the standard curve.

11.5. Estimation of Iodine Value of an Oil

The iodine value (or "iodine adsorption value" or "iodine number" or "iodine index") in chemistry is the mass of iodine in grams that is consumed by 100 grams of a chemical structure. Iodine numbers are often used to determine the amount of unsaturation in fatty acids. This unsaturation is in the form of double bonds, which react with iodine compounds. The higher the iodine number, the more C=C bonds are present in the fat. It can be seen from the table that coconut oil is very saturated, which means it is good for making soap. On the other hand, linseed oil is highlu unsaturated, which makes it a drying oil, well suited for making oil paints.

The oils contain both saturated and unsaturated fatty acids. The iodine value is a measure of the degree of unsaturation in an oil. It is constant for a perticular oil or fat. Iodine value is a useful parameter in studying oxidative rancidity of oils since higher the unsaturation the greater is possibility of the oil to go rancid. Iodine value/number is defined as the 'g' of iodine absorbed by 100 g of the oil.

A solution of iodine ions is yellow/brown in colour. When added to a complex solution however, any chemical group (usually C=C double bonds) in solution that react with iodine effectively reduce the strength, or magnitude of the colour (by taking iodine ions out of solution). Thus the amount of iodine required to make a solution retain the characteristic yellow/brown colour can effectively be used to determine the amount of iodine sensitive groups present in the solution.

In a typical procedure, the fatty acid is treated with an excess of the Hanuš or Wijs solutions, which are, respectively, solutions of iodine monobromides (IBr) and iodine monochloride (ICl) in glacial acetic acid. Unreacted iodine monobromide (or monochloride) is then allowed to react with potassium iodide, converting it to iodine, whose concentration can be determined by titration with sodium thiosulphate.

The chemical reaction associated with this method of analysis involves formation of the diiodo alkane (R and R' symbolize alkyl or other organic groups):

$$\mathbf{R - CH = CH - R + I2 \rightarrow R - CHI - CHI - R}$$

The precursor alkene (RCH=CHR') is colourless and so is the organoidine product (RCHI-CHIR').

Materials

- Hanus iodine solution – weigh 13.6 g of iodine and dissolve in 825 ml glacial acetic acid by heating. And cool. Titrate 25.0 ml of this solution against 0.1 N sodium thiosulphate. Measure another portion of 200 ml of glacial acetic acid and add 3.0 ml of bromine in it. To 5.0 ml of this solution add 10.0 ml of 15% potassium iodide solution and titrate against 0.1 N sodium thiosulphate. Calculate the value of bromine solution, to double halogen content of the remaining 800 ml of the above iodine solution as follows:
- X = B / C. where X = ml of bromine solution required to double the halogen content, B = 800 x thiosulphate equivalent of 1.0 ml of iodine solution and C = thiosulphate equivalent of 1.0 ml of bromine solution.
- 15% potassium iodide solution.
- 0.1% sodium thiosulphate.
- 1% starch.

Procedure

- Weigh 0.5 g or 0.25 g of oil into an iodine flask and dissolve in 10.0 ml of chloroform.
- Add 25.0 ml of Hanus iodine solution using a pipette, draing it in a definite time. Mix well and allow to stand in dark for exactly 30 minutes with occasional stiring.
- Add 10.0 ml of 15% potassium iodide, shake throughly and add 100 ml of freshly boiled and cooled water, washing down any free iodine on the stopper.
- Titrate against 0.1 N sodium thiosuphate until yellow solution turn almost colourless.
- Add a few drops of starch as an indicator and titrate until the blue colour completely disappear.

- Towards the end of titration, stopper the flask and shake vigorously so that any iodine remaining in solution in chloroform is taken up by potassium iodide solution.
- Run a blank without the sample.

Calculations

The quantity of thiosulphate required for blank minus quantity required for samples gives thiosulphate equivalentof iodine absorbed by the fat or oil taken for analysis.

Iodine number = ((B – S) X N X 12.69) / (g sample)

Where

B = ml thiosulphate for blank

S = ml thiosulphate for sample

N = normality of thiosulphate solution

Amount of fat / oil taken should be afdujtsted such that the excess iodine in the added 25.0 ml of Hanus iodine solution has about 60% of excess iodine of the amount added, *i.e.*, if (B – S) is greater than B/2, repeat with smaller amount of sample.

11.6. Estimation of Saponification Value of Oils

Saponification value (or "saponification number"/"Koettstorfer number", also referred to as "sap" in short) represents the number of milligrams of potassium hydroxide required to saponify 1g of fat under the conditions specified. It is a measure of the average molecular weight (or chain length) of all the fatty acids present. As most of the mass of a fat/tri-ester is in the 3 fatty acids, it allows for comparison of the average fatty acid chain length. The long chain fatty acids found in fats have a low saponification value because they have a relatively fewer number of carboxylic functional groups per unit mass of the fat as compared to short chain fatty acids. If more moles of base are required to saponify N grams of fat then there are more moles of the fat and the chain lengths are relatively small, given the following relation:

Number of moles = mass of oil/relative atomic mass

The calculated molar mass is not applicable to fats and oils containing high amounts of unsaponifiable material, free fatty acids (>0.1%), or mono- and diacylglycerols (>0.1%).

Handmade soap makers who aim for bar soap use NaOH (sodium hydroxide, lye). Because saponification values are listed in KOH (potassium hydroxide) the value must be converted from potassium to sodium to make bar soap; potassium soaps make a paste, gel or liquid soap. To convert KOH values to NaOH values, divide the KOH values by the ratio of the molecular weights of KOH and NaOH (1.403).

Fats and oils are the principle stored forms of energy in many organisms. They are highly reduced compounds and are derivatives of fatty acids. Fatty acids are carboxylic acids with hydrocarbon chains of 4 to 36 carbons, they can be saturated or unsaturated. The simplest lipids constructed from fatty acids are triacylglycerols or triglycerides. Triacylglycerols are composed of three fatty acids each in ester linkage with a single glycerol. Since the polar hydroxyls of glycerol and the polar carboxylates of the fatty acids are bound in ester linkages, triacyl glycerols are non polar, hydrophobic molecules, which are insoluble in water.

Saponification is the hydrolysis of fats or oils under basic conditions to afford glycerol and the salt of the corresponding fatty acid. Saponification literally

$$\begin{array}{l} H_2C-OH \\ HC-OH \\ H_2C-OH \end{array} + 3\,R-\overset{O}{\overset{\|}{C}}-OH \longrightarrow \begin{array}{l} H_2C-O-\overset{O}{\overset{\|}{C}}-R \\ HC-O-\overset{O}{\overset{\|}{C}}-R \\ H_2C-O-\overset{O}{\overset{\|}{C}}-R \end{array}$$

Glycerol   Fats   Triacylglycerol

means "soap making". It is important to the industrial user to know the amount of free fatty acid present, since this determines in large measure the refining loss. The amount of free fatty acid is estimated by determining the quantity of alkali that must be added to the fat to render it neutral. This is done by warming a known amount of the fat with strong aqueous caustic soda solution, which converts the free fatty acid into soap. This soap is then removed and the amount of fat remaining is then determined. The loss is estimated by subtracting this amount from the amount of fat originally taken for the test.

The saponification number is the number of milligrams of potassium hydroxide required to neutralize the fatty acids resulting from the complete hydrolysis of 1.0 g of fat. It gives information concerning the character of the fatty acids of the fat- the longer the carbon chain, the less acid is liberated per gram of fat hydrolysed. It is also considered as a measure of the average molecular weight (or chain length) of all the fatty acids present. The long chain fatty acids found in fats have low saponification value because they have a relatively fewer number of carboxylic functional groups per unit mass of the fat and therefore high molecular weight.

Materials

- Hydrochloric acid 0.5 N, accurately standardised.
- Alcoholic KOH: Dissolve 40 g of KOH in one liter of distilled alcohol keeping the temperature below 15.5^0 C while the alkali is being dissolved. This solution should remain clear.
- Phenolphthalein indicator: 1% in 95% alcohol.
- Air condenser.

Procedure

- Melt the sample, if it is not already liquid and filter through paper to remove any impurities and the last traces of moisture.
- Take 4-5 g of sample into a flask and add 50 ml of alcoholic KOH from burette by allowing it to drain for a definite period of time.
- Prepare a blank by taking only 50 ml of alcoholic KOH.
- Connect air condenser to the flask and boil them gently for about an hour.
- After the flask and condenser got cooled. Rinse down the inside of the condenser with a little distilled water and then remove the condenser.
- Add about 1 ml of indicator and titrate against 0.5 N HCl until the pink colour just disappear.

Calculation

Saponification value = (28.05 x (titer value of blank – titer value of sample) / Weight of sample

Reading

William Horowitz (Ed. (1975). Official Methods of Analysis of AOAC, 12th Edision, pp. 490.

11.7. Estimation of Peroxide Value of an Oil

Detection of Peroxide gives the initial evidence of rancidity in unsaturated fats and oils. Other methods are available but peroxide value is the most widely used. It gives a measure of the extent to which an oil sample has undergone primary oxidation, extent of secondary oxidation may be determined from p-anisidine test. The double bonds found in fats and oils play a role in autoxidation. Oils with a high degree of unsaturation are most susceptible to autoxidation. The best test for autoxidation (oxidative rancidity) is determination of the peroxide value. Peroxides are intermediates in the autoxidation reaction. Autoxidation is a free radical reaction involving oxygen that leads to deterioration of fats and oils which form off-flavours and off-odours. Peroxide value, concentration of peroxide in an oil or fat, is useful for assessing the extent to which spoilage has advanced.

Definition

The peroxide value is defined as the amount of peroxide oxygen per 1 kilogram of fat or oil. Traditionally this was expressed in units of milliequivalents, although if we are using SI units then the appropriate option would be in millimoles per kilogram (N.B. 1 milliequivalents = 0.5 millimole; because 1 mEq of O_2 =1 mmol/2=0.5 mmol of O_2, where 2 is valence). Note also that the unit of milliequivalent has been commonly abbreviated as mequiv or even as meq.

Peroxide values of fresh oils are less than 10 milliequivalents /kg, when the peroxide value is between 30 and 40 milliequivalents/kg, a rancid taste is noticeable.

Material

- Solvent mixture: Mix two volumes of glacial acetic acid with one volume of chloroform.
- 5% potassium iodide solution.
- 1% soluble starch solution.
- N/500 sodium thiosulphate solution: Prepare N/10 solution of sodium thiosulphate and dilute to N/500 on the day of use.

Procedure

- Weigh 1.0 g of oil or fat into a clean dry boiling tube and add 1.0 g of powdered potassium iodide and 20 ml of solvent mixture.

- Place the tube in boiling water bath so that the liquid boils within 30 seconds and allow tp boil vigorsousIy for not more than 30 seconds.
- Transfer the contents quickly to a conical flak containing 20 ml of 5% potassium iodide solution.
- Wash the tube twice with 25 ml of water wach time and collect into conocal flask.
- Titrate against N/500 sodium thiosulphate solution until yellow colour is almost disappears.
- A blank should also be set at the same time.

Calculation

Peroxide value (milliequivalent peroxide/kg ample) = (S x N x 100) / g of sample

Where S = ml of sodium thiosulphate (Test – Blank) and N = normality of sodium thiosulphate solution.

Reading

Cox, H.E. and Pearson, D. (1962). "*The chemical analysis of foods*", Chemical Publishing Company, Inc. New York, pp. 421.

11.8. Estimation of Lipase (E.C. 3.1.1.3) Activity

Lipase (E.C. 3.1.1.3) is an lipolytic enzyme that catalyses the hydrolysis of triglycerides, diglycerides and in some monoglycerides. In plant seeds, the free fatty acids produced during hydrolysis of the stored lipids are metabolized for energy. Several lipases have been described in plants. Lipases are a subclass of the esterases. Lipases perform essential roles in the digestion, transport and processing of dietary lipids *eg.* triglycerides, fats, oils) in most, if not all, living organisms. Genes encoding lipases are even present in certain viruses. Several other types of lipase activities exist in nature, such as phospholipases and sphingomyelinases, however these are usually treated separately from "conventional" lipases. Some lipases are expressed and secreted by pathogenic organisms during the infection. In particular, *Candida albicans* has a large number of different lipases, possibly reflecting broad lipolytic activity, which may contribute to the persistence and virulence of *C. albicans* in human tissue.

Lipases perform essential roles in the digestion, transport and processing of dietary lipids (*e.g.* fats and oils) in living organisms. In humans, pancreatic lipase is the key enzyme responsible for breaking down fats in the digestive system by converting triglyceride to monoglyceride and free fatty acid. Pancreatic lipase monitoring is also used to help diagnose Crohn's disease, cystic fibrosis and celiac disease. Damage to the pancreas can exhibit a 5-10 fold increase of

Although a diverse array of genetically distinct lipase enzymes are found in nature, and represent several types of protein folds and catalytic mechanisms, most are built on an alpha / beta hydrolase fold and employ a chymotrypsin-like hydrolysis mechanism involving a serine nucleophile, an acid residue (usually aspartic acid), and a histidine.

Lipases are involved in diverse biological processes ranging from routine metabolism of dietary triglycerides to cell signalling and inflammation. Thus, some lipase activities are confined to specific compartments within cells while others work in extracellular spaces.

- In the example of lysosomal lipase, the enzyme is confined within an organelle called the lysosome.
- Other lipase enzymes, such as pancreatic lipases, are secreted into extracellular spaces where they serve to process dietary lipids into more simple forms that can be more easily absorbed and transported throughout the body.
- Fungi and bacteria may secrete lipases to facilitate nutrient absorption from the external medium (or in examples of pathogenic microbes, to promote invasion of a new host).
- Certain wasp and bee venoms contain phospholipases that enhance the "biological payload" of injury and inflammation delivered by a sting.
- As biological membranes are integral to living cells and are largely composed of phospholipids, lipases play important roles in cell biology.
- *Malassezia globosa*, a fungus that is thought to be the cause of human dandruff, uses lipase to break down sebum into oleic acid and increase skin cell production, causing dandruff.

Several lipases, one with ab acid optima that hydrolyzes bth mono and triglycerides and other with alkaline pH that hydrolyzes monoglycerides. Most plant lipases have alkaline or neutral pH optima In some dry quiescent plant seeds such as rapeseed, wheat, corn, cotton sunflower, cucumber and tomato lipase was not detectable, but activities increased dramatically on germination. In other cases, lipases was found in ungerminated seeds of faba bean, oats, peanut. Because lipase can become active when seeds are damaged during shelling or dehulling, or in untreated flour, lipolysis can result in fatty acid rancidity and soapy flavors.

In cereals, lipase in present in rice and in unnated wheat, corn scutellum, rye, or barley. The lipid content of oats ranges from 4 to 11%, most of theis being triglycerides with a high degree of unsaturation.

Several methods are available to assay the lipase activity. The avalable methods raged from radioisotope lipase assay, colorimetric lipase assay and vlumetric lipase assay.

Radioisotopic Assay of Lipase Activity

Materials

- For radioisotpic assay of lipase activity, glycerol tri[1-^{14}C]oleate (46.1 milli Curie/ m mol) was used.

Procedure

- Dry ungerminated oat seed were manually dehaulled for 30 seconds.
- Lipid was extracted from the dehulled seeds by Soxhlet extraction with petroleum ether (40-60^0 C) for 2 hours and the defatted flour was air dried and regounded in a mortar and pestle.
- The standard assay mixture for oat flour suspensions contained 0.05 M Tris-HCl byffer, (pH 7.5), 1% Trition-X-100, 0.2% benzene, and 10 mM glycerol tri[1-^{14}C]oleate (0.5 micro Curie/ M mol) in a final volume of 10 ml in a 50-ml Erlenmeyer flask.
- The assay mixture was agitated vigorously and then placed in a sahking water bath (37^0 C, 160 cycles per minute).
- After equilibration for 15 minutes, the reaction was started by addition of 1-g samples of the defatted oat flour.
- Aliquots (1 ml) were transferred at an intervals of 0 to 30 minutes into 15-ml glass centrifuge tubes containing 2 ml chloroform/methanol (1:1). This mixture terminated the reaction and extracted the lipids.
- The tubes were then centrifuged at 3000 x g for 5 minutes, the lipid containing chloroform layer was transferred to small test tubes and the chloroform was evaporated to dryness under N_2.
- The lipd was redissolved in small volumes of chloroform and aliqots were spotted on silica gel G thin-layer plates (0.25 mm, prepared in the laboratory).

- The lipid was frantionated with petroleum ether/diethyl ether/acetic acid (85:15:1, v/v). Iodine vapour was used to locate the separated lipid components.
- The free fatty acid, triglyceride, and other lipd fractons were scraped into separate scientillation vials and radioactivity determined by liquid scintillation counter in 5 ml of Aquasol.
- Liberated fatty acid, labelled in the carbonyl group was detrmined by calculating the ratio of radioactivity recovered in the free fatty acid fraction to that in the total lipid recovered. This ratio, multiplied by the number of moles of triglyverides supplied times 3 (moles of fatty acid per mole of triglycerides), givs the mole of fatty acid hydrolysed per gram of flour for each interval.
- Rates of triglycerides hydrolysis were calculated from the linear part of the time course. One unit of lipase activity was defined as 1 μmol of fatty acid released per minute per gram of flour.

Colorimetric Assay of Lipase Activity

Procedure

- Defatted oat flour was prepared as described above from ungerminated groats and 0.5 g sample was mixed with 90 mg of glycerol trioleate.
- Buffer (0.05 M Tris-HCl, pH 7.5, containing 1% [v/v] Triton-X-100) was added to the final moisture content of 40%.
- The resulting dough was mixed with glass rod for 2 minutes in test tube and then incubated at 37^0 C in water bath.
- After intervals of 0 to 30 minutes samples were removed, 0.1 ml of 1 N HCl added and free fatty acids were extracted by homogenizing the acidified dough in 10 ml of chloroform/heptane/methanol (49:49:2, v/v).
- The extracts were filterted, the residue washed with additional aliquots of the solvnt mixture, and the combined solvent filterate made up to 50 ml.
- Aliquots (5.0 ml) were mixed with copper reagent and the resulting fatty acid copper soaps measured colorimetrically (Shipe *et al.*, 1980).
- Oat flour suspensions were prepared as described for radioisotop assay. Inone series of samples, unlabelled glycerol trioleate replaced the radioisotope.

- Samples were incubated at 37^0 C. The labelled samples were treated as described above and the unlabelled sample was acidified with 1.0 N HCl and the free fatty acids were extracted in chloroform/heptane/methanol as described above.

Enzyme extraction

- Five gram portion of Soxhlet lipid extracted oat flou were stirred at 4^0C in 10 ml of 0.05 M Tris-HCl buffer (pH 7.5). This preparation was centrifused at 6000 x g for 15 minutes and aliquots of the post 6000 x g supernatants assayed for lipase activity.

Reading

Matlashewski, G.J., Urquhart, A.A., Sahasrabudhe, M.R. and Altosaar, I. (1982). Lipase activity in oat flour suspension and soluble extracts. *Cereal Chem.*, **59 (5):** 418-422.

Shipe, W.F., Senyk, G.F. and Fountain, K.B. (1980). Modified copper soap solvent extraction method for measuring free fatty acids in milk. *J. Diary Sci.*, **63:** 193.

Volumetric Assay of Lipase Activity

Materials

- Take 2.0 ml of any clear vegetable oil, neutralize to pH 7.0, and stir well with 25 ml of water in the presence of 100 mg of bile salt (sodium taurocholate) till an emulsion is formed. Addition of 2 g gum arebica hasten emulsification.

- Grind a known quantity of sample with a mortar and pastle. Homogenize the tissue with twice the volume of ice cold acetone. Filter and wash the powder successively with acetone, acetone:ether (1:1) and ether. Air dry the sample. This acetone powder can be stored in a refrigerator. Extract 1 g of the powder in 20 ml ice cold water or a suitable buffer (phosphate buffer pH 7.0). Centrifuge at 15,000 rpm for 10 minutes and use the supernatant as enzyme source.

Procedure

- Take 20 ml of substrate in a 500 ml beaker. Add 5 ml of phosphate buffer (pH 7.0).

- Set the beaker on top of a magnetic stirrer cum hot plate and stir the content slowly. Maintain the temperature at 350 C. Dip the electrode of a pH meter in the reactioon mixture. Note the pH and adjust to pH 7.0.
- Add enzyme extract (0.5 ml), immediately record the pH and set the timer on. Let it be the pH at zero time.
- At frequent intervals or as the pH drops by about 0.2 unit add 0.1 N NaOH to bring the pH to the initial value. Continue the titration for 50-60 minute period.
- Note the volume of alkali consumed.

Calculation

The enzyme activity is defined as the amount of enzyme which releases one milliequivalent of free fatty acid per minute per g sample. Specific activity is expressed as milliequivalents/min/mg protein.

Activity meq/min/g sample = (Vol. Of alkali consumed x strength of alkali) / (Wt. Of sample x time in minute)

Reading

Jayaraman, J. (1981). In "*Laboratory Manual in Biochemistry*", Wiley Eastern Limited, New Delhi.. pp. 133.

CHAPTER 12

Plant Nutrition

Plant nutrition is the study of the chemical elements and compounds that are necessary for plant growth, and also of their external supply and internal metabolism. The criteria for an element to be essential for plant growth:

1. In its absence the plant is unable to complete a normal life cycle; or
2. That the element is part of some essential plant constituent or metabolite.

This is in accordance with Liebig's law of the minimum. There are 14 essential plant nutrients. Carbon and oxygen are absorbed from the air, while other nutrients including water are obtained from the soil. Plants must obtain the following mineral nutrients from the growing media:

- the primary macronutrients: nitrogen (N), phosphorus (P), potassium (K)
- the three secondary macronutrients: calcium (Ca), sulphur (S), magnesium (Mg)
- the micronutrients/trace minerals: boron (B), chlorine (Cl), manganese (Mn), iron (Fe), zinc (Zn), copper (Cu), molybdenum (Mo) and nickel (Ni)

The macronutrients are consumed in larger quantities and are present in plant tissue in quantities from 0.2% to 4.0% (on a dry matter weight basis). Micro nutrients are present in plant tissue in quantities measured in parts per million, ranging from 5 to 200 ppm, or less than 0.02% dry weight.

Most soil conditions across the world can provide plants with adequate nutrition and do not require fertilizer for a complete life cycle. However, humans can artificially modify soil through the addition of fertilizer to promote vigorous growth and increase yield. The plants are able to obtain their required nutrients from the fertilizer added to the soil. A colloidal carbonaceous residue, known as humus, can serve as a nutrient reservoir. Even with adequate water and sunshine, nutrient defiency can limit growth. Nutrient uptake from the soil is achieved by cation exchange, where root hair pump hydrogen ions (H^+) into the soil through proton pumps. These hydrogen ions displace cations attached to negatively charged soil particles so that the cations are available for uptake by the root.

Plant nutrition is a difficult subject to understand completely, partly because of the variation between different plants and even between different species or individuals of a given clone. An element present at a low level may cause deficiency symptoms, while the same element at a higher level may cause toxicity. Further, deficiency of one element may present as symptoms of toxicity from another element. An abundance of one nutrient may cause a deficiency of another nutrient. For example, lower availability of a given nutrient such as SO_2^{4-} can affect the uptake of another nutrient, such as NO_3^-. As another example, K^+ uptake can be influenced by the amount of NH_4^+ available.

The root, especially the root hair, is the most essential organ for the uptake of nutrients. The structure and architecture of the root can alter the rate of nutrient uptake. Nutrient ions are transported to the centre of the root, the stele in order for the nutrients to reach the conducting tissues, xylem and phloem. The Casparian strip, a cell wall outside of the stele but within the root, prevents passive flow of water and nutrients, helping to regulate the uptake of nutrients and water. Xylem moves water and inorganic molecules within the plant and phloem accounts for organic molecule transportation. Water potential plays a key role in a plants nutrient uptake. If the water potential is more negative within the plant than the surrounding soils, the nutrients will move from the region of higher solute concentration—in the soil—to the area of lower solute concentration: in the plant.

There are three fundamental ways plants uptake nutrients through the root: 1.) simple diffusion, occurs when a nonpolar molecule, such as O_2, CO_2, and NH_3 follows a concentration gradient, moving passively through the cell lipid bilayer membrane without the use of transport proteins. 2.) facilitated diffusion, is the rapid movement of solutes or ions following a concentration gradient, facilitated by transport proteins. 3.) Active transport, is the uptake by cells of ions or molecules against a concentration gradient; this requires an energy source, usually ATP, to power molecular pumps that move the ions or molecules through the membrane.

- Nutrients are moved inside a plant to where they are most needed. For example, a plant will try to supply more nutrients to its younger leaves than to its older ones. When nutrients are mobile, symptoms of any deficiency become apparent first on the older leaves. However, not all nutrients are equally mobile. Nitrogen, phosphorus, and potassium are mobile nutrients, while the others have varying degrees of mobility. When a less mobile nutrient is deficient, the younger leaves suffer because the nutrient does not move up to them but stays in the older leaves. This phenomenon is helpful in determining which nutrients a plant may be lacking.

Many plants engage in symbiosis with microorganisms. Two important types of these relationship are 1.) with bacteria such as rhizobia, that carry out biological nitrogen fixation, in which atmospheric nitrogen (N_2) is converted into ammonium (NH_4); and 2.) with mycorrhizalal, which through their association with the plant roots help to create a larger effective root surface area. Both of these mutualistic relationships enhance nutrient uptake.

Though nitrogen is plentiful in the Earth's atmosphere, relatively few plants harbor nitrogen fixing bacteria, so most plants rely on nitrogen compounds present in the soil to support their growth. These can either be supplied by decaying matter, nitrogen fixing bacteria, animal waste, or through the agricultural application of purpose-made fertilizers.

Hydroponics, is a method for growing plants in a water-nutrient solution without the use of nutrient-rich soil. It allows researchers and home gardeners to grow their plants in a controlled environment. The most common solution, is the Hogland solution and it consists of all the essential nutrients in the correct proportions necessary for most plant growth. An aerator is used to prevent an anoxic event or hypoxia. Hypoxia can affect nutrient uptake of a plant because without oxygen present, respiration becomes inhibited within the root cells. The Nutrient film technique is a variation of hydroponic technique. The roots are not fully submerged, which allows for adequate aeration of the roots, while a "film" thin layer of nutrient rich water is pumped through the system to provide nutrients and water to the plant.

12.1. Macronutrients (Primary)

Phosphorus

Phosphorus is important in plant bioenergetics. As a component of ATP, phosphorus is needed for the conversion of light energy to chemical energy (ATP) during photosynthesis. Phosphorus can also be used to modify the activity of various enzymes by phosphorylation, and can be used for cell signalling. Since ATP can be used for the biosynthesis of many plant biomolecules, phosphorus is important for plant growth and flower / seed formation. Phosphate esters make up DNA, RNA, and phospholipids. Most common in the form of polyprotic phosphoric acid (H_3PO_4) in soil, but it is taken up most readily in the form of $H_2PO_4^-$ Phosphorus is limited in most soils because it is released very slowly from insoluble phosphates. Under most environmental conditions it is the limiting element because of its small concentration in soil and high demand by plants and microorganisms. Plants can increase phosphorus uptake by a mutualism with mycorrhiza. A Phosphorus deficiency in plants is characterized by an intense green coloration in leaves. If the plant is experiencing high phosphorus deficiencies the leaves may become denatured and show signs of necrosis. Occasionally the leaves may appear purple from an accumulation of anthocyanin. Because phosphorus is a mobile nutrient, older leaves will show the first signs of deficiency. It is useful to apply a high phosphorus content fertilizer, such as bone meal, to perennials to help with successful root formation.

Potassium

Potassium regulates the opening and closing of the stomata by a potassium ion pump. Since stomata are important in water regulation, potassium reduces water loss from the leaves and increases drought tolerance. Potassium deficiency may cause necrosis or interveinal chlorosis. K^+ is highly mobile and can aid in balancing the anion charges within the plant. Potassium helps in fruit colouration, shape and also increases its brix. Hence, quality fruits are produced in Potassium rich soils. It also has high solubility in water and leaches out of rocky or sandy soils. This water solubility can result in potassium deficiency. Potassium serves as an activator of enzymes used in photosynthesis and respiration Potassium is used to build cellulose and aids in photosynthesis by the formation of a chlorophyll precursor. Potassium deficiency may result in higher risk of pathogens, wilting, chlorosis, brown spotting, and higher chances of damage from frost and heat.

Nitrogen

Nitrogen is an essential component of all proteins. Nitrogen deficiency most often results in stunted growth, slow growth, and chlorosis. Nitrogen deficient plants will also exhibit a purple appearance on the stems, petioles and underside of leaves from an accumulation of anthocyanin pigments. Most of the nitrogen taken up by plants is from the soil in the forms of NO_3^-, although in acid environments such as boreal forests where nitrification is less likely to occur, ammonium NH_4^+ is more likely to be the dominating source of nitrogen. Amino acids and proteins can only be built from NH_4^+ so NO_3^- must be reduced. Under many agricultural settings, nitrogen is the limiting nutrient of high growth.

Some plants require more nitrogen than others, such as corn (*Zea mays*). Because nitrogen is mobile, the older leaves exhibit chlorosis and necrosis earlier than the younger leaves. Soluble forms of nitrogen are transported as amines and amides.

12.2. Macronutrients (Secondary and Tertiary)

Sulphur

Sulphur is a structural component of some amino acids and vitamins, and is essential in the manufacturing of chloroplasts. Sulphur is also found in the Iron Sulphur complexes of the electron transport chains in photosynthesis. It is immobile and deficiency therefore affects younger tissues first. Symptoms of deficiency include yellowing of leaves and stunted growth.

Calcium

Calcium regulates transport of other nutrients into the plant and is also involved in the activation of certain plant enzymes. Calcium deficiency results in stunting. This nutrient is involved in photosynthesis and plant structure. Blossom end rot is also a result of inadequate calcium.

Magnesium

Magnesium is an important part of chlorophyll, a critical plant pigment important in photosynthesis. It is important in the production of ATP through its role as an enzyme cofactor. Magnesium deficiency can result in interveinal chlorosis.

Silicon

In plants, silicon strengthens cell walls, improving plant strength, health, and productivity. Other benefits of silicon to plants include improved drought and frost resistance, decreased lodging potential and boosting the plant's natural pest and disease fighting systems. Silicon has also been shown to improve plant vigor and physiology by improving root mass and density, and increasing above ground plant biomass and crop yield. Although not considered an essential element for plant growth and development (except for specific plant species - sugarcane and members of the horsetail family), silicon is considered a beneficial element in many countries throughout the world due to its many benefits to numerous plant species when under abiotic or biotic stress. Silicon is the second most abundant element in earth's crust. Higher plants differ characteristically in their capacity to take up silicon. Depending on their SiO_2content they can be divided into three major groups:

- Wetland graminae-wetland rice, horsetail (10–15%)
- Dryland graminae-sugar cane, most of the cereal species and few dicotyledons species (1–3%)
- Most of dicotyledons especially legumes (<0.5%)
- The long distance transport of Si in plants is confined to the xylem. Its distribution within the shoot organ is therefore determined by transpiration rate in the organs
- The epidermal cell walls are impregnated with a film layer of silicon and effective barrier against water loss, cuticular transpiration rate in the organs.

Si can stimulate growth and yield by several indirect actions. These include decreasing mutual shading by improving leaf erectness, decreasing susceptibility to lodging, preventing Mn and Fe toxicity.

12.3. Micro-nutrients

Some elements are directly involved in plant metabolism however, this principle does not account for the so-called beneficial elements, whose presence, while not required, has clear positive effects on plant growth. Mineral elements those either stimulates growth but are not essential, or that are essential only for certain plant species, or under given conditions, are usually defined as beneficial elements.

Iron

Iron is necessary for photosynthesis and is present as an enzyme cofactor in plants. Iron deficiency can result in interveinal chlorosis and necrosis. Iron is not the structural part of chlorophyll but very much essential for its synthesis. Copper deficiency can be responsible for promoting an iron deficiency.

Molybdenum

Molybdenum is a cofactor to enzymes important in building amino acids. Involved in Nitrogen metabolism. Mo is part of Nitrate reductase enzyme.

Boron

Boron is important for binding of pectins in the RGII region of the primary cell wall, secondary roles may be in sugar transport, cell division, and synthesizing certain enzymes. Boron deficiency causes necrosis in young leaves and stunting.

Copper

Copper is important for photosynthesis. Symptoms for copper deficiency include chlorosis. Involved in many enzyme processes. Necessary for proper photosythesis. Involved in the manufacture of lignin (cell walls). Involved in grain production. It is also hard to find in some conditions.

Manganese

Manganese is necessary for photosynthesis, including the building of chloroplasts. Manganese deficiency may result in coloration abnormalities, such as discolored spots on the foliage.

Sodium

Sodium is involved in the regeneration of phosphoenol pyruvate in CAM and C4 plants. It can also substitute for potassium in some circumstances.

Zinc

Zinc is required in a large number of enzymes and plays an essential role in DNA transcription. A typical symptom of zinc deficiency is the stunted growth of leaves, commonly known as "little leaf" and is caused by the oxidative degradation of the growth hormone auxin.

Nickel

In higher plants, Nickel is absorbed by plants in the form of Ni^{2+} ion. Nickel is essential for activation of urease, an enzyme involved with nitrogen metabolism that is required to process urea. Without Nickel, toxic levels of urea accumulate, leading to the formation of necrotic lesions. In lower plants, Nickel activates several enzymes involved in a variety of processes, and can substitute for Zinc and Iron as a cofactor in some enzymes.

Chlorine

Chlorine, as compounded chloride, is necessary for osmosis and ionic balance; it also plays a role in photosynthesis.

Cobalt

Cobalt has proven to be beneficial to at least some plants, but is essential in others, such as legumes where it is required for nitrogen fixation for the symbiotic relationship it has with nitrogen-fixing bacteria. Vanadium may be required by some plants, but at very low concentrations. It may also be substituting for molybdenum. Selenium and sodium may also be beneficial. Sodium can replace potassium's regulation of stomatal opening and closing.

1. The requirement of Co for N_2 fixation in legumes and non-legumes have been documented clearly
2. Protein synthesis of Rhizobium is impaired due to Co deficiency
3. It is still not clear whether Co has direct effect on higher plant

Aluminium

- Tea has a high tolerance for Al toxicity and the growth is stimulated by Al application. The possible reason is the prevention of Cu, Mn or P toxicity effects.
- There have been reports that Al may serve as fungicide against certain types of root rot.

12.4. Colorimetric Estimation of Phosphorus, Iron, Manganese, Aluminium and Crude silica in Plant Tissue

Sample preparation

Reagents

Acid mixture – Prepare a mixture containing 750 ml concentrated HNO_3, 150 ml concentrated H_2SO_4, and 300 ml of 60 to 62 percent $HClO_4$

Procedure

Put 1.0 g of dried, grounded plant material into a 75 ml Pyrex test tube. Add 10 ml of acid mixture to digest under a fume hood for at lest 2 hours. Then heat over a low gas flame. Gradually increase the heat until the mixture is clear. Cool and fill the tubes upto 50 ml mark with distilled water. Filter the sample extract through an acid-washed filter paper.

Note

- Phosphorus will be lost if the digestion to go to dryness.
- If aluminimum is to be determined, continue digesting the mixture until the volume has been reduced to 0.5 ml.
- If silica is to be detrmined, use ashless Whatman filter paper No. 44 and then keep the filter paper and residue for the crude silica determination.

Estimation for phosphorus

Reagents

- Molybdate-vandate solution: Dissolve 25 g of ammonium molybdate [$(NH_4)_6Mo_7O_{24}.4H_2O$] in 500 ml of distilled water. Dissolve 1.5 g ammonium vanadate (NH_4VO_3) in 500 ml of 1.0 N HNO_3. Then mix equal volumes of these two solutions. Prepare a fresh mixture each week.
- Nitric acid 2.0 N: Dilute 10 ml of concentrated HNO_3 to 80 ml with distilled water.
- Standard phosphorus solution: Dissolve 0.110 g of monobasic potassium phosphate (KH_2PO_4) in distilled water and dilure to 1.0 liter. This solution contains 25 ppm phosphorus.
- From the above standard phosphorus solutuion a gradient of phosphorus from 2.5 to 15.0 ppm is pprepared by succive dilution.

Procedure

- Put 1.0 ml of sample extract into a 10.0 ml tube and add 2 N HNO_3 and dilute to 8.0 ml with distilled water.
- Add 1.0 ml of molybdo-vanadte solution and make the final volume to 10 ml.
- Allow the tubes to stand for 20 – 25 minutes and measure the optical density at 420 nm and compare with ODs of standard phosphorus solution.

This method is best suited for samples of high phosphorus content.

Reading

Black, C.A. (1965). In "*Methods of soil analysis,*" Part 2 *Chemical and microbiological properties,* American Society of Agronomy, Inc., Medison, Wisconsin, pp. 1572.

Sekin, T. e*t. al.*, (1965). Photoelectric colorimetry in Biochemistry, (Part 2), Nanko-do Publishing Co., Tokyo, pp. 242

Estimation for Iron

Reagents

- Hydroquinone: Prepare a 1% solution in distilled water. If colour develop, discard and make a new solution.

- Sodium citrate: Dissolve 250 g of sodium citrate in water and dilute to 1.0 liter.
- Ortho-phenanthroline: dissolve 0.5 g ortho-phenanthroline in distilled water and dilute to 100 ml. Store the solution in a dark bottle or in a dark place. If coour develops, discard and make a new solution.
- Iron standard: Place 0.100 g of electrolytic iron in a 100 ml beaker, add 50 ml of 1:3 (v/v) HNO_3. Boil until the brown fumes of nitrous oxide are no longer there. Cool and dilute to 1.0 liter. This solution contain 100 ppm of Fe.
- From the above standard iron solutuion a gradient of iron from 0.4 to 4.0 ppm is pprepared by succive dilution.

Procedure

- Put 10.0 ml of sample extract or standard into a 25 ml volumetric flask. Add 1.0 ml of hydroquinone solution and 1.0 ml of ortho-phenanthroline solution.
- Add the required amount of sodium citrate to bring the pH to 3.5 and make the volume to 50 ml.
- Heat the flask in a water bath for an hour, and measure the optical density at 508 nm and compare with the ODs of standard iron solution.

Coment

An orange-red complex forms when ortho-phenanthroline reacts with ferrous ion. The olour intensity is independent of acidity between pH 2.0 to 9.0. The standard and sample solutions should have the same final pH and are therefore buffered with sodium citrate at pH 3.5. The coloured complex will remain stable for several months.

Reading

Sandell, E.B. (1950). Iron, p 522-554. In "*Colorimetric determination of traces of metals*" 3rd edition, Interscience Publishers, Inc. New York, pp. 1032.

Estimation of manganese

Reagent

- Acid solution: Mix 400 ml of concentrated HNO_3 with 200 ml of distilled water. Dissolve 75 g of H_2SO_4 in this solution and then add 200 ml of

85% H_3PO_4. Dissolve 0.035 g of $AgNO_3$ in this solution and make up to 1 liter with distilled water.

- Ammonium persulphate: Store the ammonium persulphate crystals $[(NH_4)_2S_2O_8]$ in a dessicator.
- Manganese stock solution: Place 3.0 8 g of $MnSO_4.H_20$ in a 100 ml beaker. Dissove by carefully adding 50.0 ml of 1:1 (v/v) HCl. Transfer the solution to a 1 liter volumetric flask and make up to volume with distilled water. This solution contain 1000 ppm Mn.
- From the above standard manganese solutuion a gradient of manganese from 0.2 to 12.0 ppm is pprepared by succive dilution.

Procedure

- Transfer 10 ml of the sample extract or standard to a 50 ml volumetric flask and add 2.5 ml of the acid solution and make upto 40 ml with distilled water.
- Then add 0.5 g of ammonium persulfate.
- Place the flask in boiling water for 5 minutes and then cool under tap water and make the volume up to 50 ml with distilled water.
- Measure the optical density at 530 nm against reagent blank and compare with the ODs of manganese standard.

Reading

Sandell, E.B. (1950). Manganese, p 606-620. In "*Colorimetric determination of traces of metals*" 3rd edition, Interscience Publishers, Inc. New York, pp. 1032.

Estimation of aluminum

Reagents

- Aluminum reagent: In separate beakers, dissolve 0.75 g ammonium aurine tricarboxylate, 15 g gum acacia, and 200 g ammonium acetate in distilled water. When each is dissolved, mix them together and add 190 ml concentrated HCl. Mix filter and dilute to 1.5 liters with distilled water.
- Thioglycolic acid: Add 1.0 ml of thioglucolic acid to a 100 ml volumetric flask and make the volume with distilled water.
- Phenolphthalein: Dissolve 1 g of phenolphthalein in 60 ml of absolute alcohol and 40 ml of distilled water.

- Ammonium hydroxide (1:9): Dilute 10 ml of NH_4OH to 100 ml with distilled water.
- Aluminum standard: Prepare a 1000 ppm aluminum standard stock solution by dissolving 8.95 g of $AgCl_3.6H_2O$ in 1.0 liter of distilled water.
- From the above standard aluminum solutuion a gradient of aluminum from 0.2 to 1.4 ppm is pprepared by succive dilution.

Procedure

- Transfer 1 to 5 ml of sample extract or standard solution of Al to test tubes graduated to 50 ml volume.
- Add 2 to 3 drops of phenolphthaleine and then add ammonium hydroxide (1:9) until the first pink colour develop.
- Dilute to 20 ml with distilled water and add 2.0 ml of thioglycolic acid solution. Mix throughly and then add 10 ml of aluminum reagent.
- Heat the tube in a boiling waterbath for 15-16 minutes. Cool at least for 1.5 hours and make the volme up to 50 ml with distilled water.
- Measure the optical density at 537 nm against a reagent blank and compare with the ODs of alumina standards.

Note

If the sample has too much iron (Fe precipaitate out during the neutralization with NH_4OH), proceed as follows: Take some of the sample extract and neautralize it with excess of 40% NaOH. Place the sample in a boiling waterbath for 5 minutes, then centrifuge and remove the supernatant by suction. Add 2 to 3 drops of phenolphthalein to the supernatant and then add HCl until the pink colour just disappears. Then dilute to 20 ml with distilled water and proceed as described above by adding 2 ml of thioglycolic acid solution *etc.*

Reading

Chenery, E.M. (1948). Thioglycolic acid as an inhibitor for iron in the colorimetric determination of aluminum by means of "Aluminon". *Analyst,* **73:** 501-502.

Sandell, E.B. (1950). Aluminum, p. 219-253. In "*Colorimetric determination of traces of metals*" 3rd edition, Interscience Publishers, Inc. New York, pp. 1032.

Estimation of Crude Silica

Procedure

- Dry the filter papera and residue of the sample extract in an oven at 80^0 C.
- Then char the paper with naked flame under a flame hood and allow it to turn to ash by placing it in a muffel firnace for 2 hours at 550^0 C.
- Cool the ash in a dessicator for at least 2 hours before weighing.
- This gives us an estimate of crude silica in 1 g of the dry planr sample.

Calculation

$$\text{Crude silica} = \frac{\text{Wt. Of cryde silica (g)} \times 100}{\text{Wt. Of sample (g)}}$$

12.5. EDTA method for Estimation of Calcium and Magnesium

Sample extraction

Reagents

- Concentrated sulphuric acid
- Hydrochloric acid (2.0 N): Dilute 100.0 ml of concentrted HCl to 600.0 ml with distilled water.
- Ferric chloride (1 g Fe/ml): Dissolve 1.66 g $FeCl_3.6H_2O$ in 250.0 ml of distilled water containing 10. ml of concentrated HCl.
- Sodium acetate (10%): Dissolve 100 g of sodium acetate trihydrate in distilled water and dilute to 1.0 liter.
- Sodium hydroxide (0.4 N): Dissolve 8.0 g of NaOH in distilled water and dilute to 500 ml.
- Bromine water: Prepare a saturated solution of bromine by adding bromine in distilled water until droplets of excess bromine can be seen in the solution.
- Ammonium chloride (25%): Dissolve 250.0 g of NH_4Cl in distilled water and dilute to 1.0 liter.
- Ammonium hydroxide (0.6 N): Dissolve 42.0 ml of concentrated NH_4OH (Sp. Gr. 0.89) to 1.0 liter with distilled water.

Procedure

- Place 2 g of oven dried grounded plant material in a tall 100 ml pyrex beaker and heat in a muffle furnace for 2 hours at 550^0 C.
- Add 10 ml of concentrated HCl to the ash and evaporate to dryness.
- Add 5 ml of 2.0 N HCl and dilute to 50 ml with distilled water.
- Dependig on the calcium and magnesium concentration expected in the sample, pipette 2 or 4 ml of sample etract into 15 ml centrifuge tube and add 0.5 ml of ferric chloride solution. Spin the tubes by hand and then add 2 ml of sodium acetate solution.
- Spin the tubes by hand again and add 2 ml of 0.4 N of NaOH.
- Add 1 ml of bromine water and digest for 1 hour at 95^0 C to flocculate the magnesium dioxide and to expel the excess of bromine.
- Add 2 ml of 25% ammonium chloride and spin the tube by hand and digest for 15 minutes at 80^0 C.
- While hot, centrifuge for 5 minutes and decant the sample extract in 250 ml beaker.

Estimation of calcium

Reagents

- KOH (8.0 N): Dissolve 448 g of KOH in 1 liter of distilled water.
- Dotite NN: (Wako Jun Yaku, Kogyo Co. Ltd., Tokyo, Japan.
- Calcium stock solution (5.00 mg/ml): Place 6.244 g of dried $CaCO_3$ in 500 ml beaker. Cover with Pyrex watch glass and add 100 ml of distilled water. Then slowly add 150 ml of 1.0 N HCl. When $CaCO_3$ has dissloved, boiled gently for 2-3 minutes. Allow the beaker to cool and then transfer the content to a 500 ml volumetric flask and make the volume with distilled water. This solution contains 5.00 mg/ml calcium.
- Dilute 10 ml of standard solution to 100 ml with distilled water to obtain a standard containing 0.5 mg calcium per mili liter.
- Standard EDTA (0.005 M): Dissolve 1.861 g of disodium ethylene diaminetetraacetic acid ($Na_2EDTA.2H_2O$) is distilled water and dilute to 1 liter. Store in polyethylene bottle. Standardize the EDTA solution as follows: Transfer 5 ml of the 0.5 mg/ml calcium standard solution to a 200 ml Erlenmeyer flask and add 80 ml of distilled water aml of 8.0 N

KOH to adjust the pH to 12. Add a pinch of Dotite NN to the flask and then titrate the EDTA solution against the Ca standard solution until a permanent blue colour develops. Obtain the blank titration value by titrating the EDTA solution against the blank containing 5 ml of 8.0 N KOH, distilled water and a pinch of Dotite NN unti! a permanent blue colour develops.

- Correct the titration value of the calcium standard solution by substracting the titration value of the blank Calculate the molarity of the EDTA solution. One milliliter of a 0.005 M EDTA solution is equivalent to 2.004 mg of calcium.

Procedure

- Add 80 ml of distilled water and 5 ml of 8.0 N KOH to the sample extract and adjust the pH to 12.
- Add a pinch of Dotite NN and then titrate with standard EDTA until a permanent blue colour develops.
- Correct for blank carried throughout the entire procedure.

Calculation

For a 2 ml sample

% Ca = 0.005 M EDTA titer x 0.250

For a 4 ml sample

% Ca = 0.005 M EDTA titer x 0.125

Sample analysis for magnesium plus calcium

Reagents

- Buffer solution (pH 10.0): Dissolve 67.5 g of NH_4Cl in 400 ml of water. Add 570 ml of concentrated NH_4OH and dilute to 1.o liter with distilled water.
- Superchrome Black TS or Erichrome Black: Dissolve 50 mg of Superchrome Black TS or Erichrome Black T in 20 ml of distilled water. Preparea fresh solution daily.
- Magnesium stock solution (5.00 mg/ml): Put 2.50 g of unoxidized Mg metal in a 500 ml beaker. Cover the beaker with a Pyrex watch glass and add 150 ml of distilled water. Then carefully add 20 ml 1: 1 (v/v) HCl.

Although the magnesium dissolves in the acid very rapidly, the solution should be boiled gently for 2 – 3 minutes to make sure this process is complete. Transfer the solution to a 500 ml volumetric flask and make the volume with distilled water. This solution contain 5.00 mg Mg/ml. Dilute 10 ml of the above standard solution to 100 ml with distilled water to obtain a standard solution containing 0.50 mg/ml magnesium.

- Standard EDTA solution (0.005M): Prepare as described for calcium determination. Stadardize the EDTA solution as follows: Transfer 5 ml of the 0.50 mg/ml Mg standard solution to a 200 ml flask. Then add 80 ml of distilled water and 5 ml of buffer solution (pH 10.0).
- Add six drops of superchrome Black TS or Erichrome Black T and titrate with EDTA solution until a permanent blue colour develops. Obtain the blank titration value by titrating EDTA solution against the blank containing 5 ml buffer solution (pH 10.0), 80 ml distilled water and six drops of superchrome Black TS or Erichrome Black T, until a permanent blue colour develops.
- Correct the titration value of Mg standard solution by substracting the titration value of the blank. Calculate the molarity of the EDTA solution.
- One milliliter of 0.005 M EDTA solution is equivalent to 0.1216 mg Mg.

Procedure

- Add 80 ml of distilled water and 5 ml of buffer solution (pH 10.0) to sample extract.
- Stir and add six drops of superchrome Black TS or Erichrome Black T.
- Titrate with standard EDTA until a permanent blue colour appear.
- Correct for a blank carried throughout the entire procedure.
- This titration determines the sum of calcium and magnesium.

Calculation

For a 2 ml sample

% Mg = [(0.005 M EDTA titer for Ca + Mg) – (0.005 M EDTA titer value for Ca)] x 0.152

For a 4 ml sample

% Mg = [(0.005 M EDTA titer for Ca + Mg) – (0.005 M EDTA titer value for Ca)] x 0.076

Reading

Barrows, H.L. and Simpson, E.C. (1962). An EDTA method for direct routine analysis of calcium and magnesium in soil and plants. *Soil Sci. Soc. Amer. Proc.*, **26:** 443-445.

Cheng, K.L. and Bray, R.H. (1951). Determination of calcium and magnesium in soil and plant material. *Soil Sci.*, **72:** 449-457.

Greweling, T. (1960). The chemical analysis of plant materials. Cornell University, Ithiaca, N.Y., USA.

12.6. Determination of Zinc, Copper, Manganese, Calcium, Magnesium, Potassium and Sodium in Plant

Sample extraction

Reagents

- Deionized-distilled water: Prepare by passing distilled water through the ion exchanger (maximum flow rate, 315 ml/min.).
- Hydrochloric acid (1.0 N): Add 1.5 liters of concentrated HCL to 16.5 liters of deionized water.

Procedure

- Place 1.0 g of dried, gounded plant sample into a dry 100 ml polysterene bottle that has beeen rinsed with 1.0 N HCl.
- Add 25 ml of 1.0 N HCl and be sure all the sample is wetted with the acid. Allow to stand for 24 hours. After 24 hours shake briefly and filter through Whatman filter paper No. 1.
- Collect the filterate in a 100 ml polysterene bottle.
- Prepare a 25 ml 1.0 N HCl blank at the same time using the same procedure.

Sample preparation for calcium

Reagents

- Lanthanum solution (5%): Take 56.85 g of La_2O_3 and wet it with 100 ml of deionized water and then slowly add 250 ml of 1:1 (v/v) HCl. When the lanthanum has dissolved, transfer the solution to 1 liter volumetric flask and make the volume with deionized water.

Procedure

- Take 1.0 ml of sample extract into a 10 ml graduated tube. Add 2.0 ml of 5% lanthanum solution and dilute to 10 ml with 1.0 HCl and seal the tube.

Sample preparation for magnesium, potassium and sodium

Procedure

- Take 2.0 ml of sample extract into a 40 ml vial and dilute to 40 ml with deionized water. Seal the vial with plastic lid.

Sample preparation for zinc, copper and manganese

Procedure

- The remaining sample extract can be used without further dilution for the analysis of Zn, Cu and Mn.

Stock solution and standards

Zinc 1000 ppm stock solution : Take 1.0 g of zinc metal into 1000 ml beaker and cover the beaker with watch glass. Dissolve the zinc slowly by adding 50 ml of 10% (v/v) HCl. Boil the solution for 3 minutes, allow the solution to cool and make the volume in a 1.0 liter vlumetric flask with deionized water.

From the above stock solution a gradient of zinc from 0.25 to 1.00 ppm is prepared by succive dilution with 1.0 N HCl.

Copper 1000 ppm stock solution : Take 0.1 g of copper metal into 100 ml beaker and cover the beaker with watch glass. Dissolve the copper slowly by adding 50 ml of 10% (v/v) HNO_3. Boil the solution for 3 minutes, allow the solution to cool and make the volume in a 1.0 liter vlumetric flask with deionized water.

From the above stock solution a gradient of copper from 0.5 to 1.50 ppm is prepared by succive dilution with 1.0 N HCl.

Manganese 1000 ppm stock solution : Take 3.08 g of $MnSO_4.H_2O$ into 100 ml beaker and cover the beaker with watch glass. Dissolve the salt slowly by adding 50 ml of 10% (v/v) HCl and make the volume in a 1.0 liter vlumetric flask with deionized water.

From the above stock solution a gradient of manganese from 5 to 20 ppm is prepared by succive dilution with 1.0 N HCl.

Magnesium 1000 ppm stock solution : Take 1.0 g of magnesium metal into 100 ml beaker and cover the beaker with watch glass. Dissolve the copper slowly by adding 50 ml of 10% (v/v) HCl. Boil the solution for 3 minutes, allow the solution to cool and make the volume in a 1.0 liter vlumetric flask with deionized water.

From the above stock solution a gradient of magnesium from 2 to 10 ppm is prepared by succive dilution with 1.0 N HCl.

Calcium 1000 ppm stock solution : Take 2.497 g of $CaCO_3$ into 100 ml beaker and cover the beaker with watch glass. Dissolve the copper slowly by adding 50 ml of 10% (v/v) HCl. Boil the solution for 3 minutes, allow the solution to cool and make the volume in a 1.0 liter vlumetric flask with deionized water.

From the above stock solution take 0, 5, 10, 15 and 20 ml and add 20 ml of 5% lanthanum solution in a 100 ml volumetric flasks and the volume is mafe to with 1.0 N HCl. This gives gradient of calcium from 0 to 2 ppm.

Potassium 1000 ppm stock solution : Take 1.907 g of air dried potassium chloride and dissolve it in 1000 ml of deionized water.

From the above stock solution a gradient of potassium from 20 to 100 ppm is prepared by succive dilution with 1.0 N HCl.

Sodium 1000 ppm stock solution : Take 2.542 g of air dried sodium chloride (NaCl) and dissolve it in 1000 ml of deionized water.

From the above stock solution a gradient of sodium from 20 to 100 ppm is prepared by succive dilution with 1.0 N HCl.

Specification for individual elemenal analysis

Use the atomicabsorption spectrophotometer according to the instruction manual.

Zinc

- Standards (ppm): 0, 0.25, 0.50, 0.75, 1.00
- Range: UV
- Wavelength: 214
- Source: 15 mA (aacording to lamp operating current)
- Slit: 5
- Scale: 5

- Filter: OUT
- Meter response: 2
- Zero knob position: 2.20
- Burner height: 0.4
- Air flow: 9.0
- Fuel flow: 9.0 (acetylene)

Plot percent absorption against the standards.

Copper

- Standards (ppm): 0, 0.50, 1.00, 1.50
- Range: UV
- Wavelength: 324.6
- Source: 30 mA (aacording to lamp operating current)
- Slit: 4
- Scale: 10
- Filter: OUT
- Meter response: 2
- Zero knob position: 1.00
- Burner height: 0.4
- Air flow: 9.0
- Fuel flow: 9.0 (acetylene)

Plot percent absorption against the standards.

Manganese

- Standards (ppm): 0, 5, 10, 15, 20
- Range: UV
- Wavelength: 279.9
- Source: 30 mA (aacording to lamp operating current)
- Slit: 4

- Scale: 2
- Filter: OUT
- Meter response: 2
- Zero knob position: 1.01
- Burner height: 0.4
- Air flow: 9.0
- Fuel flow: 9.0 (acetylene)

Divide the percent absorption by 2. Convert the absorption using the tables provided with the instrument, and plot percent absorption against the standards.

Calcium

- Standards (ppm): 0, 5, 10, 15, 20
- Range: VIS
- Wavelength: 211.9
- Source: 15 mA (aacording to lamp operating current)
- Slit: 4
- Scale: 2
- Filter: OUT
- Meter response: 2
- Zero knob position: 0.57
- Burner height: 0.3
- Air flow: 7.5
- Fuel flow: 9.0 (acetylene)

Divide the percent absorption by 2. Convert the absorption using the tables provided with the instrument, and plot percent absorption against the standards.

Magnesium

- Standards (ppm): 0, 2, 4, 6, 8, 10
- Range: UV
- Wavelength: 285.4

- Source: 15 mA (aacording to lamp operating current)
- Slit: 5
- Scale: 2
- Filter: OUT
- Meter response: 2
- Zero knob position: 1.05
- Burner height: 0.3
- Air flow: 8.0
- Fuel flow: 9.0 (acetylene)

Divide the percent absorption by 2. Convert the absorption using the tables provided with the instrument, and plot percent absorption against the standards.

Potassium (Flame photometer)

- Standards (ppm): 0, 20, 40, 60, 80, 100
- Use the photometer according to the instruction manual and plot the meter reading against the standards.

Sodium (Flame photometer)

- Standards (ppm): 0, 20, 40, 60, 80, 100
- Use the photometer according to the instruction manual and plot the meter reading against the standards.

Note

For samples such as grains which contain high phosphorus but low calcium, remove phosphorus interference either by an ion exchange method (Leyton, 1954, Hemingway, 1956, Hinson, 1962) or by addition of magesium and sulphuric acid (David, 1959) prior to calcium determination.

Reading

David, D.J. (1959). Determination of calcium in plant material by atomic absorption spectrophotometery. *Analyst*, **84:** 536-545.

Eiko-Seiki, Co. Ltd. Instruction manual for EKO Flame photometr.

Hemingway, R.G. (1956). The determination of calcium in plant material by flame photometery. *Analyst*, **81:** 194-198.

Hinson, W.H. (1962). Ion exchange treatment of plant ash extracts for removal of interfering anions in the determination of calcium by atomic absorption. *Spectrochim. Acta.*, **18:** 427-429.

Kushizaki, M. (1968). An extraction procedure of plant materials for the rapid determination of Mn, Cu, Zn, and Mg by the atomic absorption analysis. *J. Sci. Soil Manure*, Japan, **39:** 489-490.

Leyton, L. (1954). Phosphate interference in flame photometric determination of calcium. *Analyst*, **79:** 497-500.

Perkin-Elmer Corp. Analytical methods for atomic absorption spectrophotometry. Rev. Sept. 1968.

12.7. Estimation of Boron

Equipment

- Adjustable pipette.
- 10-ml test tubes.
- Spectrophotometer with 420 nm of wavelength.
- Boil all Pyrex glassware with a mixture of 3:1 concentrated HNO_3 and concentrated $HClO_4$ at 225°C for two hours.
- Soak them in a 2*M* HCl acid bath overnight. Rinse thoroughly with deionized water.

Reagents

- Buffer-masking solution - Dissolve 250 g of ammonium acetate and 15 g of ethylendiamine-tetraacetic acid disodium (EDTA) salt in 400-ml high quality distilled water and slowly add 125-ml of glacial acetic acid.
- Azomethine-H solution - Dissolve 0.45 g of azomethine-H in 100-ml of 1% (1 g/100 ml water) L-ascorbic acid solution. Let stand 24 hours prior to suing. This reagent will keep in refrigerator at 40°F for two weeks. This reagent is light sensitive and should be kept in a brown plastic bottle or a plastic bottle wrapped in aluminum foil.

Procedure

- Pipette 1.0 ml of filtrate from the dry ashing procedure into a test tube and add 2-ml buffer-masking solution and 2 ml of azomethine-H solution. Thoroughly mix by swirling.

- Allow the mixture to stand for 30 minutes and read on spectrophotometer at wavelength of 420 nm. Set 100% T with reagent blank, which is a 1.0 ml 1.2 N HCl solution with reagent.
- All working standards should follow steps 1 and 2.

Calibration and Standards

- 1000 mg B/L stock solution - Weigh 0.5716 g of boric acid and dilute to 100 ml with deionized water.
- 10 mg B/L working stock solution - Dilute 1 ml of 1000 mg B/L stock solution to 100 ml with deionized water.
- Working standards - Pipette the following volumes of 10 mg B/L working stock solution into 50 ml Volumetric flasks and dilute to volume with 1.2 *N* HCI solution:

ml of 10 mg B/L	mg B/L in solution	mg B/L in plant
0	0.0	0.0
2.5	0.5	50
5.0	1.0	100

Store in plastic bottles and keep in the refrigerator until ready to use.

Calculations

- mg B/L in plant = mg B/L in reading x final volume / plant weight
- mg B/L in plant = mg B/L in reading x 50 / 0.5
- mg B/L in plant = mg B/L in reading x 100

12.8. Estimation of Extractable Chloride

Reagents and Apparatus

- Extracting solution - Add about 600 ml of distilled water to a 1 liter volumetric flask. Add 20.0 ml of glacial acid in the same flask. Bring to final volume with distilled water. Mix and transfer to a plastic storage bottle
- Chloride Standard Stock Solution - Dissolve 0.2103 g of reagent grade potassium chloride in approximately 50 ml of extracting solution. Bring up to 100-ml volume with extracting solution and label "1000 ppm Cl in 2% acetic acid extracting solution."

- Working standards for chloride.- Set of five working Cl- standards: 200, 160, 120, 80, and 40 mg cl/l by volume: To five 100 ml volumetric flasks add, respectively, 20, 16, 12, 8, and 4 ml of chloride standard stock solution. Dilute each to the mark with the extracting solution and invert to mix.
- 125 ml shaking bottles.
- 10 ml testing tubes
- Whatman #2 filter paper.

Procedure

- Weigh out 0.100g of dry ground plant material into each 125 ml shaking bottle.
- Add 25 ml of extracting solution into each flask, insuring that all plant material has been wetted.
- Include at least one Blank and one Check sample per run.
- Shake vigorously for 30 minutes.
- Filter into 30 ml of beaker using Whatman #2 filter paper.
- Transfer clear filtrate into 10 ml test tubes.
- Analyze for chloride on a Flow Injection Analyzer.

Calculation

Cl ppm in plant = reading (ppm) x 25 / 0.100

12.9. Estimation of Total Sulfur

Reagents

- Buffer and $BaCl_2$ solution:
- Dissolve 5 g of gum arabic in about 500 ml of hot, D. I. Water and filter if cloudy.
- Add 50 g of $BaCl_2.2H_2O$ and 450 ml of glacial acetic acid and dilute to 1 liter.

Procedure

- Pipette 10 ml of filtrate from ashing procedure into a 50 ml Erlenmeyer flask.

- Add 10 ml of the gum arabic-$BaCl_2$-acetic acid solution.
- Shake for 10 minutes.
- Read samples on spectrophotometer with 420 nm wavelength using appropriate standards.

Calibration and Standards

- 100 ppm S standard solution - dissolve 0.544 g of oven dried (105° C) K_2SO_4 in about 500 ml of D. I. Water, add 10 ml of acetic acid as a preservative, and dilute to 1 liter with distilled water.
- Working standards – take 0, 4, and 8 ml of 100 ppm standard solution into 100 ml volumetric flasks. Add 25 ml of a 2000 ppm-P and 8N acetic acid solution (8.12 g of $Ca(H_2PO_4)_2.H_2O$ plus 460 ml of glacial acetic acid diluted to 1 liter) and dilute to volume with distilled water. Store in plastic bottles and keep in the refrigerator until ready to use.

Alternative method for total sulphur estimation

Procedure

- Take 1.0 g of dried and powdered plant sample and digest it with 10.0 ml of nitric acid in a 250 ml digestion tube for 30 minutes at 100° C.
- Cool the digested material and then 6.0 ml of 70% perchloric acid is added and digestion continued for another 60 minutes at 250° C.
- The digested material is cooled to room temperature and then 10.0 ml of 6.0 N HCl is added and final volume is made to 250 ml with water.
- To a 15.0 ml aliquot of the above 250 mg of finally grounded barium chloride is added followed by a mild shaking for 10 minutes.
- The per cent transmission of turbid suspension is measured at 420 nm.
- Standard curve for sulphur is made with potassium sulphate.

Reading

- Singh, U. (1983). *Grain Quality & Biochemistry Chickpea And Pigeonpea Amino Acid Composition. In "Grain Quality and Biochemistry Progress Report – 2".* Research Report. International Crops Research Institute for the Semi-Arid Tropics, ICRISAT Patancheru, India.

12.10. Estimation of Molybdenum

Analysis for molybdenum content of samples was carried out using a new extraction-photometric method with Tetrazolium Violet (TV). The absorption spectra of solutions of TV and the ion-association complex of 479 molybdenum (VI) show that the absorption maximum is around 230 nm. The molar absorptivity of the associate, calculated by the method of Komar-Tolmatchov, was (1.30 ± 0.08) x 105 l mol^{-1} cm^{-1}). This indicates the high sensitivity of the given reaction.

Procedure

- A wet burning of the plant samples was carried out and a mixture of sulfuric and nitric acids was used for oxidation of the organic substance. A portion of 2 g of air-dry plant material was placed into a Kjeldahl flask and moistened with 4 ml distilled water; 5 ml conc. sulphuric acid, 2 ml conc. perchloric acid and 10 ml conc. nitric acid were added. The flask was heated only slightly to avoid splashing of the solution, decomposition and fuming away of nitric acid. When all the organic material was oxidized, the solution was heated at a higher temperature for 10 minutes. After cooling, the solution was diluted with water and filtered, transferred into a volumetric flask of 50 ml and diluted up to the mark with distilled water. For determination of manganese in a separatory funnel of 100 ml, the following solutions were introduced: 1.8 ml of Toluidine Blue 0.5 x 10.3 mol l^{-1}, 0.5 ml of perchloric acid 9 mol l^{-1}, aliquote of the prepared solution of plant sample. It was diluted up to a volume of the aqueous phase of 10 ml with distilled water and extracted with 3 ml of 1,2-dichloroethane for 1 min. The organic phase was filtered through a dry paper into a 1 cm cuvette and the absorbance measured at 290 nm. A blank was run in parallel in the absence of a plant sample. A calibration graph was constructed with standards similarly treated.
- For determination of molybdenum in a separatory funnel of 100 ml, the following solutions were introduced: 1.2 ml of 1 x 10^{-4} M Tetrazolium Violet, 0.2 ml of 1 x 10 -2 M 3,4,5-trihydroxybenzoic acid (R), aliquote of the prepared solution of plant sample. Dilute to a volume of 10 ml with distilled water and extract with 3 ml of 1,2-dichloroethane for 30 seconds. The organic layer was then transferred through filter paper into a 1-cm cell and measured at 230 nm. A blank solution containing all reagents except molybdenum was prepared and treated in the same way.

CHAPTER 13

Plant Pigments

Biological pigments, also known simply as pigments or biochromes are substances produced by living organisms that have a colour resulting from selective colour absorption. Biological pigments include plant pigments and flower pigments. Many biological structures, such as skin, eyes, fur and hair contain pigments such as melanin in specialized cells called chromatophores.

Pigment colour differs from structural colour in that it is the same for all viewing angles, whereas structural colour is the result of selective reflection or iridescence, usually because of multilayer structures. For example, butterfly wings typically contain structural colour, although many butterflies have cells that contain pigment as well.

Biological Pigments

- Heme/porphyrin-based: chlorophyll, bilirubin, hemocyanine, hemoglobin, myoglobin
- Light-emitting: luciferin
- Carotenoids:
  - Hematochromes (algal pigments, mixes of carotenoids and their derivates)
  - Carotenes: alpha and beta carotene, lycopene, rhodopsin
  - Xanthophylls: canthaxanthin, zeaxamthin, lutein
- Proteinaceous: phytochrome, phycobiliproteins
- Polyene enolates: a class of red pigments unique to parrots
- Other: melanin, urochrome, flavonoids.

13.1. Pigments in Plants

The primary function of pigments in plants is photosynthesis, which uses the green pigment chlorophyll along with several red and yellow pigments that help to capture as much light energy as possible. Other functions of pigments in plants include attracting insects to flowers to encourage pollination.

Plant pigments include a variety of different kinds of molecule, including porphyrins, carotenoids, anthocyanins and betalains. All biological pigments selectively absorb certain wavelengths of light while reflecting others. The light that is absorbed may be used by the plant to power chemical reactions, while the reflected wavelengths of light determine the colour the pigment will appear to the eye.

Anthocyanin gives these pansies their purple pigmentation.

The principal pigments responsible are:

- **Chlorophyll** is the primary pigment in plants; it is a chlorin that absorbs yellow and blue wavelengths of light while reflecting green. It is the presence and relative abundance of chlorophyll that gives plants their green colour. All land plants and green algae possess two forms of this pigment: chlorophyll *a* and chlorophyll *b*. Kelps, diatoms, and other photosynthetic heterokonts contain chlorophyll *c* instead of *b*, while red algae possess only chlorophyll *a*. All chlorophylls serve as the primary means plants use to intercept light in order to fuel photosynthesis.

- **Carotenoids** are red, orange, or yellow tetraterpenoids. They function as accessory pigments in plants, helping to fuel photosynthesis by gathering wavelengths of light not readily absorbed by chlorophyll. The most familiar carotenoids are carotene (an orange pigment found in carrots), lutein (a yellow pigment found in fruits and vegetables), and lycopene (the red pigment responsible for the colour of tomatoes). Carotenoids have been shown to act as antioxidants and to promote healthy eyesight in humans.
- Anthocyanins (litreally "flower blue") are water soluble flavonoid pigments that appear red to blue, according to pH. They occur in all tissues of higher plants, providing colour in leaves, plant stem, roots, flowers, and fruits, though not always in sufficient quantities to be noticeable. Anthocyanins are most visible in the petals of flowers, where they may make up as much as 30% of the dry weight of the tissue. They are also responsible for the purple colour seen on the underside of tropical shade plants such as *Tradescantia zebriana*; in these plants, the anthocyanin catches light that has passed through the leaf and reflects it back towards regions bearing chlorophyll, in order to maximize the use of available light.
- **Betalains** are red or yellow pigments. Like anthocyanins they are water-soluble, but unlike anthocyanins they are indole-derived compounds synthesized from tyrosine. This class of pigments is found only in the Caryophyllales (including cactus and amaranth), and never co-occur in plants with anthocyanins. Betalains are responsible for the deep red colour of beets, and are used commercially as food-colouring agents.

A particularly noticeable manifestation of pigmentation in plants is seen with autum leaf colour, a phenomenon that affects the normally green leaves of many deciduous trees and shrubs whereby they take on, during a few weeks in the autum season, various shades of red, yellow, purple, and brown. Chlorophylls degrade into colourless tetrapyrroles known as *nonfluorescent chlorophyll catabolites* (NCCs). As the predominant chlorophylls degrade, the hidden pigments of yellow xanthophylls and orange beta-carotene are revealed. These pigments are present throughout the year, but the red pigments, the amthocyanins, are synthesised *de novo* once roughly half of chlorophyll has been degraded. The amino acids released from degradation of light harvesting complexes are stored all winter in the tree's roots, branches, stems, and trunk until next spring when they are recycled to re leaf the tree.

13.2. Estimation of Chlorophyll Contents

Chlorophyll is a green pigment found in cyanobacteria and the chloroplasts of algae and plants. Chlorophyll is an extremely important biomolecule, critical in photosynthesis, which allows plants to absorb energy from light. Chlorophyll absorbs light most strongly in the blue portion of the electromagnetic spectrum, followed by the red portion. Conversely, it is a poor absorber of green and near-green portions of the spectrum, hence the green colour of chlorophyll-containing tissues.

Chlorophyll is vital for photosynthesis, which allows plants to absorb energy from light. Chlorophyll molecules are specifically arranged in and around photosystem that are embedded in the thylakoid membranes of chloroplasts. In these complexes, chlorophyll serves two primary functions. The function of the vast majority of chlorophyll (up to several hundred molecules per photosystem) is to absorb light and transfer that light energy by resonance energy transfer to a specific chlorophyll pair in the reaction centre of the photosystems. The two currently accepted photosystem units are Photosystem I and Photosystem II, which have their own distinct reaction center chlorophylls, named P680 and P700, respectively. These pigments are named after the wavelength (in nanometers) of their red-peak absorption maximum. The identity, function and spectral properties of the types of chlorophyll in each photosystem are distinct and determined by each other and the protein structure surrounding them. Once extracted from the protein into a solvent (such as acetone or methanol), these chlorophyll pigments can be separated in a simple paper chromatography experiment and, based on the number of polar groups between chlorophyll a and chlorophyll b, will chemically separate out on the paper.

The function of the reaction center chlorophyll is to use the energy absorbed by and transferred to it from the other chlorophyll pigments in the photosystems to undergo a charge separation, a specific redox reaction in which the chlorophyll donates an electron into a series of molecular intermediates called an electron transport chain. The charged reaction center chlorophyll ($P680^+$) is then reduced back to its ground state by accepting an electron. In Photosystem II, the electron that reduces $P680^+$ ultimately comes from the oxidation of water into O_2 and H^+ through several intermediates. This reaction is how photosynthetic organisms such as plants produce O_2 gas, and is the source for practically all the O_2 in Earth's atmosphere. Photosystem I typically works in series with Photosystem II; thus the $P700^+$ of Photosystem I is usually reduced, via many intermediates in the thylakoid membrane, by electrons ultimately from Photosystem II. Electron transfer reactions in the thylakoid membranes are complex, however, and the source of electrons used to reduce $P700^+$ can vary. The electron flow produced by the reaction center chlorophyll pigments is used to shuttle H^+ ions across the thylakoid membrane, setting up a chemiosmotic potential used mainly to produce ATP chemical energy; and those electrons ultimately reduce $NADP^+$ to NADPH, a universal reductant used to reduce CO_2 into sugars as well as for other biosynthetic reductions. Reaction center chlorophyll–protein complexes are capable of directly absorbing light and performing charge separation events without other chlorophyll pigments, but the absorption cross section (the likelihood of absorbing a photon under a given light intensity) is small. Thus, the remaining chlorophylls in the photosystem and antenna pigment protein complexes associated with the photosystems all cooperatively absorb and funnel light energy to the reaction center. Besides chlorophyll *a*, there are other pigments, called accessory pigments, which occur in these pigment–protein antenna complexes.

Measuring absorption of the chlorophyll solution at two wavelengths allows the use of the following simultaneous equations:

$$D663 = 82.04\ Ca + 9.27\ Cb \quad (1)$$

$$D645 = 16.75\ Ca + 45.6\ Cb \quad (2)$$

Where D663 is the absorption at 663 nm, D645 is the absorption at 645 nm, Ca is the concentration of chlorophyll a in g/litre and Cb is the concentration of chlorophyll b in g/litre, and 82.04, 9.27, 16.75 and 45.6 are specific absorption coefficients of chlorophylla and b at wavelengths 663 and 645 nm, respectively.

Solving the equations (1) and (2) gives

$$Ca = (0.0127 \times D663) - (0.00269 \times D645) \quad (3)$$

$$Cb = (0.0229 \times D645) - (0.00468 \times D663) \quad (4)$$

Therefore

$$\text{Total chlorophyll } C \text{ (g/litre)} = Ca + Cb = (0.0202 \times D645) + (0.00802 \times D\ 663) \quad (5)$$

When equation (5) is expressed in mg/litre, it becomes

$$C = (20.2 \times D645) + (8.02 \times D663) \quad (6)$$

This is the equation used for calculating total chlorophyll content. For simplified procedure, the following equation can be used:

$$\text{Total chlorophyll } C = D645 / 34.5 \text{ (g/litre)} \quad (7)$$

i.e., $C = (D652 \times 1000) / 34.5$ (mg/litre)

At 652 nm, chlorophyll a and b intersect, and 34.5 is the specific absorption coefficient for both pigments at this wavelength.

Materials

Acetone 80% (v/v)

Proedure

- Fresh leaves are cut into fine pieces and 2.0 g of fresh tissue is taken in mortar. The leaves are homogenized thoroughly with pastle with 80% acetone. Enough acetone is added to allow the tissue to be thoroughly homogenized.
- Continue homogenization of the leaf tissue and take the supernatant through filter paper into a 100 ml flask. Complete homogenization till the tissue is white or gray in clour. Make the volume of supernatant to 100.0 ml with 80% acetone. This is used as chlorophyll source.
- Optical densities of the chlorophyll is recorded at 645 and 663 nm against 80% acetone.
- Contents of chlorophyll a, and b and total chlorophyll content is calculated as per equations given above.

Reading

Arnon, D.I. (1959). Copper enzymes in isolated chloroplasts.Polyphenol oxidases in *Beta vulgaris*. *Plant Physiol.* **24:** 1-15.

Chlorophyll estimation using acetone

Ziegler, R. and Egle, K. (1965). Zur quantitative analyse der Chloroplasten-pigmente. I. Kritische Uberprüfüng der spectral-photometrischen chlorophyll – Bentimmung. *Beitr. Biol. Pflanzen.*, **41:** 11-37.

Acetone 80%

Chlorophyll a = (11.78 x D664) – (2.29 x D647) (mg. l^{-1})

Chlorophyll b = (20.05 x D647) – (4.77 x D664) (mg. l^{-1})

Chlorophyll (a+b) = (7.01 x D664) + (17.76 x D667) (mg. l^{-1})

Alternative methods for chlorophyll estimation using acetone

Holm, G. (1954). Chlorophyll mutation in barley. *Acta. Agr. Scand.*, **4:** 457-471.

Acetone 100%

Chlorophyll a = (9.68 x D662) – (0.99 x D644) (mg. l^{-1})

Chlorophyll b = (21.4 x D644) – (4.65 x D662) (mg. l^{-1})

Chlorophyll (a+b) = (5.13 x D662) + (20.41 x D664) (mg. l^{-1})

Estimation of chlorophylls using alcohol

MacKinney, G. (1941). Absorption of light by chlorophyll solutions. *J. Biol. Chem.*, **140:** 315-322.

Methanol – 100%

Chlorophyll a = (16.5 x D665) – (8.3 x D650) (mg. l^{-1})

Chlorophyll b = (33.8 x D650) – (12.5 x D665) (mg. l^{-1})

Chlorophyll (a+b) = (4.0 x D665) + (25.5 x D650) (mg. l^{-1})

Alternative method for estimation of chlorophylls using alcohol

Wintermans, J.F.G.M. and deMots, A. (1965). Spectrophotometric characteristics of chlorophyll a and b and their pheophytins in ethanol. *Biochim. Biophys. Acta.*, **109:** 448-453.

Ethanol – 96%

Chlorophyll a = (13.7 x A665) – (5.76 x A649) (mg. l^{-1})

Chlorophyll b = (25.8 x A645) – (7.6 x A665) (mg. l^{-1})

Chlorophyll (a+b) = (6.1 x A665) + (20.04 x A649) (mg. l^{-1})

The calculation of chlorophyll contents for ethanol extracts can be simplified by using following tables of Wintermanns (1969). The ratio of measured absorbances A_{max} a (665 nm) and A_{max} b (649 nm) is established first. Contents of chlorophyll a, band (a+b) are then found by multiplication of A_{max} a or A_{max} b by coefficients found on the appropriates line of the table.

A_{665}/A_{649}	$C_a = A_{665}$ x	$C_b = A_{649}$ x	$C_{(a+b)} = A_{645}$ x	C_a / C_b =
2.00	10.82	5.30	16.12	2.04
2.05	10.88	5.00	15.88	2.18
2.10	10.96	4.69	15.65	2.33
2.15	11.02	4.40	16.42	2.50
2.20	11.08	4.13	15.21	2.68
2.25	11.14	3.87	15.01	2.88
2.30	11.20	3.61	14.81	3.10
2.35	11.25	3.36	14.61	3.33
2.40	11.30	3.14	14.44	3.59
2.45	11.35	2.92	14.27	3.88
2.50	11.40	2.72	14.12	4.19

Ethanol 96%, cuvette 10 mm pathlength, chlorophyll = $mg.l^{-1}$, ratio of chlorophyll a.b is read directly

Wintermanns, J.F.G.M. (1969). Comparative chlorophyll determinations by spectrophotometry of leaf extracts in different solvants. *Photosynthetica,* **3:** 112-119.

Estimation of chlorophyll a, b and (a+b) in diethyl eather

Transfer of pigments from extraction solvants into diethyl ether is recommended by some workers. This removes the hydrophilic pigments when executed carefully, clear turbid extracts. Moreover, specific absorption coefficients measurd in diethyl ether are taken as standard values of chlorophyll analysis. These advantages are balanced by the longer time needed for analysis and risk of pigment losses, specially in experiments with small extracts. The biggest error is introduced by the washing out and the loss of chlorophyllides from chlorophyllase action during the extraction. Diethyl ether must be peroxide free and eather solution should not contain water.

There are no general agreement on the size and wavelength of the absorption peaks of diethyl ether solutions of chlorophylls. The two most frequently used sets of equations are:

Chl a = (9.93 x A 660) – (0.78 x A642.5) (mg/litre)

Chl b = (17.60 x A 642.5) – (2.81 x A660) (mg/litre)

Chl (a+b) = (7.12 x A 660) – (16.80 x A642.5) (mg/litre)

Comar, C.L. and Zscheile, F.P. (1942). Analysis of plant extracts for chlorophyll a and b by a photoelectric spectrophotometric method. *Plant Physiol.*, **17:** 198-209.

Chl a = (10.1 x A 662) – (1.01 x A644) (mg/litre)

Chl b = (16.4 x A 644) – (2.57 x A662) (mg/litre)

Chl (a+b) = (7.53 x A 662) – (15.39 x A644) (mg/litre)

Smith, JHC and Benitez, A. (1955). Chlorophyll analysis in plant materials. In "*Modern Methods of Plant Analysis*" Ed. K. Prach and MV Tracey, **6**: 142-196, Springer Verlag, Berlin.

Wintermans (1969) proposed the following factors for easy calculations of the chlorophyll contents in diethyl ether extracts and absorption in Smith and Benitez (1955) method.

A_{662}/A_{644}	$C_a = A_{662}$ x	$C_b = A_{662}$ x	$C_{(a+b)} = A_{662}$ x	$C_a / C_b =$
2.20	9.64	4.88	14.52	1.97
2.30	9.66	4.56	14.22	2.12
2.40	9.68	4.27	13.95	2.26
2.50	9.69	3.99	13.68	2.43
2.60	9.71	3.73	13.44	2.60
2.70	9.72	3.50	13.22	2.78
2.80	9.74	3.28	13.02	2.97
2.90	9.75	3.08	12.83	3.16
3.00	9.76	2.89	12.65	3.38
3.10	9.77	2.72	12.49	3.60
3.20	9.78	2.56	12.34	3.82
3.30	9.79	2.40	12.19	4.08

Estimation of Chlorophyll a and b with one Wavelength Using Hydroxylamine

The main source of error in the two wavelength spectrophotometric determination of chlorophylls is in the adjustment of one or both wavelengths when they are interchanged. Inaccurate adjustement of the wavelength for the chlorophyll a and b peaks is most serious because this wavelengths correspond to a point laying on the steep part of the absorption curve of chlorophyll mixture. It has been reported that 1 nm shift of wavelength results in only 1.6% error in the chlorophyll a determination but 15.2% error in the total chlorophyll determination.

These errors can be reduced by the procedure of Ogawa and Shibata (1965). A methanolic reaction mixture is devided into two parts, the aborbance of which are measured at one wavelebgth with and without addition of hydroxylamine reagent. Hudroxylamine does not change the character of the absorption curve of chlorophyll a, but react with aldehydic group of chlorophyll b. The red maxima of chlorophyll b in methanol is shifted from 653 to 663 nm and the blue band from 471 to 453 nm. The resulting absorption band of the mixture is relatively wide in the red so that the amount of of chlorophyll a and b are determined from estimation of absorption at 660 nm.

The reagent is prepared by dissolving 4 g of hydroxylamine hydrochloride in 6 ml of water and pH is adjusted to 4.0 ± 0.5 with NaOH solution and adding water to make the final volume upto 10.0 ml.

Two 9.5 ml aliquats are piptted from methanolic solution of pigment. To one 0.5 ml of water is added and to the other 0.5 ml of the hydroxylamine hydrochloride reagent is added. The reagent mixtures are allowed to stand for 5 minutes and thereafter optical densities are recorded at 660 nm. The amount of chlorophyll ($mg.l^{-1}$) are calculated using the following equations:

Chlorophyll a = 7.2 x (2 A666 – Δ A666)

Chlorophyll b = 25.4 x Δ A666

Chlorophyl (a+b) = (14.4 x A666) + (18.2 x Δ A666)

Where Δ A666 is the differences in absorbance of the solutions with and without the reagent.

The advantage of this method is the stability of the reaction product, the absorption characteristics of which do not change during 24 hours storage in darkness and cold. A disadvantage is the necessity to extract with mrthanol, the extraction power of which is often smaller than acetone, particularly with well hydrated tissues.

Reading

Ogawa, T. and Shibata, K. (1965). A sensitive method for determining chlorophyll b in plant extracts. *Photochem. Photobiol.*, **4:** 193-200.

Estimation of chlorophyll using dimethyl sulphoxide (DMSO)

Use of dimethyl sulphoxide (DMSO) for chlorophyll extraction technique has two principal advantages over other extractions (*e.g.* methanol, ethanol, or acetone). First, the method is faster, largely because grinding and centrifuging is not required. Second, the chlorophyll extracts are more stable in DMSO, and do not break down as quickly as those in acetone: It has been reported that DMSO extracts are stable for up to 5 days, whereas with acetone extracts the measured level of chlorophyll begins to fall off immediately. For the extractions 7.0 ml of DMSO were preheated to 65° C in a water bath. Chlorophyll is extracted from approximately 100 mg od fresh leaf sample. The extraction of chlorophyll is done at 65° C for 15 – 20 minutes. After chlorophyll extraction, samples were removed from the water bath and 10.0 ml of DMSO is added. Absorbance of both reagent blank (DMSO) and sample were measured at 645 and 663 nm. Chlorophyll content was calculated as per the following equations suggested by Hiscox and Israelstam (1979):

Chl a ($g\ l^{-1}$) = (0.0127 x A663) – (0.00269 x A645)

Chl b ($g\ l^{-1}$) = (0.0229 x A645) – (0.00468 x A663)

Total Chlorophyll ($g\ l^{-1}$) = (0.0202 x A645) + (0.00802 x A663).

The Chl concentration of the extract calculated from these equations was then converted to leaf Chl content (mg Chl cm^{-2} leaf area).

Reading

Hiscox JD, Israelstam GF. 1979. A method for the extraction of chlorophyll from leaf tissue without aceration. *Canadian Journal of Botany,* **57:** 1332 – 1334.

13.3. Chlorophyll Fluorescence

Chlorophyll fluorescence is light that has been re-emitted after being absorbed by chlorophyll molecules of plant leaves. By measuring the intensity and nature of this fluorescence, plant ecophysiology can be investigated.

Assessing physiology of plant with chlorophyll fluorescence

Light energy that has been absorbed by a leaf will excite electrons in chlorophyll molecules. Energy in photosystem II can be converted to chemical energy to drive photosynthesis (photochemistry). If photochemistry is inefficient, excess energy can damage the leaf. Energy can be emitted (known as energy quenching) in the form of heat (called non-photochemical quenching) or emitted as chlorophyll fluorescence. These three processes are in competition, so fluorescence yield is high when less energy is emitted as heat or used in photochemistry. Therefore, by measuring the amount of chlorophyll fluorescence, the efficiency of photochemistry and non-photochemical quenching can be assessed. The fluorescence emitted from a leaf has a longer wavelength than the light absorbed by the leaf. Therefore, fluorescence can be measured by shining a defined wavelength of light onto a leaf and measuring the level of light emitted at longer wavelengths.

By measuring chlorophyll fluorescence, plant ecophysiology can be investigated. Chlorophyll fluorometers are used by plant researchers to assess plant stress. Gitelson (1999) states, "The ratio between chlorophyll fluorescence at 735 nm and the wavelength range 700nm to 710 nm, F735/F700 was found to be linearly proportional to the chlorophyll content (with determination coefficient, r^2, more than 0.95) and thus this ratio can be used as a precise indicator of chlorophyll content in plant leaves.

Measuring fluorescence

Usually the initial measurement is the minimal level of fluorescence, F_0. This is the fluorescence in the absence of photosynthetic light. To use measurements of chlorophyll fluorescence to analyse photosynthesis, researchers must distinguish between photochemical quenching and non-photochemical quenching (heat dissipation). This is achieved by stopping photochemistry, which allows researchers to measure fluorescence in the presence of non-photochemical quenching alone. To reduce photochemical quenching to negligible levels, a high intensity, short flash of light is applied to the leaf. This transiently closes all PSII reaction centres, which prevents energy of PSII being passed to downstream electron carriers. Non-photochemical quenching will not be effected if the flash is short. During the flash, the fluorescence reaches the level reached in the absence of any photochemical quenching, known as maximum fluorescence F_m.

The efficiency of photochemical quenching (which is a proxy of the efficiency of PSII) can be estimated by comparing F_m to the steady yield of fluorescence in the light F_t and the yield of fluorescence in the absence of photosynthetic light F_0. The efficiency of non-photochemical quenching is altered by various internal and external factors. Alterations in heat dissipation mean changes in F_m. Heat dissipation cannot be totally stopped, so the yield of chlorophyll fluorescence in the absence of non-photochemical quenching cannot be measured. Therefore, researchers use a dark-adapted point (F^0_m) with which to compare estimations of non-photochemical quenching.

Common fluorescence parameters

- $\mathbf{F_v}$: Minimal fluorescence (arbitrary units). Fluorescence level when all antenna pigment complexes associated with the photosystem are assumed to be open (dark adapted).
- $\mathbf{F_m}$: Maximal fluorescence (arbitrary units). Fluorescence level when a high intensity flash has been applied. All antenna sites are assumed to be closed.
- $\mathbf{F_{tr}}$: Terminal fluorescence (arbitrary units). Fluorescence quenching value at the end of the test.
- $\mathbf{T_{1/2}}$: Half rise time from $\mathbf{F_0}$ **to** $\mathbf{F_m}$.

Calculated parameters

- $\mathbf{F_v}$ is variable fluorescence. Calculated as $\mathbf{F_v = F_m\ F_0}$
- $\mathbf{F_v / F_m}$ is the ratio of variable fluorescence to maximal fluorescence. Calculated as $\mathbf{(F_m - F_0) / F_m}$. This is a measure of the maximum efficiency of PSII (the efficiency if all PSII centres were open). $\mathbf{F_v / F_m}$ can be used to estimate the potential efficiency of PSII by taking dark-adapted measurements.
- φ_{PSII} measures the efficiency of Photosystem II. Calculated as $\mathbf{(F_m - F_{tr}) / F_m}$. This parameter measures the proportion of light absorbed

by PSII that is used in photochemistry. As such, it can give a measure of the rate of linear electron transport and so indicates overall photosynthesis.

- qP (photochemical quenching). Calculated as $(F_m - F_{tr}) / (F_m - F_0)$. This parameter approximates the proportion of PSII reaction centres that are open.
- Whilst φ_{PSII} gives an estimation of the efficiency, qP and F_v / F_m tell us which processes which have altered the efficiency. Closure of reaction centers as a result of a high intensity light will alter the value of qP. Changes in the efficiency of non-photochemical quenching will alter the ratio F_v / F_m.

Applications of the theory

PSII yield as a measure of photosynthesis

Chlorophyll fluorescence appears to measure of photosynthesis, but this is an over-simplification. Fluorescence can measure the efficiency of PSII photochemistry, which can be used to estimate the rate of linear electron transport by multiplying by the light intensity. However, researchers generally mean carbon fixation when they refer to photosynthesis. Electron transport and CO_2 fixation can correlate well, but may not correlate in the field due to processes such as photorespiration, nitrogen metabolism and the Mehler reaction.

- Upon examining the banana peel degreening after treatment with 1-methylcyclopropane (1-MCP). Many parameters were measured, including chlorophyll fluorescence since fluorescense is a measure of chloroplast photochemical efficiency. 1-MCP treatment was shown to delay the decrease in chlorophyll fluorescence. Greenness and photochemical efficiency decreased simultaneously after ethylene treatment, suggesting that changes in chlorophyll fluorescence F_m were responsible for the patterns of de-greening. After 7 and 8 days, F_m values of 1-MCP peel were ca. 3-fold the values of the other treatments, showing that 1-MCP treated fruit maintain their photochemical efficiency when the other treatment regimes do not.

Relating electron transport to carbon fixation

A powerful research technique is to simultaneously measure chlorophyll fluorescence and gas exchange to obtain a full picture of the response of plants to their environment. One technique is to simultaneously measure CO_2 fixation and PSII photochemistry at different light intensities, in non-photorespiratory conditions. A plot of CO_2 fixation and PSII photochemistry indicates the electron requirement per molecule CO_2 fixed. From this estimation, the extent of photorespiration may be estimated. This has been used to explore the significance of photorespiration as a photoprotective mechanism during drought.

Fluorescence analysis can also be applied to understanding the effects of low and high temperatures.

- Gas exchange and chlorophyll *a* fluorescence responses to high intensity light, of pioneer species and forest species was investigated. Midday leaf gas exchange was measured using a photosynthesis system, which measured net photosynthetic rate, gs, and intercellular CO_2 concentration (C_i). In the same leaves used for gas exchange measurements, chlorophyll *a* fluorescence parameters (initial, F_0; maximum F_m; and variable, F_v) were measured using a fluorometer. The results showed that despite pioneer species and forest species occupying different habitats, both showed similar vulnerability to midday photoinhibition in sun-exposed leaves.

Measuring stress and stress tolerance

Chlorophyll fluorescence can measure most types of plant stress. Chlorophyll fluorescence can be used as a proxy of plant stress because environmental stresses, *e.g.* extremes of temperature, light and water availability, can reduce the ability of a plant to metabolise normally. This can mean an imbalance between the absorption of light energy by chlorophyll and the use of energy in photosynthesis.

- Favaretto *et a*l. (2011) investigated adaptation to a strong light environment in pioneer and late successional species, grown under 100% and 10% light. Numerous parameters, including chlorophyll *a* fluorescence, were measured. A greater decline in $\mathbf{F_v / F_m}$ under full sun light in the late-successional species than in the pioneer species was observed. Overall, their results show that pioneer species perform better under high-sun light than late- successional species, suggesting that pioneer plants have more potential tolerance to photo-oxidative damage.

- Neocleous and Vasilakakis (2009) investigated the response of rasberry to boron and salt stress. An chlorophyll fluorometer was used to measure $\mathbf{F_0}$; $\mathbf{F_m}$; and $\mathbf{F_v}$ and. The leaf chlorophyll fluorescence was not significantly affected by NaCl concentration when B concentration was low. When B was increased, leaf chlorophyll fluorescence was reduced under saline conditions. It could be concluded that the combined effect of B and NaCl on raspberries induces a toxic effect in photochemical parameters.

Reading

Favaretto, V.F., Martinez, C.A., Soriani, H.H. and Furriel, R.P.M. (2011). Differential responses of antioxidant enzymes in pioneer and late-succession tropical tree species grown under sun and shade conditions. *Environmental and Experimental botany*, **70(1):** 20-28.

13.4. Isolation of Chloroplasts

Chloroplasts are organelles, specialized subunits, in plant and algal cells. Their main role is to conduct photosynthesis, where the photosynthetic pigment chlorophyll captures the energy from sunlight, and stores it in the energy storage molecules ATP and NADPH while freeing oxygen from water. They then use the ATP and NADPH to make organic molecules from carbon dioxide in a process known as the Calvin cycle. Chloroplasts carry out a number of other functions, including fatty acid synthesis, much amino acid synthesis, and the immune response in plants.

A chloroplast is one of three types of plastid, characterized by its high concentration of chlorophyll. (The other two types, the leucoplast and the chloroplast, contain little chlorophyll and do not carry out photosynthesis.) Chloroplasts are highly dynamic—they circulate and are moved around within plant cells, and occasionally pinch in two to reproduce. Their behaviour is strongly influenced by environmental factors like light colour and intensity. Chloroplasts, like mitochondria, contain their own DNA, which is thought to be inherited from their ancestor—a photosynthetic cyanobacterium that was engulfed by an early eukaryotic cell. Chloroplasts cannot be made by the plant cell, and must be inherited by each daughter cell during cell division.

Isolated chloroplasts are required to study the electron transport system of the phoitosynthetic apparatus. There are many techniques available to isolate the chloroplasts from leaf tissues. A general procedure is given below:

Materials

- Isolation medium – Take 750 ml of distilled water and to it add 2.42 g Tris (20 mM), 72.8 g sorbitol (0.4 M), 1.168 g NaCl (20 mM), and 0.610 g $MgCl_2.6H_2O$ (3.0 mM). Make the final volume to 1000 ml and adjust the pH to 7.8.

Procedure

- Take 5- 10 g of fresh leaf tissue and cut it into fine pieces. Add 20.0 ml of pre chilled isolation medium.
- Homogenise in a blender for 3 – 5 seconds.

- Filter the mixture through eight layers of cheesecloth followed by centrifugation at 3000 g for 2 minutes.
- Discard the supernatant and re-suspend the pellet in a small volume of the isolation medium and store on ice.
- Estimate the chlorophyll by diluting 0.1 – 0.2 ml of chloroplast suspension to a total volume of 4.0 ml with 80% acetone. Measure the optical density at 663 and 645 nm against 80% acetone.
- Calculate the chlorophyll content as follows:

  $(12.7 \times A_{663}) - (2.69 \times A_{645})$ = Chlorophyll 'a' (μg/ml)

  $(22.9 \times A_{645}) - (4.68 \times A_{663})$ = Chlorophyll 'b' (μg/ml)

  $(20.2 \times A_{645}) - (8.02 \times A_{663})$ = Total chlorophyll (μg/ml)

Notes

- An alternate isolation medium, most useful for many purpose is a mixture of 330 mM sorbitol, 10 mM $Na_2P_2O_7$ (sodium pyrophosphate), 5 mM $MgCl_2$ and 2 mM Na isoascorbate. Adjust the pH at 6.5 with HCl.
- Most leaves yield better chloroplasts if freshly harvested, except spinach leaves which can be stored for four weeks in cold for better yield. If leaves are brightly illuminated for 20 - 30 minutes prior to grinding, the chloroplast yield is increased.

Reading

Procedure Manual, Plant Science Division, School of Biological Sciences, MK Universitym Madurai (1982).

Walker, D.A. (1980). In "*Methods in Enzymology*" **69**, (Eds. S.P. Colowick and N.O. Kaplan), Academic Press, pp. 94.

An alternate method for chloroplast isolation

Materials

- Fresh spinach leaves
- Grinding solution
  - 0.33 M Sorbitol
  - 10 mM Sodium pyrophosphate ($Na_4P_2O_7$)
  - 4 mM MgCl

    - 2 mM Ascorbic Acid
    - Adjust pH to 6.5 with HCl
- Chopping board and knife
- Chilled mortar and pestle
- Cheesecloth
- Refrigerated preparative centrifuge
- Suspension solution
    - 0.33 M Sorbitol
    - 2 mM EDTA
    - 1 mM MgCl
    - 50 mM HEPES
    - Adjust pH to 7.6 with NaOH
- Hemacytometer and microscope

Procedure

- Prepare an ice bath and pre-cool all glassware to be used.
- Select several fresh spinach leaves and remove the large veins by tearing them loose from the leaves. Weigh out 4.0 grams of deveined leaf tissue.
- Chop the tissue as fine as possible. Add the tissue to an ice-cold mortar containing 15 ml of grinding solution and grind to a fine paste.
- Filter the solution through double layered cheesecloth into a beaker and squeeze the tissue pulp to recover all of the suspension.
- Transfer the green suspension to a cold 50 ml. centrifuge tube and centrifuge at 200 xg for 1 minute at 4° C to pellet the unbroken cells and fragments.
- Decant the supernatant into a clean centrifuge tube and recentrifuge at 1000 xg for 7 minutes. The pellet formed during this centrifugation contains chloroplasts. Decant and discard the supernatant.
- Resuspend the chloroplast pellet in 5.0 ml. of cold suspension solution or 0.035 M NaCl. Use a glass stirring rod to gently disrupt the packed pellet. This is the chloroplast suspension for use in subsequent procedures.
- Enclose the tube in aluminum foil and place it in an ice bucket.

- Determine the number of chloroplasts/ml of suspension media using a hemocytometer.

Reading

Heidcamp, W.H. In "*Cell Biology Laboratory Manual*", Biology Department, Gustavus Adolphus College, St. Peter

13.5. Anthocyanin

Anthocyanins are water soluble vascular pigments that may appear red, purple, or blue depending on the pH. They belong to a parent class of molecules called flavonoids synthesized via the phenylpropanoid pathway; they are odorless and nearly flavorless, contributing to taste as a moderately astringent sensation. Anthocyanins occur in all tissues of higher plants, including leaves, stems, roots, flowers and fruits. Anthoxanthins are clear, white to yellow counterparts of anthocyanins occurring in plants. Anthocyanins are derived from anthocyanidins by adding pendant sugars.

In flowers, bright-reds and -purples are adaptive for attracting pollinators. In fruits, the colourful skins also attract the attention of animals, which may eat the fruits and disperse the seeds. In photosynthetic tissues (such as leaves and sometimes stems), anthocyanins have been shown to act as a "sunscreen", protecting cells from high-light damage by absorbing blue-green and ultraviolet light, thereby protecting the tissues from photoinhibition, or high-light stress. This has been shown to occur in red juvenile leaves, autumn leaves, and broad-leaf evergreen leaves that turn red during the winter. The red colouration of leaves has been proposed to possibly camouflage leaves from herbivores blind to red wavelengths, or signal unpalatability, since anthocyanin synthesis often coincides with synthesis of unpalatable phenolic compounds.

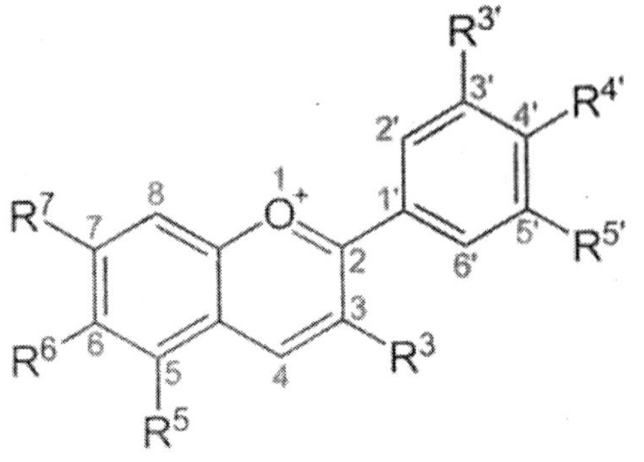

Anthocyanins are glucosides of anthicyaninins and its basic chemical structure

In addition to their role as light-attenuators, anthocyanins also act as powerful antioxidants. However, it is not clear whether anthocyanins can significantly contribute to scavenging of free radicals produced through metabolic processes in leaves, since they are located in the vacuole and, thus, spatially separated from metabolic reactive oxygen species. Some studies have shown hydrogen peroxide produced in other organelles can be neutralized by vacuolar anthocyanin.

Anthocyanins are found in the cell vacuole, mostly in flowers and fruits but also in leaves, stems, and roots. In these parts, they are found predominantly in outer cell layers such as the epidermis and peripheral mesophyll cells. Most frequently occurring in nature are the glycosides of cyanidin, delphindin, malvidin, pelargonidin, peonidin, and petunidin. Roughly 2% of all hydrocarbons fixed in photosynthesis are converted into flavonoids and their derivatives such as the anthocyanins. No fewer than 10^9 tons of anthocyanins are produced in nature per year. Not all land plants contain anthocyanin; in the Caryophyllales (including cactus, beets, and amaranth), they are replaced by betalains. Anthocyanins and betalains have never been found in the same plant. Plants with abnormally high anthocyanin quantities are popular as ornamental plants.

The anthocyanins, anthocyanidins with sugar group(s), are mostly 3-glucosides of the anthocyanidins. The anthocyanins are subdivided into the sugar-free anthrocyandin aglycones and the anthocyanin glycosides. As of 2003, more than 400 anthocyanins had been reported while more recent litreature (early 2006), puts the number at more than 550 different anthocyanins. The difference in chemical structure that occurs in response to changes in pH is the reason why anthocyanins are often used as pH indicators, as they change from red in acids to blue in bases.

Anthocyanins are thought to be subject to physiochemical degradation *in vivo* and *in vitro*. Structure, pH, temperature, light, oxygen, metal ions, intramolecular association, and intermolecular association with other compounds (copigments, sugars, proteins, degradation products, etc.) are generally known to affect the colour and stability of anthocyanins. B-ring hydroxylation status and pH have been shown to mediate the degradation of anthocyanins to their phenolic acid and aldehyde constituents. Indeed, significant portions of ingested anthocyanins are likely to degrade to phenolic acids and aldehyde *in vivo*, following consumption. This characteristic confounds scientific isolation of specific anthocyanin mechanisms *in vivo*.

Materials

- Ethyl acetate, iso-amyl alcohol, methanol, ethanol, concentrated hydrochloric acid, potassium chloride 0.025 M (in distilled water) and sodium acetate 0.4 M (in distilled water).

Preparation of Plant Extracts

Six grams of the dried plant sample was extracted with 150 ml of three different extracting solvents: methanol (E-1), water (E-2) and acidified ethanol (40% ethanol with 1% HCl, E-3), using reflux apparatus at 50^{0}C for one hour. The resulted filtrate was made off up to 150 ml with the same extracting solvent (sol. A for each of the three extracting solvents E-1, E-2 and E-3). Solution A was then divided in three parts, as follows:

Part one : 50 ml of sol. A was hydrolyzed with a solution of 4 N HCl in a ratio of (1:1) for 30 minutes. The resulted hydrolyzed solution was first partitioned with ethyl acetate (15 ml x 3) to extract the flavonoids; the ethyl acetate layer was washed with distilled water (10 ml x 3), evaporated to dryness under reduced pressure and re-dissolved in ethanol to a final volume of 25 ml. This solution is denoted as sol. 1. The remained hydrolyzed layer was secondly partitioned with iso-amyl alcohol (15 ml. x 3) to extract the anthocyanidins. The iso-amyl alcohol layer was evaporated to dryness and finally the residue was re-dissolved in 0.01 % HCl in ethanol. This solution was denoted as sol. 2

Part two: 50 ml of sol. A was hydrolyzed with 2 N HCl in a ratio of (1:1) for 8 hours. The resulted hydrolyzed solution was treated using the same method mentioned in part one, resulting in sol. 3 (the red anthocyanidins part).

Part three: 50 ml of sol A was centrifuged at 2000 rpm for 10 min at 4^{0}C. The collected aliquots were evaporated to dryness using a rotary evaporator at 40^{0}C under vacuum condition, then re-dissolved in 0.01% HCl in distilled water to a final volume of 25 ml. This solution is sol.4.

Quantitative analysis of anthocyanins

Two methods with their modifications are examined in order to select the best procedure followed for the determination of anthocyanin content in different plant samples.

The pH-Differential method

To perform this method, two types of buffers should be prepared:

- **pH = 1 buffer**

This buffer is prepared by dissolving 1.86g of KCl in 980 ml distilled water. The pH is adjusted to pH=1 by the drop wise addition of HCl, then the volume is completed to 1L with distilled water.

- **pH = 4.5 buffer**

54.43g of sodium acetate is dissolved in 970 ml distilled water. The pH is adjusted to 4.5 through the addition of HCl drop by drop, and then the volume is completed to 1L using distilled water.

Procedure

1 ml of sol. 4 was leveled off to 10 ml with the buffer solution of pH 1. A second 1 ml of sol. 4 was diluted to 10 ml using buffer solution of pH 4.5. This procedure was performed on each of the three extracting solvents. The flasks are left at room temperature for 15 min., and then the absorbance was read at ë = 520 nm, and ë =700 nm. Distilled water was used as a blank. The anthocyanins are calculated as cyanidin-3-glucoside equivalents, mg/l, using the equation:

$$\frac{A \times MW \times DF \times 10^3}{\varepsilon \times l}$$

where A = (*A* 520 nm – *A* 700 nm) pH 1.0 – (*A* 520 nm – *A* 700 nm) pH 4.5; MW (molecular weight) = 449.2 g/mol for cyanidin-3-glucoside (cyd-3-glu); DF = dilution factor (1:10); l = path length in cm; ε = 26900 molar extinction coefficient, in $L \cdot mol^{-1} \cdot cm^{-1}$, for cyd-3-glu; and 10^3 = factor for conversion from g to mg.

The Standard Curve method

This method depends on the preparation of cyanidin chloride standard curve. The anthocyanins are calculated as cyanidin chloride equivalents by means of the standard curve equation obtained.

Preparation of the standard curve

The first step here is the preparation of the standard cyanidin chloride by dissolving 0.1g of quercetin, using 5 ml of ethanol in a 100 ml volumetric flask. 2.5 g of Magnesium strips, and 16 ml of conc. HCl (added step wise). The colour will change from yellow to dark violet-red colour. The volume was completed to 100 ml using 50% ethanol; this is sol B.

The second step was the preparation of the standard curve. Take 0.5 ml, 0.75 ml, 1 ml, 1.25 ml, 1.5 ml, 1.75 ml, 2 ml, 2.25 ml, 2.5 ml, 2.75 ml, and 3 ml of sol. B, and we put them separately in a series of 10 ml volumetric flasks, completing the volumes to 10 ml with 50% ethanol. We left the flasks for 10 minutes and read the absorbance at λ = 520 nm.

Measurement of samples

First: 5 ml of sol. 2 (for each of the three extracts) were diluted with 0.01 % HCl in ethanol to 25 ml. Secondly: 5 ml of of diluted with 0.01% HCl in ethanol to 25 ml. Finally, 2 ml of sol. 4 were diluted with 2 % HCl in distilled water to a volume of 25 ml. All the flasks were left for 10 min, and then the absorbance of each was read at $\lambda = 520$ nm. The blanks used were 0.01%HCl in ethanol, and 2% HCl in water respectively.

Reading

Humad, S.S. and Istudor, V. (2009). Quantitative analysis of bio-active compounds in *Hibiscus sabdariffa* l. extracts. ii. quantitative analysis and biological activities of anthocyanins, *Farmacia,* **LVII** (1): 74-81.

13.6. Analysis of the Flavonoid Content in Plant Extracts

Determination of flavonoids according to the Christ-Mullers method

Procedure

20 ml of acetone, 2 ml of 25% HCl and 1 ml of 0.5% hexamethylenetetramine were added to 25 ml extract (corresponding to 2g of plant material) and refluxed at 56^{0}C for 30 minutes. The extract was filtered and re-extracted twice with 20 ml acetone for 10 min. After cooling and filtration, the extract was made up to 100.0 ml with acetone (basic sample solution, BSS). 20 ml of BSS was mixed with 20.0 ml of water and then extracted with ethyl acetate (first with 15.0 ml and then three times with 10.0 ml). Ethyl acetate extracts were rinsed twice with water then filtered and made up to 50.0 ml with ethyl acetate (S1). In 10 ml of S1 0.5 ml of 0.5% solution of sodium citrate and 2 ml of $AlCl_3$ (prepared by dissolving 2 g of $AlCl_3$ in 100 ml of 5% acetic acid in methanol) were added and then made up to 25.0 ml with 5% methanolic solution of acetic acid (sample solution, SS). The same procedure was performed with blank sample solution but without $AlCl_3$. After 45 minutes, yellow solutions were filtered and absorbance at 425 nm was measured. The content of total flavonoids was evaluated upon three independent analyses. The yield was calculated as quercetin percent using the following expression:

g % = A × 0.772 / b, where A is absorbance and b represents the mass of dry herbal material in grams.

Determination of flavonoids by Chang *et al.* method

Flavonols are expressed as quercetin equivalent. Quercetin was used to perform the calibration curve (standard solutions of 6.25, 12.5, 25.0, 50.0, 80.0 and 100.0 $\mu g\ ml^{-1}$ in 80 % ethanol *(v/v)*. Sample extracts (1g plant material in 25 ml extract) were all evaporated to dryness and re-dissolved in 80 % ethanol to be ready for the analytical test.

1 ml of a sample (ethanolic solutions *Hibiscus*) was mixed with 3 ml 95 % ethanol *(v/v)*, 0.2 ml 10 % aluminum chloride *(m/V)*, 0.2 ml of 1 mol l^{-1} potassium acetate and 5.6 ml water. A volume of 10 % *(m/V)* aluminum chloride was substituted by the same volume of distilled water in blank. After incubation at room temperature for 30 minutes, the absorbance of the reaction mixture was measured at 415 nm.

13.7. Estimation of Carotenes and Xanthophylls

Carotenes: The term carotene is used for several related unsaturated hydrocarbons substances having the formula $C_{40}H_x$, which are synthesized by plants but cannot be made by animals. Carotene is an orange photosynthetic pigment important for photosynthesis. Carotenes are all coloured to the human eye. They are responsible for the orange colour of the carrot, for which this class of chemicals is named, and for the colours of many other fruits and vegetables (for example, sweet potatoes, chanterelle and orange cantalouge melon). Carotenes are also responsible for the orange (but not all of the yellow) colours in dry foliage. They also (in lower concentrations) impart the yellow colouration to milk-fat and butter. Omnivorous animal species which are relatively poor converters of coloured dietary carotenoids to colourless retinoids have yellowed-coloured body fat, as a result of the carotenoid retention from the vegetable portion of their diet. The typical yellow-coloured fat of humans and chickens is a result of fat storage of carotenes from their diets.

Carotenes contribute to photosynthesis by transmitting the light energy they absorb to chlorophyll. They also protect plant tissues by helping to absorb the energy from singlet oxygen, an excited form of the oxygen molecule O_2 which is formed during photosynthesis. β-carotene is composed of two retinyl groups, and is broken down in the mucosa of the human small intestine by β-carotene 15.15'-monooxygenase to retinal, a form of vitamin A. β-Carotene can be stored in the liver and body fat and converted to retinal as needed, thus making it a form of vitamin A for humans and some other mammals. The carotenes α-carotene and λ-carotene, due to their single retinyl group (β-ionone ring), also have some vitamin A activity (though less than β-carotene), as does the xanthophyll carotenoid β-cryptoxanthin. All other carotenoids, including lycopene, have no beta-ring and thus no vitamin A activity (although they may have antioxidant activity and thus biological activity in other ways).

Chemically, carotenes are polyunsaturated hydrocarbons containing 40 carbon atoms per molecule, variable numbers of hydrogen atoms, and no other elements. Some carotenes are terminated by hydrocarbon rings, on one or both ends of the molecule. All are coloured to the human eye, due to extensive systems of conjugate double bonds. Structurally carotenes are tetraterpenes, meaning that they are synthesized biochemically from four 10-carbon terpene units, which in turn are formed from eight 5-carbon isoprene units.

Carotenes are found in plants in two primary forms designated by characters from the Greek alphabet: alpha carotene (α-carotene) and beta carotene (β-carotene). Gamma-, Delta-, Epsilon-, and zeta-carotene (γ, δ, ε, and ς-carotene) also exist. Since they are hydrocarbons, and therefore contain no oxygen, carotenes are fat-soluble and insoluble in water (in contrast with other cartotenoids, the xanthophylls, which contain oxygen and thus are less chemically hydrophobic).

Xanthophylls (originally phylloxanthins) are yellow pigments that form one of two major divisions of the cartenoid group. The name is due to their formation of the yellow band seen in early chromatography of leaf pigments. Their molecular structure is similar to carotenes, which form the other major carotenoid group division, but xanthophylls contain oxygen atoms, while *carotenes* are purely hydrocarbons with no oxygen. Xanthophylls contain their oxygen either as hydroxyl groups and/or as pairs of hydrogen atoms that are substituted by oxygen atoms acting as a bridge (epoxide). For this reason, they are more polar than the purely hydrocarbon carotenes, and it is this difference that allows their separations from carotenes in many types of chromatography. Typically, carotenes are more orange in colour than xanthophylls.

Like other carotenoids, xanthophylls are found in highest quantity in the leaves of most green plants, where they act to modulate light energy and perhaps serve as a non-photochemical quenching agent to deal with triplet chlorophyll (an excited form of chlorophyll), which is overproduced at high light levels in photosynthesis. The xanthophylls found in the bodies of animals, and in dietary animal products, are ultimately derived from plant sources in the diet. For example, the yellow colour of chicken egg yolks, fat, and skin comes from ingested xanthophylls (primarily lutein, which is often added to chicken feed for this purpose).

The yellow colour of the human macula lutea (litreally, *yellow spot*) in the retina of the eye comes from the lutein and zeaxanthin it contains, both xanthophylls again requiring a source in the human diet to be present in the eye.

These function in eye protection from ionizing blue light, which they absorb. These two specific xanthophylls do not function in the mechanism of sight, since they cannot be converted to retinal (also called retinaldehyde or vitamin A aldehyde).

The group of xanthophylls includes (among many other compounds) lutein, zeaxanthin, neoxanthin, violaxanthin, and α- and β-cryptoxanthin. The latter compound is the only known xanthophyll to contain a beta-ionone ring, and thus β-cryptoxanthin is the only xanthophyll that is known to possess pro-vitamin A activity for mammals. Even then, it is a vitamin only for plant-eating mammals that possess the enzyme to make retinal from carotenoids that contain beta-ionone (some carnivores lack this enzyme). In species other than mammals, certain xantho-phylls may be converted to hydroxylated retinal-analogues that function directly in vision. For example, with the exception of certain flies, most insects use the xanthophyll derived R-isomer of 3-hydroxyretinal for visual activities, which means that β-cryptoxanthin and other xanthophylls (such as lutein and zeaxanthin) may function as forms of visual "vitamin A" for them, while carotenes (such as beta carotene) do not.

Materials

- The chromatographic column is constructed of borosilicate glass, 12 mm id X 33 cm length, fitted with an adapter containing a replaceable fritted disc. The column and adapter are fitted with an "0" ring and held together with a spring clamp. The adapter is fitted with a stopcock containing a needle valve. The column is attached to a Fisher Filtrator through a rubber stopper
- The column adsorbent is a mixture of magnesium oxide, diatomaceous earth 1 :1 (w/w), used as received, mixed.
- The extractants are hexane-acetone, 7:3 (v/v), and acetone. Carotenes are eluted with hexane-acetone, 9 :1 (v/v).
- Xanthophylls are eluted with hexane-acetone-methanol, 8 :1 :1 (v/v/v).

Procedure for dry samples

- Grind plant sample to pass a No . 50 sieve.
- Place 4 g of accurately weighed sample in a 100 ml volumetric flask.
- Add 40 ml of hexane-acetone (7:3) and 1 ml distilled water.
- Extract either by overnight infusion or by shaking for one hour.

- Dilute the sample to 100 ml with hexane; mix well and set aside in a dark cabinet until the tobacco particles settle.
- Dry pack the chromatographic column under reduced pressure. Tamp firmly to a final height of 7 cm and add a 2-cm layer of anhydrous sodium sulfate to the top of the column.
- With a 25-ml volumetric flask in place beneath the column, transfer 15 ml of sample extract to the column and apply a vacuum. As the last of the sample extract is entering the sodium sulfate layer, add 10 ml of hexane-acetone (9 :1) to the column.
- When all of this solvent mixture has entered the sodium sulfate layer, add 10 ml of hexane-acetone-methanol (8:1:1).
- The 25-ml volumetric receiving flask is changed when the orange carotene band is completely eluted and the xanthophyll band, also orange, is about halfway down the column.
- Elute the xanthophyll band completely with small additional volumes of hexane-acetone-methanol.
- The eluates are adjusted to 25 ml with hexane-acetone (9:1) for carotenes and with acetone for the xanthophylls.
- Keep eluates out of strong light. Storage in a dark cabinet is ideal .
- Measure absorbances at 436 nm for carotenes and at 475 nm for xanthophylls. Use distilled water as the reference solution.

Procedure for fresh sample

- Freeze the green leaf as quickly as possible after picking and store in a freezer until the analysis, performed under reduced lighting, is completed.
- Blend 4 g of green leaf with approximately 12 ml acetone for one minute.
- Filter the extract into a 100-ml volumetric flask using a water pump. Collect the residue on a course fritted-glass crucible.
- Wash the residue with a small volume of acetone (<10 ml).
- Transfer the residue to the mini-container and repeat this extraction. After 4 or 5 such extractions the extract should be colourless. Do not exceed the 100-ml capacity of the receiving flask.
- Add 1.5 ml of potassium hydroxide solution (40% KOH in CH_30H) to the flask; dilute to 100 ml with acetone and mix well.

- Allow this solution to stand at least 10 minutes .
- Pipet a 15-ml aliquot of extract into a 125-ml separatory funnel. Add 15 ml of an ammonium sulfate solution (20%); mix well. Extract with 10 ml hexane. Drain the aqueous phase into a second 125-mi separatory funnel.
- Extract the aqueous phase with a second 10-ml portion of hexane. Repeat this hexane extraction until the hexane is colourless.
- Combine all hexane extracts in a 125-ml separatory funnel and extract twice with 15-ml
- portions of water and once with sodium sulfate solution (2% w/v). Do not extract too vigorously because of the ease of emulsion formation.
- Transfer the hexane quantitatively to a small beaker containing 1 to 2 g sodium sulfate (anhydrous). Stir well to remove all traces of water from the hexane.
- Quantitatively transfer the hexane extract to the chromatographic column. Apply a vacuum, collecting the hexane in a 50-ml volumetric flask.
- When the carotenoid bands have formed on the column and as the last of the hexane extract enters the sodium sulfate layer, add hexane-acetone (9:1) to elute the carotene band. As the carotene band nears the bottom of the column, add hexane-acetone-methanol (8:1:1) to elute the xanthophylls band. Collect the carotene band eluate in a 25-ml volumetric flask.
- Collect the xanthophyll band in a 25-ml volumetric flask containing 2 ml acetone. Dilute the carotene eluate to 25 ml with hexane-acetone (9:1) .
- Dilute the xanthophyll eluate to 25 ml with acetone. Measure absorbances at 436 nm for carotenes and at 475 nm for xanthophylls. Use distilled water as the reference solution.

Preparation of Calibration Curves

- Weigh 0.1000 g α-carotene in a 100-ml beaker. Add 30 ml of benzene; dissolve the g-carotene by stirring. Add 30 ml of hexane-acetone (9:1) and mix well.
- Carefully transfer this solution into a 250-ml volumetric flask. Thoroughly rinse the beaker with hexane-acetone (9:1) and bring the combined solutions to a total volume of 250 ml with hexane-acetone (9:1) ; mix well.

- Transfer exactly 5 ml of this solution to a second 250-ml volumetric flask and dilute to volume with hexane-acetone (9:1).
- Transfer 15, 30, and 45-ml aliquots of this solution to individual 100-ml volumetric flasks and dilute to volume with hexane-acetone (9:1). These standard solutions contain 1 .2, 2 .4, and 3 .6 μg β-carotene per millilitre, respectively.
- To prepare standard curves, measure the absorbances of these working standards at 436 nm for carotene and at 475 nm for xanthophylls.

Notes

- All of the carotenoids are extremely sensitive to acids, oxidation, and heat.
- Any reliable procedure must be as short and simple as possible to avoid these causes of decomposition and possible enzyme action.
- Extraction of dry samples for carotenoids have been carried out using hot (one-hour reflux) or cold (overnight infusion) techniques.
- Cold extraction is the preferred technique since hot refluxing may result in oxidation or isomerization.
- To verify the efficiency of one-hour shaking (wrist-action) compared with overnight infusion, a series of eight ground, dry samples were analyzed for total carotenes and total xanthophylls. The procedure previously given was used.

13.8. Estimation of Lycopene

Lycopene is a bright red carotene and carotenoids pigment and phytochemicals found in tomatoes and other red fruits and vegetables, such as red carrots, watermelons, gac, and papayas (but not strawberries, red bell peppers, or cherries). Although lycopene is chemically a carotene, it has no vitamin A activity. Foods that are not red may contain lycopene as well. In plants, algae, and other photosynthetic organisms, lycopene is an important intermediate in the biosynthesis of many carotenoids, including beta carotene, responsible for yellow, orange or red pigmentation, photosynthesis, and photo-protection. Like all carotenoids, lycopene is a polyunsaturated hydrocarbon (an unsubstituted alkenes). Structurally, it is a tetraprene assembled from eight isoprene units, composed entirely of carbon and hydrogen, and is insoluble in water. Lycopene's eleven conjugated double bonds give it its deep red colour and are responsible for its antioxidant activity. Due to its strong colour and non-toxicity, lycopene is a useful food colouring and is approved for usage in the USA, Australia and New Zealand and the EU.

Lycopene is not an essential nutrient for humans, but is commonly found in the diet, mainly from dishes prepared from tomatoes. When absorbed from the intestine, lycopene is transported in the blood by various lipoproteins and accumulates in primarily the blood, adipose, skin, liver, adrenal glands, prostate and testes, but can be found in most tissues.

Because preliminary research has shown that people who consume tomatoes have a lower cancer risk, lycopene has been considered a potential agent for prevention of some types of cancers, particularly prostate cancer. However, this area of research and the relationship with prostate cancer have been deemed insufficient of evidence for health claim approval. Lycopene is a symmetrical tetraterpene assembled from 8 isoprene units. It is a member of the carotenoid family of compounds, and because it consists entirely of carbon and hydrogen, is also a carotene. Isolation procedures for lycopene were first reported in 1910, and the structure of the molecule was determined by 1931. In its natural, all-*trans* form, the molecule is long and straight, constrained by its system of eleven conjugated double bonds. Each extension in this conjugated system reduces the energy required for electrons to transition to higher energy states, allowing the molecule to absorb visible light of progressively longer wavelengths. Lycopene absorbs all but the longest wavelengths of visible light, so it appears red.

Plants and photosynthetic bacteria naturally produce all-*trans* lycopene, but a total of 72 geometric isomers of the molecule are sterically possible. When exposed to light or heat, lycopene can undergo isomerisation to any of a number of these *cis*-isomers, which have a bent rather than linear shape. Different isomers were shown to have different stabilities due to their molecular energy (highest stability: 5-cis e" all-trans > 9-cis > 13-cis > 15-cis > 7-cis > 11-cis: lowest). In the human bloodstream, various *cis*-isomers constitute more than 60% of the total lycopene concentration, but the biological effects of individual isomers have not been investigated.

Carotenoids like lycopene are important pigments found in photosynthetic pigment-protein complexes in plants, photosynthetic bacteria, fungi, and algae. They are responsible for the bright colours of fruits and vegetables, perform various functions in photosynthesis, and protect photosynthetic organisms from excessive light damage. Lycopene is a key intermediate in the biosynthesis of many important carotenoids, such as beta-carotene, and xanthophylls.

Lycopene is non-toxic and is commonly found in the diet, but cases of excessive carotenoid intake have been reported. In a middle-aged woman who had prolonged and excessive consumption of tomato juice, her skin and liver were coloured orange-yellow and she had elevated levels of lycopene in her blood. After three weeks on a lycopene-free diet her skin colour returned to normal. This discolouration of the skin is known as lycopenodermia and is non-toxic. There are also cases of intolerance or allergic reaction to dietary lycopene, which may cause diarrhea, nausea, stomach pain or cramps, gas, vomiting, and loss of appetite.

The general chemical formula of lycopene is $C_{40}H_{56}$ with a molar mass of 536.87 g mol^{-1}. It has a density of 0.889 g/ml and a melting point of 173^0 C. Lycopene is insoluble in water and appear deep red in colour.

Estimation of Lycopene

The carotenoids in the sample are extracted in acetone and then taken up in petroleum ether. Lycopene has absoption maxima at 473 nm and 503 nm. One mole of lycopene whwn dissolved in one litre light petroleum ($40 - 60^0$ C) and measured in a spectrophotometer at 503 nm in 1 cm light path gives an absorbance of 17.2 x 104. Therefore a concentration of 3.1206 µg lycopene/ml gives ubit absobance.

Material

- Acetone
- Petroleum ether ($40 - 60^0$ C)
- Anhydrous sodium sulphate
- Sodium sulphate – 5%

Procedure

- Freshly prepared tomato pulp (5 – 10g) is extracted repeatedly with acetone, until the residue is colourless.
- Pool the acetone extracts and transfer quantitatively in separating funnel having 20.0 ml petroleum ether and mix gently.
- Add 20 ml of sodium sulphate solution and shake the separating funnel gently. Add 20.0 ml of petroleum ether for clear separation of two distinct layers.

- Separate the two phases and re-extract the lower aqueous phase with additional 20.0 ml petroleum ether until aqueous phase is colourless.
- Pool the petroleum ether extract and keep in brown coloured bottle having 10.0 g of anhydrous sodium sulphate. Keep it for 30 – 45 minutes.
- Decant the extract in 100 ml volumetric flask by passing through glass wool and observe the optical density at 503 nm against pure petroleum ether.

Calculation

Absobance (1 unit) = 3.1206 µg lycopene/ml

Lycopene (mg) in 100 g sample = (31.206 X Absorbance)/ Weight of sample (g)

Reading

Ranganna, S. (1976). In "*Manual of Analysis of Fruits and Vegetable Products*" McGraw Hill Publication, New Delhi, pp. 77.

13.9. Curcumin

Curcumin is a diarylheptanoid. It is the principal curcuminoid of the popular South Asian spice turmeric, which is a member of the ginger family (Zingiberaceae). Turmeric's other two curcuminoids are desmethoxycurcumin and bis- desmethoxycurcumin. The curcuminoids are natural phenols that are responsible for the yellow colour of turmeric. Curcumin can exist in several tautomeric forms, including a 1,3-diketo form and two equivalent enol forms. The enol form is more energetically stable in the solid phase and in solution. Curcumin can be used for boron quantification in the curcumin method. It reacts with boric acid to form a red-colour compound, rosocyanine.

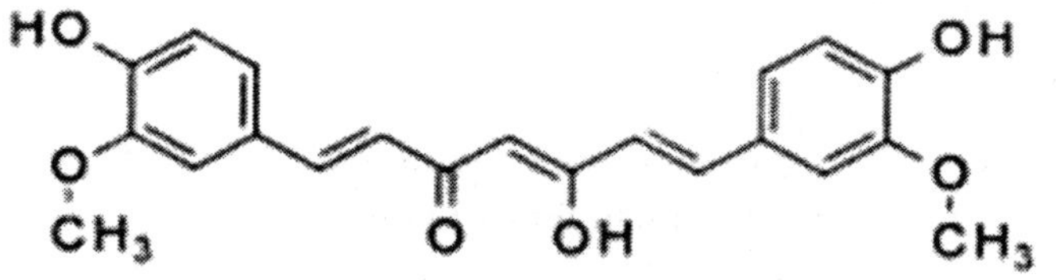

Curcumin is a bright-yellow colour and may be used as a food colouring. Its molecular formula is $C_{21}H_{20}O_6$ with molar mass of 368.38 g mol^{-1} and appear bright yellow-orange in colour.

Curcumin molecule

Curcumin incorporates several functional groups. The aromatic ring systems, which are phenols, are connected by two α, β-unsaturated carbonyl groups. The diketones form stable enol and are readily deprotonated to form enolates; the α, β-unsaturated carbonyl group is a good Michael acceptor and undergoes nucleophilic addition. Research has identified curcumin as the agent responsible for most of the biological activity of turmeric. Laboratory research shows that curcumin is a pleiotropic molecule possibly capable of interacting with molecular targets involved in inflammation. *In vitro*, curcumin modulates the inflammatory response by down-regulating the activity of cyclooxygenase-2, lipoxygenase, and inducible nitric oxide synthase enzymes; and inhibits several other enzymes involved in inflammation mechanisms.

A systematic review of the use of curcumin based supplements for treating diabetic wounds found no significant positive outcome for human use. The effectiveness of curcumin has neither been confirmed in sufficient preliminary research nor been conclusively demonstrated in randomized, placebo-controlled, double blind clinical trials. A survey of the litreature shows a number of other potential uses and that daily doses over a 3-month period of up to 12 grams proved safe. Clinical trials in humans are studying the effect of curcumin on various diseases, including multiple myeloma, pancreatic cancer, myelodysplastic syndromes, colon cancer, psoriasis, arthritis, and Alzheimer's disease. A number of trials studying curcumin efficacy and safety revealed poor absorption and low bioavailability. Methods to possibly increase absorption and systemic bioavailability are under study, including combined administration with piperine and quercetin.

Curcumin has also been shown to be a vitamin D receptor ligand "with implications for colon cancer chemoprevention." Curcumin is used as an indicator for boron. Curcumin is quantitatively extracted by refluxing the material in alcohol and is estimated spectrophotometrically at 425 nm.

Procedure

- Dissolve 0.2 – 0.5 g of moisture free turmeric powder in 250 ml of absolute alcohol. Reflux the content for 4–5 hours.
- Cool the material and decant the extract into 250 ml volumetric flask and make the volume with absolute alcohol.
- Dilute a suitable aliquot (1–2 ml) to 0 ml with absolute alcohol and optical density is recorded at 425 nm against absolute alcohol.

Calculation

Curcumin content g/100 g =

(0.0025 x A425 x Volume of sample x Dilution factor x 100) / (0.42 x Weight of sample (g) x 1000)

CHAPTER 14

Phenols

In organic chemistry, phenols, sometimes called phenolics, are a class of chemical compounds consisting of a hydroxyl group (—OH) bonded directly to an aromatic hydrocarbon group. The simplest of the class is phenol, which is also called carbolic acid C_6H_5OH. Phenolic compounds are classified as simple phenols or polyphenols based on the number of phenol units in the molecule.

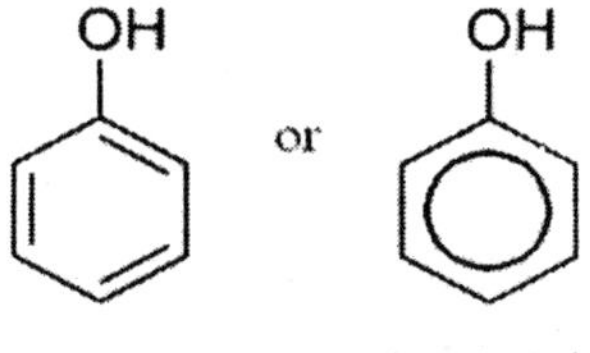

OH or OH

Phenol - the simplest of the phenols.

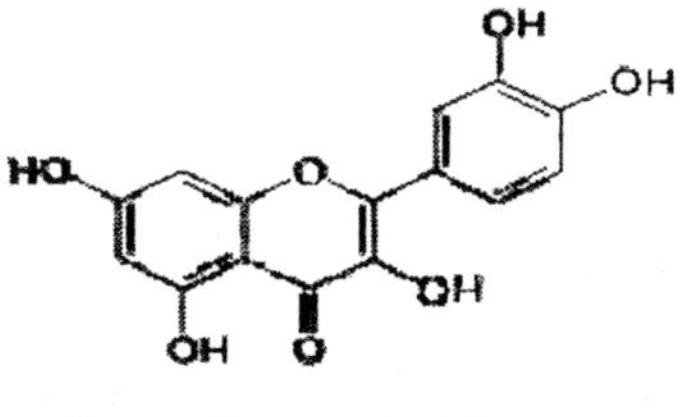

Quercetin, a typical flavonoid, is a polyphenol

Phenolic compounds are synthesized industrially; they also are produced by plants and microorganisms, with variation between and within species. Although similar to alcohols, phenols have unique properties and are not classified as alcohols (since the hydroxyl group is not bonded to a *saturated* carbon atom). They have higher acidities due to the aromatic ring's tight coupling with the oxygen and a relatively loose bond between the oxygen and hydrogen. The acidity of the hydroxyl group in phenols is commonly intermediate between that of aliphatic alcohols and carboxylic acids (their pK_a is usually between 10 and 12).

Loss of a positive hydrogen ion (H^+) from the hydroxyl group of a phenol forms a corresponding negative phenolate ion or phenoxide ion, and the corresponding salts are called phenolates or phenoxides, although the term aryloxides is preferred according to the IUPAC Gold Book. Phenols can have two or more hydroxy groups bonded to the aromatic ring(s) in the same molecule. The simplest examples are the three benzenediols, each having two hydroxy groups on a benzene ring. Organisms that synthesize phenolic compounds do so in response to ecological pressures such as pathogen and insect attack, UV radiation and wounding. As they are present in food consumed in human diets and in plants used in traditional medicines of several cultures, their role in human health and disease is a subject of research. Some phenols are germicidal and are used in formulating disinfectants. Others possess estrogenic or endocrine disrupting activity. Phenolic compounds are mostly found in vascular plants (trachea-phytes) *i.e.* Lycopodiophyta (lycopods), Pteridophyta (ferns and horsetails), Angiosperms (flowering plants or Magnoliophyta) and Gymnosperms (conifers, cycads, Ginkago and Gnetales). In ferns, compounds such as kaempferol and its glucoside can be isolated from the methanolic extract of fronds of *Phegopteris connectilis* or kaempferol-3-O-nutinoside, a known bitter-tasting flavonoid glycoside, can be isolated from the rhizomes of *Selliguea feei*. Hypogallic acid, caffeic acid, paeoniflorin and pikuroside can be isolated from the freshwater fern *Salvinia molesta*. In conifers (Pinophyta), phenolics are stored in polyphenolic parenchyma cells, a tissue abundant in the phloem of all conifers. The aquatic plant *Myriophyllum spicatum* produces ellagic, gallic and pyrogallic acids and (+)-catechin.

Occurrences in Monocotyledons - Alkylresorcinols can be found in cereals. 2,4-Bis(4-hydroxybenzyl) phenol is a phenolic compound found in the orchids Gastrodia elata and Galeola faberi.

14.1. Estimation of Total Phenolic Compounds

The quantitative estimation of total phenols in biological extracts can be accompanied in a number of different ways. Most phenolic compounds are readily oxidised by oxidising agents, many react with diazitised amines and like electrophilic substances, form coloured compounds, some form colured complexes with certain metals; all show a definite absorption peaks in ultra-violet, and methods for their estimation have been deviced based on such properties. Since, however, individual compounds vary widely in their ability to react with (a) oxidising agents such as phosphomolybdate or permanganate, (b) coupling reaction such diazotized p-nitroaniline, (c) metals such as iron, (d) in their ultra-violet or visible abroption spectra; all procedures for estimation of total phenols are necessarily emperical. It was found that methods based on the use of oxidising agents are the most useful as the variation between estimations of individual compounds is much less than that obtained using methods using the other properties.

The addition of extra saturated sodium carbonate to the solution (2.0 ml instead of 1.0 ml), although increasing the likelyhood of precipitation, was found to make no differences to the reading obtained, but the use of the bases such as K_2CO_3, NaOH or Na_2PO_4 buffer of various pH gave less reproducible results. The transmission curve of the complex showed a very broad maximum between 620 nm and 740 nm with a pea at 725 nm and hence simple colorimeters can be used with a suitable red filter without much loss of precision. All the phenolic compounds tested showed an almost linear relationship between absorption and concentration, altough the slopes of the curves were different, the useful range using 1 cm cells being of the order of 10 – 100 µg of substnces in the aliquote taken for examination.

Material

- Folin-Dennis reagent – To 750 ml of distilled water, 100 g of sodium tungstate ($Na_2WO_4 \cdot 2H_2O$), 20 g of phoshpho molybdic acid and 50 ml of phosphoric acid (H_3PO_4) is added. The chemical mixture is refluxed for 2 hours, cooled and final volume is made to 1000 ml with distilled water.
- Saturated sodium carbonate solution – To each 100 ml of distilled water add 35 g of anhydrous sodium carbonate (Na_2CO_3) and dissolve it at 70 – 80^0 C and it is cooled for over night period. Seed the super saturated solution with the crystals of $Na_2CO_3.10H_2O$ and after crystalization filter through glass wool.
- Phloroglucinol – A stock solution of 10 µmole/0.1 ml is prepared by dissolving 126.11 mg of phloroglucinl in 10.0 ml of distilled water. From this stock solution, with appropriate dilution a gradient of phloroglucinol from 1.0 to 10.0 µmole is prepared.

Procedure

- 5 g of fresh plant sample are homoginated in 10 ml of 80% (v/v) of ethanol / methanol in mortar and pastle.
- The homogenates are centrifuged at 10,000 rpm for 10 minutes at room temperature.
- The supernatants are collected and the residue is washed 3 ties with 80% ethanol / methanol and centrifuged. All the supernatants are pooled and made to a volume of 50 ml with 80% ethanol / methanol.
- To 0.5 ml of saple extract is diluted to 7.0 ml with distilled water and 0.5 ml of Follin-Dennis reagent is added. Reagent mixture was allowed to

stan for 5 minutes and then 1.0 ml of saturated sodium carbonate solution is added.

- Final volume is made to 10 ml with distilled water and the reaction mixture is allowed to stand for an hour at room temperature.
- Optical density is recorded at 725 nm against a reagent blank.
- Standard curve of phloroglucinol is prepared as described above.

14.2. Estimation of Chlorogenic Acid Content

Chlorogenic acid (CGA) is a natural chemical compound which is the ester of caffeic acid and (-)-quinic acid. It is an important biosynthetic intermediate. Chlorogenic acid is an important intermediate in lignin biosynthesis. This compound, known as an antioxidant, may also slow the release of glucose into the bloodstream after a meal.

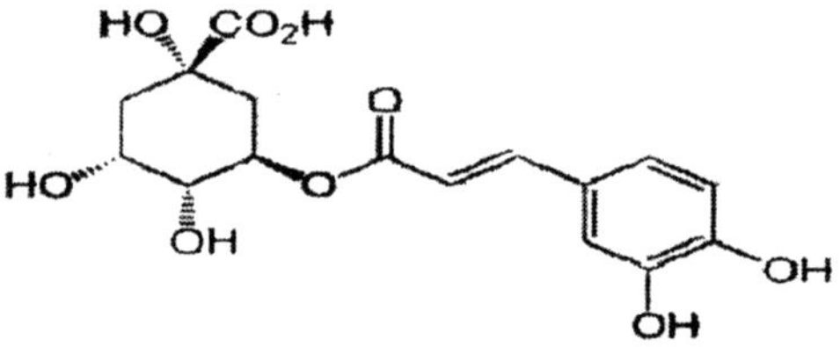

Chlorogenic acid - (1*S*,3*R*,4*R*,5*R*)-3-{[(2*Z*)-3-(3,4-dihydroxyphenyl) prop-2-enoyl]oxy}-1,4,5-trihydroxycyclohexanecarboxylic acid

The term chlorogenic acids can also refer to a related family of esters of hydroxycinnamic acids (caffeic acid, ferulic acid and p-coumaric acid) with quinic acid. Chlorogenic acids contain no chlorine.

Structurally, chlorogenic acid is the ester formed between caffeic acid and L-quinic acid. Isomers of chlorogenic acid include 4-*O*-caffeoylquinic acid (cryptochlorogenic acid or 4-CQA), 5-*O*-caffeoylquinic acid (neochlorogenic acid or 5-CQA). The epimer at position 1 has not yet been reported. Isomers containing two caffeic acid molecules are called isochlorogenic acid. It can be found in coffee. There are several isomers such as 3,4-dicaffeoylquinic acid and 3,5-dicaffeoylquinic acid. Cynanine (1,5-dicaffeoylquinic acid) is an other isomer with two caffeic acid molecules..Chlorogenic acid is freely soluble in ethanol and acetone.

Isomers of chlorogenic acid are found in potatoes. Chlorogenic acid can be found in bamboo *Phyllostachys edulis* as well as in many other plants. It is one of the major phenolic compounds identified in peach and in prunes. It also is one of the phenol found in green coffee bean extract. Chlorogenic acid, its 3-

O-glucoside, 3-*O*-galactocide and 3-*O*-arabinoside can be found in the shoots of *Calluna vulgaris* (heather).

Chlorogenic acid is extracted with alcohol, dried and dissolved in acetone. It is reacted with titanium ion to form a coloured complex which can be measured at 450 nm.

Materials

- Titanium reagent – 20% titanium chloride ($TiCl_4$) in concentrated HCl.
- Standard 25 – 200 μg/ml chlorogenic acid in acetone.
- Acetone
- Ethanol 80%
- HCl 2.5 N

Procedure

- Reflux twice a known quantity of defatted sunflower meal in 80% ethanol (adjusted to pH 4.0 with 2.5 N HCl) for 30 minutes (125 ml to 1.0 g meal).
- Discard the precipitate and collect 250 ml of extract.
- Remove 0.5 ml of sample and dry in a vaccum oven at 50^0 C and 700 mm pressure for 2 hours.
- Redissolve the dried extract in 4.75 ml of acetone.
- Add 0.25 ml of $TiCl_4$ and observe the optical density at 450 nm against a reagent blank.
- Similarly treat the standards with TiCl4 and measure the optical ensity at 450 nm and draw the standar curve.
- Find the amount of cholorogenic acid in the sample through the standard curve.
- Express the chlorogenic acid content as mg per 100 g of sample.

14.3. Estimation of Tannin Content

A tannin (also known as *vegetable tannin, natural organic tannins* or sometimes *tannoid, i.e.* a type of biomolecule, as opposed to modern synthetic tannin) is an astrigent, bitter plant polyphenolic compound that binds to and precipitates proteins and various other organic compounds including amino acids and alkaloids. The term tannin (from *tanna*, an Old High German word for oak or fir tree, as in Tannenbaum) refers to the use of wood tannins from oak in tanning animal hides into leather; hence the words "tan" and "tanning" for the treatment of leather. However, the term "tannin" by extension is widely applied to any large polyphenolic compound containing sufficient hydrolysis and other suitable groups (such as carboxyls) to form strong complexes with various macromolecules.

The tannin compounds are widely distributed in many species of plants, where they play a role in protection from predation, and perhaps also as pesticides, and in plant growth regulation. The astringency from the tannins is what causes the dry and puckery feeling in the mouth following the consumption of unripened fruit or red wine. Likewise, the destruction or modification of tannins with time plays an important role in the ripening of fruits and the aging of wine. Tannins have molecular weights ranging from 500 to over 3,000 (gallic acid esters) and up to 20,000 (proanthocyanidins).

Tannins are distributed in species throughout the plant kingdom. They are commonly found in both gymnosperms as well as angiosperms. Most families of dicot contain tannin-free species (tested by their ability to precipitate proteins). The best known families of which all species tested contain tannin are: Aceracea, Actinidaceae, Anacardiaceae, Bixaceae, Burseraceae, Combertaceae, Dipterocarpaceae, Ericaceae, Grossulariaceae, Myricaceae for dicot and Najadaceae and Typhaceae in Monocot. To the family of the oak, Faquaceae, 73% of the species tested (N = 22) contain tannin. For those of Acacias, Mimosaceae, only 39% of the species tested (N = 28) contain tannin, among Solanaceae rate drops to 6% and 4% for the Asteraceae. Some families like the Boraginaceae, Cucurbitaceae, Papaveraceae contain no tannin-rich species.

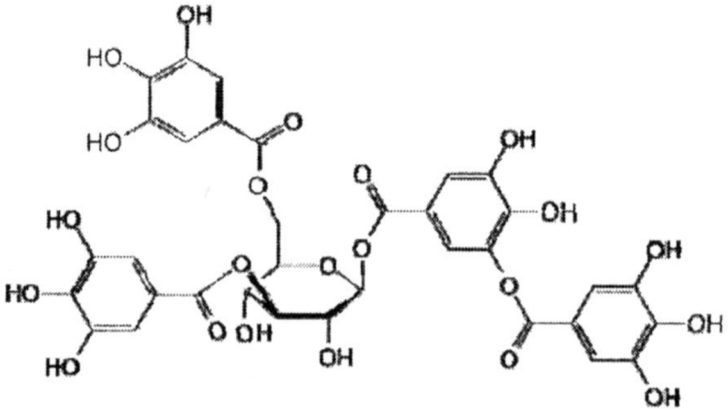

Tannic acid, a type of tannin

The most abundant polyphenols are the condensed tannins, found in virtually all families of plants, and comprising up to 50% of the dry weight of leaves. Tannins of tropical woods tend to be of a cathetic nature rather than of the gallic type present in temperate woods. There may be a loss in the bio-availability of still other tannins in plants due to birds, pests, and other pathogens.

Tannins are found in leaf, bud, seed, root, and stem tissues. An example of the location of the tannins in stem tissue is that they are often found in the growth areas of trees, such as the secondary phloem and xylem and the layer between the cortex and epidermis. Tannins may help regulate the growth of these tissues. In all vascular plants studied so far, tannins are manufactured by a chloroplast-derived organelle, the tannpsome. Tannins are mainly physically located in the vacuoles or surface wax of plants. These storage sites keep tannins active against plant predators, but also keep some tannin from affecting plant metabolism while the plant tissue is alive; it is only after cell breakdown and death that the tannins are active in metabolic effects.

Tannins are classified as ergastic substances, *i.e.,* non-protoplasm materials found in cells. Tannins, by definition, precipitate proteins. In this condition, they must be stored in organelles able to withstand the protein precipitation process. Idioblasts are isolated plant cells which differ from neighbouring tissues and contain non-living substances. They have various functions such as storage of reserves, excretory materials, pigments, and minerals. They could contain oil, latex, gum, resin or pigments etc. They also can contain tannins. In Japanese persimmon (*Diospyrous kaki*) fruits, tannin is accumulated in the vacuole of tannin cells, which are idioblasts of parenchyma cells in the flesh.

The tannins are quantitatively estimated by following teo method:

- **Folin-Danis method** – This is based on the non-stichiometric oxidation of the molecules containing a phenol hydroxyl group.
- **Vanillin hydrochloride method** – Vanillin method is specific for dihydroxy phenols and is particularly sensitive for meta-substituted, di- and tri-hydroxy benzene containing molecules.

Estimation of Tannin Content by Folin-Dennis Method

Colorimetric reactions are widely used in the UV/VIS spectrophotometric method, which is easy to perform, rapid and applicable in routine laboratory use, and low-cost. However, it is important that colorimetric assay need to use a reference substance, then this method mensures the total concentration of phenolic hydroxyl groups in the plant extract. Polyphenols in plant extracts react with specific redox reagents (Folin-Ciocalteu reagent) to form a blue complex that can be quantified by visible-light spectrophotometry. The reaction forms a blue chromophore constituted by a phosphortungsti-phosphomolybdenum complex, where the maximum absorption of the chromospheres depends on the alkaline solution and the concentration of phenolic compounds. However, this reagent rapidly decomposes in alkaline solutions, which makes it necessary to use an enormous excess of the reagent to obtain a complete reaction.

Materials

- Folin-Dennis reagent – To 750 ml of distilled water, 100 g of sodium tungstate (Na_2WO_4. $2H_2O$), 20 g of phoshpho molybdic acid and 50 ml of phosphoric acid (H_3PO_4) is added. The chemical mixture is refluxed for 2 hours, cooled and final volume is made to 1000 ml with distilled water.
- Saturated sodium carbonate solution – To each 100 ml of distilled water add 35 g of anhydrous sodium carbonate (Na_2CO_3) and dissolve it at 70 – 80^0 C and it is cooled for over night period. Seed the super saturated solution with the crystals of $Na_2CO_3.10H_2O$ and after crystalization filter through glass wool.
- Standard tannin acid solution – Dissolve 100 mg of tanninc acid in 100 ml of water
- Working standard solution – Dilute 5 ml of stock solution to 100 ml of water and this solution contain 50 mg of tnnins per ml.

Procedure

- About 0.5 g of powdered plant materials are taken in conical flask and 75 ml of water is added. Flask is heated gently to boiling for 30 minutes. Cool the solution and centrifuge at 2000 rpm for 20 minutes and supernatant is collected and the final volume is made to 1000 ml with distilled water.
- To 1.0 ml of the sample extract 75 ml of water is added and 5.0 ml of Folin Dennis reagent and 10.0 ml of sodium carbonate solution are added and final volume is made to 100 ml with distilled water.
- Shake well and let it stand for 30 minutes and thereafter optical density is recorded at 700 nm against a reagent blank.
- In case the optical density is high then diiute the reaction mixture by 1:4 with distilled water.

- A standard graph is made by using 0 – 100 μg.

Calculation

Calculate the tannin content of the sample as tanninc acid equivalents from from the standard graph.

Estimation of Tannin Content by Vani!lin Hydrochloride Method

The vanillin reagent reacts with any phenol that has an unsubstituted resorcinol or phloroglucinol nucleus and forms a coloured substituted product which is measured at 500 nm.

Materials

- Vanillin Hydrochloride Reagent – Mix equal volumes of 8% hydrochloric acid in methanol and 4% vanillin in methanol. The solutions must be mixed just before use and avoid using it if it is slightly coloured.
- Catechin stock standard solution – A standard stock solution containing 1.0 mg catachol/ ml methanol.
- Working standard solution – Dilute the above standard solution ten times *i.e.*, 10.0 ml is taken and final volume is made to 100 ml.

Procedure

- Extract 1.0 g of dried plant material by adding 50.0 ml of methanol. Allow to stand for 20 – 24 hours with occasional stirring. Centrifuge and supernatant is used for tannin estimation.
- To 1.0 ml of extract, quickly add 5.0 ml of vanillin hydrochloride reagent.
- Observe the optical density at 500 nm after 20 minutes against reagent blank.
- Prepare a standard curve with 20 – 100 μg of catechin using the diluted stock solution.
- From the standard curve calculate the amount of catechin *i.e.*, tannins in the sample as per the optical density value and express the results as catehin equivalent.

14.4. Estimation of Lignin Content

Lignin or lignen is a complex polymer of aromatic alcohols known as monolignols. It is most commonly derived from wood, and is an integral part of the secondary cell walls of plants and some algae. The term was introduced in 1819 by de Condole and is derived from the Latin word *lignum*, meaning wood. It is one of the most abundant organic polymers on Earth, exceeded only by cellulose. Lignin constitutes 30% of non-fossil organic carbon and a quarter to a third of the dry mass of wood. As a biopolymer, lignin is unusual because of its heterogeneity and lack of a defined primary structure. Its most commonly noted function is the support through strengthening of wood (xylem cells) in trees. The molecular formula of lignin is $C_9H_{10}O_2$, $C_{10}H_{12}O_3$, $C_{11}H_{14}O_4$. Global production of lignin is around 1.1 million metric tons per year and is used in a wide range of low volume, niche applications where the form but not the quality is important.

Lignin fills the spaces in the cell wall between cellulose, hemicellulose, and pectin components, especially in xylem tracheids, vessel elements and sclereid cells. It is covalently linked to hemicellulose and, therefore, crosslinks different plant polysaccharides, conferring mechanical strength to the cell wall and by extension the plant as a whole. It is particularly abundant in compression wood but scarce in tension woods, which are types of reaction wood.

Lignin plays a crucial part in conducting water in plant stems. The polysaccharide components of plant cell walls are highly hydrophilic and thus permeable to water, whereas lignin is more hydrophobic. The crosslinking of polysaccharides by lignin is an obstacle for water absorption to the cell wall. Thus, lignin makes it possible for the plant's vascular tissue to conduct water efficiently. Lignin is present in all vascular plants, but not in bryophytes, supporting the idea that the original function of lignin was restricted to water transport. However, it is present in red algae, which seems to suggest that the common ancestor of plants and red algae also synthesised lignin. This would suggest that its original function was structural; it plays this role in the red alga *Calliathron*, where it supports joints between calcified segments. Another possibility is that the lignin in red algae and in plants are result of convergent evolution, and not of a common origin.

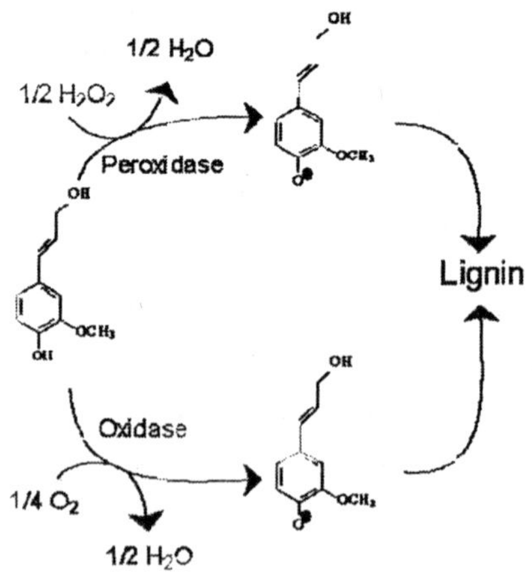

Polymerisation of coniferyl alcohol to lignin. The reaction has two alternative routes catalysed by two different oxidative enzymes, peroxidases or oxidases.

Lignin biosynthesis begins in the cytosol with the synthesis of glycosylated monolignols from the amino acid phenylalanine. These first reaction are shared with the phenylpropanoid pathway. The attached glucose renders them water soluble and less toxic. Once transported through the cell membrane to the apoplast, the glucose is removed and the polymerisation commences. Much about its anabolism is not understood even after more than a century of study. The polymerisation step, that is a radical-radical coupling, is catalysed by oxidative emzymes. Both peroxidase and laccase enzymes are present in the plant cell walls, and it is not known whether one or both of these groups participates in the polymerisation. Low molecular weight oxidants might also be involved. The oxidative enzyme catalyses the formation of monolignol radicals. These radicals are often said to undergo uncatalyzed coupling to form the lignin polymer, but this hypothesis has been recently challenged. The alternative theory that involves an unspecified biological control is however not widely accepted.

A solution of hydrochloric acid and phloroglucinol is also used for the detection of lignin (Weisner test). A brilliant red colour develops, owing to the presence of coniferaldehyde groups in the lignin.

Gravimetric method for estimation of lignin content

Materials

- Acid detergent solution – Dissolve 20.0 g of cetyl trimethyl ammonium bromide in 1000 ml of 1.0 N sulphuric acid
- Slphuric acid 72%
- Acetone
- Refluxing set
- Muffle furnace
- Sintered glass crucible

Procedure

A.Acid detergent fibre (ADF)

- One g of powdered plant sample is taken with 100.0 ml of acid detergent solution and refluxed for 60 minutes on low heating heater.
- Quantitatively transfer the refluxed sample in pre weighed cintered glass crucible with 2 – 3 washing with hot water.
- Wash with acetone until the filterarate is colourless.

- Dry at 100^0 C overnight
- Weight after cooling in a desicator.
- Express ADF content in percentage *i.e.*, W / S x 100, where W is the weight of fibre and S in the weight of sample.

B. Determination of acid detergent lignin (ADL)

- Transfer ADF in 100 ml beaker with 20-25 ml of 72% sulphuric acid, add 1.0 g of asbestos and allow it to stand for 3 hours with occasional stirring.
- Dilute the acid with water and quantitatively filer with pre weighed filter paper with 2 – 3 washings with water.
- Dry the filter paper at 100^0 C and weigh after cooling in a desiccator.
- Transfer the filter paper to a pre weighed silica crucible and ash the filter paper with its content in a muffle furnace at 550^0 C for about 3 hours.
- Cool the crucible in a desiccators and weigh. Calculate the ash content.
- For blank take 1.0 g of asbestos, add 72% H_2SO_4 and follow the procedure.

Calculation

$$\text{ADL \%} = \frac{\text{Weight 72\% } H_2SO_4 \text{ washed fibre (Test – Asbestos blank)} - \text{Ash (Test – Asbestos blank)} \times 100}{\text{Weight of sample}}$$

Reading

Goering, HD, and van Soest, PJ. (1975). Forage fibre analysis, US department of Agriculture , Agriculture Reearch Service, Washington DC.

Colorimetric method for lignin estimation

- Lignin is determined with acetyl bromide procedure and for it 10 – 15 mg of dried sample is taken with 2.0 ml of acetyl bromide in glacial acetic acid (1:3, v/v) containg perchloric acid (78%, 0.08 ml) in brown vials of 4.0 ml capacity.
- After digestion, the samples are transferred with the aid of acetic acid, to a 50 ml volumetric flask containg 2.0 M NaOH (5.0 ml) and acetic acid (12.0 ml). The flasks are made to mark with acetic acid.

- The lignin content are calculated from the following equation:

$$\text{Absorbance value} = \frac{OD_{sample} - OD_{blank}}{\text{Concentration of sample (g } l^{-1})}$$

Lignin content = 33.6 absorbance – 11.1 g kg^{-1} dry matter

Reading

Morrison, LM, Asibdu, EA, Stuchbury, T and Powell, AA. (1995). Determination of lignin and tannin content in cowpea seed coats. *Annals of Botany,* **76:** 287-290.

14.5. Estimation of Capsaicin Content

Capsaicin is the active component of chilli peppers, which are plants belonging to the genus *Capsicum.* It is an irritant for mammals, including humans, and produces a sensation of burning in any tissue with which it comes into contact. Capsaicin and several related compounds are called capsaicinoids and are produced as secondary metabolites by chili peppers, probably as deterrents against certain mammals and fungi. Pure capsaicin is a volatile, hydrophobic, colourless, odourless, and crystalline to waxy compound.

Capsaicin is the main capsaicinoid in chilli peppers, followed by dihydrocapsaicin. These two compounds are also about twice as potent to the taste and nerves as the minor capsaicinoids norhydrocapsaicin, homodihydrocapscaine, and homocapsaicin. Dilute solutions of pure capsaicinoids produced different types of pungency; however, these differences were not noted using more concentrated solutions.

Capsaicin is believed to be synthesized in the interlocular spectrum of chili peppers by addition of a branched-chain fatty acid to vanillylamine; specifically, capsaicin is made from vanillylamine and 8-methyl-6-nonenoyl CoA. Biosynthesis depends on the gene *AT3*, which resides at the *pun1* locus, and which encodes a putative acetyltransferase.

Besides the six natural capsaicinoids, one synthetic member of the capsaicinoid family exists. Vanillylamide of n-nonanioc (VNA, also PAVA) is used as a reference substance for determining the relative pungency of capsaicinoids.

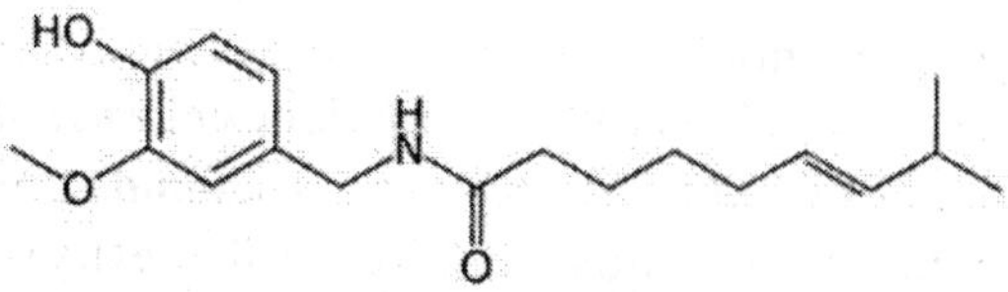

Capsaicin (8-Methyl-*N*-vanillyl-*trans*-6-nonenamide)

Capsaicin is present in large quantities in the placental tissue (which holds the seeds), the internal membranes and, to a lesser extent, the other fleshy parts of the fruits of plants in the genus *Capsicum*. The seeds themselves do not produce any capsaicin, although the highest concentration of capsaicin can be found in the white pith of the inner wall, where the seeds are attached. The seeds of *Capsicum* plants are dispersed predominantly by birds: in birds, the TRPV channel does not respond to capsaicin or related chemicals (avian vs mammalian TRPV1 show functional diversity and selective sensitivity). This is advantageous to the plant, as chili pepper seeds consumed by birds pass through the digestive tract and can germinate later, whereas mammals have molar teeth which destroy such seeds and prevent them from germinating. Thus, natural selection may have led to increasing capsaicin production because it makes the plant less likely to be eaten by animals that do not help it reproduce. There is also evidence that capsaicin may have evolved as an anti-fungal agent: the fungal pathogen *Fusarium*, which is known to infect wild chilies and thereby reduce seed viability, is deterred by capsaicin, which thus limits this form of predispersal seed mortality. In 2006, it was discovered that the venom of a certain tarantula species activates the same pathway of pain as is activated by capsaicin; this was the first demonstrated case of such a shared pathway in both plant and animal anti-mammal defence.

Because of the burning sensation caused by capsaicin when it comes in contact with mucous membranes, it is commonly used in food products to give them added spice or "heat" (piquancy). In high concentrations, capsaicin will also cause a burning effect on other sensitive areas of skin. The degree of heat found within a food is often measured on the Scoville scale. In some cases, people enjoy the heat; there has long been a demand for capsaicin-spiced food and beverages. There are many cuisines and food products featuring capsaicin such as hot sauce, salsa, and beverages. It is common for people to experience pleasurable and even euphoriant effects from ingesting capsaicin. Folklore among self-described "chiliheads" attributes this to pain-stimulated release of endorphins, a different mechanism from the local receptor overload that makes capsaicin effective as a topical analgesic. In support of this theory, there is some evidence that the effect can be blocked by nolaxone and other compounds that compete for receptor sites with endorphins and opiates.

Capsaicinoids are soluble in moderate polar organic solvents *e.g.* chloroform, acetone, ethyl acetate, methylene chloride, methanol, ethanol, acetonitrile, and among others. The major capsaicinoids present in most varieties of the chilli pepper are capsaicin (tran-8-methyl-N-vanillyl-6-nonenamide)and dihydrocapsaicin (8-methyl-N-vanillylnonanamide). In addition, other minor ones are also found such as nordihydrocapsaicin, norcapsaicin, homocapsaicin, nornorcapsaicin, nornornorcapsaicin and nonivamide. Capsaicin has been used in neurological research to stimulate sensory nerves and also to treat bladder inflammation. It is also found in topical ointments used for arthritis and neuralgia, and exerts its effect on the sensory nerves by interacting with the vanilloid receptor, promoting the release of substance P as well as other cytokines.

The phenolic group of capasaicin reduces phosphomolybdic acid to lower acids of molybdenum. The resulting componenet is blue in colour and read at 650 nm. The colour intensity is directly proportional to the concentration of capasaicin.

Materials

- Sodium hydroxide (0.40%)
- Phoshomolybdic acid (3%)
- Dry acetone – about 25.0 g of anhydrous sodium sulphate is added to 500 ml of acetone and left for over night to stand.
- Standard capasaicin solution – 50.0 mg of capasaicin is dissolved in 50.0 ml of 0.4% of NaOH and it gives 1000 µg/ml.
- Working standard solution – 10.0 ml of standard solution is diluted to 50.0 ml with 0.4% of NaOH to get a solution of 200 µg/ml.

Procedure

- Approximately 0.5 g of chilli powder is taken with 10.0 ml of dry acetone and shaken for 3 hours in a mechanical shaker. After 3 hours of shaking the solution is allowed to stand for 30 minutes and centrifuged at 10,000 rpm for 10 minutes.
- With the help of boiling waterbath, acetone is removed from extract and 1.0 ml of supernatant and the residue is dissolved in 5.0 ml of 0.4% NaOH.
- To the reaction mixture 3.0 ml of 3.0% phosphomolybdic acid is mixed and the content is shaken thoroughly and let it stand for an hour.
- Centrifuge at 5,000 rpm for 10 – 15 minutes. And the optical density of the clear supernatant is measuredat 650 nm against reagent blank.
- Standard graph for capasaicin is prepared using 0 – 200 µg/ml capasaicin.

Reading

Quagliotti, L., 1971: Effects of soil moisture and nitrogen level on the pungency of berries of *Capsicum annuum* L. *Horticultural Research*, **11:** 2, 93 - 97.

Balasubramanium, T., Raj, D., Kasthuri, R. and Rengaswami, P. (1982). Capsaicin and plant cacteristic in chilli. *Indian Journal of Horticulture*, **19:** 239

Alternate method for capsaicin estimation

Capsaicin is extracted with ethyl acetate and made to react with ethyl acetate solution of vanadium oxychloride. The developed colour is read at 720 nm. This method is very sensitive and can measure small quantity (less than 0.05%).

Materials

- Vanadium oxychloride (0.5%) in ethyl acetate.
- Pure capsaicin (1.0% in ethyl acetate (10.0 mg in 100.0 ml)

Procedure

- Approximately 2.0 g of plant sample is mixed with 70.0 ml of ethyl acetate and let it stand for 24 hours and then volume (100 ml) is made up with ethyle acetate.
- The extract is further diluted (1.0 ml to 5.0 ml) with ethyl acetate and 0.5 ml of vanadium oxychloride solution is added.
- Optical density is measured at 720 nm against a reagent blank.
- Standard curve for capsaicin is prepared by using 0.5, 1.0, 1.5, 2.0, 2.5, and 3.0 ml of standard capsaicin solution containing 50, 100, 150, 200, 250, 300 μg capsaicin respectively.

Calculation

$$\text{Capasaicin (\%)} = (\mu g \text{ capsaicin} / 1000 \times 1000) \times (100 / 1) \times (100 / 2)$$

$$= \mu g \text{ capsaicin} / 200$$

Reading

Palacio, JJR. 1977. Biochemical readings, *J. Agri. Org. App. Chem.*, **60:** 970

14.6. Estimation of Solanine Content

Solanine is a glycoalkaloid poison found in species of the nightshade family (Solanaceae), such as the potato (*Solanum tuberosum*) and the tomato (*Solanum lycopersicum*). It can occur naturally in any part of the plant, including the leaves, fruit, and tubers. Solanine has fungicidal and pesticidal properties, and it is one of the plant's natural defences. Solanine was first isolated in 1820 from the berries of the European black nightshade (*Solanum nigrum*), after which it was named.

Solanine occurs naturally in many species of the genus *Solanum*, including the potato (*Solanum tuberosum*), tomato (*Solanum lycopersicum*), eggplant (*Solanum melongena*), and bittersweet nightshade (*Solanum dulcamara*).

Potatoes naturally produce solanine and chaconine, a related glycoalkaloid, as a defence mechanism against insects, diseases, and predators. Potato leaves, stems and shoots are naturally high in glycoalkaloids. When potato tubers are exposed to light, they turn green and increase glycoalkaloid production. This is a natural defense to help prevent the uncovered tuber from being eaten. The green colour is from chlorophyll, and is itself harmless. However, it is an indication that increased level of solanine and chaconine may be present.

Some diseases, such as late blight, can dramatically increase the levels of glycoalkaloids present in potatoes. Mechanically damaged potatoes also produce increased levels of glycol-alkaloids. This is believed to be a natural reaction of the plant in response to disease and damage. In potato tubers, 30–80% of the solanine develops in and close to the skin. In the 70s, Solanine poisoning affected 78 school boys in Britain. Due to immediate and effective treatments, no one died.

Showing green under the skin strongly suggests solanine build-up in potatoes, although each process can occur without the other. A bitter taste in a potato is another, potentially more reliable indicator of toxicity. Because of the bitter taste and appearance of such potatoes, solanine poisoning is rare outside conditions of food shortage. The symptoms are mainly vomiting and diarrhoea, and the condition may be misdiagnosed as gastroenteritis. Most potato poisoning victims recover

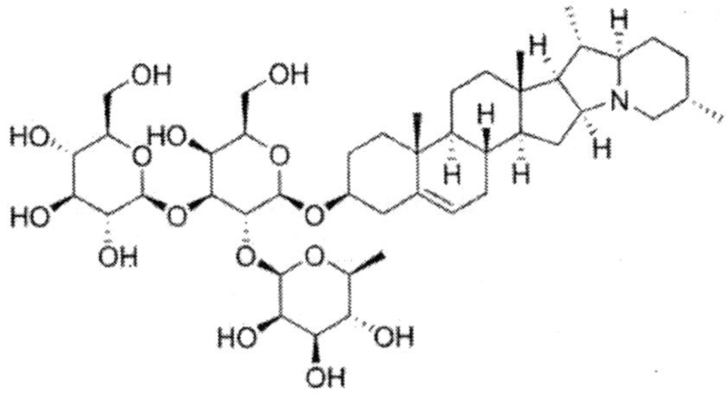

α-Solanine

fully, although fatalities are known, especially when victims are undernourished or do not receive suitable treatment. Fatalities are also known from solanine poisoning from other plants in the nightshade family, such as the berries of *Solanum dulcamara* (woody nightshade).

The United States National Institute of Health's information on solanine says to never eat potatoes that are green below the skin.

Deep drying potatoes at 170°C (338°F) is known to effectively lower glycoalkaloid levels (because they move into the frying fat), as does boiling (because solanine is water soluble), while microwaving is only somewhat effective, and freeze drying or dehydration has little effect. Some have claimed that tomatoeas and tomato leaves contain solanine. However, some others have contradicted this claim stating that tomatine, a relatively benign alkaloid, is the tomato alkaloid while solanine is found in potatoes.

Solanine has fungicidal and pesticidal properties, and solanine hydrochloride (a salt of solanine) has been used as a commercial pesticide, but never on a large scale. Solanine has sedative and anticonvulsant properties, and has been used as a treatment for asthama, as well as for cough and cold medicines. However, its effectiveness for either use is questionable. Solanine poisoning is primarily displayed by gastrointestinal and neurological disorders. Symptoms include nausea, diarrhea, vomiting, stomach cramps, burning of the throat, cardiac dysrhythmia, nightmare, headache and dizziness. In more severe cases, hallucinations, loss of sensation, paralysis, fever, jaundice, dilated pupils, hypothermia and death have been reported.

In large quantities, solanine poisoning can cause death. One study suggests that doses of 2 to 5 mg per kilogram of body weight can cause toxic symptoms, and doses of 3 to 6 mg per kilogram of body weight can be fatal. Symptoms usually occur 8 to 12 hours after ingestion, but may occur as rapidly as 30 minutes after eating high-solanine foods. The lowest dose to cause symptoms of nausea is about 25 mg solanine for adults, a life-threatening dose for a regular-weight adult ranges about 400 mg solanine.

Solanine is extracted from plant into a week acid solution and then precipated with concentrated NH_4OH. The precipitate is dissolved in ethanol: H_2SO_4 mixture and treated with formaldehyde reagent. The intensity of greenish blue colour is measured colorimetrically between 565-570 nm.

Materials

- Acetic acid 15%
- Concentrated and 1% ammonium hydroxide solution.

- Ethanol 96%
- Sodium sulphate 60% solution.
- Formaldehyde 0.5% solution in 60% hydrogen peroxide.
- Sulphuric acid 1.0 M
- Standard solution – 0.2 mM – 30 mM solanine in 96% ethanol: 20% H_2SO_4 mixture (1:1).

Procedure

- Extract the tissue by macerating with 7% acetic acid (15 – 20 ml/g of plant tissue) and filter through cheese cloth.
- Warm the filterate to 70^0 C in a water bath and add conc. NH_4OH dropwise until pH 10.0 is attained.
- Centrifuge at low speed (2000 rpm) and discard the supernatant.
- Wash the precipitate with 1% NH_4OH and centrifuge. Collect the precipitate, dry and weigh the crude solanine.
- Dissolve a known amount (50.0 mg) of crude solanine in a 1:1 mixture of 96% ethanol and 20% H_2SO_4.
- To 1.0 ml of alkaloid solution add 5.0 ml of 60% Na_2SO_4 and after 15 minutes, add 5.0 ml of formaldehyde reagent. Incubate for 60 minutes at room temperature.
- Measure the optical density at 565-570 nm against reagent blank.
- Calculate the amount of solanine in the sample using standard graph obatained through pure solanine.

Notes

- It is easier to isolate solanine from the sprouts, berry or flowers of potato since these plant parts have higher concentration of solanine (0.04%/g fresh weight).
- Yomatin from unripe tomatoes and ajmalicine from *Vicea rosea* can also be extracted using the similar method.

Reading

Clarke, E.G.C. (1958). Identification of solanine. *Nature*, **181:** 1152-1153.

14.7. Estimation of Catalase (EC 1.11.1.6.) Activity

Catalase is a common enzyme found in nearly all living organisms exposed to oxygen. It catalyses the decomposition of hydrogen peroxide to water and oxygen. It is a very important enzyme in protecting the cell from oxidative damage by reactive oxygen species (ROS). Likewise, catalase has one of the highest turnover numbers of all enzymes; one catalase molecule can convert millions of molecules of hydrogen peroxide to water and oxygen each second. Catalase is a tetramer of four polypeptide chains, each over 500 amino acids long. It contains four porphyrin heme (iron) groups that allow the enzyme to react with the hydrogen peroxide. The optimum pH for human catalase is approximately 7, and has a fairly broad maximum (the rate of reaction does not change appreciably at pHs between 6.8 and 7.5). The pH optimum for other catalases varies between 4 and 11 depending on the species. The optimum temperature also varies by species.

The reaction of catalase in the decomposition of living tissue:

$$2\ H_2O_2 \rightarrow 2\ H_2O + O_2$$

The presence of catalase in a microbial or tissue sample can be tested by adding a volume of hydrogen peroxide and observing the reaction. The formation of bubbles, oxygen, indicates a positive result. This easy assay, which can be seen with the naked eye, without the aid of instruments, is possible because catalase has a very high specific activity, which produces a detectable response. Catalase can also catalyze the oxidation, by hydrogen peroxide, of various metabolites and toxins, including formaldehyde, formic acid, phenols, acetyldehyde and alcohol. It does so according to the following reaction:

$$H_2O_2 + H_2R \rightarrow 2H_2O + R$$

The exact mechanism of this reaction is not known. Any heavy metal ion (such as copper cations in copper(II) suphate) can act as a non-competitive inhibitor of catalase. Also, the poison cyanide is a competitive inhibitor of catalase, strongly binding to the heme of catalase and stopping the enzyme's action.

Hydrogen peroxide is a harmful byproduct of many normal metabolic processes; to prevent damage to cells and tissues, it must be quickly converted into other, less dangerous substances. To this end, catalase is frequently used by cells to rapidly catalyze the decomposition of hydrogen peroxide into less-reactive gaseous oxygen and water molecules.

The true biological significance of catalase is not always straightforward to assess. Mice genetically engineered to lack catalase are phenotypically normal, indicating this enzyme is dispensable in animals under some conditions. A catalase deficiency may increase the likelihood of developing type 2 diabetes. Some humans have very low levels of catalase (acatalasia), yet show few ill effects. The predominant scavengers of H_2O_2 in normal mammalian cells are likely peroxiredoxins rather than catalase.

Catalase is usually located in a cellular, bipolar environment organelle called the peroxisome. Peroxisomes in plant cells are involved in photorespiration (the use of oxygen and production of carbon dioxide) and symbiotic nitrogen fixation (the breaking apart of diatomic nitrogen (N_2) to reactive nitrogen atoms). Hydrogen peroxide is used as a potent antimicrobial agent when cells are infected with a pathogen. Catalase-positive pathogens, such as *Mycobacterium tuberculosis*, *Legionelle pneumophila*, and *Campylobacter jeiuni*, make catalase to deactivate the peroxide radicals, thus allowing them to survive unharmed within the host. Catalase contributes to ethanol metabolism in the body after ingestion of alcohol, but it only breaks down a small fraction of the alcohol in the body.

Materials

- Phosphate buffer 0.1 M phosphate buffer (pH 7.5) – Dissolve 6.8045 g of potassium dihydrogen phosphate in 500 ml of water and 8.709 g of potassium hydrogen phosphate in 500 ml of water. The pH 7.5 is made by adding 16.0 ml of potassium dihydrogen phosphate and 84 ml of potassium hydrogen phosphate. Final pH is addusted by pH meter.
- Hydrogen peroxide solution – Dissolve 10.0 ml of 30% (8.82 M) of H_2O_2 in water to make the final volume 88.2 ml. This gives approximately 1.0 M solution (1 mM / 1 ml).
- Titanium reagent – 1.0 g of titanium dioxide (TiO_2) and 10.0 g of potassium sulphate (K_2SO_4) is mixed and is digested with 150.0 ml of concentrated H_2SO_4 for 2 to 3 hours on a mantel heater. The digested mixture is then cooled, diluted to 1.5 liter with distilled water and used as titanium reagent.

Procedure

- Weighed leaf sample is cut for grinding in a prechilled mortar with 0.1 M phosphate buffer (pH 7.5). Homogenates are centrifuged at 0° C for 20 minutes at 12,000 g. Supernatant is used for the enzyme assay.
- The assay mixture consists of the following
- Enzyme extract 0.1 ml (For control, boiled enzyme extract was used)
- Hydrogen peroxide solution 0.2 ml.

- Phosphate buffer (pH 7.5) 2.9 ml
- Incubation in done at 30^0 C in a water bath for one minute and the reaction is stopped by adding 0.1 ml of 1.0 M zinc acetate.
- To the above tube 0.9 ml of distilled water is added and is centrifiged at 3000 g for 10 minutes. Three ml of supernatant is taken and to it 2.0 ml of titanium reagent is mixed to develop colour. Optical density is recorded at 410 nm for residual hydrogen peroxide content.
- For making a standard curve H_2O_2 is first standarised using a standar solution of potassium permangante (0.1 M). From this solution several concentrations *viz.*, 5, 10, 15, 20......70 μ moles of H_2O_2 is atken. To this 2.0 ml of titanium reagent is added and the final volume is made up to 5.0 ml and OD is taken at 410 nm.

14.8. Estimation of Polyphenol Oxidase (Monophenol, Dihydroxyphe -nylalanine: Oxygen Axidoreductase EC 1.14.18.1) Activity

Polyphenol oxidase (PPO or monophenol monooxygenase) is a tetramer that contains four atoms of copper per molecule, and binding sites for two aromatic compounds and oxygen. The enzyme catalyses the *o*-hydroxylation of monophenols (phenol molecules in which the benzene ring contains a single hydroxyl substituent) to *o-diphenols* (phenol molecules containing two hydroxyl substituent). They can also further catalyse the oxidation of *o*-diphenols to produce *o-quinones*. It is the rapid polymerisation of *o*-quinones to produce black, brown or red pigments (polyphenols) that is the cause of fruit browning. The amino acid tyrosine contains a single phenolic ring that may be oxidised by the action of PPOs to form *o*-quinone. Hence, PPOs may also be referred to as tyrosinase. Enzyme nomenclature differentiates between monophenol oxidase enzymes (tyrosinase) and *o*-diphenol:oxygen oxidoreductase enzymes (catechol oxidases).

A mixture of monophenol oxidase and catechol oxidase enzymes is present in nearly all plant tissues, and can also be found in bacteria, animals, and fungi. In insects, cuticular polyphenol oxidases are present and their products are responsible for desiccation tolerance. In fact, browning by PPO is not always an undesirable reaction; the familiar brown colour of tea and cocoa is developed by PPO enzymatic browning during product processing. Tentoxin has also been used in recent research to eliminate the polyphenol oxidase activity from seedlings of higher plants. Tropolone is a grape polyphenol oxidase inhibitor. Another inhibitor of this enzyme is potassium pyrosulphite ($K_2S_2O_5$).

Materials

- Citrate-phosphate buffer 0.2 M (pH 6.0)
- 2-Nitro-5-thiobenzoic acid anion (TNB): Add 30 mg sodium borohydride to a suspension of Ellman's reagent, *i.e.* 5,5-dithiobis (2-nitrobenzoic acid) (19mg) in 10 ml water. Within 1h, the disulphide is quantitatively reduced to the intensely yellow, water soluble thiol. This solution is stable for at least one week when stored at 4^0C.
- Quinine Solutions: Dissolve 4-methyl-1,2-benzoquinone in double-distilled water in a 50 ml. volumetric flask by bubbling nitrogen gas until quinine is completely dissolved. Prepare p-benzoquinone solution also in a similar manner. Both solution are stable for 30 minutes. Each time, add 100 ml cold ethanol and collect the precipitate. Pool the precipitate, dissolve in 1.5 M NaCl and use as enzyme source.
- Methyle catachol solution – 4-methyl carachol (2 mM) for catachol oxidase assay quinol (1,4-dihydroxybebzene, 2 mM) for laccase assay.

Procedure

- Freshly prepared acetone powder of fresh plant tissue is prepared and 100 mg of acetone powder is mixed with 2.5 ml of 0.2 mM citrate-phosphate buffer (pH 6.0), and is mixed with 1.0 ml of 1% Triton X-100, 6.5 ml of water and 500 mg polyamide. This mixture is shaken for 60 minutes and filter.The filterate is used as enzyme source.
- In a 1 cm cuvette 1.4 ml of citrate-phoshphate buffer (pH 6.0), 0.5 ml of TNB and 1.0 ml of methyl catachol solution is taken.
- The reaction is initiated by the addition of 0.1 ml of enzyme preparation and immediately optical density at 412 nm is recorded.
- Decrease of optical density at 412 nm for 30 seconds is recorded.

Calculation

Read the change of OD per minute from linear part of the curve. Calculate the enzyme units according to the equation:

$$\textbf{Unit in the test} = \textbf{K x } (\Delta\textbf{A/minute})$$

Where K is 0.272 for catachol oxidase and 0.242 for laccase.

One unit of either catachol oxidase or laccase is defined as the enzyme which transforms 1 μ mol of dihydriciphanol to 1 μ mol of quinine per minute under the assay conditions. One unit is equivalent to the consumption of 1 μ mol of TNB.

Notes

- The method described above is rapid and sensitive than many other methods.
- The enzyme concentration is required to get satisfactory linearity of time or decrease in optical density have to be standarised.
- Another simple method of assaying polyphenol oxidase is given below:
- The enzyme extract is prepared by grinding 5.0 g of fresh leaves with about 20 ml of extraction medium having 50 mM Tris-HCl buffer (pH 7.2), 0.4 M sorbitol and 10 mM NaCl. Cetrifuge at 20,000 g for 19 minutes and use the supernatant for assay.
- Add 2.5 ml of 0.1 M phosphate buffer pH 6.5, 0.5 ml of catachol solution (0.01 M) into cuvette and set the spectrophotometer at 495 nm. Now add 0.2 ml of enzyme extract and start recording the change in absoption for every 30 seconds upto 5 minutes.

14.9. Estimation of Peroxidase (Donor: H_2O_2 Oxidoreductase EC. 1.11.1.7) Activity

Peroxidases (E.C. 1.11.1.x) are a large family of enzymes that typically catalyze a reaction of the form:

$$\textbf{ROOR' + electron donor (2 } e^-\textbf{) + 2H}^+ \rightarrow \textbf{ROH + R'OH}$$

For many of these enzymes the optimal substrate is hydrogen peroxide, but others are more active with organic hydroperoxides such as lipid peroxides. Peroxidases can contain a heme cofactor in their active sites, or alternately redox-active cysteine or selenocystein residues.

The nature of the electron donor is very dependent on the structure of the enzyme.

- For example, horse radish peroxidase can use a variety of organic compounds as electron donors and acceptors. Horse radish peroxidase has an accessible active site, and many compounds can reach the site of the reaction.
- Because there is a very closed active site, for an enzyme such as cytochrome c peroxidase, the compounds that donate electrons are very specific.

While the exact mechanisms have yet to be elucidated, peroxidases are known to play a part in increasing a plant's defenses against pathogens. Peroxidases are sometimes used as histological marker. Cytochrome c peroxidase is used as a soluble, easily purified model for cutochrome c oxidase. The glutathione peroxidase family consists of 8 known human isoforms. Glutathione peroxidases use glutathion as an electron donor and are active with both hydrogen peroxide and organic hydroperoxide substrates. Gpx1, Gpx2, Gpx3, and Gpx4 have been shown to be selenium-containing enzymes, whereas Gpx6 is a selenoprotein in humans with cysteine-containing homologues in rodents.

The enzyme horse radish peroxidase (HRP), found in the roots of plant horseradish, is used extensively in biochemistry applications primarily for its ability to amplify a weak signal and increase detectability of a target molecule. It is a metalloenzyme with many isoforms and the most studied type is C.

Lignin is highly resistant to biodegradation and only higher fungi are capable of degrading the polymer via an oxidative process. This process has been studied extensively in the past twenty years, but the actual mechanism has not yet been fully elucidated. The complicated structure of the lignin polymer and major difficulties in analysis are responsible for the relatively slow progress. Lignin is found to be degraded by an enzyme lignin peroxidases produced by some fungi like *Phanerochaete chrysosporium*. The mechanism by which lignin peroxidase (Lip) interacts with the lignin polymer which involves Veratryl alcohol (Valc), a secondary metabolite of white rot fungi, acts as a cofactor for the enzyme. In enzymology, a lignin peroxidase (E.C. 1.11.1.14) is an enzyme that catalyses the chemical reaction

1,2-bis (3,4-dimethoxyphenyl)propane-1,3-diol + H_2O_2 $\rightleftharpoons$ 3, 4-dimethoxybenzaldehyde + 1-(3,4-dimethoxyphenyl)ethane-1,2-diol + H_2O

Thus, the two substrates of this enzyme are 1,2-bis (3,4-dimethoxyphenyl) propane-1,3-diol and H_2O_2, whereas its 3 products are 3,4-dimeth oxybenzaldehyde, 1-(3,4-dimethoxyphenyl) ethane-1,2-diol, and H_2O.

This enzyme belongs to the family of oxidoreductase, specifically those acting on a peroxide as acceptor (peroxidases) and can be included in the broad category of ligninases. The systematic name of this enzyme class is 1,2-bis(3,4-dimethoxyphenyl) propane-1,3-diol:hydrogen-peroxide oxidoreductase. Other names in common use include diarylpropane oxygenase, ligninase I, diarylpropane peroxidase, LiP, diarylpropane:oxygen,hydrogen-peroxide oxidoreductase (C-C-bond-cleaving). It employs one cofactor, heme. A typical group of peroxidases are the haloperoxidases. This group is able to form reactive halogen species and, as a result, natural organohalogen substances. A majority of peroxidase protein sequences can be found in the PeroxiBase database.

To assay peroxidase activity guaiacol is used as substrate

Guaiacol H_2O_2 $\rightarrow$ Oxidised guiacol + $2H_2O$

The resulting oxidized (dehydrogenated) guaiacol is probably more than one compound and depends on the reaction conditions. The rate of formation of guaiacol dehydragenation product is a measure of the peroxidase activity and can be assayed spectrophotometrically.

Materials

- Phosphate buffer 0.1 M pH 7.0
- Guaiacol solution 20 mM – 240 mg of guaiacol is dissolved in water and volume is made to 100 ml. It can be stored for many months as freezed.
- Hydrogen peroxide solution (0.042% = 12.3 mM) Dilute 0.14 ml of 30% (v/v) H_2O_2 to 100 ml with water. The extinction of this solution should be 0.485 at 240 nm.

Procedure

- Fresh leaves (~ 1 g) is extracted with pre chilled 3.0 ml of 0.1 M phosphate buffer pH 7.0. Centrifuge the homogenate at 18,000 x g at 5°C for 15 minutes. Supernatant is used as the enzyme source.
- To 3.0 ml of 0.1 M phosphate buffer (pH 7.0) 0.05 ml of guaiacol solution and 0.1 ml of enzyme extract and 0.03 ml of hydrogen peroxide solution is added in a 1.0 cm cuvette.
- Place the cuvette in spectrophotometer, wait until the OD at 436 nm increases by 0.05. Note the time required in minutes (Δt) to increase optical density by 0.1.

Calculation

Since the extinction coefficient of guaiacol dehydrogenation product at 436 nm is 6.39 per micromole, the enzyme activity per liter of extract is calculated as:

Enzyme activity = (3.18 x 0.1 x 1000) / (6.39 x 1 x 0.1 x Δt) = 500/Δt

Reading

Putter, J. (1974). In "*Methods of Enzymatic Analysis*" Ed. Bergmeyer, Academic Press, New York, pp. 655.

Malik, CP. And Singh, MB. (1980). In "*Plant Enzymology and Histoenzymology*", Kalyani Publishers, New Delhi, pp. 53.

14.10. Estimation of Glutathion Peroxidaes (EC. 1.11.1.9) Activity

Glutathione peroxidase (GPx) (E.C. 1.11.1.9) is the general name of an enzyme family with peroxide activity whose main biological role is to protect the organism from oxidative damage. The biochemical function of glutathione peroxidase is to reduce lipid hydrogenperoxide to their corresponding alcohols and to reduce free hydrogen peroxide to water.

Several isozymes are encoded by different genes, which vary in cellular location and substrate specificity. Glutathion peroxidise 1 (GPx1) is the most abundant version, found in the cytoplasm of nearly all mammalian tissues, whose preferred substrate is hydrogen peroxide. Glutathion peroxidise 4 (GPx4) has a high preference for lipid hydroperoxides; it is expressed in nearly every mammalian cell, though at much lower levels. Glutathion peroxidise 2 is an intestinal and extracellular enzyme, while glutathione peroxidase 3 is extracellular, especially abundant in plasma. So far, eight different isoforms of glutathione peroxidase (GPx1-8) have been identified in humans. The main reaction that glutathione peroxidase catalyses is:

$$\mathbf{2GSH + H_2O_2 \rightarrow GS\text{–}SG + 2H_2O}$$

where GSH represents reduced monomeric glutathion, and GS–SG represents glutathione disulphide. The mechanism involves oxidation of the selenol of a selenocysteine residue by hydrogen peroxide. This process gives the derivative with a selenic acid (RSeOH) group. The selenenic acid is then converted back to the selenol by a two step process that begins with reaction with GSH to form the GS-SeR and water. A second GSH molecule reduces the GS-SeR intermediate back to the selenol, releasing GS-SG as the by-product. A simplified representation is shown below:

$$RSeH + H_2O_2 \rightarrow RSeOH + H_2O$$

$$RSeOH + GSH \rightarrow GS\text{-}SeR + H_2O$$

$$GS\text{-}SeR + GSH \rightarrow GS\text{-}SG + RSeH$$

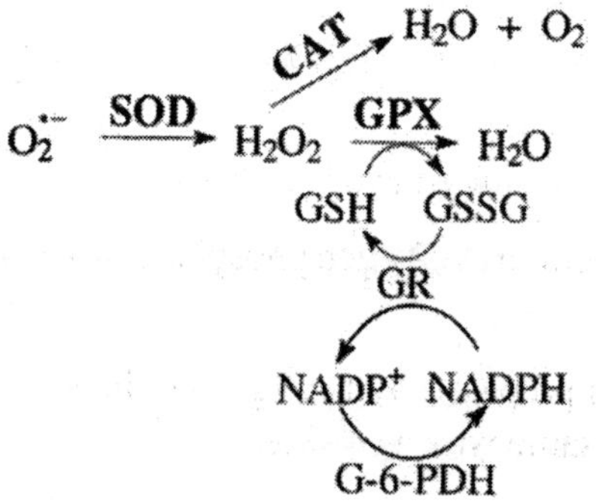

$$O_2^{\bullet -} \xrightarrow{SOD} H_2O_2 \xrightarrow{CAT} H_2O + O_2$$

$$H_2O_2 \xrightarrow{GPX} H_2O$$

GSH GSSG

GR

NADP$^+$ NADPH

G-6-PDH

Glutathion reductase then reduces the oxidized glutathione to complete the cycle:

$$\textbf{GS–SG + NADPH + H}^{+} \rightarrow \textbf{2 GSH + NADP}^{+}.$$

Materials

- Sodium phosphate buffer 0.4 M pH 7.0
- Sodium azide 10 mM
- Glutathion reduced 4 mM
- Hydrogen peroxide 2.5 mM
- Trichloroacetic acid (TCA) 10% (w/v)
- Disodium hydrogen phosphate buffer 0.3 M
- 5,5'-Dithio bis 2-nitrobenzoic (DTNB) acid – 40.0 mg of 5,5'-Dithio bis 2-nitrobenzoic acid is dissolved in 100.0 ml of 1% sodium citrate solution.
- Standard reduced glutathione 20% mg solution

Procedure

- Fresh tissue (approximately 10.0 g) is homogenated in 100.0 ml of phosphate buffer pH 7.0. Homogenate is centrifused at 10,000 rpm for 15 minutes at 2^0 C. Supernatant is used as enzyme extract.
- To 0.4 ml of buffer addition are done with 0.1 ml of sodium azide, 0.2 ml of reduced glutathione, 0.5 ml of enzyme extract and 0.1 ml of H_2O_2.
- Final volume of the above reaction mixture is made to 2.0 ml with water.
- The reaction mixture is incubated for 30 minutes at 37° C and after incubation period, reaction is stopped by the addition of 0.5 ml 10% TCA solution.
- The reaction mixture is centrifuged at 10,000 rpm at room temperature and to the supernatant 3.0 ml of disodium hydrogen phosphate and 1.0 ml of DTNB solution is added and colour is allowed to develop for 15 minutes.
- Optical density is measured at 412 nm against a reagent blank.
- For blank only disodium hydrogen phosphate and 1.0 ml of DTNB reagent is used.
- Standard curve is prepared by taking a gradient of reduced glutathione and treating it in same way as above.

Reading

Rotruck, JT., Pope, AL., Ganther, HE., Swanson, AB., Hafeman, DG. And Hoekstra, WG. (1973). Selenium: Biochemical role as a component of glutathione peroxidase purification and assay. *Science*, **179:** 588.598.

14.11. Estimation of Super Oxide Dismutase (EC. 1.15.1.1) Activity

Superoxide dismutases (SOD, E.C. 1.15.1.1) are enzymes that catalyze the dismutation of superoxide (O_2^-) into oxygen and hydrogen peroxide. Thus, they are an important antioxidant defence in nearly all cells exposed to oxygen. One anomaly is *Lactobacillus plantarum* and related lactobacilli, which use a different mechanism.

The SOD-catalysed dismutation of superoxide may be written with the following half-reactions :

- $M^{(n+1)+}$-SOD + O_2^- → M^{n+}-SOD + O_2
- M^{n+}-SOD + O_2^- + $2H^+$ → $M^{(n+1)+}$-SOD + H_2O_2.

where M = Cu (n=1) ; Mn (n=2) ; Fe (n=2); Ni (n=2).

In this reaction the oxidation state of the metal cation oscillates between n and n+1.

Superoxide dismutase (SODs) was previously known as a group of metalloproteins with unknown function; for example, CuZnSOD was known as erythrocuprein and as the veterinary anti-inflammatory drug "Orgotein". Later it became known as superoxide dismutase as an indophenol oxidase by protein analysis of starch gels using the phenazine-tetrazolium technique.

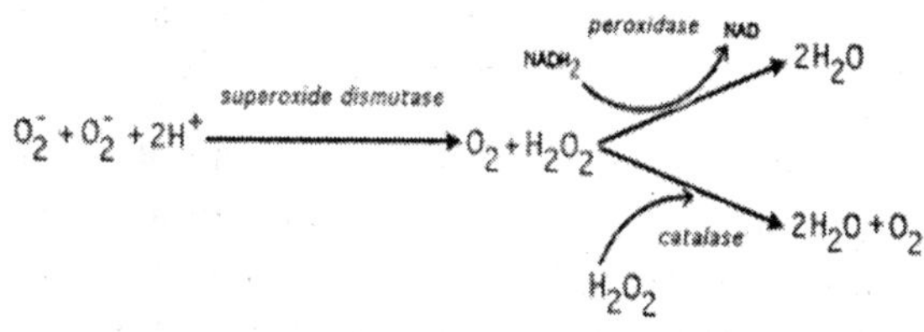

Several common forms of SOD exist: they are proteins cofactored with copper and zinc, or manganese, iron or nickel. Thus, there are three major families of superoxide dismutase, depending on the metal cofactor: Cu/Zn (which binds both copper and zinc), Fe and Mn types (which bind either iron or manganese), and the Ni type, which binds nickel.

- Copper and zinc – most commonly used by eukaryotes. The cytosis of virtually all eukaryotic cells contain an SOD enzyme with copper and zinc (Cu-Zn-SOD). For example, Cu-Zn-SOD available commercially is normally purified from the bovine erythrocytes: The Cu-Zn enzyme is a homodimer of molecular weight 32,500. The bovine Cu-Zn protein was the first SOD structure to be solved, in 1975. It is an 8-stranded "Greek key" beta-barrel, with the active site held between the barrel and two surface loops. The two subunits are tightly joined back-to-back, mostly by hydrophobic and some electrostatic interactions. The ligands of the copper and zinc are six histidine and one aspartate side-chains; one histidine is bound between the two metals.

- Iron or manganese – used by prokaruotes and protists, and in mitochondria
  - Iron – *E. coli* and many other bacteria also contain a form of the enzyme with iron (Fe-SOD); some bacteria contain Fe-SOD, others Mn-SOD, and some contain both. Fe-SOD can be found in the plastids of plants. The 3D structures of the homologous Mn and Fe superoxide dismutases have the same arrangement of alpha-helices, and their active sites contain the same type and arrangement of amino acid side-chains.
  - Manganese – Chicken liver (and nearly all other) mitochondria, and many bacteria (such as *E. coli*), contain a form with manganese (Mn-SOD): For example, the Mn-SOD found in human mitochondria. The ligands of the manganese ions are 3 histidine side-chains, an aspartate side-chain and a water molecule or hydroxyl ligand, depending on the Mn oxidation state (respectively II and III).
- Nickel – prokaryotic. This has a hexameric structure built from right-handed 4-helix bundles, each containing N-terminal hooks that chelate a Ni ion. The Ni-hook contains the motif His-Cys-X-X-Pro-Cys-Gly-X-Tyr; it provides most of the interactions critical for metal binding and catalysis and is, therefore, a likely diagnostic of NiSODs.

In higher plants, SOD isozymes have been localized in different cell compartments. Mn-SOD is present in mitochondria and peroxisomes. Fe-SOD has been found mainly in chloroplasts but has also been detected in peroxisomes, and CuZn-SOD has been localized in cytosol, chloroplasts, peroxisomes, and apoplast. In higher plants, superoxide dismutase enzymes (SODs) act as antioxidants and protect cellular components from being oxidized by reactive oxygen specices (ROS). ROS can form as a result of drought, injury, herbicides and pesticides, ozone, plant metabolic activity, nutrient deficiencies, photoinhibition, temperature above and below ground, toxic metals, and UV or gamma rays. To be specific, molecular O_2 is reduced to O_2^- (an ROS called superoxide) when it absorbs an excited electron released from compounds of the electron transport chain. Superoxide is known to denature enzymes, oxidize lipids, and fragment DNA. SODs catalyze the production of O_2 and H_2O_2 from superoxide (O_2^-), which results in less harmful reactants.

When acclimating to increased levels of oxidative stress, SOD concentrations typically increase with the degree of stress conditions. The compartmentalization of different forms of SOD throughout the plant makes them counteract stress very effectively. There are three well-known and -studied classes of SOD metallic coenzymes that exist in plants. First, Fe SODs consist of two species, one homodimer (containing 1-2 g Fe) and one tetramer (containing 2-4 g Fe). They are thought to be the most ancient SOD metalloenzymes and are found within both prokaryotes and eukaryotes. Fe SODs are most abundantly localized inside plant chloroplasts, where they are indigenous. Second, Mn SODs consist of a homodimer and homotetramer species each containing a single Mn(III) atom per subunit. They are found predominantly in mitochondrion and peroxisomes. Third, Cu-Zn SODs have electrical properties very different from those of the other two classes. These are concentrated in the chloroplast cytosol, and in some cases the extracellular space. Note that Cu-Zn SODs provide less protection than Fe SODs when localized in the chloroplast.

Superoxide is one of the main reactive oxygen species in the cell. As a consequence, SOD serves a key antioxidant role. The physiological importance of SODs is illustrated by the severe pathologies evident in mice genetically engineered to lack these enzymes. Mice lacking SOD2 die several days after birth, amid massive oxidative stress. Mice lacking SOD1 develop a wide range of pathologies, including hepatocellular carcinoma, an acceleration of age-related muscle mass loss, an earlier incidence of cataracts and a reduced lifespan. Mice lacking SOD3 do not show any obvious defects and exhibit a normal lifespan, though they are more sensitive to hyperoxic injury. Knockout mice of any SOD enzyme are more sensitive to the lethal effects of superoxide-generating drugs, such as paraquat and diquat.

Drosophila lacking SOD1 have a dramatically shortened lifespan, whereas flies lacking SOD2 die before birth. SOD knockdowns in *C. elegans* do not cause major physiological disruptions. Knockout or null mutations in SOD1 are highly detrimental to aerobic growth in the yeast Sacchormyces cerevisiae and result in a dramatic reduction in post-diauxic lifespan. SOD2 knockout or null mutations cause growth inhibition on respiratory carbon sources in addition to decreased post-diauxic lifespan.

Several prokaryotic SOD null mutants have been generated, including *E. Coli*. The loss of periplasmic CuZnSOD causes loss of virulence and might be an attractive target for new antibiotics. Mutations in the first SOD enzyme (SOD1) can cause familial amyotrophic lateral sclerosis (ALS, a form of motor neuron disease). The most common mutation in the U.S. is A4V, while the most intensely studied is G93A. The other two isoforms of SOD have not been linked to any human diseases, however, in mice inactivation of SOD2 causes perinatal lethality and inactivation of SOD1 causes heptacellular carcinoma. Mutations in SOD1 can cause familial ALS (several pieces of evidence also show that wild-type SOD1, under conditions of cellular stress, is implicated in a significant fraction of sporadic ALS cases, which represent 90% of ALS patients.), by a mechanism that is presently not understood, but not due to loss of enzymatic activity or a decrease in the conformational stability of the SOD1 protein. Overexpression of SOD1 has been linked to the neural disorders seen in Down syndrome. In recent years, it has become more apparent that in mice the extracellular superoxide dismutase (SOD3, ecSOD) is critical in the development of hypertension. In other studies, diminished SOD3 activity was linked to lung diseases such as Acute Respiratory Distress Syndrome (ARDS) or Chronic obstructive pulmonary disease (COPD). Superoxide dismutase is also not expressed in neural crest cells in the developing foetus. Hence, high levels of free radicals can cause damage to them and induce dysraphic anomalies (neural tube defects).

The assay of super oxide dismutase is based on the inhibition of the formation of NADH-phenazine methosulphate-nitroblue tetrazolium formazon. The colour formed at the end of the reaction can be extracted into butanol and measured at 560nm.

Materials

- Sodium phosphate buffer (0.025 M, pH 8.3)
- Phenazine methosulphate (PMS) (186 μM)
- Nitroblue tetrazolium (NBT) (300 μM)
- NADH (780 μM)
- Glacial acetic acid
- n-butanol
- Potassium phosphate buffer (50 mM, pH 6.4)

Procedure

- 0.20-1.50 g of tissue are minced and homogenized in 4 vols. of 0.2 M phosphate buffer, pH 7.8, 2 min at 1,000 rpm in a ice bath in a homogenizer.

- The homogenate was sonicated under cooling, 3 minutes. The sonicated homogenate was centrifuged for 30 min at 20,000 x g and the supernatant is used as enzyme source.
- The assay mixture contained 1.2 ml of sodium phosphate buffer, 0.1 ml of PMS, 0.3 ml of NBT, 0.2 ml of the enzyme preparation and water in a total volume of 2.8 ml.
- The reaction was initiated by the addition of 0.2ml of NADH.
- The mixture was incubated at 30°C for 90 seconds and arrested by the addition of 1.0ml of glacial acetic acid.
- The reaction mixture was then shaken with 4.0 ml of n-butanol, allowed to stand for 10 minutes and centrifuged.
- The intensity of the chromogen in the butanol layer was measured at 560 nm.
- One unit of enzyme activity is defined as the amount of enzyme that gave 50% inhibition of NBT reduction in one minute.

Reading

Elstner EF, Youngman R, Obwald W (1995). Superoxide dismutase. In Bergmeyer J, Grabl BM (eds) Methods of Enzymatic Analysis vol. III. Enzymes oxidoreductases, 3rd ed. Weinheim: Verlag-Chemie. pp: 293-302.

Alternative method for SOD

Materials

- Potassium phosphate buffer 250 mM, pH 7.8.
- Methionine 100 mM in distilled water
- Riboflavin 10 mM
- EDTA 5 mM
- Nitroblue tetrazolium (NBT) salt 750 μ M

Procedure

- Fresh tissue (approximately 1.0) is grinded with 10.0 ml of pre chilled 50.0 mM phosphate buffer pH 7.8. The homogenate is centrifused at 10,000 x g for 10 minutes at 4^0 C. The supernatant is used as enzyme extract.

- The assay mixture have 50 mM phosphate buffer pH 7.8, 13 mM methionine, 2.0 μM riboflavin, 0.1 mM EDTA, 75.0 μM NBT and 50 μl of crude enzyme extract.
- In the blank, enzyme and NBT are omitted.
- The assay mixture is exposed to 400 W bulb (4 x 100 w bulbs) for 15 minutes.
- , Measure the optical density at 560 nm immediately.
- Calculate the percentage inhibition.
- The 50% inhibition of the reaction between riboflavin and NBT in the presence of methionine is taken as 1 unit of SOD activity.
- The enzyme activity is expressed as units/mg of protein.

Reading

Beauchamp, C and Fridovich, I. (1971). Super oxide dismutase. Improved assay and an assay applicable to acrylamide gels. *Anal Biochem.*, **44:** 276-287.

14.12. Estimation of Phenyl Ammonia Lyase (L-Phenylalanine Ammonia lyase EC. 4.3.1.24) Activity

Phenylalanine ammonia lyase (E.C. 4.3.1.24) is an enzyme that catalyses a reaction converting L-phenylalanine to ammonia and trans-cinnamic acid. Phenylalanine ammonia lyase (PAL) is the first and committed step in the ohenyl propanoid pathway and is therefore involved in the biosynthesis of the polyphenol compounds such as flavonoids, phenylpropanoids, and lignin in plants. Phenylalanine ammonia lyase is found widely in plants, as well as some yeast and fungi, with isoenzymes existing within many different species. It has a molecular mass in the range of 270-330 kDa. The activity of PAL is induced dramatically in response to various stimuli such as tissue wounding, pathogenic attack, light, low temperatures, and hormones. PAL has recently been studied for possible therapeutic benefits in humans afflicted with phenylketonuria. It has also been used in the generation of L-phenylalanine as precursor of the sweeter aspartame.

The enzyme is a member of the ammonia lyase family, which cleaves carbon-nitrogen bonds. Like other lyases, phenylalanine requires only one substrate for the forward reaction, but two for the reverse. It is thought to be mechanistically similar to the related enzyme histidine ammonia-lyase (E.C. 4.3.1.3 HAL). The systematic name of this enzyme class is L-phenylalanine ammonia-lyase (trans-cinnamate-forming). Previously, it was designated E.C.

4.3.1.5, but that class has been redesignated as EC 4.3.1.24 (phenylalanine ammonia-lyases), E.C. 4.3.1.25 (tyrosine ammonia-lyases), and E.C. 4.3.1.26 (phenylalanine/tyrosine ammonia-lyases). Other names in common use include tyrase, phenylalanine deaminase, tyrosine ammonia-lyase, L-tyrosine ammonia-lyase, phenylalanine ammonium-lyase, PAL, and L-phenylalanine ammonia-lyase.

Phenylalanine ammonia lyase can perform different functions in different species. It is found mainly in some plants, fungi, and yeast. In fungal and yeast cells, PAL plays an important catabolic role, generating carbon and nitrogen. In plants it is a key biosynthetic enzyme that catalyzes the first step in the synthesis of a variety of polyphenyl compounds and is mainly involved in defense mechanisms. PAL is involved in 5 metabolic pathways: tyrosine metabolism, phenylalanine metabolism, nitrogen metabolism, phenyl propanoid biosynthesis, and alkaloid biosynthesis. Phenylalanine ammonia lyase activity is determined spectrophotometrically by following the formation of trans-cinnamic acid which exhibit an increase in absorbance at 290 nm.

Materials

- Borate buffer 0.2 M pH 8.7
- L-Phenylalanine 0.1 M – dissolve 165 mg of L-phenylalanine in 10 ml of water and adjust the pH at 8.7 with 0.1 M KOH.
- Trichloro acetic acid 1.0 M – dissolve 16.3 g of TCA in water and raise the volume to 100.0 ml.

Procedure

- Fresh plant material, approximately 1.0 g, is homogenized with 10.0 ml of chilled 5.0 m Mole borate-HCl buffer pH 8.2 containing 5.0 m Mole β-mercaptoethanol (0.4 ml/l). The homogenate is centrifuged at 12,000 x g for 20 minutes at 4^0 C. The supernatant is used as the enzyme extract.
- To 0.2 ml of borate buffer (pH 8.2) add 1.3 ml of water and 0.2 ml of enzyme extract.
- The reaction is initiated by the addition of 1.0 ml of L-phenylalanine solution.
- The reaction mixture is incubated for 60 minutes at 32^0 C.
- After incubation period, the reaction is terminated by the addition of 0.5 ml of trichloroacetic acid. In control, 1.0 ml of L-phenyl alanine solution is added after the addition of 0.5 ml of trichloro acetic acid.

- From the reaction mixture, the product is extracted with 15.0 ml of ethyl acetate followed by evaporation to remove extracting solvent.
- The solid residue is suspended in 3.0 ml of 0.05 NaOH and cinnamic acid concentration by measuring the optical density at 290 nm.
- Standard graph is prepared through trans-cinnamic acid.
- One unit of PAL activity is equal to 1.0 μ mole of cinnamic acid produced per minute.

Reading

Brueske, C.H. (1980). *Physiol. Plant Pathol.*, **16: 409.**

Wang, JW, Zheng, LP, Wu, JY and Tan, RX. (2006). Involvement of nitric oxide in oxidation burst phenylalanine ammonia lyase activation and taxol production induced by low energy ultrasound in *Taxus yunnanensis* cell suspension cultures. *Nitric Oxide-Biol. Ch.*, **15:** 351-358.

14.13. Estimation of Pectin Methyl Esterase (Pectin pectyl hydrolase EC. 3.1.1.11) Activity

Pectinesterase (PE) (E.C. 3.1.1.11) is a ubiquitous cell-wall-associated enzyme that presents several isoforms that facilitate plant cell wall modification and subsequent breakdown. It is found in all higher plants as well as in some bacteria and fungi. Pectinesterase functions primarily by altering the localised pH of the cell wall resulting in alterations in cell wall integrity. Pectin-esterase catalyses the de-esterification of pectin into pectate and methanol. Pectin is one of the main components of the plant cell wall. In plants, pectinesterase plays an important role in cell wall metabolism during fruit ripening. In plant bacterial pathogens such as *Erwinia carotovora* and in fungal pathogens such as *Aspergillus niger*, pectinesterase is involved in maceration and soft-rotting of plant tissue. Plant pectinesterases are regulated by pectinesterase inhibitors, which are ineffective against microbial enzymes.

$$\textbf{Pectin} + H_2O \rightarrow \textbf{Pectic acid} + CH_3OH$$

Recent studies have shown that the manipulation of pectinesterase expression can influence numerous physiological processes. In plants, pectinesterase plays a role in the modulation of cell wall mechanical stability during fruit ripening, cell wall extension during pollen germination and pollen tube growth, abscission, stem elongation, tuber yield and root development. Pectinesterase has also been shown to play a role in a plants response to pathogen attack. A cell wall-associated pectinesterase of *Nicotiana tabacum* is involved in host cell receptor recognition for the tobacco mosaic virus movement protein and it has been shown that this interaction is required for cell-to-cell translocation of the virus. Pectinesterase action on the components of the plant cell wall can produce two diametrically opposite effects. The first being a contribution to the stiffening of the cell wall by producing blocks of unesterified carboxyl groups that can interact with calcium ions forming a pectate gel. The other being that proton release may stimulate the activity of cell wall hydrolases contributing to cell wall loosening.

Pectins form approximately 35% of the dry weight of dicot cell walls. They are polymerised in the cis Golgi, methylesterified in the medial Golgi and substituted with side chains in the trans Golgi cisternae. Pectin biochemistry can be rather complicated but put simply, the pectin backbone comprises 3 types of polymer: homogalacturonan (HGA); rhamnogalacturonan I (RGI); rhamnogalacturonan II (RGII).

Homogalacturonan is highly methyl-esterified when exported into cell walls and is subsequently de-esterified by the action of pectinesterase and other pectic enzymes. Pectinesterase catalyses the de-esterification of methyl-esterified D-galactosiduronic acid units in pectic compounds yielding substrates for depolymerising enzymes, particularly acidic pectins and mrthanol. Most of the purified plant pectinesterases have neutral or alkaline isoelectric points and are bound to the cell wall via electrostatic interactions. Pectinesterases can however display acidic isoelectric points as detected in soluble fractions of plant tissues. Until recently, it was generally assumed that plant pectinesterases remove methyl esters in a progressive block-wise fashion, giving rise to long contiguous stretches of un-esterified GalA residues in homogalacturonan domains of pectin. Alternatively it was thought that fungal pectinesterases had a random activity resulting in the de-esterification of single GalA residues per enzyme/substrate interactions. It has now been shown that some plant pectinesterase isoforms may exhibit both mechanisms and that such mechanisms are driven by alterations in pH. The optimal pH of higher plants is usually between pH 7 and pH 8 although the pH of pectinesterase from fungi and bacteria is usually much lower than this.

The enzyme activity of demethylation of the pectin results in the appearance of free carboxyl groups associated with the galacturonic acids making up the chains. The pectin methyl-esterase activity is estimated by determination of the methanol. The alcohol oxydase of *Pichia pastoris* is specific to primary alcohols with a low molecular weight and catalyses the oxidation of the methanol into formaldehyde. 2,4- Pentanedione condenses exclusively with aldehydes of low molecular weight such as formaldehyde, forming a chromophore absorbing at 412 nm.

Materials

- Sodium acetate buffer 50 mM, pH 4.5. (Solution A) Dissolve 4.10 g of sodium acetate into 1 liter of water, (Solution B) dissolve 2.8 ml of acetic acid into 1 liter of water. Mix 39.2% of solution A + 60.8% of solution B, check that the pH equals 4.5 using a pH-meter. Maintain at 4°C
- Citrus fruit pectin solution at 0.5% (p/v) – Diussolve 0.5 g of citrus fruit pectin into 100 ml of sodium acetate buffer. The solution must be prepared as needed.
- Acetic acid solution 0.05 M – Add 0.283 ml of acetic acid into 100 ml of water.
- Ammonium acetate solution 2 M - Dissolve 15.4 g of ammonium acetate in 100.0 ml of acetic acid.
- 2,4-Pentanedione 0.02 M – Dissolve 40.8 µl 2,4-pentanedione in 20.0 ml of ammonium acetate solution. The solution must be prepared as needed.
- Sodium phosphate buffer (0.25 M; pH 7.5) – (Solution A) Dissolve 34.015 g of potassium dihydrogen phosphate in 1 liter of water. (Solution B) Disslove 44.5125 g of disodium hydrogen phosphate in 1 liter of water. Mix 16.25 % of solution A + 83.75% of solution B to obtain a pH of 7.5. Check the pH using a pH-meter. Maintain at 4°C, for a maximum of one week
- Stock solution of methanol at 40 µg/ml – Mix 5 µl of methanol into 100.0 ml of sodium phosphate buffer.
- Alcohol oxydase at 1U/ml - Dilute alcohol oxydase of *Pichia pastoris* in a phosphate buffer in order to obtain a solution at 1U/ml. The solution must be prepared as needed.

Procedure

- To solubilize and extract pectin methylesterase, green bean pod or tomato pericarp material is homogenized in a mortar and pestle in an equal weight of a solution containing 2.0 M NaCl and 10 mM phosphate buffer (pH 7.5). The homogenate was clarified by centrifugation at 16100 x g for 2 min. The supernatant is used as source of pectin methyl esterase.
- To assay for pectin methyl esterase activity, 2.5 ml of a solution containing 0.5% apple pectin, 0.2 M NaCl, and 0.1 mM phosphate buffer (pH 7.5) is taken in a stirred, water-jacketed cell and preheated to the desired temperature (30-60° C).

- After starting the assay by the addition of the supernatant, the pH of the incubation was adjusted to 7.5 by the addition of small aliquots of 0.1 M NaOH.
- The reaction is stopped by placing the tubes immediately in the water bath at 100°C for 10 min. The tubes are then cooled under running cold water.
- Methanol produced through enzymic action is quantified using alcohol oxidase and purpald. Our standard reaction contained 90 µl of 200 mM phosphate buffer (pH 7.2), 10 µl of alcohol oxidase at 0.01 unit/µl in 200 mM hosphate buffer (pH 7.2), 25 – 50 µl of the sample solution, and H_2O to give a final volume of 200 µl.
- In sample solution contained added calcium, 100 mM citrate buffer (pH 6.5) was used in place of phosphate buffer to avoid the formation of insoluble $Ca_3(PO_4)_2$. For samples that were homogenized in trichloroacetic acid an aliquot of 0.5 N NaOH sufficient to neutralize the acid was added and the amount of H_2O in the assay reduced to maintain a final volume of 200 µl. Assays were started by the addition of the alcohol oxidase solution and then incubated in a water bath at 30° C. After 10 min, 200 µl of a freshly prepared 5 mg/ml Purpald solution in 0.5 N NaOH was added, and the samples were vigorously vortexed to ensure oxygenation. After an additional 30 min at 30° C, the samples were removed from the water bath and 0.6 ml of H_2O was added for a final volume of 1.0 ml.
- Absorbance at 550 nm was then determined. Methanol concentrations were determined from a standard curve between 0 and 50 nmol of methanol.

Reading

Anthon, GE and Barrett, DM. (2006). Characterization of the temperature activation of pectin methylesterase in green beans and tomatoes. *J. Agric. Food Chem.*, **54:** 204-211.

14.14. Estimation of Cinnamic Acid Content

Cinnamic acid is an organic compound with the formula $C_6H_5CHCHCO_2H$. It is a white crystalline compound that is slightly soluble in water. Classified as an unsaturated carboxylic acid, it occurs naturally in a number of plants. It is freely soluble in many organic solvents. It exists as both a cis and a trans isomer, although the latter is more common. Cinnamic acid and its phenolic analogues are natural substances. Chemically, in cinnamic acids the 3-phenyl acrylic acid functionality offers three main reactive sites; substitution at the phenyl ring, addition at the α,β- unsaturation and the reactions of the carboxylic acid functionality. Cinnamic acid is also known as phenylacrylic acid, cinnamylic acid, 3-Phenylacrylic acid, (E)-cinnamic acid, benzenepropenoic acid, and isocinnamic acid

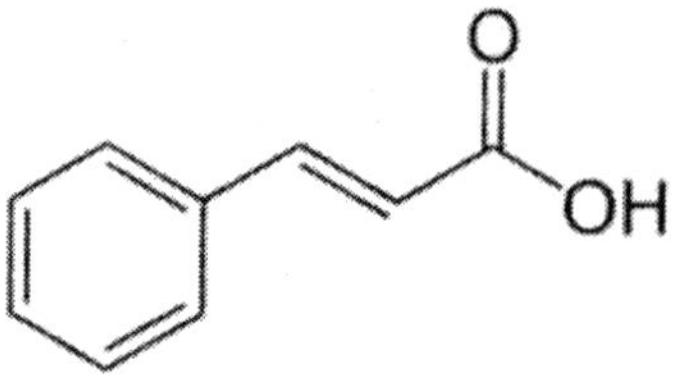

Cinnamic acid

Cinnamic acid is used in flavors, synthetic indigo, and certain pharmaceuticals. A major use is in the manufacturing of the methyl, ethyl, and benzyl esters for the perfume industry. Cinnamic acid is a precursor to the sweetener aspartate via enzyme-catalysed amination to phenylalanine. Cinnamic acid is also a kind of self-inhibitor produced by fungal spore to prevent germination.

Owing to the unique chemical aspects cinnamic acid derivatives received much attention in medicinal research as traditional as well as recent synthetic antitumor agents. Inspite of their rich medicinal tradition, cinnamic acid derivatives and their anticancer potentials remained underutilized for several decades since the first published clinical use in 1905. In last two decades, there has been huge attention towards various cinnamoyl derivatives and their antitumor efficacy.

Cinnamic acid is obtained from oil of cinnamon, or from balsams such as storax. It is also found in shea butter and is the best indication of its environmental history and post-extraction conditions. Cinnamic acid has a honey-like odor; it and its more volatile ethyl ester (rthyl cinnamate) are flavor components in the essential oil of cinnamon, in which related cinnamaldehyde is the major constituent. Cinnamic acid is also part of the biosynthetic shikimate and phenylpropanoid pathways. Its biosynthesis is performed by action of the enzyme phenylalanine ammonia-lyase (PAL) on phenylalanine.

Materials

- Isotachophoresis apparatus with (1) high voltage (30 kV) supply, (2) Fused silica capillary tubes, (3) UV monitor. Detection is carried out by on-colum measurement of UV absorption at 254 nm at a position of 30 cm from the cathode.
- 3,4-Dimethoxycinnamic acid (DMCA)
- 4-Methoxycinnamic acid (MCA)
- Cinnamic acid (CA)
- 3-methoxy-4-hydroxycinnamic acid (ferullic acid, FA)

- 4-hydroxycinnamic acid (HCA)
- Phosphate buffer used for carrier and electrode liquid are prepared by mixing an aquous solution of disodium hydrogen phosphate of the same concentration of potassium dihydrogen phosphate (pH 6-9)

Procedure for Electrophoresis

- The capillary tube is filled with carrier by dipping one end of the tube into a buffer solution and allowing the liquid to flow downhill.
- After the tub is filled completely, both ends of the tube is dipped in separate buffer having same salt concentration and pH as well as the same level.
- To introduce a sample into the tube, the anodic end of the tube is rapidly moved into the same solution, whose level is raised by 5.0 cm higher than the other buffer, and maintained for a specified period between 5 to 15 seconds. The end of the tube is returned to the buffer.
- The a high voltage is applied in the constant current mode.
- To determine the actual amount of sample introduced, the liquid in the tube is drained, after electrophoresis separation, into a small test tube and an aliquot of the drainage is analysed by using high performance liquid chromatography equipped with UV-254 detector.

CHAPTER 15

Secondary Metabolites

Plant secondary metabolism produces products that aid in the growth and development of plants but are not required for the plant to survive. Secondary metabolism facilitates the primary metabolism in plants. This primary metabolism consists of chemical reactions that allow the plant to live. In order for the plants to stay healthy, secondary metabolism plays a pinnacle role in keeping all the of plants' systems working properly. A common role of secondary metabolites in plants is defense mechanisms. They are used to fight off herbivores, pests, and pathogens. Although researchers know that this trait is common in many plants it is still difficult to determine the precise role each secondary metabolite. Secondary metabolites are used in anti-feeding activity, toxicity or acting as precursors to physical defense systems.

There is no fixed, commonly agreed upon system for classifying secondary metabolites. Based on their biosynthetic origins, plant secondary metabolites can be divided into three major groups:

1. Flavonoids and allied phenolic and polyphenolic compounds,
2. Terpenoids and
3. Nitrogen-containing alkaloids and sulphur-containing compounds.

Other researchers have classified secondary metabolites into following, more specific types

Class	Type	Example
Alkaloids	Nitrogen containing	Atropin, Berberine, Cocaine, Caffeine, Ephedrine, Galantamine, Morphine, Nicotine, Psilocin, Quinidine, Quinine, Reserpine, Vincristine, Vicamine,
Non protein amino acids	Nitrogen containing	
Amines	Nitrogen containing	
Cyanogenic glycosides	Nitrogen containing	Amygdalin, Dhurrin, Linamarin, Lotaustralin, Prunasin
Glusinolates	Nitrogen containing	
Alkamides	Nitrogen containing	
Lectins, peptides and polypeptides	Nitrogen containing	
Terpenes	Without nitrogen	Azadirachtin, Artemisinin, Tetrahydrocannabinol
Steroids and saponins	Without nitrogen	These are terpenoids with a particular ring structure. Cycloartenol
Flavonoids and Tannins	Wthout nitrogen	
Phenylpropanoids, Lignins, Coumarins and Lignans	Without nitrogen	
Polyacetylenes, fatty acids and waxes	Without nitrogen	
Polyketides	Without nitrogen	
Carbohydrates and organic acids	Without nitrogen	

Some of the secondary metabolites are discussed below:

Atropine

Atropine is a naturally occurring tropane alkaloid extracted from deadly nightshade (*Atropa belladona*), Jimson weed (*Datura stramonium*), mandrake (*Mandragora officinarum*) and other plants of the family Solanaceae. It is a secondary metabolite of these plants and serves as a drug with a wide variety of effects.

In general, atropine counters the "rest and digest" activity of glands regulated by the parasympathetic nervous system. This occurs because atropine is a competitive antagonist of the muscrinic acetylcholine receptor (acetylcholine being the main neurotransmitter used by the parasympathetic nervous system). Atropine dilates the pupils, increases heart rate, and reduces salivation and other secretions.

Atropine is a *core* medicine in the World Health Organization's "Essential Drug List", which is a list of minimum medical needs for a basic health care system.

Atropine is found in many members of the Solanaceae family. The most commonly found sources are *Atropa belladonna, Datura inoxia, D. metel*, and *D. stramonium*. Other sources include members of the *Brugmansia* and *Hyoscyamus* genera. The *Nicotiana* genus (including the tobacco plant, *N. tabacum*) is also found in the Solanaceae family, but these plants do not contain atropine or other tropane alkaloids.

Atropine is a type of secondary metabolite called a tropane alkaloid. Alkaloids contain nitrogens, frequently in a ring structure, and are derived from aminoacids. Tropane is an organic compound containing nitrogen and it is from tropane that atropine is derived. Atropine is synthesized by a reaction between tropine and tropate, catalyzed by atropinase. Both of the substrates involved in this reaction are derived from amino acids, tropine from pyridine (through several steps) and tropate directly from phenylalamine. Within *Atropa belladona* atropine synthesis has been found to take place primarily in the root of the plant. The concentration of synthetic sites within the plant is indicative of the nature of secondary metabolites. Typically, secondary metabolites are not necessary for normal functioning of cells within the organism meaning the synthetic sites are not required throughout the organism. As atropine is not a primary metabolite, it does not interact specifically with any part of the organism, allowing it to travel throughout the plant.

Flavonoids

Flaonoids are one class of secondary plant metabolites that are also known as Vitamin P or citrin. These metabolites are mostly used in plants to produce yellow and other pigments which play a big role in coloring the plants. In addition, Flavonoids are readily ingested by humans and they seem to display important anti-inflammatory, anti-allergic and anti-cancer activities. Flavonoids are also found to be powerful anti-oxidants and researchers are looking into their ability to prevent cancer and cardiovascular diseases. Flavonoids help prevent cancer by inducing certain mechanisms that may help to kill cancer cells, and researches believe that when the body processes extra flavonoid compounds, it triggers specific enzymes that fight carcinogens. Good dietary sources of Flavonoids are all citrus fruits, which contain the specific flavanoids hesperidins, quercitrin, and rutin, apples, apricot, blueberries, pears, raspberries, strawberries, black beans, cabbage, onions, parsley, pinto beans, and tomatoes. green tea, dark chocolate and red wine. Flavonoids are synthesized by the phenylpropanoid metabolic pathway where the amino acid phenylalanine is used to produce 4-coumaryol-CoA, and this is then combined with malonyl-

CoA to produce chalcones which are backbones of Flavonoids Chalcones are aromatic ketones with two phenyl rings that are important in many biological compounds. The closure of chalcones causes the formation of the flavonoid structure. Flavonoids are also closely related to flavones which are actually a sub class of flavonoids, and are the yellow pigments in plants. In addition to flavones, 11 other subclasses of Flavonoids including, isoflavones, flavans, flavanones, flavanols, flavanolols, anthocyanidins, catechins (including proanthocyanidins), leukoanthocyanidins, dihydrochalcones, and aurones.

Flavonoids help us help protect blood vessels from rupture or leakage, enhance the power of vitamin C, protect cells from oxygen damage and prevent excessive inflammation throughout your body.

15.1. Estimation of Flavonoids in Plant Samples

Flavonoids (or bioflavonoids) (from the Latin word *flavus* meaning yellow, their colour in nature) are a class of plant secondary metabolites. Flavonoids were referred to as Vitamin P (probably because of the effect they had on the permeability of vascular capillaries) from the mid-1930s to early 50s, but the term has since fallen out of use.

According to the IUPAC nomenclature, they can be classified into:

- *flavonoids* or *bioflavonoids*.
- *isoflavonoids*, derived from 3-phenylchromen-4-one (3-phenyl-1,4-benzopyrone) structure
- *neoflavonoids*, derived from 4-phenylcoumarine (4-phenyl-1,2-benzopyrone) structure.

The three flavonoid classes above are all ketone-containing compounds, and as such, are anthoxanthins (flavones and flavonoids). This class was the first to be termed *bioflavonoids*. The terms *flavonoid* and *bioflavonoid* have also been more loosely used to describe non-ketone polyhydroxy polyphenol compounds which are more specifically termed *flavanoids*. The three cycle or heterocycles in the flavonoid backbone are generally called ring A, B and C. Ring A usually shows a phloroglucinol substitution pattern.

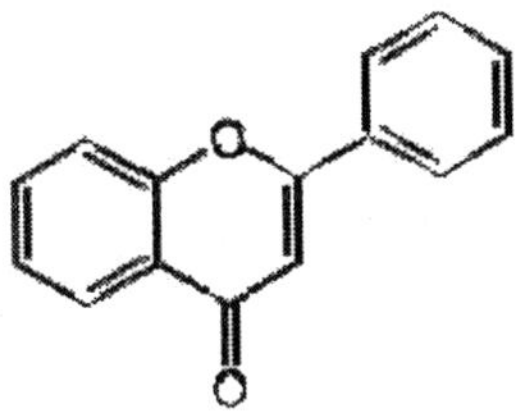

Molecular structure of Flavone 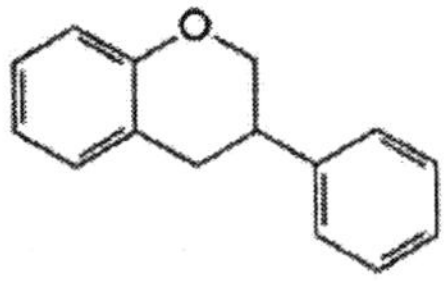 Isoflavon structure 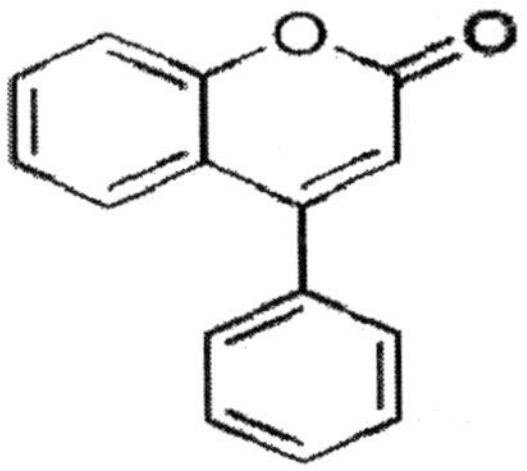 Neoflavonoid structure

Flavonoids are widely distributed in plants, fulfilling many functions. Flavonoids are the most important plant pigments for flower colouration, producing yellow or red/blue pigmentation in petals designed to attract pollinator animals. In higher plants, flavonoids are involved in UV filtration, symbiotic nitrogen fixation and floral pigmentation. They may also act as chemical messengers, physiological regulators, and cell cycle inhibitors. Flavonoids secreted by the root of their host plant help *Rhizobia* in the infection stage of their symbiotic relationship with legumes like peas, beans, clover, and soy. Rhizobia living in soil are able to sense the flavonoids and this triggers the secretion of Nod factors, which in turn are recognized by the host plant and can lead to root hair deformation and several cellular responses such as ion fluxes and the formation of a root nodule. In addition, some flavonoids have inhibitory activity against organisms that cause plant diseases, *e.g. Fusarium oxysporum*.

Tests for detection

Shinoda test

Four pieces of magnesium fillings (ribbon) are added to the ethanolic extract followed by few drops of concentrated hydrochloric acid. A pink or red colour indicates the presence of flavonoid. Colours varying from orange to red indicated flavones, red to crimson indicated flavonoids, crimson to magenta indicated flavonones.

Sodium hydroxide test

About 5 mg of the compound is dissolved in water, warmed and filtered. 10% aqueous sodium hydroxide is added to 2 ml of this solution. This produces a yellow colouration. A change in colour from yellow to colourless on addition of dilute hydrochloric acid is an indication for the presence of flavonoids.

p-Dimethylaminocinnamaldehyde test

A colorimetric assay based upon the reaction of A-rings with the chromogen p-dimethylaminocinnamaldehyde (DMACA) has been developed for flavanoids in beer that can be compared with the vanillin procedure.

Quantification

Lamaison and Carnet have designed a test for the determination of the total flavonoid content of a sample ($AlCl_3$ method). After proper mixing of the sample and the reagent, the mixture is incubated for 10 minutes at ambient temperature and the absorbance of the solution is read at 440 nm. Flavonoid content is expressed in mg/g of quercetin.

15.2. Cyanogenic Glycoside

In chemistry, a glycoside is a molecule in which a sugar is bound to another functional group via a glycosidic bond. Glycosides play numerous important roles in living organisms. Many plants store chemicals in the form of inactive glycosides. These can be activated by enzyme hydrolysis, which causes the sugar part to be broken off, making the chemical available for use. Many such plant glycosides are used as medications. In animals and humans, poisons are often bound to sugar molecules as part of their elimination from the body.

In formal terms, a glycoside is any molecule in which a sugar group is bonded through its aromatic carbon to another group via a glycosidic bond. Glycosides can be linked by an O- (an *O-glycoside*), N- (a *glycosylamine*), S-(a *thioglycoside*), or C- (a *C-glycoside*) glycosidic bond. According to the IUPAC the name "*C*-glycoside" is a misnomer, the preferred term is "*C*-glycosyl compound". The given definition is the one used by IUPAC, which recommends the Haworth projection to correctly assign stereochemical configurations. Many authors require in addition that the sugar be bonded to a *non-sugar* for the molecule to qualify as a glycoside, thus excluding polysaccharides. The sugar group is then known as the glycone and the non-sugar group as the aglycone or genin part of the glycoside. The glycone can consist of a single sugar group (monosaccharide) or several sugar groups (oligosaccharide).

Many plants have adapted to iodine-deficient terrestrial environment by removing iodine from their metabolism, in fact iodine is essential only for animal cells. An important antiparasitic action is caused by the block of the transport of iodide of animal cells inhibiting sodium-iodide symporter (NIS). Many plant pesticides are cyanogenic glycoside which liberate cyanide, which, blocking cytochrome c oxidase and NIS, is poisonous only for a large part of parasites and herbivores and not for the plant cells in which it seems useful in seed dormancy phase. To get a better understanding of how secondary metabolites play a big role in plant defense mechanisms we can focus on the recognizable defense-related secondary metabolites, cyanogenic glycosides. The compounds of these secondary metabolites are found in over 2000 plant species. Its structure allows the release of cyanide, a poison produced by certain bacteria, fungi, and algae that is found in numerous plants. Animals and humans possess the ability to detoxify cyanide from their systems naturally. Therefore cyanogenic glycosides can be used for positive benefits in animal systems always. For example, the larvae of the southern armyworm consumes plants that contain this certain metabolite and have shown a better growth rate with this metabolite in their diet, as opposed to other secondary metabolite-containing plants. Although this example shows cyanogenic glycosides being beneficial to the larvae many still argue that this metabolite can do harm. To help in determining whether cyanogenic glycosides are harmful or helpful researchers look closer at its biosynthetic pathway. Past research suggests that cyanogenic glucosides stored in the seed of the plant are metabolized during germination to release nitrogen for seedling to grow. With this, it can be inferred that cyanogenic glycosides play various roles in plant metabolism. Though subject to change with future research, there is no evidence showing that cyanogenic glycosides are responsible for infections in plants.

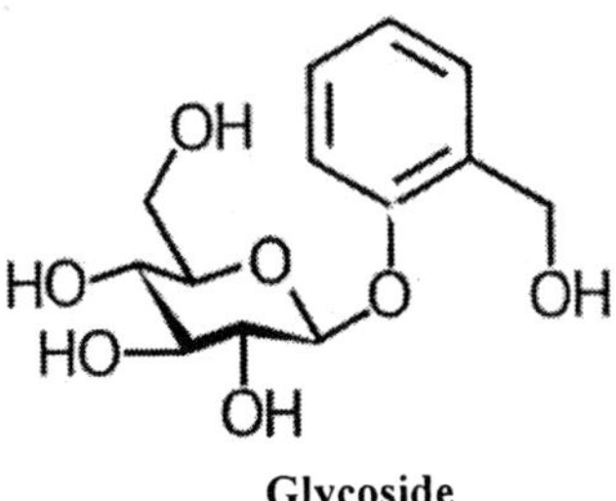

Glycoside

Estimation of Cyanogens

Materials

- Chloroform
- Whatman No. 1 filter paper – Cut filter paper into strips 10-12 cm long and 0.5 cm wide and saturate them with alkaline picrate solution.
- Alkaline picrate solution – 25.0 g of sodium carbonate is mixed with 5.0 g of picric acid and dissolved in 1000 ml of water.
- Standard hydrogen cyanide solution – 0.241g of KCN/litre of water. This gives a solution containing 100 µg hydrogen cyanide/ml.

Procedure

- Fresh plant sample (approximately 1.0 g) is homogenized in 25.0 ml of water along with 3-4 drops of chloroform.

- Place the homogenates in a flask and attach the filter paper strips in a hanging position along the mouth of the flask. Put the stop corck on the flask.
- Incubate the mixture at room temperature (20^0 C) for 20-24 hours.
- The sodium picrate in filter paper strips produce a reddish compound in proportion to the amount of hydrocyanic acid present in plant homogenate.
- Elute the colour with 10.0 ml of water and measure the optical density at 625 nm against water.
- For standard curve, 5.0 ml of alkaline picrate solution is mixed with 5.0 ml of potassium cyanide solution. Heat the mixture for 5 minutes in a boiling waterbath. From this solution a gradient of is taken as 0.1, 0.2, 0.4, 0.6, 0.8, 1.0 is taken and total volume is made to 10.0 ml with water. Optical density at 625 nm. The amount of hydrogen cyanide in these volumes is equal to 5, 10, 20, 30, 40 and 50 µg, respectively.

Calculation

Calculate the hydrogen cyanide content of the sampe from the standard curve.

Reading

Hogg, PG and Ahlgren, HL. (1942). A rapid method for determining hydrogen cyanic acid content of single plant of sudan grass. *J. Ame. Soc Agron.*, **34:** 199 – 200.

Indira, P and Sinha, SK. (1969). Colorimetric method for determination of HCN in tubers and leaves of Cassava (*Manihot. esculenta* Crantz). *Ind. J. Agric. Sci.*, **39:** 1021 – 1023.

15.3. Phytic Acid

Phytic acid (known as inositol hexakisphosphate (IP6), or phytate when in salt form), a saturated cyclic acid, is the principal storage form of phosphorus in many plant tissues, especially bran and seeds. Phytate is not digestible to humans or nonruminant animals, so it is not a source of either inositol or phosphate if eaten directly. Moreover, phytic acid chelates and thus makes unabsorbable certain important minor minerals such as zinc and iron, and to a lesser extent, also macro minerals such as calcium and magnesium; phytin refers specifically to the calcium or magnesium salt form of phytic acid. Catabolites of phytic acid are called lower inositol polyphosphates. Examples are inositol penta- (IP5), tetra- (IP4), and triphosphate (IP3). Phosphorus and inositol in phytate form are not, in general, bioavailable to nonruminant animals because these animals lack the digestive enzyme phytase required to remove phosphate from the inositol in the phytate molecule. On the other hand, ruminants readily digest phytate because of the phytase produced by rumen microorganisms.

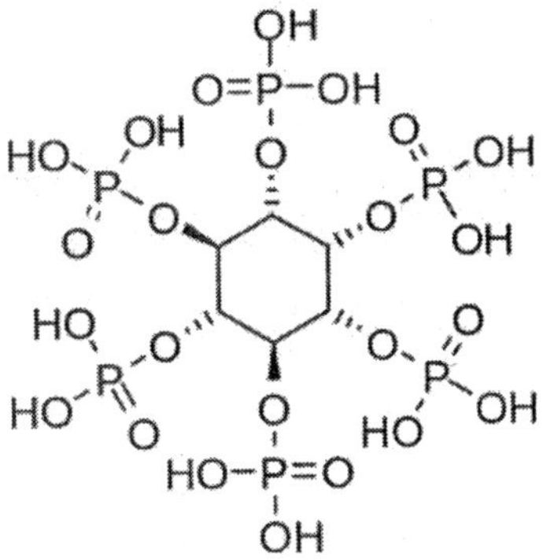

Molecular structure of Phytic acid

In most commercial agriculture, nonruminant livestock, such as swine, fowl, and fish, are fed mainly grains, such as maize, and legumes, such as soybeans. Because phytate from these grains and beans is unavailable for absorption, the unabsorbed phytate passes through the gastrointestinal tract, elevating the amount of phosphorus in the manure. Excess phosphorus excretion can lead to environmental problems, such as eutrophication.

The bioavailability of phytate phosphorus can be increased by supplementation of the diet with the enzyme phytase. Also, viable low-phytic acid mutant lines have been developed in several crop species in which the seeds have drastically reduced levels of phytic acid and concomitant increases in inorganic phosphorus. However, reported germination problems have hindered the use of these cultivars thus far. The use of sprouted grains will reduce the quantity of phytic acids in feed, with no significant reduction of nutritional value. Phytates also have the potential to be used in soil remediation, to immobilize uranium, nickel and other inorganic contaminants.

Although undigestable for many animals, phytic acid and its metabolites as they occur in seeds and grains have several important roles for the seedling plant. Most notably, phytic acid functions as a phosphorus store, as an energy store, as a source of cations and as a source of myoinositol (a cell wall precursor). Phytic acid is the principal storage forms of phosphorus in plant seeds.

In animal cells, myoinositol polyphosphates are ubiquitous, and phytic acid (myoinositol hexakisphosphate) is the most abundant, with its concentration ranging from 10 to 100 uM in mammalian cells, depending on cell type and developmental stage. This compound is not obtained from the animal diet, but must be synthesized inside the cell from phosphate and inositol. The interaction of intracellular phytic acid with specific intracellular proteins has been investigated *in vitro*, and these interactions have been found to result in the inhibition or potentiation of the physiological activities of those proteins. The best evidence from these studies suggests an intracellular role for phytic acid as a cofactor in DNA repair by nonhomologous end-joining. Other studies using yeast mutants have also suggested intracellular phytic acid may be involved in mRNA export from the nucleus to the cytosol. There are still major gaps in the understanding of this molecule, and the exact pathways of phytic acid and lower inositol phosphate metabolism are still unknown. As such, the exact physiological roles of intracellular phytic acid are still a matter of debate.

Phytic acid is the main method of phosphorus storage in plant seeds, but is not readily absorbed by many animals. Not only is phytic acid a phosphorus storage unit, but it also is a source of energy and cations, a natural antioxidant for plants, and can be a source of myinositol which is one of the preliminary pieces for cell walls. Phytic Acid is also known to bond with many different minerals, and by doing so prevents those minerals from being absorbed; making Phytic Acid an anti-nutrient. There is a lot of concern with phytic acids in nuts and seeds because of its anti-nutrient characteristics. In preparing foods with high phytic acid concentrations, it is recommended they be soaked in after being ground to increase the surface area. Soaking allows the seed to undergo germination which increases the availability of vitamins and nutrient, while reducing phytic acid and protease inhibitors, ultimately increasing the nutritional value. Cooking can also reduce the amount of phytic acid in food but soaking is much more effective.

Phytic Acid is an antioxidant found in plant cells that most likely serves the purpose of preservation. This preservation is removed when soaked, reducing the phytic acid and allowing the germination and growth of the seed. When added to foods it can help prevent discolouration by inhibiting lipid peroxidation. There is also some belief that the chelating of phytic acid may have potential use in the treatment of cancer.

Estimation of Phytic acid

Materials

- Trichloroacetic acid (3%)
- Sodium sulphate (3%) in 3% trichloroacetic acid
- Sodium hydroxide 1.5 N
- Nitric acid 3.2 N
- Ferric chloride solution – 583.0 mg of ferric chloride ($FeCl_3$) in 100.0 ml of 3% TCA
- Potassium thiocyanate – 29.15 g of potassium thiocyanate (KSCN) in 200.0 ml of water
- Standard $Fe(NO_3)_3$ solution

Procedure

- To 5-30 mg of plant sample 50.0 ml of TCA is added and shaken mechanically for 45 minutes and then centrifuged at 10,000 rpm for 10 minutes.
- To 10.0 ml of extract, 4.0 ml of ferric chloride solution is added and the mixtures is kept in boiling water bath for 30 minutes, few drops of 3% sodium sulphate in 3% TCA is added and continue the heating for another 30 minutes in boiling water bath.
- Cool the content and centrifuge the conten at 10,000 rpm for 10 minutes. Decant the supernatant.
- Wash the precipitate twice with 20-25 ml of 3% TCA, heat in boiling water bath for 5-10 minutes and centrifuge. Repeat the washing with water.
- The precipitate is dispesed in a few ml of water and 3.0 ml of 1.5 N HNO_3.
- Bring te volume to 30.0 ml with water and heat in boiling water for 30 minutes. Filter hot through Whatman Filter paper # 2..
- Wash the precipitate with 60-70 ml water and discard filterate.
- The precipitate in filterpaper in dissolved in 40.0 ml hot 3.2 N HNO_3, and precipitate is washed with water. All the washings and dissolved precipitate is pooled. The content is cooled and volume is made to 100 ml with water.
- To a 5.0 ml above aliqote about 70.0 ml of water is added followed by 20.0 ml of 1.5 M potassium thiocyanate solution.
- The optical density of the solution is measured immediately (within 1 minute) at 480 nm against a reagent blank.
- For standard curve 433.0 mg of $Fe(NO_3)_3$ is dissolved in 100.0 ml of water. 2.5 ml of this solution is diluted to 250.0 ml with water and from this solution a gradient of 2.5, 5.0 7.5, 10.0, 15.0, 20.0 and 25.0 ml solution is taken and to these solutions, 20.0 ml of KSCN solution is added and optical density is measured at 480 nm against a reagent blank.

Calcultions

Through standard curve find the μg of iron present in sample solution and calculate the phytate P content as:

Phytate P mg/100 g of sample = (µg Fe x 15) / (Weight of sample (g))

Reading

Wheeler, EL and Ferrel, RE. (1971). A method for phytic Acid determination in wheat and wheat fractions. *Cereal Chem.*, **38:** 312-320.

15.4. Gossypol

Gossypol is a natural phenol derived from the cotton plant (genus *Gossipium*). Gossypol is a phenolic aldehyde that permeates cells and acts as an inhibitor for several dehydrogenase enzymes. It is a yellow pigment.

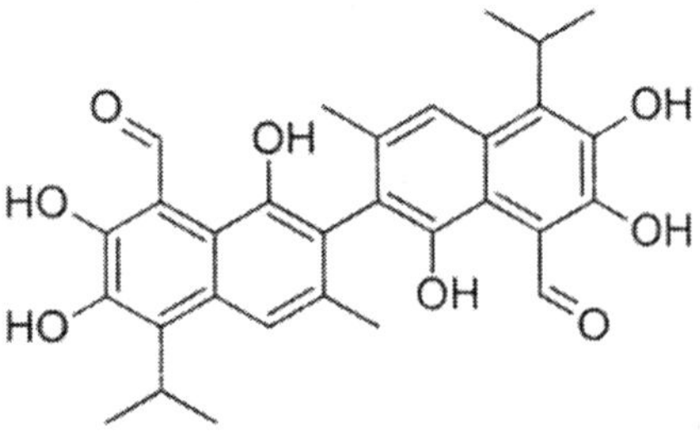

Molecular structure of Gossypol

Among other things, it has been tested as a male oral contraceptive in China. In addition to its contraceptive properties, gossypol has also long been known to possess antimalarial properties. Other researchers are investigating the anticancer properties of gossypol

It has proapopototic properties, probably due to the regulation of the Bax and Bcl2. It also reversibly inhibits calcineurin and binds to calmodlin. It inhibits replication of the HIV-1 virus. It is an effective protein kinase C inhibitor It also causes low potassium levels, and thus causes temporary paralysis.

Gossypol has a yellow pigment and is found in cotton plants. It occurs mainly in the root and/or seeds of different species of cotton plants. Gossypol can have various chemical structures. It can exist in three forms: gossypol, gossypol acetic acid, and gossypol formic acid. All of these forms have very similar biological properties. Gossypol is a type of aldehyde, meaning that it has a formyl group. The formation of gossypol occurs through an isoprenoid pathway. Isoprenoid pathways are common among secondary metabolites. Gossypol's main function in the cotton plant is to act as an enzyme inhibitor. An example of gossypol's enzyme inhibition is its ability to inhibit nicotinamide adenine dinucleotide-linked enzymes of *Trypanosoma cruzi. Trypanosoma cruzi* is a parasite which causes Chaga's disease.

For some time it was believed that gossypol was merely a waste product produced during the processing of cottonseed products. Extensive studies have shown that gossypol has other functions. Many of the more popular studies on gossypol discuss how it can act as a male contraceptive. Gossypol has also been linked to causing hypokalemic paralysis. Hypokalemic paralysis is a disease characterized by muscle weakness or paralysis with a matching fall in potassium levels in the blood. Hypokalemic paralysis associated with gossypol in-take usually occurs in March, when vegetables are in short supply, and in September, when people are sweating a lot. This side effect of gossypol in-take is very rare however. Gossypol induced hypokalemic paralysis is easily treatable with potassium repletion.

Estimation of Gossypol

Materials

- Ethanol (96% (v/v)
- Phloroglucinol reagent – 5.0 g of phloroglucinol is dissolved in 100.0 ml of 80% (v/v) ethanol.

Procedure

- Fresh plant samples (approximately 5.0 g) is repeatedly extracted with 15-20 ml of boiling 95% ethanol, filter the and all the filterates are pooled. Dilute the extract with 40% ethanol and pH of the extract is adjusted to 3.0 with I N HCl.
- The content is mixed with 1.5 volumes of diethyl ether at 10^0 C in a separating funnel and ether phase is saved and washed two to three times with water.
- Using a flash evaporator, ether is evaporated from the ether extract, and residue is redissolved in known volume of 95% ethanol and is used as gossypol extract.
- Gossypol extract (1.2 ml) in ethanol is mixed with 0.5 ml of phloroglucinol reagent followed by 1.0 ml of concentrated HCl.
- Incubate for 30 minutes with occasional stirring.
- Make the volume of the reaction mixture to 10.0 ml with 80% ethanol and measure the optical density at 550 nm against a reagent blank.
- Standard curve is prepared with gossypol acetate.

Reading

Bell, AA. (1967). Formation of gossypol in infected or chemically irritated tissues of *Gossypium* species. *Phytopath*, **57:** 759-764.

Sood, DR. and Singh, A. (1973). *Cotton Dev.* **3:** 15.

15.5. Phytoestrogens

Phytoestrogens are plant-derived xenostrogens not generated within the endocrine system but consumed by eating phytoestrogenic plants. Also called "dietary estrogens", they are a diverse group of naturally occurring nonsteroidal plant compounds that, because of their structural similarity with estradiol (17-β-estradiol), have the ability to cause estrogenic or/and antiestrogenic effects, antiestrogenic effects by sitting in and blocking receptor sites against estrogen.

The similarities, at molecular level, of estrogens and phytoestrogens allow them to mildly mimic and sometimes act as antagonists of estrogen. Phytoestrogens were noticed for the first time that red clover (a phytoestrogens-rich plant) pastures had effects on the fecundity of grazing sheep. Researchers are exploring the nutritional role of these substances in the regulation of chloestrol, and the maintenance of proper bone density post-menopause. Evidence is accruing that phytoestrogens may have protective action against diverse health disorders, such as prostrate, breast, browel, and other cancers, cardiovascular disease, brain function disorders and osteoporosis,

Phytoestrogens cannot be considered as nutrients, given that the lack of these in diet does not produce any characteristic deficiency syndrome, nor do they participate in any essential biological function. Phytoestrogens are chemicals which act like the hormone estrogen. Estrogen is important for women's bone and heart health, but high amounts of it has been linked to breast cancer. In the plant, the phytoestrogens are involved in the defense system against fungi. Phytoestrogens can do two different things in a human body. At low doses it mimics estrogen, but at high doses it actually blocks the body's natural estrogen. The estrogen receptors in the body which are stimulated by estrogen will acknowledge the phytoestrogen, thus the body may reduce its own production of the hormone. This has a negative result, because there are various abilities of the phytoestrogen which estrogen does not do. Its effects the communication pathways between cells and has effects on other parts of the body where estrogen normally does not play a role. It has also been found to induce tumor growth of the estrogen receptor cells in the breast. But, one role of estrogens which phytoestrogens mimic is its protective behavior for the heart. So, an intake of phytoestrogens has also been seen to reduce the risk of cardiovascular disease. Resveratrol, a phytoestrogen found in grapes is responsible for this. For example, the French suffer relatively little heart disease despite the average French diet being relatively high in fat. One proposed reason for this is the resveratrol found in red wine, which has been linked to decreased risk of cardiovascular disease.

The most widely used techniques for measurement of phytoestrogens are:

- reversed phase high performance liquid chromatography with ultraviolet detection (HPLC-UV)
- gas chromatography with mass spectrometric detection (GC-MS)
- liquid chromatography with mass spectrometric detection (LC-MS)

The most appropriate analytical method is dependent on the type of biological matrix and compound to be analysed. The analytical methods used in phytoestrogen analysis require reference standards, which are pure, samples of the analyte and are used to calibrate the analytical method to ensure it performs in the way that is anticipat. As methods have developed, the limits of detection and quantification have decreased so that phytoestrogens can now be measured in foodstuffs and biological samples down to concentrations of parts per billion.

High performance liquid chromatography with ultraviolet detection (HPLC-UV) for Phytoestrogen

HPLC-UV is a relatively rapid way of measuring phytoestrogens compared to mass spectrometric (MS) based methods. Following isolation of phytoestrogens from the matrix, they are directly separated and quantified. This method allows simultaneous purification and measurement of complex mixtures. UV detection is generally less sensitive than MS detection and the reported detection limits can be variable. The analytes are quantified by comparison with calibration curves derived from reference standards. However, substances present in the sample, but not in the reference standard, which may co-elute with the analyte during chromatography, may lead to falsely high measurements of the analyte. Reference standards are not available for many phytoestrogens, such as the acetyl- and malonyl isoflavone glucosides. Therefore, measurements of these compounds are based on calibration curves of isoflavone glucosides and aglucones. Research indicates that this approach could introduce errors in measurements. However, when reference materials of the acetyl- and malonylglucosides are available such assumptions can be tested and results re-calculated.

Gas chromatography with mass spectrometric detection (GC-MS) for Phytoestrogen

GC-MS is sufficiently sensitive to measure concentrations of phytoestrogens of less than parts per million. Measurement is usually done with the use of internal standards (preferably isotopically labelled analogues of the analytes), but unlike HPLC-UV, the standard and the analyte are measured under exactly the same conditions. Samples must be treated to remove conjugating groups prior to analysis by GC-MS and as a result phytoestrogens cannot be measured as they appear in the matrix. As such, the analytical results obtained using this method are expressed as quantity of 'total phytoestrogen'. Chemical and enzymatic methods have been developed to remove these groups. Enzymatic methods are preferable for isoflavones as they can be unstable under acidic conditions. However, lignan glucosides are resistant to enzymatic hydrolysis and require strong acid to remove the sugars.

Liquid chromatography with mass spectrometric detection (LC-MS) for Phytoestrogen

LC-MS is also sufficiently sensitive to measure concentrations of phytoestrogens of less than parts per million. However, in contrast to GC-MS, removal of conjugating groups is not required prior to analysis. Therefore, different forms of phytoestrogens can be measured and expressed directly rather than as a value of 'total phytoestrogen'. However, measurement is dependent on the availability of internal standards for each analyte and many of these materials are unavailable at present. Laboratories using LC-MS have to use the available aglucone standards and hydrolyse samples prior to analysis. As such, the advantage of non-destructive sample preparation that LC-MS offers over GC-MS has not yet been realised. More recently, methods using LC coupled to mass spectrometers in tandem (MS/MS) have been developed for the analysis of phytoestrogens.

15.6. Carotenoids

Carotinoids are organic pigments found in the chloroplasts and chromoplasts of plants. They are also found in some organisms such as algae, fungi, some bacteria, and certain species of aphids. There are over 600 known carotenoids. They are split into two classes, xanthophylls and carotenes. Xanthophylls are carotenoids with molecules containing oxygen, such as lutein and zeaxanthin. Carotenes are carotenoids with molecules that are unoxygenated, such as α-carotene, β-carotene and lycopene. In plants, carotenoids can occur in roots, stems, leaves, flowers, and fruits. Carotenoids have two important functions in plants. First, they can contribute to photosynthesis. They do this by transferring some of the light energy they absorb to chlorophyll, which then uses this energy for photosynthesis. Second, they can protect plants which are over-exposed to sunlight. They do this by harmlessly dissipating excess light energy which they absorb as heat. In the absence of carotenoids, this excess light energy could destroy proteins, membranes, and other molecules. Some plant physiologists believe that carotenoids may have an additional function as regulators of certain developmental responses in plants. tetraterpenes are synthesized from DOXP precursors in plants and some bacteria. Carotenoids involved in photosynthesis are formed in chloroplasts; Others are formed in plastids. Carotenoids formed in fungi are presumably formed from mevalonic acid precursors. Carotenoids are formed by a head-to-head condensation of geranylgeranyl pyrophosphate or diphosphate (GGPP) and there is no NADPH requirement.

Alkaloids

Alkaloids are a group of naturally occurring chemical compounds that contain mostly basic nitrogen atoms. This group also includes some related compounds with neutral and even weakly acidic properties. Some synthetic compounds of similar structure are also attributed to alkaloids. In addition to carbon, hydrogen and nitrogen, alkaloids may also contain oxygen, sulphur and more rarely other elements such as chlorine, bromine, and phosphorus. Alkaloids are produced by a large variety of organisms, including bacteria, fungi, plants, and animals, and are part of the group of ntural products (also called secondary metabolites). Many alkaloids can be purified from crude extracts by acid-base extraction. Many alkaloids are toxic to other organisms. They often have pharmacological effects and are used as medication, as recreational drugs, or in entheogenic rituals. Examples are the local anesthetic and stimulant cocaine, the psychedelic psilocin, the stimulant caffeine, nicotine, the analgesic morphine, the antibacterial berberine, the anticancer compound vincristine, the antihypertension agent reserprine, the cholinomimeric galantamine, the spasmolysis agent atropine, the vasodilator vinacamine, the anti-arrhythmia compound quinidine, the anti-asthma therapeutic ephedrine, and the antimalarial drug quinine. Although alkaloids act on a diversity of metabolic systems in humans and other animals, they almost uniformly invoke a bitter taste. The boundary between alkaloids and other nitrogen-containing natural compounds is not clear-cut. Compounds like amino acid peptides, proteins, nucleotides, nucleic acid, amines, and antibiotics are usually not called alkaloids. Natural compounds containing nitrogen in the exocyclic position (mescaline, serotonin, dopamine, *etc.*) are usually attributed to amines rather than alkaloids. Some authors, however, consider alkaloids a special case of amines.

Compared with most other classes of natural compounds, alkaloids are characterized by a great structural diversity and there is no uniform classification of alkaloids. First classification methods have historically combined alkaloids by the common natural source, *e.g.*, a certain type of plants. This classification was justified by the lack of knowledge about the chemical structure of alkaloids and is now considered obsolete.

More recent classifications are based on similarity of the carbon skeleton (*e.g.*, indole-isoquinoline, and pyridine-like) or biogenetic precursor (ornithine, lysine, tyrosine, tryptophane, *etc.*). However, they require compromises in borderline cases; for example, nicotine contains a pyridine fragment from nicotinamide and pyrrolidine part from ornithine and therefore can be assigned to both classes.

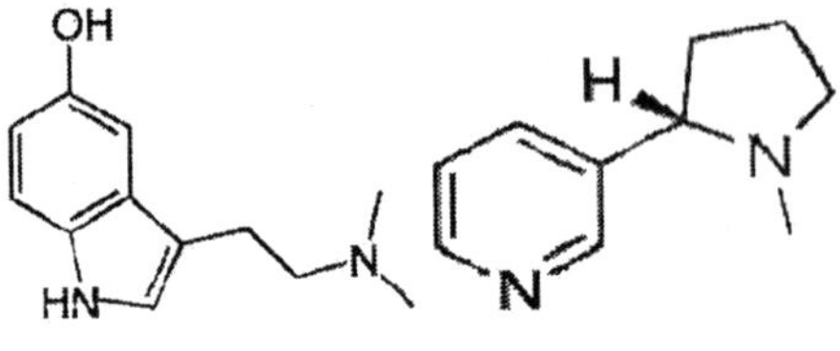

Some alkaloids

Alkaloids are often divided into the following major groups:

1. "True alkaloids", which contain nitrogen in the heterocycle and originate from amino acids. Their characteristic examples are atropine, nocotine, and morphine. This group also includes some alkaloids that besides nitrogen heterocycle contain terpene (*e.g.*, evonine) or peptide fragments (*e.g.* ergotamine). This group also includes piperidine alkaloids coniine and conicerine although they do not originate from amino acids.
2. "Protoalkaloids", which contain nitrogen and also originate from amino acids. Examples include mescaline, adrenalline and ephidrine.
3. Polyamine alkaloids – derivatives of putrescine, spermidine, and spermine.
4. Peptide and cyclopeptide alkaloids.
5. Pseudalkaloids – alkaloid-like compounds that do not originate from amino acids. This group includes, terpene-like and steroid-like alkaloids, as well as purine-like alkaloids such as caffeine, theobromine, theacrine and theophylline. Some authors classify as pseudoalkaloids such compounds such as ephidrine and cathinone. Those originate from the amino acid phenylalanine, but acquire their nitrogen atom not from the amino acid but through transamination.

Some alkaloids do not have the carbon skeleton characteristic of their group. So, galantamine and homoaporphines do not contain isoquinoline fragment, but are, in general, attributed to isoquinoline alkaloids.

Most alkaloids contain oxygen in their molecular structure; those compounds are usually colourless crystals at ambient conditions. Oxygen-free alkaloids, such as nicotine or coniine, are typically volatile, colourless, oily liquids. Some alkaloids are coloured, like berberine (yellow) and sanguinarine (orange).

Most alkaloids are weak bases, but some, such as theobromine and theophylline, are amphoteric. Many alkaloids dissolve poorly in water but readily dissolve in organic solvents, such as diethyl ether, chloroform or 1,2-dichloroethane. caffeine, cocaine, codeine and nicotine are water soluble (with a solubility of ≥ 1g/l), whereas others, including morphine and yohimbine are highly water soluble (0.1–1 g/l). Alkaloids and acids form salts of various strengths. These salts are usually soluble in water and alcohol and poorly soluble in most organic solvents. Exceptions include scopolamine hydrobromide, which is soluble in organic solvents, and the water-soluble quinine sulfate.

Most alkaloids have a bitter taste or are poisonous when ingested. Alkaloid production in plants appeared to have evolved in response to feeding by herbivorous animals; however, some animals have evolved the ability to detoxify alkaloids. Some alkaloids can produce developmental defects in the offspring of animals that consume but cannot detoxify the alkaloids. One example is the alkaloid cyclopamine, produced in the leaves of corn lily. During the 1950s, up to 25% of lambs born by sheep that had grazed on corn lily had serious facial deformations. These ranged from deformed jaws to cyclopia. After decades of research, in the 1980s, the compound responsible for these deformities was identified as the alkaloid 11-deoxyjervine, later renamed to cyclopamine. Alkaloids are generated by various living organisms, especially by higher plants – about 10 to 25% of those contain alkaloids. Therefore, in the past the term "alkaloid" was associated with plants. The alkaloids content in plants is usually within a few percent and is in homogeneous over the plant tissues. Depending on the type of plants, the maximum concentration is observed in the leaves (black henbane), fruits or seeds (*Strycnine tree*), root (*Rauwofia serpentina*) or bark (*Cinchona*). Furthermore, different tissues of the same plants may contain different alkaloids.

Beside plants, alkaloids are found in certain types of fungi, such as psilocybin in the fungus of the genus *Psilocybe*, and in animals, such as bufotenin in the skin of some toads. Many marine organisms also contain alkaloids. Some amines, such as adrenaline and serotonin, which play an important role in higher animals, are similar to alkaloids in their structure and biosynthesis and are sometimes called alkaloids.

The role of alkaloids for living organisms that produce them is still unclear. It was initially assumed that the alkaloids are the final products of nitrogen metabolism in plants, as urea in mammals. It was later shown that alkaloid concentrations varies over time, and this hypothesis was refuted. Most of the known functions of alkaloids are related to protection. For example, aporphine alkaloid liriodenine produced by the tulip tree protects it from parasitic mushrooms. In addition, presence of alkaloids in the plant prevents insects and chordate animals from eating it. However, some animals adapted to alkaloids and even use them in their own metabolism. Such alkaloid-related substances as serotonin, dopamine and histamine are important neurotransmitters in animals. Alkaloids are also known to regulate plant growth. Another example of an organism that uses alkaloids for protection is the *Utetheisa ornatrix*, more commonly known as the Ornate Moth. Pyrrolizidine alkaloids render these larvae and adult moths unpalatable to many of their natural enemies like coccinelid beetles, green lacewings, insectivorous hemiptera and insectivorous bats.

Qualitative analysis of alkaloids

Qualitative test for alkaloid was carried out on the plant materials that were obtained from the sample locations. An alcoholic extract of each material was prepared separately by adding 50 ml of 70 % ethanol to 1g of each powdered material of the various plant organs collected. To 2 ml portions of each alcoholic extract of the various organs in separate test tubes was added the following test reagents:

(a) Mayer's reagent (b) Dragendorff's reagent (c) N /10 Iodine solution

Quantitative analysis of alkaloids

Quantitative determination of the alkaloid content in the ethno-plant materials of *C. sanguinolenta* (Lindl.) Schtr. *M. lucida* Benth and *V. africana* Stapl obtained from the sample location were carried out. Three replicates were prepared for each plant sample and the mean values computed. The plant materials were air-dried for 30 days and pulverized into powder using the Manesty disintegrator. 50 g of the powdered material obtained in each case were Soxhlet extracted with hexane for 12 hours in order to defat it. The defatted powder in each case was taken and the alkaloid extracted with 500 ml of ethanol. The extracts were filtered and concentrated under reduced pressure using a rotary evaporator. The residue was mixed with 200 ml of 10 % aqueous a cetic acid and allowed to stand overnight. The mixture was filtered and the pH of the resulting filtrate adjusted to 10 using drops of dilute ammonium solution. The alkaloids were extracted with two equal volumes of 200 ml of chloroform. The chloroform extract was dried with anhydrous sodium sulphate, the solvent evaporated and the residue weighed. The percentage of alkaloid was calculated using the following formula:

$$\text{Total alkaloid (\%)} = (W / Y) \times 100$$

W = Weight of alkaloid content extracted and Y = Weight of powdered plant material

Reading

Y. Ameyaw and G. Duker-eshun (2009). The alkaloid contents of the ethno-plant organs of three antimalarial medicinal plant species in the eastern region of Ghana. *Int. J. Chem. Soc.*, **7(1):** 48-58.

Alternative Method for Estimation of Total Alkaloid Content

Materials

- Bromocresol Green Solution (BCG) ($1x10^{-4}$) - Warm 69.8 mg bromocresol green with 3 ml of 2N NaOH and 5 ml distilled water until completely dissolved and dilute to 1000 ml with distilled water.
- Phosphate Buffer (pH 4.7) Adjust pH of 2 M sodium phosphate (71.6 g Na_2HPO_4 in 1 L distilled water) to 4.7 with 0.2 M citric acid (42.02 g citric acid in 1 L distilled water).
- Atropine Standard Solution - Dissolve 1 mg pure atropine in 10 ml distilled water.
- Preparation of Standard Curve - Accurately measure aliquots (0.4, 0.6, 0.8, 1 and 1.2 ml) of atropine standard solution and transfer each to different separatory funnels. Add 5 ml pH 4.7 phosphate buffer and 5 ml BCG solution. Shake mixture with 1, 2, 3 and 4 ml of chloroform. The extracts were collected in a 10 ml volumetric flask and then diluted to volume with chloroform. The absorbance of the complex in chloroform was measured at 470 nm against blank prepared as above but without atropine.
- Extraction Preparation - The dry seeds of Syrian rue (100 g) were grinded and then extracted with methanol for 24 h in a continuous extraction (soxhlet) apparatus. The extract was filtered and methanol was evaporated on a rotary evaporator under vaccum at a temperature of 45°C to dryness. A part of this residue was dissolved in 2 N HCl and then filtered. One milliliter of this solution was transferred to a separatory funnel and washed with 10 ml chloroform (3 times). The pH of this solution was adjusted to neutral with 0.1 N NaOH. Then 5 ml BCG solution and 5 ml phosphate buffer were added to this solution. The mixture was shaken and the complex formed was extracted with 1, 2, 3 and 4 ml chloroform by vigorous shaking. The extracts were collected in a 10 ml volumetric flask and diluted to volume with chloroform. The absorbance of the complex in chloroform was measured at 417 nm vs. similarly prepared blank.

Reading

Shamsa Fadhil, Monsef Hamid Reza, Ghamooshi Rouhollah and Verdian Rizi Mohammad Reza, (2007). Spectrophotometric Determination of Total Alkaloids in *Peganum harmala* L. Using Bromocresol Green. *Research. Journal of Phytochemistry,* **1:** 79-82.

15.7. Saponins

Saponins are a class of chemical compounds found in particular abundance in various plant species. More specifically, they are ampipathic glycosides grouped phenomenologically by the soap-like foaming they produce when shaken in aqueous solutions, and structurally by having one or more hydrophilic glycoside moieties combined with a lipophilic triterpene derivative.

The aglycone (glycoside-free) portions of the saponins are termed saponenins. The number of saccharide chains attached to the sapogenin/aglycone core can vary – giving rise to another dimension of nomenclature (monodesmosidic, bidesmosidic, *etc.*) – as can the length of each chain. A somewhat dated compilation has the range of saccharide chain lengths being 1–11, with the numbers 2-5 being the most frequent, and with both linear and branched chain saccharides being represented. Dietary monosaccharides such as D-glucose and D-galactose are among the most common components of the attached chains.

The lipophilic aglycone can be any one of a wide variety of polycyclic organic structures originating from the serial addition of 10-carbon (C10) terpene units to compose a C30 triterpene skeleton, often with subsequent alteration to produce a C27 steroidal skeleton. The subset of saponins that are steroidal have been termed saraponins; Aglycone derivatives can also incorporate nitrogen, so some saponins also present chemical and pharmacologic characteristics of alkaloid natural products. The figure at right above presents the structure of the alkaloid phytotoxin solanine, a monodesmosidic, branched-saccharide steroidal saponin. (The lipophilic steroidal structure is the series of connected six- and five-membered rings at the right of the structure, while the three oxygen-rich sugar rings are at left and below.)

Saponins have historically been understood to be plant-derived, but they have also been isolated from marine organisms. Saponins are indeed found in many plants, and derive their name from the soapwort plant (genus *Saponaria*, family Caryophyllaceae), the root of which was used historically as a soap. Saponins are also found in the botanical family Sapinadaceae, with its defining genus *Spindus* (soapberry or soapnut), and in the closely related families Aceraceae (maples) and Hippocastanaceae (horse chestnuts). It is also found heavily in

Gynostemma pentaphyllum (*Gynostemma*, cucurbitaceae) in a form called gypenosides, and ginseng or red ginseng (*Panax Araliaceae*) in a form called quinsenosides. Within these families, this class of chemical compounds is found in various parts of the plant: leaves, stems, roots, bulbs, blossom and fruit. Commercial formulations of plant-derived saponins, *e.g.*, from the soap bark (or soapbark) tree, *Quillaja saponaria*, and those from other sources are available via controlled manufacturing processes, which make them of use as chemical and biomedical reagents

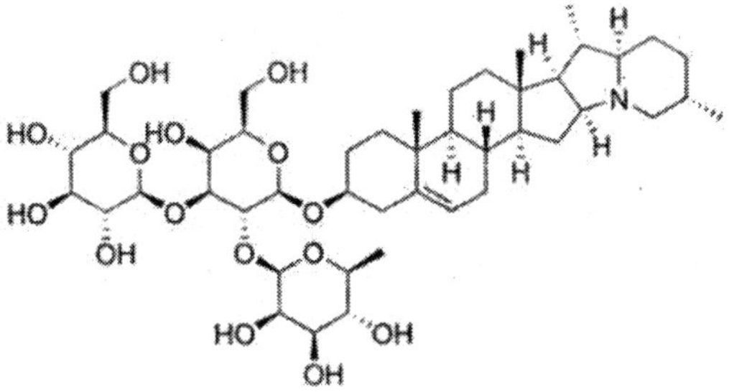

Chemical structure of saponins

Estimation of saponins

- Plant sample (~ 5.0 g) is weighed and was dispersed in 100 ml of 20% ethanol.
- The suspension is heated over a hot water bath for 4 h with continuous stirring at about 55°C. The filtrate and the residue is re-extracted with another 100 ml of 20% ethanol.
- The combined extracts is reduced to 40 ml over water bath at about 90°C.
- The concentrate is transferred into a 250 ml separating funnel and 20 ml of diethyl ether was added and shaken vigorously.
- The aqueous layer was recovered while the ether layer is discarded.
- The purification process was repeated and about 30 ml of n-butanol is added.
- The combined n-butanol extracts is washed twice with 10 ml of 5% aqueous sodium chloride. The remaining solution was heated in a water bath.
- After evaporation, the samples is dried in the oven to a constant weight.
- The saponin content was calculated in percentage.

CHAPTER 16

Plant Growth Hormones

Plant hormones (also known as **phytohormones**) are chemicals that regulate plant growth, which are termed 'plant growth substances'.

Plant hormones are signal molecules produced within the plant, and occur in extremely low concentrations. Hormones regulate cellular processes in targeted cells locally and, when moved to other locations, in other locations of the plant. Hormones also determine the formation of flower, stems, leaves, the shedding of leaves, and the development and ripening of fruit. Plants, unlike animals, lack glands that produce and secrete hormones. Instead, each cell is capable of producing hormones. Plant hormones shape the plant, affecting seed growth, time of flowering, the sex of flowers, senescence of leaves, and fruits. They affect which tissues grow upward and which grow downward, leaf formation and stem growth, fruit development and ripening, plant longevity, and even plant death. Hormones are vital to plant growth, and, lacking them, plants would be mostly a mass of undifferentiated cells. So they are also known as growth factors or growth hormones..

Phytohormones are found not only in higher plants, but in algae too, showing similar functions, and in microorganisms, like fungi and bacteria, but, in this case, they play no hormonal or other immediate physiological role in the producing organism and can, thus, be regarded as secondary metabolites.

The word hormone is derived from Greek, meaning *set in motion*. Plant hormones affect gene expression and transcription levels, cellular division, and growth. They are naturally produced within plants, though very similar chemicals are produced by fungi and bacteria that can also affect plant growth. A large number of related chemical compounds are synthesised by humans. They are used to regulate the growth of cultivated plants, weeds, and *in vitro-*

grown plants and plant cells; these man made compounds are called **Plant Growth Regulators** or **PGRs** for short. Early in the study of plant hormones, "phytohormone" was the commonly used term, but its use is less widely applied now.

Plant hormones are not nutrients, but chemicals that in small amounts promote and influence the growth, development, and differentiation of cells and tissues. The biosynthesis of plant hormones within plant tissues is often diffuse and not always localized. Plants lack glands to produce and store hormones, because, unlike animals which have two circulatory systems (lymphatic and cardiovascular) powered by a heart that moves fluids around the body plants use more passive means to move chemicals around the plant. Plants utilize simple chemicals as hormones, which move more easily through the plant's tissues. They are often produced and used on a local basis within the plant body. Plant cells produce hormones that affect even different regions of the cell producing the hormone.

Hormones are transported within the plant by utilizing four types of movements. For localized movement, cytoplasmic streaming within cells and slow diffusion of ions and molecules between cells are utilized. Vascular tissues are used to move hormones from one part of the plant to another; these include sieve tubes or phloem that move sugars from the leaves to the roots and flowers, and xylem that moves water and mineral solutes from the roots to the foliage.

Not all plant cells respond to hormones, but those cells that do are programmed to respond at specific points in their growth cycle. The greatest effects occur at specific stages during the cell's life, with diminishing effects occurring before or after this period. Plants need hormones at very specific times during plant growth and at specific locations. They also need to disengage the effects that hormones have when they are no longer needed. The production of hormones occurs very often at sites of active growth within the meristems, before cells have fully differentiated. After production, they are sometimes moved to other parts of the plant, where they cause an immediate effect; or they can be stored in cells to be released later. Plants use different pathways to regulate internal hormone quantities and moderate their effects; they can regulate the amount of chemicals used to biosynthesize hormones. They can store them in cells, inactivate them, or cannibalise already-formed hormones by conjugating them with carbohydrates, amino acids, or peptides. Plants can also break down hormones chemically, effectively destroying them. Plant hormones frequently regulate the concentrations of other plant hormones. Plants also move hormones around the plant diluting their concentrations.

The concentration of hormones required for plant responses are very low (10^{-6} to 10^{-5} mol/l). Because of these low concentrations, it has been very difficult to study plant hormones, and only since the late 1970s have scientists been able to start piecing together their effects and relationships to plant physiology. Much of the early work on plant hormones involved studying plants that were genetically deficient in one or involved the use of tissue-cultured plants grown *in vitro* that were subjected to differing ratios of hormones, and the resultant growth compared. The earliest scientific observation and study dates to the 1880s; the determination and observation of plant hormones and their identification was spread-out over the next 70 years.

16.1. Classes of Hormones

In general, it is accepted that there are five major classes of plant hormones, some of which are made up of many different chemicals that can vary in structure from one plant to the next. The chemicals are each grouped together into one of these classes based on their structural similarities and on their effects on plant physiology. Other plant hormones and growth regulators are not easily grouped into these classes; they exist naturally or are synthesized by humans or other organisms, including chemicals that inhibit plant growth or interrupt the physiological processes within plants. Each class has positive as well as inhibitory functions, and most often work in tandem with each other, with varying ratios of one or more interplaying to affect growth regulation. The five major classes are:

Abscisic acid

Abscisic acid (also called ABA) is one of the most important plant growth regulators. It was discovered and researched under two different names before its chemical properties were fully known, it was called *dormin* and *abscicin II*. Once it was determined that the two compounds are the same, it was named abscisic acid. The name "abscisic acid" was given because it was found in high concentrations in newly abscissed or freshly fallen leaves.

This class of PGR is composed of one chemical compound normally produced in the leaves of plants, originating from chloroplasts, especially when plants are under stress. In general, it acts as an inhibitory chemical compound that affects bud growth, and seed and bud dormancy. It mediates changes within the apical meristem, causing bud dormancy and the alteration of the last set of leaves into protective bud covers. Since it was found in freshly abscissed leaves, it was thought to play a role in the processes of natural leaf drop, but further research has disproven this. In plant species from temperate parts of the world, it plays a role in leaf and seed dormancy by inhibiting growth, but, as it is dissipated from seeds or buds, growth begins. In other plants, as ABA levels decrease, growth then commences as gibberellin levels increase. Without ABA, buds and seeds would start to grow during warm periods in winter and be killed when it froze again. Since ABA dissipates slowly from the tissues and its effects take time to be offset by other plant hormones, there is a delay in physiological pathways that provide some protection from premature growth. It accumulates within seeds during fruit maturation, preventing seed germination within the fruit, or seed germination before winter. Abscisic acid's effects are degraded within plant tissues during cold temperatures or by its removal by water washing in out of the tissues, releasing the seeds and buds from dormancy.

In plants under water stress, ABA plays a role in closing the stomata. Soon after plants are water-stressed and the roots are deficient in water, a signal moves up to the leaves, causing the formation of ABA precursors there, which then move to the roots. The roots then release ABA, which is translocated to the foliage through the vascular system and modulates the potassium and sodium uptake within the guard cells, which then lose turgidity, closing the stomata. ABA exists in all parts of the plant and its concentration within any tissue seems to mediate its effects and function as a hormone; its degradation, or more properly catabolism, within the plant affects metabolic reactions and cellular growth and production of other hormones. Plants start life as a seed with high ABA levels. Just before the seed germinates, ABA levels decrease; during germination and early growth of the seedling, ABA levels decrease even more. As plants begin to produce shoots with fully functional leaves, ABA levels begin to increase, slowing down cellular growth in more "mature" areas of the plant. Stress from water or predation affects ABA production and catabolism rates, mediating another cascade of effects that trigger specific responses from targeted cells. Scientists are still piecing together the complex interactions and effects of this and other phytohormones.

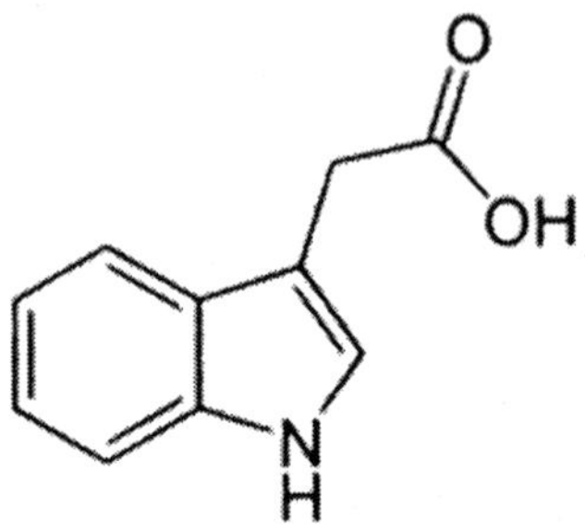

O
OH
N
H

Auxins

The Auxin-indole-3-acetic acid

Auxins are compounds that positively influence cell enlargement, bud formation and root initiation. They also promote the production of other hormones and in conjunction with Cytokinins, they control the growth of stems, roots, and fruits, and convert stems into flowers. Auxins were the first class of growth regulators discovered. They affect cell elongation by altering cell wall plasticity. They stimulate cambium, a subtype of meristem cells, to divide and in stems cause secondary xylem to differentiate. Auxins act to inhibit the growth of buds lower down the stems (apical dominance), and also to promote lateral and adventitious root development and growth. Leaf abscission is initiated by the growing point of a plant ceasing to produce auxins. Auxins in seeds regulate specific protein synthesis, as they develop within the flower after pollination, causing the flower to develop a fruit to contain the developing seeds. Auxins are toxic to plants in large concentrations; they are most toxic to dicots and less so to monocots. Because of this property, synthetic auxin herbicides including 2,4-D and 2,4,5-T have been developed and used for weed control. Auxins, especially 1-napthaleneacetic acid (NAA) and Indole-3-butyric acid (IBA), are also commonly applied to stimulate root growth when taking cuttings of plants. The most common auxin found in plants is indole-3-acetic acid or IAA. The correlation of auxins and cytokynins in the plants is a constant (A/C = const.)

Cytokinin

The cytokinin zeatin, the name is derived from *Zea*, in which it was first discovered in immature kernels. Cytokinines or CKs are a group of chemicals that influence cell division and shoot formation. They were called kinins in the past when the first cytokinins were isolated from yeast cells. They also help delay senescence of tissues, are responsible for mediating auxin transport throughout the plant, and affect inter nodal length and leaf growth. They have a highly synergistic effect in concert with auxins, and the ratios of these two groups of plant hormones affect most major growth periods during a plant's lifetime. Cytokinins counter the apical dominance induced by auxins; they in conjunction with ethylene promote abscission of leaves, flower parts, and fruits. The correlation of auxins and cytokinins in the plants is a constant (A/C = const.).

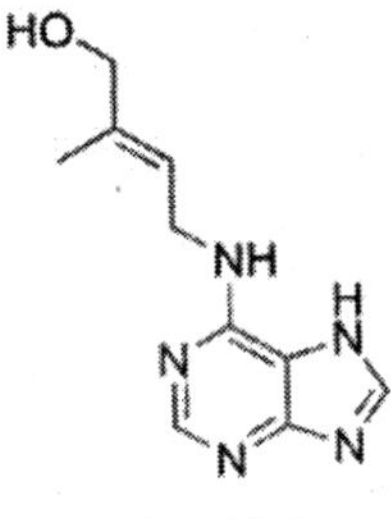

Cytokinins

Ethylene

Ethylene is a gas that forms through the breakdown of methionine, which is in all cells. Ethylene has very limited solubility in water and does not accumulate within the cell but diffuses out of the cell and escapes out of the plant. Its effectiveness as a plant hormone is dependent on its rate of production versus its rate of escaping into the atmosphere. Ethylene is produced at a faster rate in rapidly growing and dividing cells, especially in darkness. New growth and newly germinated seedlings produce more ethylene than can escape the plant, which leads to elevated amounts of

H H
C=C
H H

Ethylene

ethylene, inhibiting leaf expansion. As the new shoot is exposed to light, reactions by phytochrome in the plant's cells produce a signal for ethylene production to decrease, allowing leaf expansion. Ethylene affects cell growth and cell shape; when a growing shoot hits an obstacle while underground, ethylene production greatly increases, preventing cell elongation and causing the stem to swell. The resulting thicker stem can exert more pressure against the object impeding its path to the surface. If the shoot does not reach the surface and the ethylene stimulus becomes prolonged, it affects the stem's natural genotropic response, which is to grow upright, allowing it to grow around an object. Studies seem to indicate that ethylene affects stem diameter and height: When stems of trees are subjected to wind, causing lateral stress, greater ethylene production occurs, resulting in thicker, more sturdy tree trunks and branches. Ethylene affects fruit-ripening: Normally, when the seeds are mature, ethylene production increases and builds-up within the fruit, resulting in a climacteric event just before seed dispersal. The nuclear protein Ethylene Insensitive2 (EIN2) is regulated by ethylene production, and, in turn, regulates other hormones including ABA and stress hormones.

Gibberellins

Gibberellins or GAs, include a large range of chemicals that are produced naturally within plants and by fungi. They were first discovered when Japanese researchers, including Eiichi Kurosawa, noticed a chemical produced by a fungus called *Giberella fujikuroi* that produced abnormal growth in rice plants. Gibberellins are important in seed germination, affecting enzyme production that mobilizes food production used for growth of new cells. This is done by modulating chromosomal transcription. In grain (rice, wheat, corn, *etc.*) seeds, a layer of cells called the aleurone layer wraps around the endosperm tissue. Absoption of water by the seed causes production of GA. The GA is transported to the aleurone layer, which responds by producing enzymes that break down stored food reserves within the endosperm, which are utilized by the growing seedling. GAs produce bolting of rosette-forming plants, increasing internodal length. They promote flowering, cellular division, and in seeds growth after germination. Gibberellins also reverse the inhibition of shoot growth and dormancy induced by ABA.

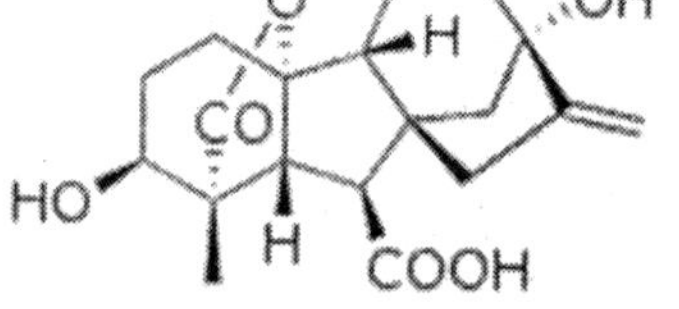

HO, O, CO, H, OH, H, COOH

Gibberellin A1

Other known hormones

Other identified plant growth regulators include:

- **Brassinosteroids**, are a class of polydroxysteroids, a group of plant growth regulators. Brassinosteroids have been recognized as a sixth class of plant hormones, which stimulate cell elongation and division, gravitropism, resistance to stress, and xylem differentiation. They inhibit root growth and leaf abscission. Brassinolide was the first identified brassinosteroid and was isolated from extracts of rapeseed (*Brassica napus*) pollen in 1979.
- **Salicylic acid** — activates genes in some plants that produce chemicals that aid in the defense against pathogenic invaders.
- **Jasmonates** — are produced from fatty acids and seem to promote the production of defense proteins that are used to fend off invading organisms. They are believed to also have a role in seed germination, and affect the storage of protein in seeds, and seem to affect root growth.
- **Plant peptide hormones** — encompasses all small secreted peptides that are involved in cell-to-cell signaling. These small peptide hormones play crucial roles in plant growth and development, including defense mechanisms, the control of cell division and expansion, and pollen self-incompatibility.
- **Polyamines** — are strongly basic molecules with low molecular weight that have been found in all organisms studied thus far. They are essential for plant growth and development and affect the process of mitosis and meiosis.
- **Nitric oxide (NO)** — serves as signal in hormonal and defense responses (*e.g.* stomatal closure, root development, germination, nitrogen fixation, cell death, stress response). NO can be produced by a yet undefined NO synthase, a special type of nitrite reductase, nitrate reductase, mitochondrial cytochrome c oxidase or non enzymatic processes and regulate plant cell organelle functions (*e.g.* ATP synthesis in chloroplasts and mitochondria).
- **Strigolactones**, implicated in the inhibition of shoot branching.
- **Kamikins,** a group of plant growth regulators found in the smoke of burning plant material that have the ability to stimulate the germination of seeds.

16.2. Colorimetric Estimation of Indole Acetic Acid

A sensitive colorimetric method for estimation of indole acetic acid was developed by Anthony and Street (1970) and this method can estimate IAA down to 1-2 µg/5 ml.

Materilas

- Stock solution of IAA in methanol is prepared and was diluted with distilled water to give test solution containing 1 – 50 µg/ml.
- Ehrlich's reagent : also known as the "DMAB test" is prepared by dissolving 2.0 g of p-dimethylaminobenjaldehyde (DMAB) in 100 ml of 2.5 N hydrochloric acid. It is best prepared fresh.
- Trichloroacetic acid (TCA) – 25% (w/v) in water.

Procedure

- The standard reaction mixture consists of 1.0 ml of indole acetic acid and added with 2.0 ml of TCA solution and 2.0 ml of Ehrlich's reagent.
- The reaction mixture is incubated at room temperature for 30 minutes.
- After incubation period, optical density was measured at 540 nm against a reagent blank.

Reading

Anthony, A. and Street, H.E. (2006). A colorimetric method for estimation of certain indoles. *New Phytol.* **69:** 47-50.

16.3. Colorimetric Estimation of Gibberellic Acid

Gibberellic acid reacts with phosphomolybdic acid reagent to give stable colour, intensity of which is proportional to the amount of gibberellic acid present when measured at 780 nm. The temperature of heating the gibberellic acid with phosphomolybdic acid reagent is very critical. At boiling temperature longer periods are required for maximum colour development. The effect of several variables on the reaction was thoroughly investigated and conditions for reproducibility established. Indoleacetic acid, 2,4-dichlorophenoxyacetic acid and kinetin failed to give the typical wine-red colour when heated at 100° for five minutes under identical conditions.

Material

- Phosphomolybdic acid reagent – 35 g of molybdic acid and 5 g of sodium tungstate were taken in 1 liter beaker. 200 ml of 10% NaOH solution is added to it along with 200 ml of distilled water. The content is boiled vigorously for 40 minutes and then cooled and to this mixture 125 ml of phosphoric acid (85%) is added and finally the volume is made to 500 ml with distilled water.

Procedure

- 5.0 ml of plant extract (methanol extract in 8:2 v/v methanol and water mixture) is taken in a 25 ml volumetric flask and 15.0 ml of phosphomolybdic reagent is added and mixed thoroughly.
- The flasks are kept in a boiling water bath for 60 minutes. Zero time sample is drawn after keeping the flasks in boiling water bath for 2 minures.
- After 60 minutes the flasks are kept in icecold water and volume is amde to 25 ml with distilled water.
- Measure the optical density at 780 nm against distilled water
- To draw standard curve the same process is repeated with known quantity of gibberelic acid and the ODs are plotted against gibberelic acid concentration.

Reading

Graham, H.D. and Handerson, J.H.M. (1961). Reaction of gibberellic acid & gibberellins with Folin-Wu phosphomolybdic acid reagent & its use for quantitative assay. *Plant Physiol.*, **36**(4): 405–408.

16.4. Estimation of Abscisic Acid

Kinetin is a type of Cytokinins, a class of plant hormone that promotes cell division. Kinetin is often used in plant tissue culture for inducing formation of callus (in conjunction with auxin) and to regenerate shoot tissues from callus (with lower auxin concentration). For a long time, it was believed that kinetin was an artifact produced from the deoxyadenosine residues in DNA, which degrade on standing for long periods or when heated during the isolation procedure. Therefore, it was thought that kinetin does not occur naturally, but, since 1996, it has been shown by several researchers that kinetin exists naturally in the DNA of cells of almost all organisms tested so far, including human and various plants. The mechanism of production of kinetin in DNA is thought to be via the production of furfural, an oxidative damage product of deoxyribose sugar in DNA and its quenching by the adenine base's converting it into N^6-furfuryladenine, kinetin.

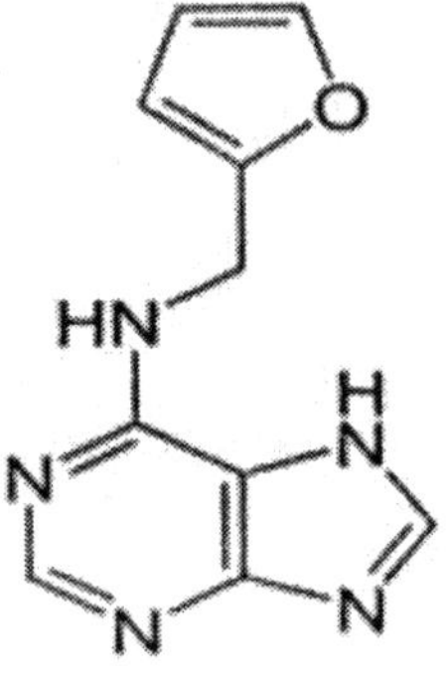

N^6-furfuryladenine

Kinetin is currently sold commercially under the trade name "Bonide Tomato and Blossom Set Spray", and can be used to increase yields of various fruits and vegetables, produce seedless fruits, and increase 'budding' of various herbs. Kinetin is also widely used in producing new plants from tissue cultures.

Determination of ABA

Powdered plant material (1 g) was extracted with 80% aqueous methanol (three replicate extractions) and the extracts were passed through a C18 reversed phase prepacked column For ABA, the effluent was concentrated in vacuo, dissolved in 3 ml double distilled water, acidified to pH 2.5 with 1 M HCl and partitioned three times into ethyl acetate (3ml). The organic solvent was evaporated to dryness under a N_2 stream and samples were redissolved in 2 ml of 5% methanol in 0.1 M acetic acid. Separation of ABA in a 1 ml aliquot of the extract was performed by HPLC on a reverse phase Nucleosil 120–5 µm C_{18} column (250×10 mm, Machery Nagel, Düren, Germany) using a linear gradient from 5% methanol in 0.1 M acetic acid to 95% methanol. The fractions containing ABA were collected and the quantitative determination was performed by enzyme immunoassay (ELISA). Monoclonal antibodies for ABA were used for analyses according to a standard procedure.

Alternative Method for Estimating abscisic acid

Procedure

For extraction of ABA, leaves were homogenized with a pestle and mortar in extraction solution (80% methanol containing 2% glacial acetic acid). To remove plant pigments and other non-polar compounds which could interfere in the immunoassay extracts were first passed through a polyvinylpyrrolidone column and C18 cartridges. The eluates were concentrated to dryness by vacuum evaporation and resuspended in Tris-buffered saline before enzyme-linked immunosor-bent assay (ELISA). ABA was quantified by ELISA. The ABA immunoassay detection kit was specific for (+)-ABA. By evaluating [^{3}H] ABA recovery, [^{3}H] ABA loss was <3% by the method described here. The content of ABA was expressed on the fresh weight basis.

Reading

Chun-Hsin LIU, Yun-Yang CHAO, and Ching Huei KAO (2012). Abscisic acid is an inducer of hydrogen peroxide production in leaves of rice seedlings grown under potassium deficiency. *Botanical Studies*, **53:** 229-237

16.5. Estimation of Ascorbic Acid Oxidases (E.C. 1.10.3.3) Activity

In enzymology, a L-ascorbate oxidase (E.C. 1.10.3.3) is an enzyme that catalyses the chemical reaction

$$\textbf{2 L-ascorbate} + \textbf{O}_2 \rightleftharpoons \textbf{2 dehydroascorbate} + \textbf{2 H}_2\textbf{O}$$

Thus, the two substrates of this enzyme are L-ascorbate and O_2, whereas its two products are dehydroascorbate and H_2O.

This enzyme belongs to the family of oxidoreductases, specifically those acting on diphenols and related substances as donor with oxygen as acceptor. The systematic name of this enzyme class is L-ascorbate:oxygen oxidoreductase. Other names in common use include ascorbase, ascorbic acid oxidase, ascorbate oxidase, ascorbic oxidase, ascorbate dehydrogenase, L-ascorbic acid oxidase, AAO, L-ascorbate:O_2• oxidoreductase, and AA oxidase. This enzyme participates in ascorbate metabolism. It employs one cofactor, copper. The role of this enzyme is to regulate the levels of oxidised and reduced glutathione and NADH

Materials

- Phosphate buffer 0.1 M (pH5.6 and 6.5 seperately)
- Substrate solution: Dissolve 8.8 mg of ascorbic acid in 300 ml of phosphate buffer (pH5.6).
- Enzyme extract: Macerate one part of plant tissue with five part of (w/v) of 0.1 M phosphate buffer (pH 6.5) in a homogenizer. Centrifuge the homogenate at 3000 x g for 15 minutes. Use the supernatant as enzyme source.

Procedure

- Take 3.0 ml of substrate solution to each of sample and referece cuvettes of a spectrophotometer.
- Add 0.1 ml of enzyme extract to the reference cuvette.

- Measure the change in optical density at 265 nm in 30 seconds intervals for 5 minutes.

Calculation

- From the linear phase of reaction, compute the change in absorption per minute.
- Alternatively, activity may be expressed in enzyme units. One enzyme unit is equivalent to 0.405 μ mol oxygen per minute. Ascorbic acid requires only one atom of oxygen per mole oxidised. Therefore, 0.81 μ mol of ½ oxygen per minute is utilised per enzyme unit.
- Ascorbic acid has an $E^{1\%}_{1\ cm}$ of 760 at 265 nm. From this it is calculated that ascorbic acid has an absorbance of 4.4 per μ mol in a 3 ml volume.
- Hence, one enzyme unit (0.81 μ mol ½ oxygen per minute) would would be equivalent to an absorbance change of 3.58 per minute.

Reading

Oberbacher, M.F.and Vines, H.M. (1963). Spectrophotometric assay of ascorbic acid oxidase. *Nature (London)*, **197:** 1203-1204.

16.6. Estimation of Indole Acetic Acid Oxidase Activity

Indole acetic acid (IAA) is best known naturally occuring plant auxin. It participate in controlling many phases of plant growth and differentiation. Levels of free IAA are in turn egulated *via* synthesis, binding, esterification and enzyme degradation. Indole acetic acid oxidase (IAAO) is the enzyme involved in the catabolic degradtion of IAA to 3-methylene oxindole. The IAAO activity is determined by measuring residual IAA following dark incubation with shaking at 30^0 C. The IAA is determined by Salkowski reaction. Since monophenols act as cofactors of IAAO, and 0- and p-dihydroxy and polyphenols acrs as inhibitors of this enzyme, the monophenolic compounds, para-coumaric acid is added in the enzyme assay for activation.

Materials

- Phosphate buffer 0.071 M pH 6.2.
- Para coumaric acid solution – Dissolve 25 mg p-coumaric acid in 50 ml of water.

- IAA solution- Dissolve 10 mg IAA in 40 ml water.
- Manganese chloride solution – Dissolve 118 mg of $MnCl_2.4H_2O$ in 20 ml of water.
- Perchloric acid 5 M.
- Ferric nitrate 0.1 M – Dissolve 24.18 g of $Fe(NO)_3$ in 100 ml of water.

Procedure

- Prepare acetone powder from frozen plant tissue (25 g) by blending in two successive 100 ml aliquots of ice dold acetone. The homogenates are filtered through Whatman No. 1 filter paper. The homogenate was air dried until traces of acetone is removed and freeze stored.
- One g of acetone powder is grinded in two successive 20 ml aliquots of 25 mM phosphate buffer (pH 6.2) in a chilled mortar. The extracts are collected and filtered throgh Whatman No. 1 filter paper. The extracts are pooled and dilute to 50 ml with phosphate buffer. This is used as enzyme extract.
- To 2.0 ml of phosphate buffer (pH 6.2) 1.0 ml each of para coumaric acid solution and manganese chloride solution is added and lastly 2.0 ml of enzyme extrac is also added.
- The reaction is started by the addition of 4.0 ml of IAA solution and allowed to proceed in dark with shaking at 30° C.
- Withdraw 2 ml of assay mixture after 0, 50 minutes of incubation and add 5.2 ml of perchloric acid and 0.5 ml of ferric nitrate solution and the mixture is diluted to 10.0 ml with water.
- After incubating the reaction mixture in dark for 60 minutes measure the absorbance at 535 nm.
- Protein cntent of the enzyme extract is estimated by Lowry method.
- The enzyme activity is expressed as micromole IAA oxidised per min/ per mg protein.

Reading

Byrant, S.D. and Lane, F.E. (1979). Indole-3-acetic Acid Oxidase from Peas: I. Occurrence and Distribution of Peroxidative and Nonperoxidative Forms. *Plant Physiol.*, **63** (4): 696-699.

CHAPTER 17

Nucleic Acids

Nucleic acids are polymeric macromolecules, or large biological molecules, essential for all known forms of life. Nucleic acids, which include DNA (deoxyribonucleic acid) and RNA (ribonucleic acid), are made from monomers known as nucleotides. Each nucleotide has three components: a 5-carbon sugar, a phosphate group, and a nitrogenous base. If the sugar is deoxiribose, the polymer is DNA. If the sugar is ribose, the polymer is RNA. Together with proteins, nucleic acids are the most important biological macromolecules; each is found in abundance in all living things, where they function in encoding, transmitting and expressing genetic information—in other words, information is conveyed through the nucleic acid sequence, or the order of nucleotides within a DNA or RNA molecule. Strings of nucleotides strung together in a specific sequence are the mechanism for storing and transmitting hereditary, or genetic, information via protein synthesis.

The term *nucleic acid* is the overall name for DNA and RNA, members of a family of biopolymers, and is synonymous with polynucleotide. Nucleic acids were named for their initial discovery within the nucleus, and for the presence of phosphate groups (related to phosphoric acid). Although first discovered within the nucleus of eukaryotic cells, nucleic acids are now known to be found in all life forms as well as some nonliving entities, including within bacteria, archaea, mitochondria, chloroplasts, viruses and viroids. All living cells contain both DNA and RNA (except some cells such as mature red blood cells), while viruses contain either DNA or RNA, but usually not both. The basic component of biological nucleic acids is the nucleotide, each of which contains a pentose sugar (ribose or deoxyribose), a phosphate group, and a nucleobase. Nucleic acids are also generated within the laboratory, through the use of enzymes (DNA and RNA polymerases) and by solid-phase chemical synthesis. The chemical methods also enable the generation of altered nucleic acids that are not found in nature, for example peptide nucleic acids.

Nucleic acids are generally very large molecules. Indeed, DNA molecules are probably the largest individual molecules known. Well-studied biological nucleic acid molecules range in size from 21 nucleotides (small interfering RNA) to large chromosomes (human chromosome is a single molecule that contains 247 million base pairs).

In most cases, naturally occurring DNA molecules are double-stranded and RNA molecules are single-stranded. There are numerous exceptions, however, some viruses have genomes made of double-stranded DNA and other viruses have single stranded DNA genomes, and, in some circumstances, nucleic acid structures with three or four strands can form.

Nucleic acids are linear polymers (chains) of nucleotides. Each nucleotide consists of three components: a purine or pyrimidine nucleobase (sometimes termed *nitrogenous base* or simply *base*), a pentose sugar, and a phosphate group. The substructure consisting of a nucleobase plus sugar is termed a nucleoside. Nucleic acid types differ in the structure of the sugar in their nucleotides - DNA contains 2'-deoxyribose while RNA contains ribose (where the only difference is the presence of a hydroxyl group). Also, the nucleobases found in the two nucleic acid types are different: adenine, cytosine, and guanine are found in both RNA and DNA, while thymine occurs in DNA and uracil occurs in RNA.

The sugars and phosphates in nucleic acids are connected to each other in an alternating chain (sugar-phosphate backbone) through phosphodiester linkages. In conventional nomenclature, the carbons to which the phosphate groups attach are the 3'-end and the 5'-end carbons of the sugar. This gives nucleic acids directionality, and the ends of nucleic acid molecules are referred to as 5'-end and 3'-end. The nucleobases are joined to the sugars via an N-glycosidic linkage involving a nucleobase ring nitrogen (N-1 for pyrimidines and N-9 for purines) and the 1' carbon of the pentose sugar ring.

Non-standard nucleosides are also found in both RNA and DNA and usually arise from modification of the standard nucleosides within the DNA molecule or the primary (initial) RNA transcript. Transfer RNA (tRNA) molecules contain a particularly large number of modified nucleosides.

17.1. Types of Nucleic Acids

Deoxyribonucleic acid

Deoxyribonucleic acid (DNA) is a molecule that encodes the genetic instructions used in the development and functioning of all known living organisms and many viruses. DNA is a nucleic acid; alongside proteins and carbohydrates, nucleic acids compose the three major macromolecules essential for all known forms of life. Most DNA molecules consist of two biopolymer strands coiled around each other to form a double helix. The two DNA strands are known as polynucleotides since they are composed of simpler units called nucleotides. Each nucleotide is composed of a nitrogen-containing nucleobase, either guanine (G), adenine (A), thymine (T), or cytosine (C), as well as a monosaccharide sugar called deoxyribose and a phosphate group. The nucleotides are joined to one another in a chain by covalent bonds between the sugar of one nucleotide and the phosphate of the next, resulting in an alternating sugar-phosphate backbone. According to base pairing rules (A with T and C with G), hydrogen bonds bind the nitrogenous bases of the two separate polynucleotide strands to make double-stranded DNA.

DNA is well-suited for biological information storage. The DNA backbone is resistant to cleavage, and both strands of the double-stranded structure store the same biological information. Biological information is replicated as the two strands are separated. A significant portion of DNA (more than 98% for humans) is non-coding, meaning that these sections do not serve a function of encoding proteins.

The two strands of DNA run in opposite directions to each other and are therefore anti-parallel, one backbone being 3' (three prime) and the other 5' (five prime). This refers to the direction the 3rd and 5th carbon on the sugar molecule is facing. Attached to each sugar is one of four types of nucleobases (informally, *bases*). It is the sequence of these four nuclèobases along the backbone that encodes biological information. Under the genetic code, RNA strands are translated to specify the sequence of amino acids within proteins.

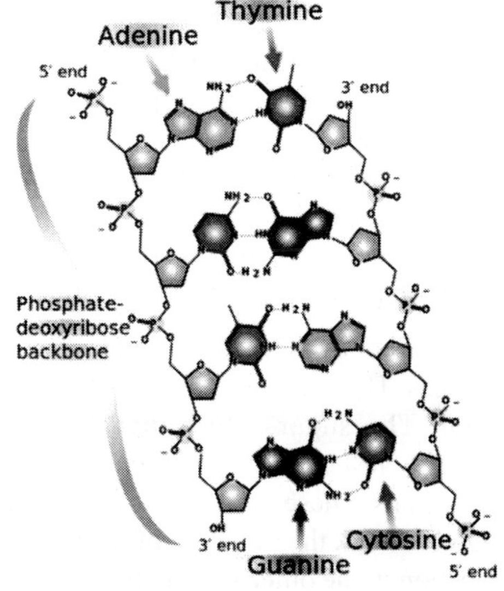

These RNA strands are initially created using DNA strands as a template in a process called transcription.

Within cells, DNA is organized into long structures called chromosomes. During cell division these chromosomes are duplicated in the process of DNA replication, providing each cell its own complete set of chromosomes. Eukaryotic organisms (animals, plants, fungi, and protists) store most of their DNA inside the cell nucleus and some of their DNA in organelles, such as mitochondria or chloroplasts. In contrast, prokaryotes (bacteria and archaea) store their DNA only in the cytoplasm. Within the chromosomes, chromatin proteins such as histones compact and organize DNA. These compact structures guide the interactions between DNA and other proteins, helping control which parts of the DNA are transcribed.

DNA is a long polymer made from repeating units called nucleotides. DNA was first identified and isolated by Friendrich Miescher and the double helix structure of DNA was first discovered by James Watson and Francis Crick. The structure of DNA of all species comprises two helical chains each coiled round the same axis, and each with a pitch of 34 ångströms (3.4 namo metres) and a radius of 10 ångströms (1.0 nano metres). According to another study, when measured in a particular solution, the DNA chain measured 22 to 26 ångströms wide (2.2 to 2.6 nano metres), and one nucleotide unit measured 3.3 Å (0.33 nm) long. Although each individual repeating unit is very small, DNA polymers can be very large molecules containing millions of nucleotides. For instance, the largest human chromosomes, chromosome number 1, consists of approximately 220 million base pairs and is 85 nm long.

In living organisms DNA does not usually exist as a single molecule, but instead as a pair of molecules that are held tightly together. These two long strands entwine like vines, in the shape of a double helix. The nucleotide repeats contain both the segment of the backbone of the molecule, which holds the chain together, and a nucleobase, which interacts with the other DNA strand in the helix. A nucleobase linked to a sugar is called a nucleoside and a base linked to a sugar and one or more phosphate groups is called a nucleotide. A polymer comprising multiple linked nucleotides (as in DNA) is called a polynucleotide.

The backbone of the DNA strand is made from alternating phosphate and sugar residues. The sugar in DNA is 2-deoxyribose, which is a pentose (five-carbon) sugar. The sugars are joined together by phosphate groups that form phosphodiester bonds between the third and fifth carbon atoms of adjacent sugar rings. These asymmetric bonds mean a strand of DNA has a direction. In a double helix the direction of the nucleotides in one strand is opposite to their direction in the other strand: the strands are *antiparallel*. The asymmetric ends of DNA strands are called the 5' (*five prime*) and 3' (*three prime*) ends, with the 5' end having a terminal phosphate group and the 3' end a terminal hydroxyl group. One major difference between DNA and RNA is the sugar, with the 2-deoxyribose in DNA being replaced by the alternative pentose sugar ribose in RNA.

The DNA double helix is stabilized primarily by two forces: hydrogen bonds between nucleotides and base-stacking interactions among aromatic nucleobases. In the aqueous environment of the cell, the conjugated Π bonds of nucleotide bases align perpendicular to the axis of the DNA molecule, minimizing their interaction with the salvation shell and therefore, the Gibbs free energy. The four bases found in DNA are adenine (abbreviated A), cytosine (C), quanine (G) and thymine (T). These four bases are attached to the sugar/ phosphate to form the complete nucleotide, as shown for adenosine monophosphate.

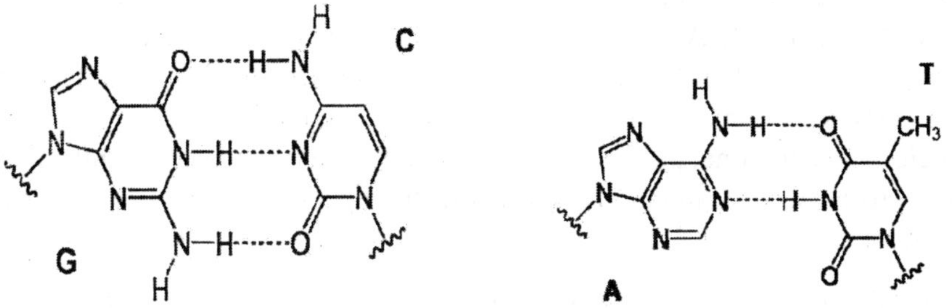

A GC base pair with three hydrogen bonds. and, an AT base pair with two hydrogen bonds. Non-covalent hydrogen bonds between the pairs are shown as dashed lines.

Deoxyribonucleic acid (DNA) is a nucleic acid containing the genetic instructions used in the development and functioning of all known living organisms (with the exception of RNA viruses). The DNA segments carrying this genetic information are called genes. Likewise, other DNA sequences have structural purposes, or are involved in regulating the use of this genetic information. Along with RNA and proteins, DNA is one of the three major macromolecules that are essential for all known forms of life. DNA consists of two long polymers of simple units called nucleotides, with backbones made of sugars and phosphate groups joined by ester bonds. These two strands run in opposite directions to each other and are, therefore, anti-parallel. Attached to each sugar is one of four types of molecules called nucleobases (informally, bases). It is the sequence of these four nucleobases along the backbone that encodes information. This information is read using the genetic code, which specifies the sequence of the amino acids within proteins. The code is read by copying stretches of DNA into the related nucleic acid RNA in a process called transcription. Within cells, DNA is organized into long structures called chromosomes. During cell division these chromosomes are duplicated in the process of DNA replication, providing each cell its own complete set of chromosomes. Eukaryotic organisms (animals, plants, fungi, and protists) store most of their DNA inside the cell nucleus and some of their DNA in organelles, such as mitochondria or chloroplasts. In contrast, prokaryotes (bacteria and archaea) store their DNA only in the cytoplasm. Within the chromosomes, chromatin proteins such as histones compact and organize DNA. These compact structures guide the interactions between DNA and other proteins, helping control which parts of the DNA are transcribed.

Ribonucleic acid

Ribonucleic acid (RNA) is a ubiquitous family of large biological molecules that perform multiple vital roles in the coding, decoding, regulation, and expression of genes. Together with DNA, RNA comprises the nucleic acids, which, along with proteins, constitute the three major macromolecules essential for all known forms of life. Like DNA, RNA is assembled as a chain of nucleotides, but is usually single-stranded. Cellular organisms use messenger RNA (mRNA) to convey genetic information (often notated using the letters G, A, U, and C for the nucleotides guanine, adenine, uracil and cytosine) that directs synthesis of specific proteins, while many viruses encode their genetic information using an RNA genome.

Some RNA molecules play an active role within cells by catalyzing biological reactions, controlling gene expression, or sensing and communicating responses to cellular signals. One of these active processes is protein synthesis, a universal function whereby mRNA molecules direct the assembly of proteins on ribosomes. This process uses transfer RNA (tRNA) molecules to deliver amino acids to the ribosome, where ribosomal RNA (rRNA) links amino acids together to form proteins

The chemical structure of RNA is very similar to that of DNA, but differs in three main ways:

- Unlike double-stranded DNA, RNA is a single-stranded molecule in many of its biological roles and has a much shorter chain of nucleotides. However, RNA can, by complementary base pairing, form intrastrand double helixes, as in tRNA.
- While DNA contains *deoxyribose*, RNA contains *ribose* (in deoxyribose there is no hydroxyl group attached to the pentose ring in the 2' position). These hydroxyl groups make RNA less stable than DNA because it is more prone to hydrolysis.

- The complementary base to adenine is not thymine, as it is in DNA, but rather uracil, which is an unmethylated form of thymine.

Like DNA, most biologically active RNAs, including mRNA, tRNA, rRNA, snRNA, and other non-coding RNAs, contain self-complementary sequences that allow parts of the RNA to fold and pair with itself to form double helices. Analysis of these RNAs has revealed that they are highly structured. Unlike DNA, their structures do not consist of long double helices but rather collections of short helices packed together into structures akin to proteins. In this fashion, RNAs can achieve chemical catalysis, like enzymes. For instance, determination of the structure of the ribosome—an enzyme that catalyzes peptide bond formation—revealed that its active site is composed entirely of RNA.

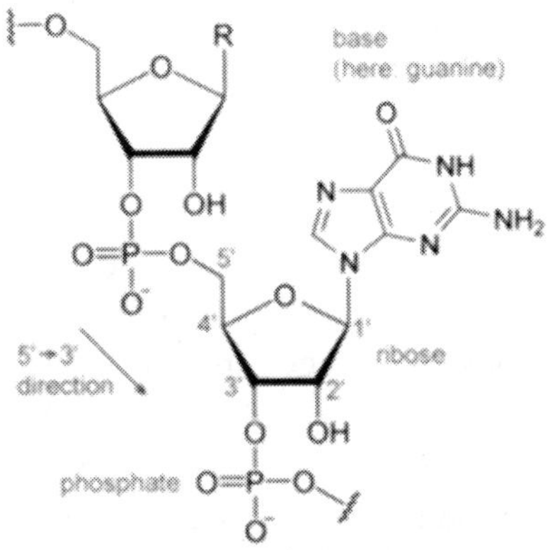

Chemical structure of RNA

Each nucleotide in RNA contains a ribose sugar, with carbons numbered 1' through 5'. A base is attached to the 1' position, in general, adenine (A), cytosine (C), guanine (G), or uracil (U). Adenine and guanine are purines, cytosine and uracil are pyrimidines. A phosphate group is attached to the 3' position of one ribose and the 5' position of the next. The phosphate groups have a negative charge each at physiological pH, making RNA a charged molecule (polyanion). The bases form hydrogen bonds between cytosine and guanine, between adenine and uracil and between guanine and uracil. However, other interactions are possible, such as a group of adenine bases binding to each other in a bulge, or the GNRA tetraloop that has a guanine–adenine base-pair. An important structural feature of RNA that distinguishes it from DNA is the presence of a hydroxyl group at the 2' position of the ribose sugar. The presence of this functional group causes the helix to adopt the A-form geometry rather than the B-form most commonly observed in DNA. This results in a very deep and narrow major groove and a shallow and wide minor groove. A second consequence of the presence of the 2'-hydroxyl group is that in conformationally flexible regions of an RNA molecule (that is, not involved in formation of a double helix), it can chemically attack the adjacent phosphodiester bond to cleave the backbone. There are several types of RNAs is present in cell of any living organism and they differ in their function.

Types of RNA.

Types	Abbriviation	Function	Distribution
Messenger RNA	mRNA	Codes of protein	All organism
Ribosomal RNA	rRNA	Translation	All organism
Signal recognition particle RNA	7SL RNA or SRP RNA	Membrane integration	All organism
Transfer RNA	tRNA	Translation	All organism
Transfer messenger RNA	tmRNA	Rescuing stalled ribosomes	Bacteria

RNA is transcribed with only four bases (adenine, cytosine, guanine and uracil), but these bases and attached sugars can be modified in numerous ways as the RNAs mature. Pseudourinine (Ψ), in which the linkage between uracil and ribose is changed from a C–N bond to a C–C bond, and ribothymidine (T) are found in various places (the most notable ones being in the TψC loop of tRNA). Another notable modified base is hypoxanthine, a deaminated adenine base whose nucleoside is called inosine (I). Inosine plays a key role in the wobble hypothesis of the genetic code.

There are more than 100 other naturally occurring modified nucleosides, The greatest structural diversity of modifications can be found in tRNA, while pseudouridine and nucleosides with 2'-O-methylribose often present in rRNA are the most common. The specific roles of many of these modifications in RNA are not fully understood. However, it is notable that, in ribosomal RNA, many of the post-transcriptional modifications occur in highly functional regions, such as the peptidyl transferase center and the subunit interface, implying that they are important for normal function.

Ribonucleic acid (RNA) functions in converting genetic information from genes into the amino acid sequences of proteins. The three universal types of RNA include transfer RNA (tRNA), messenger RNA (mRNA), and ribosomal RNA (rRNA). Messenger RNA acts to carry genetic sequence information between DNA and ribosomes, directing protein synthesis. Ribosomal RNR is a major component of the ribosome, and catalyzes peptide bond formation. Transfer RNA serves as the carrier molecule for amino acids to be used in protein synthesis, and is responsible for decoding the mRNA. In addition, many other classes of RNA are now known.

Plasmids

A plasmid is a small DNA molecule that is physically separate from, and can replicate independently of, chromosomal DNA within a cell. Most commonly found as small circular, double-stranded DNA molecules in bacteria, plasmids are sometimes present in archaea and eukaryotic organisms. In nature, plasmids carry genes that may benefit survival of the organism (*e.g.* antibiotic resistance), and can frequently be transmitted from one bacterium to another (even of another species) via horizontal gene transfer. Artificial plasmids are widely used as vectors in molecular cloning, serving to drive the replication of recombinant DNA sequences within host organisms. The term *plasmid* was first introduced in 1952. Plasmid sizes vary from 1 to over 1,000 kbp. The number of identical plasmids in a single cell can range anywhere from one to thousands under some circumstances. Plasmids can be found in all three major domains: Archaea, Bacteria, and Eukarya.

Plasmids are considered *replicons*, capable of replicating autonomously within a suitable host. However, plasmids, like viruses, are not considered by some to be a form of life. Plasmids can be transferred between bacterial hosts through a process known as bacterial conjugation. Because conjugation is a mechanism of horizontal gene transfer, plasmids can be considered part of the mobilome. Unlike viruses (which encase their genetic material in a protective protein coat called a capsid), plasmids are "naked" DNA and do not encode genes necessary to encase the genetic material for transfer to a new host. However, some classes of plasmids encode the conjugative "sex" plus necessary for their own transfer. Plasmid host-to-host transfer can also require changes in incipient host gene expression allowing the intentional uptake of the genetic element by transformation. Microbial transformation with plasmid DNA is neither parasitic nor symbiotic in nature, because each implies the presence of an independent species living in a detrimental or commensal state with the host organism. Rather, plasmids provide a mechanism for horizontal gene transfer within a population of microbes and typically provide a selective advantage under a given environmental state. Plasmids may carry genes that provide resistance to naturally occurring antibiotics in a compe-titive environmental niche, or the proteins produced may act as toxins under similar circumstances. Plasmids can also provide bacteria with the ability to fix nitrogen or to degrade recalcitrant organic compounds that provide an advantage when nutrients are scarce.

17.2. Extraction of Plant DNA

Plant DNA extraction has unique challenges that require procedures specifically designed to deal with carbohydrates, phenolics, and other compounds abundant in plant tissues. Plant cell walls can be very difficult to disrupt, and lysates often contain significant amounts of compounds such as tannins, phenolics, and complex polysaccharides that can affect DNA quality and inhibit downstream reactions. To make DNA isolation from plant samples easier while achieving highyield and high-purity results a number of methods are available.

The efficiency and recovery of extraction depends on the sample material, ionic conditions of the extraction medium, type of lysing agent used *etc.* In general, the commonly used method for DNA extraction is that of Murmur's method. Extraction of DNA is accomplished by rupturing of cell walls and nuclear membrane followed by deproteinization and precipitation of the nuclic acid using ethanol.

Materials

- Liquid nitrogen
- Extraction buffer – 100 mM Tris-HCl buffer (pH 7.8) is prepared by dissolveing 6.06 g of Tris-HCl along with 9.30 g of sodium salt of EDTA (10 mM) and 14.61 NaCl (500 mM) in water and volume is made to 500 ml.
- Potassium acetate 5.0 M – Dissolve 24.55 g of potassium acetate in water make the volume to 50 ml.
- Sodium acetate 3.0 M - Dissolve 2.46 g of sodium acetate in water make the volume to 10 ml.
- Suspension buffer - 50 mM Tris-HCl pH (8.0) is prepared by dissolve 0.61 g of Tris-HCl along with 0.37 g of sodium salt of EDTA (10 mM) in water and volume is made to 100 ml.
- TRis-EDTA (TE) buffer – 10 mM pH (7.5) is prepared by dissolving 0.12 g of Tris-HCl and 0.37 g of sodium salt of EDTA (1.0 mM) in water and volume is made to 100 ml.

Procedure

- Approximately 5.0 g of plant samle is taken and freezed by dipping in liquid nitrogen and then grounded in fine powder followed by extraction with 75.0 ml of extraction buffer.
- The homogenates are transferred in 250 ml flask and 5.0 ml 20 SDS solution is added and mixed thoroughly using magnetic stirrer for 20 minutes and then incubate at 65° C for 10 minutes.
- Add 50 ml of potassium acetate solution, mix and incubate at 0° C for 30 minutes for precipitation of proteins and polysaccharides.
- Centrifuge out the precipitate at 25,000 x g for 15 minutes. To the supernatant add six tenth volume of isopropyl alcohol and allow it to stand for 30 minutes at -20° C.

- DNA is pelleted at 20,000 x g for 15 minutes and supernatant is discarded.
- DNA is redissolved in 3.0 ml of suspension buffer and add 1.8 ml of isopropyl alcohol and 180 µl of 3.0 M sodium acetate and allow it stand for 60 minutes at -20^0 C.
- DNA is repelleted and washed with ice-cold 80% ethanol and gently dried under vaccume.
- Redissolve the DNA pallet in a suitable volume (0.5 – 5.0 ml) of TE buffer.
- Estimate the DNA content and check the purity by UV spectroscopy.

Purification of DNA through cesium chloride

- To 1.0 ml of DNA solution add 0.75 g of cesium chloride (CsCl) and after dissolving completly add 10 µl of ethidium bromide (10 mg/ml) in TE buffer.
- Transfer the content to a suitable ultracentrifuge tube and centrifuge at 1,80,000 x g at 15^0 C for 16 – 24 hours.
- Collect the DNA band usin a 'Pasture pipette or syringe' while viewing in a UV transilluminator.
- Extract the ehidium bromide with a small volume of water saturated *n*-butanol until the extract is colourless. Mix the aqueous extract with equal volume of TE buffer.
- Pricipitate the DNA by adding an exualvolume of isopropanol at -20^0 C.
- Pallet the DN by centrifugation, wash with ice-cold 80% ethanol, dry and resuspende in a small volume of TE buffer.

Reading

Murmur, J. (1961). A procedure for isolation of deoxyribose nuclic acid from microganism. *J. Mol. Biol.*, **3:** 208.

17.3. Extraction of Plant Total DNA

Material

- Extraction buffer – 50 mM Tris-HCl buffer (pH 8.0) is prepared by dissolveing 50 mM of Tris-HCl along with 50 mM of sodium salt of EDTA and 250 mM of NaCl in 15% sucrose solution.

- Ethanol.
- Isopropanol.
- Potassium acetate 5.0 M – Dissolve 24.55 g of potassium acetate in water make the volume to 50 ml.
- Sodium acetate 3.0 M - Dissolve 2.46 g of sodium acetate in water make the volume to 10 ml.
- TRis-EDTA (TE) buffer – 10 mM pH (7.5) is prepared by dissolving 0.12 g of Tris-HCl and 0.37 g of sodium salt of EDTA (1.0 mM) in water and volume is made to 100 ml.

Procedure

- Wash the plant material under running water followed by sterile water, plant materials are dried with blotting paper and about 3 – 5 g is freezed in liquid nitrogen and grounded to fine powder.
- The frozen powdered material is transferred to a 50 ml tube having 15 ml of extraction buffer at 65^0 C. Material is mixed well and incubated at 65^0 C for 15 minutes.
- Add 5.0 ml of potassium acetate (5.0 M) mixed vigorously and incubated on ice for 20 minutes.
- Centrifuge the content at 4000 rpm for 20 minutes.
- Filter the supernatant through two layers of fine cloth and collect the filterate. If the filterate is green or brown add another 5.0 ml of potassium acetate and repeat the process.
- Add 2/3 volumes of isopropyl alcohol to the filterate (2 ml of isopropyl alcohol to 3.0 ml of filterate) slowly aand shake gently. Incubate the tubes at -70^0 C for 30 minutes.
- Separate the DNA by using a 'Pasture pipette or syringe' or pellet the DNA by centrifugation at 10,000 rpm for 15 minutes. Wash the pellet with ice-cold 70% ethanol followed by absolute ethanol. Dry the pellet in vaccume.
- Suspende the DNA in TE buffer.
- This DNA preparation can be further purified by adding 10 μl of RNase (10 mg/ml) and incubate at room temperature for 15 – 30 minutes to remove RNA impurities.

- Add 1/10 volume of 3.0 M sodium acetate solution and 2 volumes of 95% ethanol. Mix gently to precipitate the DNA. Incubate at -20^0 C for 60 minutes.
- Collect the DNA by centrifugation.
- Resuspende the DNA in TE buffer for further use. It can be stored at 4^0 C for some weeks.

Reading

Dellaporta, S.L., Wood, J. And Hicks, J.B. (1983). A plant DNA minipreparation: version II. *Plant Molecular Biology reporter* I **(4):** pp. 19-21.

17.4. A Simple Method for DNA Isolation from Plant Sample

DNA from the samples was isolated as follows. Around 0.5 g of leaf tissue was placed in a mortar and homogenized with 2 ml of extraction buffer. The extraction buffer (pH 8.0) consisted of 100 mM Tris, 20 mM EDTA, 0.5 M NaCl, 7 M Urea, 0.1% β-mercaptoethanol and 2% SDS. Long fibers of the tissue were retained back after crushing and the homogenate was transferred to a 2 ml-microfuge tube. An equal volume of phenol:chloroform :Isoamlyalcohol (25:24:1) was added to the tubes and mixed well by gently shaking the tubes. The tubes were centrifuged at room temperature for 15 min at 15,000 rpm. The upper aqueous phase was collected in a new tube and an equal volume of chloroform: Isoamlyalcohol (24:1) was added and mixed.

The upper aqueous phase obtained after centrifuging at room temperature for 10 min at 15,000 rpm was transferred to a new tube. The DNA was precipitated from the solution by adding 0.1 volume of 3 M Sodium acetate pH 7.0 and 0.7 volume of Isopropanol. After 15 min of incubation at room temperature the tubes were centrifuged at 4°C for 15 min at 15,000 rpm. The DNA pellet was washed twice with 70% ethanol and then very briefly with 100% ethanol and air-dried. The DNA was dissolved in TE (Tris-Cl 10 mM pH 8.0, EDTA 1 mM). To remove RNA 5 μl of DNAse free RNAse A (10 mg/ml) was added to the DNA.

Reading

E. Nalini, E., Jawali, N. and Bhagwat, S.G. (2004). A simple method for isolation of DNA from plants suitable for long term storage and dna marker analysis, *BARC News Letter* No. **249**: 208-214

17.5. Extraction of RNA from Plant Samples

Plants are diverse, and individual species and organs or tissues of plants can behave differently during extraction of RNA (and DNA) for use in molecular studies. Hence, a range of extraction methods has also been devised, depending on the tissue or genotype being extracted. Problems encountered include the presence of large quantities of polysaccharides; high levels of RNases; various different kinds of phenolics, including tannins; low concentrations of nucleic acids (high water content); tissue, such as lignin (wood), that is difficult to break up; and so on. In addition, sampling techniques can have an effect on yield and lack of degradation, recognizing also that most tissues extracted are generally composed of a range of cell types and, hence, functions. In some instances, kits obtained from biotechnology supply companies are sufficient to perform the task, but, in other instances, especially with a new plant or tissue that has not had RNA extracted from it before, methods may need to be modified to suit the particular characteristics of that material. There is no simple indication that a tissue will be difficult. Depending on the use of the RNA extracted, further purification of messenger RNA (mRNA) may be required. In general, commercial kits will perform this job more than adequately, and any of a range of methods of quantification will be successful. The biggest problem encountered in RNA extraction usually originates from the initial sampling and extraction protocols, and from personal technique and care taken. Three extraction protocols will, therefore, be outlined that have had widespread success in plant RNA extractions in our laboratory.

Materials

Sampling

- Liquid nitrogen (and container).
- Polystyrene box and/or second liquid nitrogen-proof container.
- Sharp knife, scalpel, razor blade, tweezers, cork borer, metal needle/probe, and flame source.
- Eppendorf tubes, tinfoil, and plastic bottles of various sizes.
- Analytical balance.
- Plant material.
- –80°C freezer or liquid nitrogen storage container and/or dry ice.

Arabidopsis Extractions: Trizol™ or Guanidine Isothiocyanate-Based Procedure (see Note 1)

- Trizol reagent (Invitrogen). Refer to the manufacturer's instructions and guidelines for stability and storage, and handle with eye and glove protection.
- Chloroform.
- Isopropyl alcohol.
- 75% ethanol in RNase-free water.
- RNase-free water (made by adding 0.01% DEPC [v/v], standing or stirring overnight, then autoclaving; or made by using Barnstead™ Ultrapure RNase-free water.
- NaOH-washed (0.1 M) and UV-treated plasticware, oven-baked sterile glassware, sterilized Eppendorf tubes, or clean sterile Falcon tubes (conical bottom).
- Liquid nitrogen and mortar and pestle.
- Benchtop centrifuges (refrigerated or access to cold room).
- mRNA purification kit (Amersham Biosciences, now GE).

Extractions From Problem Tissues: CTAB Method (see Notes 1 and 2)

- Sterile Falcon tubes (conical bottom, 25 or 50 ml).
- Oakridge tubes (round bottom, sterile, and RNase-free).
- Mira-Cloth® (Calbiochem).
- Liquid nitrogen and mortar and pestle.
- Benchtop centrifuge, vortex machine, refrigerator, and freezer.
- RNase-free water in a baked storage bottle (in an oven at >150°C, for > 4 h).
- Chloroform:isoamyl alcohol (24:1).
- 12 *M* LiCl (use RNase-free water and autoclave or filter through a Nalgene™
- Extraction buffer: 2% hexadecyl trimethyl-ammonium bromide (CTAB); 2% polyvinylpyrrolidone K30, 100 m*M* Tris-HCl, pH 8.0, 25 m*M* EDTA, sodium form, pH 8.0; 2 *M* NaCl, 0.5 g/L spermidine, and 2%

β-mercaptoethanol. Use RNase-free water for dissolving, and autoclave before using.

- Sodium dodecyl sulfate–Tris-HCl–EDTA (SSTE) buffer: 1 *M* NaCl; 0.5% sodium dodecyl sulfate (SDS), 10 m*M* Tris-HCl, pH 8.0, and 1 m*M* EDTA, sodium form, pH 8.0. Use RNase-free water and autoclave before using.

Extractions From Problem Tissues: Non-CTAB-Based

- Eppendorf and/or Falcon tubes (sterile and RNase-free).
- Liquid nitrogen and mortar and pestle.
- Polytro homogenizer.
- Oakridge tubes (sterile and RNase-free) and Corex tubes (sterile and RNase-free).
- Benchtop centrifuge, vortex machine, refrigerator, and freezer.
- RNase-free water in a baked storage bottle (in an oven at >150°C, for >4 h).
- Preheated (65°C) lysis buffer: 150 m*M* Tris-HCl, 50 m*M* EDTA, 4% SDS, pH 7.5 titrated with boric acid, 1% β-mercaptoethanol, and 1% w/w polyvinylpolypyrrolidone (PVPP). Use RNase-free water and autoclave before use.
- 5 *M* potassium acetate; use RNase-free water and autoclave or filter
- Cold absolute ethanol.
- Chloroform:isoamyl alcohol (24:1).
- Tris-HCl-equilibrated phenol, pH 8.0. Keep phenol in dark bottles in cold room (or –20°C); do not use old phenol that has been opened for a long time and is discoloured. Make the Tris-HCl buffer RNase-free by adding DEPC to make buffer in a baked (or sterile) bottle, do not autoclave buffer; otherwise,
- Equilibrate by melting 500 ml phenol at 65°C and adding 100 ml of RNase-free water, mixing, and leaving to partition overnight (can last for 4–6 wk). Discard the top, aqueous phase. Repeat two more times, but with 0.5 *M* Tris-HCl, pH 8.5, the first time, and 0.1 *M* Tris-HCl, pH 8.5, the second time. Phenol can now be used (some buffer can be left on top, but prevent carrying it over while pipetting). Store in the dark at 4°C for up to 2 to 3 wk.
- 12 *M* LiCl (use RNase-free water and autoclave or filter.

Extraction of mRNA

Use a kit from a biotechnology supplier.

Quantification, Degradation, and Storage (see Note 3)

- RNase-free water.
- Two UV-capable glass spectrophotometer cuvets.
- 10X stock Tris-base–boric acid–EDTA buffer: 108 g Tris-base, 55 g boric acid, and 40 ml of 0.5 *M* EDTA, pH 8.0, in 1 L deionized water.
- 37% formaldehyde.
- Ultrapur agarose (Invitrogen).
- 10X MePS buffer: 0.2 *M* MOPS (3-[*N*-morpholino] propanesulfonic acid, 50 m*M* Na acetate, and 10 m*M* EDTA.
- Loading buffer (store in aliquots at –20°C): 0.75 ml deionized formamide, 0.15 ml of 10X MOPS buffer (autoclaved or filtered), 0.24 ml formaldehyde, 0.1 ml RNase-free water, 0.1 ml glycerol (autoclaved), and 10% w/v bromophenol blue dye. Add 3 μl ethidium bromide to 300 μl loading buffer before using.
- Ethidium bromide as a 10% solution (Caution: ethidium bromide is toxic, handle with gloves; (*see* Note 4).
- RNase-free electrophoresis gel boxes, beds, and combs.
- –80°C freezer.
- Agilent™ chip.

Procedure

Extraction of excellent quality plant RNA starts with good practice tissue sampling and storage. To reflect mRNA present in a snapshot moment in the growing intact plant, tissues need to be treated in a manner very similar to that for analysis of metabolic intermediates or active enzymes. Partial or complete degradation of mRNA can occur because of tissue sampling and storing techniques. Successful extraction may require alternative techniques, and we outline:

- A standard method now used for *Arabidopsis* (the model plant in which the genome has been fully sequenced), tomato, maize, tobacco, and other commonly researched plants.

- Two methods for use in more difficult tissues in which guanidine-based methods result in zero yield—the CTAB (1) and the hot phenol/chloroform methods (2)—which have had much greater success in tissues with low yields and/or high polysaccharide, secondary product, or RNase contents.
- The need for care and for use of RNase-free solutions and equipment, including during quantification and storage, is common across all methods.

Sampling

Ample liquid nitrogen supply is essential. In general, tissue from a plant is sampled by plucking and covering immediately with liquid nitrogen in a polystyrene container, such as those in which chemicals are dispatched on dry ice by laboratory suppliers. For example, with leaves, whole leaves, rapidly hand shredded leaves, or cork borer discs of leaves can be sampled and killed in a short time, equilibrating to liquid nitrogen temperatures within less than 1 min. It is essential to have a minimal time between removal from the plant and immersion in liquid nitrogen, to minimize the expression of new mRNAs because of tissue wounding or detachment from the plant. Some mRNA has been shown to be upregulated within 5 minutes of tissue detachment from a plant. Other tissues are more bulky, *e.g.*, fruit or tubers; these tissues take longer to equilibrate to liquid nitrogen temperatures if immersed whole. Bulky tissues hold more heat, and exchange is slower with liquid nitrogen. This leads to tissue damage (altering osmoticum leading to leaky cells) and degradation of the mRNA present, because RNases gain access to the mRNA. Hence, for bulky tissues, it is better to rapidly remove the tissue from the plant and to sub sample quickly (preferably a minute between detachment and immersion of sub samples in liquid nitrogen). Slicing and/or dicing with cork borers is very effective. Take care to sample the tissue in which you are interested in a representative manner. Tissue samples can be added directly to preweighed Eppendorf tubes or storage containers or to homemade tinfoil pouches immersed in the larger liquid nitrogen container. If using Eppendorf tubes, prepare the tubes with a small hole (heat a needle over a flame and pierce the lid) to prevent explosions because of the remnants of liquid nitrogen inside the sealed tube when the lids are closed. This can also be done with other containers, or else the container can be drained before placing the lid on the container. The amount of tissue can be calculated if the containers or tubes are preweighed, then weighed again after the tissue has been killed and the liquid nitrogen evaporated from the container. Care needs to be taken, however, that the tissue does not thaw during weighing. After the tissue has been sampled, store the sample in a liquid nitrogen storage container or a –80°C freezer. Take care not to remove tissue or allow it to reach subzero temperatures, by keeping the tissue in liquid nitrogen as much as is practical during sub sampling and weighing before grinding (if not performed before storage) to extract the mRNA. Repeated removal from storage and sub sampling of tissues has often led to reduced quality of mRNA. To extract the RNA, grind the weighed material in a mortar and pestle, under liquid nitrogen, to a fine powder and transfer the powder to the extraction buffer.

Caution: do not ever let the plant tissue thaw after killing the tissue in liquid nitrogen and before complete mixing in extraction buffers after grinding the tissue to a powder. The amount of tissue required to achieve acceptable yields of RNA varies according to the material. Tissues with a high water content require higher amounts of tissue to be extracted. For example, 2 g *Arabidopsis* leaf tissue yields approx 60 to 200 µg RNA (Trizol method); and 5 to 8 g fruit tissue (high water) yields approx 400 µg RNA (non-CTAB/non-guanidine-based method).

Arabidopsis Extractions

- Grind approx 0.1 g tissue in liquid nitrogen.
- Add 1 ml of Trizol reagent to the ground powder (*see* Note 5).
- Transfer into Eppendorf tubes.
- Centrifuge at 12,000*g* for 5 min at 2 to 8°C.
- Remove supernatant to new Eppendorf tube.
- Add 200 µL of chloroform and shake vigorously by hand for approx 15 seconds.
- Let stand at room temperature (~20–25°C) for 3 min.
- Centrifuge at 12,000*g* for 15 min at 2 to 8°C.
- Carefully transfer the upper aqueous phase to a new Eppendorf tube (ensure no interface debris is transferred, *see* Note 1).
- Add 0.5 ml of isopropyl alcohol. Mix.
- Let stand at room temperature for 10 min.
- Centrifuge at 12,000*g* for 10 min, at 2 to 8°C.
- Carefully discard supernatant (tip Eppendorf with the pellet position angled up and away from you and pipet out the supernatant). The pellet may be slightly glassy and transparent or may not be clearly visible at all.
- Add 1 ml of 75% ethanol.

- Vortex briefly and centrifuge at 12,000*g* for 5 min at 2 to 8°C.
- Discard the supernatant and allow pellet to air-dry for 10 min.
- Dissolve the pellet in 20 μl of RNase-free water by very gently sucking the liquid up and down with a pipet.
- Quantify the RNA, check the purity and degradation, and either store at –20 or –80°C until used, or extract the mRNA using commercial kits.

Extractions From Problem Tissues

CTAB-Based Method

- Pipet 15 ml of extraction buffer (minus β-mercaptoethanol) into an RNase-free Falcon tube and add 300 μl of β-mercaptoethanol. Warm in a water bath to 65°C.
- Grind the tissue in liquid nitrogen and add the tissue gradually to the heated buffer so that no powder coagulates (freezes into a lump) and, therefore, thaws before mixing fully with the buffer. Vortex after each small addition to ensure that the powder is fully dispersed and thawing in the heated buffer.
- Leave the sample sitting at room temperature while processing the next samples.
- Mix the samples using the Polytron homogenizer for approx 1 min at full speed until the sample foams close to the top of the tube. Wash the Polytron homogenizer with distilled water after each sample.
- Add an equal volume of chloroform: isoamyl alcohol, mix (vortex), transfer to an RNase-free Oakridge tube, balance the samples with buffer, and centrifuge at 11,984 x *g* for 10 min at room temperature to separate the phases.
- Filter the upper aqueous phase through an autoclaved Mira-Cloth into a new RNase-free Oakridge tube (or carefully pipet off the top aqueous phase, ensuring no transfer of any interface material to a new tube). (**Caution:** it is better to leave some aqueous phase behind than to transfer contaminants.
- Add an equal volume of chloroform:isoamyl alcohol, mix, and centrifuge to separate the phases.

- Remove the top aqueous phase to an RNase-free Falcon tube and estimate the volume to the nearest milliliter. Add an appropriate volume of LiCl solution to give a final concentration of 2 *M* LiCl (1 volume of 4 *M*, 0.5 volumes of 6 *M*, 0.33 volumes of 8 *M*, 0.25 volumes of 10 *M*, or 0.2 volumes of 12 *M*).
- Leave at 5°C (refrigerator) overnight.
- Centrifuge at 4°C and 11,984*g* for 20 min.
- Pour off the supernatant, and invert the tubes to drain onto a tissue.
- Preheat SSTE buffer to 65°C. Dissolve the pellet in 200 µl of heated SSTE, and transfer to a 1.5-ml, RNase-free Eppendorf tube.
- If the SDS in the SSTE buffer precipitates (a white cloudiness), place the Eppendorf with the sample in a heating block (37°C) until it has dissolved again before continuing.
- Add an equal volume of chloroform:isoamyl alcohol. Vortex immediately before adding and immediately after to mix completely.
- Add 2 volumes of absolute ethanol to precipitate the RNA (>30 min at –70°C, or >2 h at –20°C).
- Centrifuge the tube in a microcentrifuge in a cold room (or a in a temperature-controlled microcentrifuge) for 20 min at maximum speed.
- Discard the supernatant, allow the pellet to air-dry, and resuspend in 20 µl RNasefree water.

Non-CTAB-Based or Non-Guanidine-Based Method

- Very slowly add 5 g of ground powder to 15 ml of preheated lysis buffer containing PVPP and freshly added β-mercaptoethanol, and vortex between additions (do not allow powder to thaw or form lumps).
- Homogenize the suspension using the Polytron homogenizer at maximum speed for 20 s, or until the froth reaches the top of the tube.
- Add 0.1 volumes (1.5 ml) of 5 *M* potassium acetate and 0.25 volumes (4 ml) of cold absolute ethanol to the tube and vortex for 30 s.
- Put 1 volume of chloroform:isoamyl alcohol into each of two Oakridge tubes and put half of the homogenate into each tube, vortex, and centrifuge at 2000*g* for 10 min at room temperature.

- Remove the top aqueous phase to an RNase-free Falcon tube and add 10 ml of buffered phenol and 10 ml of chloroform:isoamyl alcohol.
- Vortex to mix, and centrifuge at 2000*g* for 10 min to separate the phases.
- Remove the top aqueous phase to an Oakridge tube and add one-third volume of 12 *M* LiCl. Incubate overnight at –20°C.
- Centrifuge at 20,000*g* for 20 min to precipitate the pellet.
- Pour off the supernatant and resuspend the pellet (by vortexing) in 10 ml of 3 *M* LiCl (12 *M* LiCl diluted with RNase-free water). Centrifuge at 20,000*g* for 20 min.
- Pour off the supernatant, and resuspend the pellet (by vortexing) in 2 ml of RNase-free water, transfer to a 30-ml Corex tube.
- Add 180 µl of 5 *M* potassium acetate and 6 ml of cold absolute ethanol. Cover with Parafilm and leave at –20°C for 1 h.
- Centrifuge at 12,000*g* for 10 min.
- Pour off the supernatant and dry pellet in air for 10 min.
- Dissolve the pellet in 200 µl of sterile RNase-free water. Store at –20°C until use, with or without aliquoting into different tubes.

Notes

1. Generally, the method used by *Arabidopsis* researchers is adequate for a wide range of plant tissues. However, many plant tissues also contain other compounds that interfere with the extractions. In particular, some tissues (*e.g*., algae, some fruits, some leaves, and woody material) have high concentrations of polysaccharides, derived either from the plant cell wall or present as mucilages. These generally entrap the RNA during extraction and, if they are not removed in the first steps and partitioned into a discarded phase, they will remain through the rest of the extraction. Hence, it is important not to take any debris or interface material during the chloroform partitioning. Heat is a good way of removing polysaccharides, by making them more soluble. Many plant tissues also have high levels of RNases and, generally, the best way to remove these RNases is to increase the SDS present in the extraction buffer. The result of insufficient SDS is partially or fully degraded RNA. The PVPP helps to bind polyphenolics, which can also be a problem in some tissues that have high levels of polyphenolics. With very low RNA/high water-containing tissue, more material to extract in a given volume of buffer is

usually required. Otherwise, there is insufficient RNA present to partition properly during purification.

2. For reasons it is not understood fully, both CTAB and guanidine can cause precipitation problems or other mixing problems when extracting some tissues. A way have been found to predict this occurrence (other than the presence of polysaccharides).

3. To minimize the presence of RNases, it is important to keep equipment aside for use only with RNA extractions. Pipet tips and Eppendorf tubes should be used only for RNA work and not mixed. Gloves should be used at all times. Keep one set of gel electrophoresis equipment for use just with RNA. Keep solutions RNase free by not using the solutions in other procedures. Always use RNase-free water for solutions. The solid chemicals are NOT RNase free, therefore, buffers should be autoclaved. Tris-HCl buffers cannot be autoclaved.

4. RNA gel matrices can be prepared in advance by dissolving the agarose in buffer in a Sorvall or similar bottle (able to be autoclaved). Melt the aliquot and add formaldehyde. Take care to fill the bottles only partially full (*e.g.*, 500 ml in a 1 l bottle). This can be stored with the lid on until use. If you have a microwave oven, loosen the lid so that air can escape and heat in a microwave oven until melted. Remove with care (it will be hot) and pour out whatever is required for the gel. Leave the remainder to resolidify in the bottle and store again. Take care when handling ethidium bromide, it is a mutagen and toxic, and gloves should always be used and surfaces wiped down after use.

5. Do not let ground (or intact) plant tissue thaw without being in the presence of extraction buffer. Small amounts of material can thaw when taking samples in and out of the freezer, weighing out aliquots, or during grinding in liquid nitrogen. It can be particularly important to ensure that the tissue does not form a lump, where the outside is in contact with the buffer, but the inside is thawing directly. When adding tissue to a hot extract, add only a little at a time, using a spoon or spatula that has also been precooled in liquid nitrogen.

6. Always add β-mercaptoethanol fresh on the day of extraction. It will become ineffective in solutions within 12 to 24 h.

7. A ratio of approx 1.8 to 2.0 (A260/A280 nm) means that the RNA is sufficiently pure and without polysaccharide contamination for use in most applications and is soluble. A lower ratio generally means

polysaccharide contamination and/or insolubility. A high reading at 240 nm also suggests polysaccharide contamination.

17.6. Isolation of Plasmid DNA

Plasmid DNA is generally isolated by the alkaline lysis method. In this method cells are first lysed with lysozyme proteins and chromosomal DNA is denatured by SDS and high alkaline pH respectively. On decreasing the pH in the subsequent step, plasmid DNA is renatured but chromosomal DNA remains denatured and forms a complex of DNA protein and SDS which gets precipitated. Plasmid DNA is recovered from the supernatant by ethanol precipitation. This method can be scaled up for use with larger volumes. The plasmid obtained may further be deproteinized by phenol: chloroform extraction followed by ethanol precipitation or may be subjected to caesium chloride equilibrium gradient centrifugation.

Materials

- Solution: 1. 50 mM glucose, 10mM EDTA, 25mM tris-Hcl (pH8.0), 4mg/ml lysozyme (lysozyme is added just before use)
- Solution: 2. 0.2 N NaOH (1.0% SDS freshly prepared each time)
- Solution: 3. Potassium acetate (pH 4.8). To 60 ml of 5 M potassium acetate, add 11.5 ml of glacial acetic acid and 28.5 ml of water. The resulting solution is 3M with respect to potassium and 5 M with respect to acetate (3M sodium acetate, pH 4.8 can also be used in place of potassium acetate).
- RNase A - 10 mg/ml in 10 mM Tris-Cl (pH7.5) containing 15 mM NaCl. Heat to 100^{0}C for 15 min and cool slowly to room temperature. Store in small aliquots at -20^{0}C.
- LB medium
- TE buffer – 10 mM Tris-HCl (pH8.0), 1mM EDTA.
- Saturated phenol: chloroform iso amyl alcohol mixture (25:24:1).

Procedure

- Inoculate 5 ml of LB medium containing the appropriate antibiotic with a single bacterial colony carrying the plasmid. Incubate at 37^{0}C overnight with vigorous shaking.
- Pour 1.5 ml of the culture into microfuge tube. Centrifuge for 1 min in a microfuge at 5000 rpm.

- Remove the medium by aspiration leaving the bacterial pellet as dry as possible.
- Resuspend the pellet in 100μl of ice cold solution 1. Vortex thoroughly. No lumps should be seen.5. Incubate in ice for 5 min.
- Add 200 μl of freshly prepared solution 2, Mix gently by inverting tube. The contents at this stage become clear and viscous. Do not vortex. Leave on ice for 5 min.
- Add 150 μl of ice cold solution 3. Mix by inverting the tube slowly several times. Keep on ice for 5 min.
- Centrifuge for 5 min at 14,000 x g 4^{0}C.
- Transfer the supernatant carefully to a fresh tube.
- Add an equal volume of phenol: chloroform: isoamyl alcohol mixture. Mix by vortexing, centrifuge for 2 min in a microfuge at 12,000 rpm. Collect supernatant in fresh tube (Optional step).
- Add two volume of ethanol at room temperature, Mix by vortexing. Allow to stand at room temperature for 2 min.
- Centrifuge for 5 minutes in a microfuge at 12,000 rpm. Discard supernatant completely by keeping the tube inverted on paper towels and save the pellet.
- Add 1 ml of 70% ethanol. Vortex briefly and recentrifuge at 12,000 rpm.
- Remove ethanol and dry the pellet.
- Dissolve the pellet in 50 μl TE and analyze an aliquot on agarose gel.
- RNase treatment of plasmid - To 20 μl of the DNA from the above step add 2 μl of RNAse A (10 mg/ml) and incubate at 37^{0}C for 30 min. Dilute DNA to 50 μl with TE and add an equal volume of phenol: chloroform : isoamyl alchohol mixture. Mix well and centrifuge in microfuge at full speed for 2-5 min. Collect the aqueous phase and repeat phenol/ chloroform extraction until the interface is clear. To the aqueous phase add 1/10 vol. of 3 M sodium acetate, pH 4.8 and 2 vol. of 100% ethanol. Mix well and keep at -70^0 C for 15 min. Centrifuge to collect the DNA, wash with 70% ethanol, dry the pellet and dissolve it in 20 μl TE.

17.7. Quantitative Estimation of DNA from Plant Leaves

Estimation of DNA is possible by a number of methods based on the physical or chemical property of the nuclic acid. A convenient and easy colorimetric method is available on the basis of quantitative reaction of deoxysugar with diphenylamine reagent. Under extreme acid condition, DNA is initially depurinated quantitatively followed by the dehydration of sugar to hydroxylevulinylaldehyde. This aldehyde condense, in alkali medium, with diphenylamine to produce a deep blue coloured condensation products with absorption maxima at 595 nm.

Materials

- DNA Standard (0.5 mg/ml).
- Saline citrate (0.15 NaCl, 0.015 M Na_3 citrate) solution.
- Diphenylamine reagent – Mix 5 g fresh or recrystallized diphenylamine, 500 ml of glacial acetic acid and 13.75 ml of concentrated H_2SO_4. Stable for months at 2° C, warm to room temperature and swirl to remix before use.

Procedure

- Isolated DNA in saline citrate solution is taken in a series *viz.*, 1.0, 2.0 and .0 ml along with standar DNA solution (5 mg/ml). The total volume of the solutions in tubes are made to 3.0 ml with water.
- Add 6.0 ml of diphenylamine solution to each tube, mix thoroughly and heat the tubes in boiling water-bath for 10 minutes. Cool the tubes.
- Measure the OD of the solutions at 600 nm against a reagent blank and the DNA content is quantified by comparing the ODs against a standard curve of DNA.

Reading

Burton, K. (1956). A study of the conditions and mechanism of the diphenylamine reaction for the colorimetric estimation of deoxyribonucleic acid. *Biochem. J.*, **62:** 315-323.

Ashwell, G. (1957). In "Methods in Enzymology", **3:** *(Eds. S.p> Colowick and N.O. Kaplan),* Academic Press, New Yorlk, pp. 99.

Himesh, S., Singhai A. K. and Sharma S. (2011). Quantitative estimation of DNA isolated from various parts of *Annona squamosa. Int. Res. J. Pharmacy,* **2(12):** 169-171.

17.8. Quantitative Estimation of RNA from Plant Samples

RNA molecules play essential roles in living cells. Many important and diverse functions of RNA molecules, including catalysis of chemical reactions and control of gene expression, have only recently come to light. Outside of the cell, novel nucleic acids have been selected using directed molecular evolution techniques *in vitro*, which can function as enzymes or aptamers with high binding specificity for target proteins. Because of the importance of RNA molecules, and because structure is key to the function of RNA molecules in their diverse roles, there is a need to have a simple and efficient method for estimating the RNA content in any plant sample.

This is a general reaction for pentoses and depends on the formation of furfural when the pentose is heated with concentrated hydrochloric acid. Orcinol reacts with the furfural in the presence of ferricchloride as a catalyst to give a green colour, which can be measured at 665 nm.

Materials

- Standard RNA solution - 200μ g/ml in 1 N perchloric acid/buffered saline.
- Orcinol Reagent- Dissolve 0.1g of ferric chloride in 100 ml of concentrated HCl and add 3.5 ml of 6% w/v orcinol in alcohol.
- Buffered Saline - 0.5 mol/litre NaCl; 0.015 mol/litre sodium citrate, pH 7.

Procedure

- Pipette out 0.0, 0.2, 0.4, 0.6, 0.8 and 1 ml of working standard in to the series of labeled test tubes.
- Pipette out 1 ml of the given sample in another test tube.
- Make up the volume to 1 ml in all the test tubes. A tube with 1 ml of distilled water serves as the blank.
- Now add 2 ml of orcinol reagent to all the test tubes including the test tubes labeled 'blank' and 'unknown'.
- Mix the contents of the tubes by vortexing / shaking the tubes and heat on a boiling water bath for 20 minutes.
- Then cool the contents and record the absorbance at 665 nm against blank.
- Then plot the standard curve by taking concentration of RNA along X-axis and absorbance at 665 nm along Y-axis.

- Then from this standard curve calculate the concent ration of RNA in the given sample.

Reading

Mahesha H B, Yuvaraja's College, Mysore - Estimation of RNA by Orcinol Reaction

Alternative colorimetric method for RNA estimation

The method depends on conversaion of the pentose, ribose in the presence of hot acid to furfural, which then reacts with orcinol to yield green colour. The colour formed largely depends on the concentration of HCl, ferric chloride, orcinol, the duration of heating at 100^0 C *etc.* upto certain maxima.

Materials

- Standard RNA and sample RNA solutions.
- Orcinol reagent – Mix 2.0 ml of 10% solution (w/v) of ferric chloride [$FeCl_3.6H_2O$] to 400.0 ml of of concentrated HCl.
- Alcohol orcinol (6%) – Mix 6.0 g orcinol to 100.0 ml of 95% ethanol. Refrigerate in a brown bottle until use. It is stable for 30 days.

Procedure

- Prepare a standard RNA (50 µg/ml) solution in ice-chilled 10 mM Tris-acetate, 1.0 mM EDTA buffer (pH 7.2) or any other buffer by dissolving RNA completely.
- Dissolve the isolated RNA in the above buffer solution to an approximate concentration (50 µg/ml).
- Make a gradient of RNA by taking 0.5, 1.0 .....3.0 ml of RNA standard as well as isolated RNA solution and make the final volume of 3.0 ml with water. For blank take only 3.0 ml of water and add 6.0 ml orcinol solution to each tube.
- Mix 0.4 ml of 6.0% alcoholic orcinol in each tube. Mix the solutions thoroughly and heat the tube for 20 minutes in boiling water bath.
- Cool the tubes and measure the absorption at 660 nm against reagent blank.
- Findout the concentration of isolated RNA solution by comparing the ODs from standard RNA solutions.

Reading

Bial, M. (1902). Estimation of RNA nucleic acid. *Den. Med. Woch.*, **28:** 253.

Ashwell, G. (1957). In "Methods in Enzymology", *3: (Eds. S.P. Colowick and N.O. Kaplan), Academic Press, New Yorlk, pp. 87.*

17.9. Separation of DNA and RNA from Plant Samples and Their Colorimetric Estimation

Procedure

- Take the plant material (0.2 – 0.5 g) and extract in 5.0 ml of 70% ethanol, centrifuge at 10,000 rpm for 10 minutes. Discard the supernatant and preserve the pellet.
- Add 5.0ml of ethanol-ether (3:1) and boil for 5-10 minutes. Centrifuge at 10,000 rpm for 10 minutes and take the residue.
- Wash the residue with 5.0 ml of 80% acetone for 2 – 3 times to decolourise it. Centrifuge at 10,000 rpm for 10 minutes.
- Wash the residue with 5.0 ml of 0.2 M perchloric acid and then centrifuge at 12,000 rpm for 15 minutes.
- Take the residue and add 5.0 ml of 1.0 N perchloric acid and allow it to stand for overnight at 4^0 C, then centrifuge at 12,000 rpm for 15 minutes, the supernatant will be RNA.
- Take the residue and to it add 5.0 ml of 0.5 N perchloric acid and incubate at 70^0 C for 15 minutes, the supernatant will be DNA.
- To the residue add 5 ml of 2.0 N HCl, centrifuge at 15,000 rpm for 15 minutes. The supernatant will be DNA. Pool both the DNA fractions. The residue would be carbohydrates and crude protein.
- Optical density is measured at 260 nm for both RNA and DNA against reagent blank and the ODs are compared with that of standard RNA and DNA to quantify the quantity.

Reading

Ouger, M. and Rosen, G. (1950). The nucleic acids of plant tissues; the extraction and estimation of desoxypentose nucleic acid and pentose nucleic acid. *Arch. Biochem.*, **25(2):** 262-276.

CHAPTER 18

Vitamins

A vitamin is an organic compound required by an organism as a vital nutrient in limited amounts. An organic chemical compound (or related set of compounds) is called a vitamin when it cannot be synthesised in sufficient quantities by an organism, and must be obtained from the diet. Thus, the term is conditional both on the circumstances and on the particular organism. For example, ascorbic acid (vitamin C) is a vitamin for humans, but not for most other animals. Supplementation is important for the treatment of certain health problems but there is little evidence of benefit when used by those who are otherwise healthy.

By convention, the term *vitamin* includes neither other essential nutrients, such as dietary minerals, essential fatty acids, or essential amino acids (which are needed in larger amounts than vitamins) nor the large number of other nutrients that promote health but are otherwise required less often. Thirteen vitamins are universally recognized at present. Vitamins are classified by their biological and chemical activity, not their structure. Thus, each "vitamin" refers to a number of *vitamer* compounds that all show the biological activity associated with a particular vitamin. Such a set of chemicals is grouped under an alphabetized vitamin "generic descriptor" title, such as "vitamin A", which includes the compounds retinal, retinol, and four known carotenoids. Vitamers by definition are convertible to the active form of the vitamin in the body, and are sometimes inter-convertible to one another, as well.

Vitamins have diverse biochemical functions. Some, such as vitamin D, have hormone-like functions as regulators of mineral metabolism, or regulators of cell and tissue growth and differentiation (such as some forms of vitamin A). Others function as antioxidants (*e.g.*, vitamin E and sometimes vitamin C).

The largest number of vitamins, the B complex vitamins, function as precursors for enzyme cofactors, that help enzymes in their work as catalysts in metabolism. In this role, vitamins may be tightly bound to enzymes as part of prosthetic groups: For example, biotin is part of enzymes involved in making fatty acids. They may also be less tightly bound to enzyme catalysts as coenzymes, detachable molecules that function to carry chemical groups or electrons between molecules. For example, folic acid may carry methyl, formyl, and methylene groups in the cell. Although these roles in assisting enzyme-substrate reactions are vitamins' best-known function, the other vitamin functions are equally important.

Until the mid-1930s, when the first commercial yeast-extract vitamin B complex and semi-synthetic vitamin C supplement tablets were sold, vitamins were obtained solely through food intake, and changes in diet (which, for example, could occur during a particular growing season) usually greatly altered the types and amounts of vitamins ingested. However, vitamins have been produced as commodity chemicals and made widely available as inexpensive semi-synthetic and synthetic-source multivitamin dietary and food supplements and additives, since the middle of the 20th century.

Classification of Vitamins

Vitamins can be classified in accordance to their solubility either in water or in fat.

Fat soluble Vitamin – Retinol (vitamin A), Vitamin D, Vitamin E, Vitamin K

Water Soluble Vitamin – Ascorbic acid (vitamin C), Thiamin (vitamin B_1), Riboflavin (vitamin B_2), Niacin (vitamin B_3), Pantothanic acid (vitamin B_5), Pyrodoxine (vitamin B_6), Folic acid (vitamin B_9), Cobalamine (vitamin B_{12}), and biotin.

18.1. Vitamin A (Retinol)

Vitamin A is a group of unsaturated nutritional organic compounds, that includes retinol, retinal, retinoic acid, and several provitamin A carenoids, among which beta-carotene is the most important. Vitamin A has multiple functions: it is important for growth and development, for the maintenance of the immune system and good vision. Vitamin A is needed by the retina of the eye in the form of retinal, which combines with protein opsin to form rhodopsin, the light-absorbing molecule necessary for both low-light (scotopic vision) and colour vision. Vitamin A also functions in a very different role as retinoic acid (an irreversibly oxidized form of retinol), which is an important hormone-like growth factor for epithelial and other cells. In foods of animal origin, the major form of vitamin A is an ester, primarily retinyl palmitate, which is converted to retinol (chemically an alcohol) in the small intestine. The retinol form functions as a storage form of the vitamin, and can be converted to and from its visually active aldehyde form, retinal.

All forms of vitamin A have a beta-ionone ring to which an isoprenoid chain is attached, called a *retinyl group*. Both structural features are essential for vitamin activity. The orange pigment of carrots (beta-carotene) can be represented as two connected retinyl groups, which are used in the body to contribute to vitamin A levels. Alpha-carotene and gamma-carotene also have a single retinyl group, which give them some vitamin activity. None of the other carotenes have vitamin activity. The carotenoid beta-cryptoxanthin possesses an ionone group and has vitamin activity in humans.

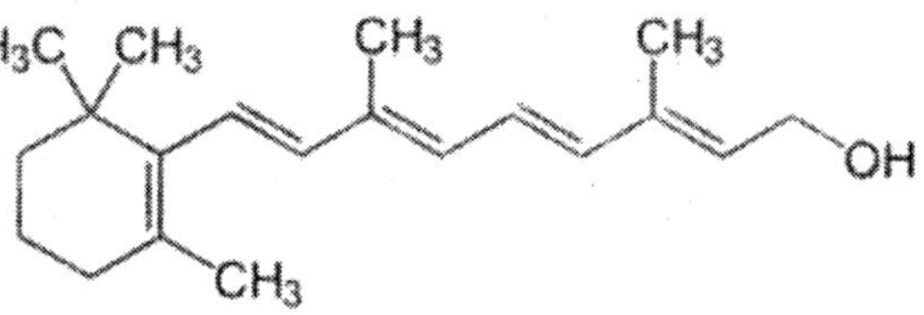

(2*E*,4*E*,6*E*,8*E*)-3,7-Dimethyl-9-(2,6,6-trimethyl-1-cyclohexen-1-yl)-2,4,6,8-nonatetraen-1-ol (Retinol)

Vitamin A can be found in two principal forms in foods:

- Retinol, the form of vitamin A absorbed when eating animal food sources, is a yellow, fat-soluble substance. Since the pure alcohol form is unstable, the vitamin is found in tissues in a form of retinyl ester. It is also commercially produced and administered as esters such as retinyl acetate or palmitate.

- The carotenes alpha-carotene, beta-carotene, gamma-carotene; and the xanthophyll beta-cryptoxanthin (all of which contain beta-ionone rings), but no other carotenoids, function as provitamin A in herbivores and omnivore animals, which possess the enzyme (15-15'-dioxygenase) which cleaves beta-carotene in the intestinal mucosa and converts it to retinol. In general, carnivores are poor converters of ionone-containing carotenoids, and pure carnivores such as cats and ferrets lack 15-15'-dioxygenase and cannot convert any carotenoids to retinal (resulting in *none* of the carotenoids being forms of vitamin A for these species).

Estimation of Vitamin A

The calorimetric method depends upon formation of a coloured derivative of vitamin A aldehyde by reaction with thiobarbituric acid. No colour is obtained with vitamin A or α-carotene. The procedure permits direct estimation of vitamin A aldehyde in tissues. Applications are described in studies on the properties of the visual pigment (rhodopsin) and on enzymatic transformations of vitamin A aldehyde.

Materials

- Thiourea Reagent - Dissolve 4 g of thiourea in 100 ml of glacial acetic acid and filter the solution through glass wool.
- Thiobarbituric Acid Reagent - Dissolve 600 mg of thiobarbituric acid in 100 ml of absolute alcohol. Filter the solution and store in refrigerator.
- Vitamin A Aldehyde Stock Solution - Dissolve 10 mg of all- trans vitamin A aldehyde in 100 ml of absolute ethanol. Shield the solution from light with aluminum foil and store it in a refrigerator.
- Vitamin A Aldehyde X standard Solution - Mix 2 ml of the stock solution with 2.5 ml of water and dilute to 25 ml with 90% ethanol. The solution is prepared each day.
- All solutions may be stored for 30 days except as noted.

Procedure

- Dilute an aliquot of the standard solution or tissue extract to contain 2 to 20 μg of vitamin A aldehyde in 3 ml of 90% ethanol.
- Add 1 ml of thiourea reagent, 1 ml of thiobarbituric acid reagent, and mix well. Prepare a reagent blank containing 3 ml of 90% ethanol.
- Allow colour development to proceed at room temperature in the dark for 30 minutes. Shield the solutions from bright light after colour development.
- Measure the absorbancy of the red solution at 530 nm agaist a reagent blank.

Reading

Futterman, S. and Saslaw, L.D. (1961). The estimation of vitamin A aldehyde with thiobarbituric acid. *J. Biol. Chem.*, **236 (6):** 1652 – 1657.

Alternative Method for Estimation of Vitamin A

Materials

- Potassium hydroxide in 90% ethanol 2.0 N
- Petroleum ether

- Sodium sulphate
- Trichloroacetic acid (TCA) – dissolve 15 g of TCA crystal in 25 ml of chloroform (alcohol free) and store in dark
- Standard vitamin A – 1.5 mg of vitamin A palmitate id dissolved in 10 ml of chloroform.

Procedure

- Grind 1-5 g of sample in a fine paste and add 1.0 ml of saponification mixture (2 N KOH in 90% alcohol).
- Refux the tube for 20 minutes at 60^0 C. Cool the tubes at room temperature then add 20 ml of water and mix thoroughly.
- Extract vitamin A with 10 ml of petroleum ether in a separating funnel. Repeat the separation twice and pool the extracts.
- To the organic layer, add sodium sulphate (amhydrous) to remove moisture for 30 – 60 minutes. Evaporate 5 ml aliquot of the ether extract to dryness at 60^0 C.
- Dissolve the dried residue in 1.0 ml of chloroform and use it as vitamin source.
- Take aliquots of standard (vitamin A acetae) in the concentration range of 1.5 – 7.5 µg and make the volume to 1.0 ml with chloroform.
- Add 2.0 ml of TCA solution and quickly mix the mixture.
- Immidiately measure the OD at 620 nm against chloroform. With the OD values construct the standard curve.
- Determine the vitamin A content of the sample by comparing its OD from that of standard curve.

Reading

Bayfield R.F. and Cole R.R. (1980). Colourimetric estimation of vitamin A with trichloroacetic acid. *Methods of Enzymology*, **67:** 180-185.

18.2. Thiamine (Vitamin B_1)

Thiamine or thiamin or vitamin B_1, named as the "thio-vitamine" ("sulfur-containing vitamin") is a water-soluble vitamin of the B complex. First named *aneurin* for the detrimental neurological effects if not present in the diet, it was eventually assigned the generic descriptor name vitamin B_1. Its phosphate derivatives are involved in many cellular processes. The best-characterized form is thiamine pyrophosphate (TPP), a coenzyme in the catabolism of sugars and amino acids. Thiamine is used in the biosynthesis of the neurotransmitter acetylcholine and gamma-aminobutyric acid (GABA). In yeast, TPP is also required in the first step of alcoholic fermentation.

All living organisms use thiamine, but it is synthesized only in bacterial, fungi, and plants. Animals must obtain it from their diet, and thus, for them, it is an essential nutrient. Insufficient intake in birds produces a characteristic polyneuritis. In mammals, deficiency results in Korsakoff's syndrome, optic neuropathy, and a disease called beriberi that affects the peripheral nervous system (polyneuritis) and/or the cardiovascular system. Thiamine deficiency has a potentially fatal outcome if it remains untreated. In less severe cases, nonspecific signs include malaose, weight loss, irritability and confusion. It is on the World Health Organisation's List of Essential Medicines, a list of the most important medication needed in a basic health system.

Thiamine is a colourless organo-sulphur compound with a chemical formula $C_{12}H_{17}N_4OS$. Its structure consists of aminopyrimidine and a thiazole ring linked by a methylene bridge. The thiazole is substituted with methyl and hydroxyethyl side chains. Thiamine is soluble in water, methanol, and glycerol and practically insoluble in less polar organic solvents. It is stable at acidic pH, but is unstable in alkaline solutions. Thiamine, which is a N-heterocyclic carbene, can be used in place of cyanide as a catalyst for benzoin condensation. Thiamine is unstable to heat, but stable during frozen storage. It is unstable when exposed to ultraviolet light and gamma irradiation. Thiamine reacts strongly in Maillard-type reactions

Thiamine is found in a wide variety of foods at low concentrations. Yeast, yeast extract, and pork are the most highly concentrated sources of thiamine. In general, cereal grains are the most important dietary sources of thiamine, by virtue of their ubiquity. Of these, whole grains contain more thiamine than refined grains, as thiamine is found mostly in the outer layers of the grain and in the germ (which are removed during the refining process). For example, 100 g of whole-wheat flour contains 0.55 mg of thiamine, while 100 g of white flour contains only 0.06 mg of thiamine. In the US, processed flour must be enriched with thiamine

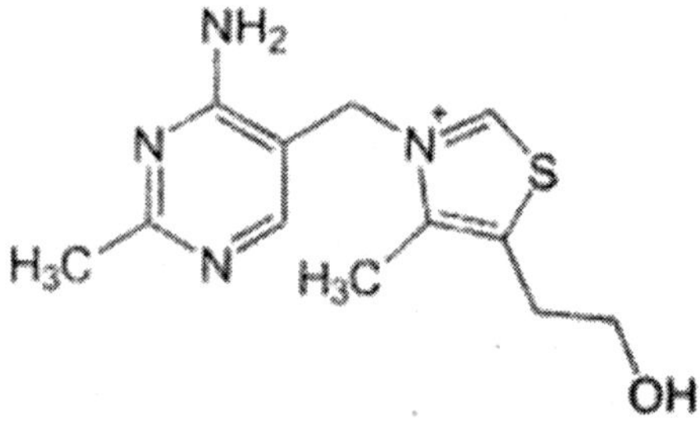

2-[3-[(4-Amino-2-methyl-pyrimidin-5-yl)methyl]-4-methyl-thiazol-5-yl] ethanol

mononitrate (along with niacin, ferrous iron, riboflavin, and folic acid) to replace that lost in processing. In Australia, thiamine, folic acid, and iodised salt are added for the same reason. A whole foods diet is therefore recommended for deficiency.

Some other foods rich in thiamine are oatmeal, flax, and sunflower seeds, brown rice, whole grain rye, asparagus, kale, cauliflower, potatoes, oranges, liver (beef, pork, and chicken), and eggs. Pasta, legumes and watermelon are also good sources of Thiamine.

Thiamine hydrochoride (Betaxin) is a (when by itself) white, crystalline hygroscopic food-additive used to add a brothy/meaty flavor to gravies or soups. It is a natural intermediary resulting from a thiamine-HCl reaction, which precedes hydrolysis and phosphorylation, before it is finally employed (in the form of TPP) in a number of enzymatic amino, fatty acid, and carbohydrate reactions.

Thiamine in foods can be degraded in a variety of ways. Sulfities, which are added to foods usually as a preservative, will attack thiamine at the methylene bridge in the structure, cleaving the pyrimidine ring from the thiazole ring. The rate of this reaction is increased under acidic conditions. Thiamine is degraded by thermolabile thiaminases (present in raw fish and shellfish). Some thiaminases are produced by bacteria. Bacterial thiaminases are cell surface enzymes that must dissociate from the membrane before being activated; the dissociation can occur in ruminants under acidotic conditions. Rumen bacteria also reduce sulfate to sulfite, therefore high dietary intakes of sulfate can have thiamine-antagonistic activities.

Plant thiamine antagonists are heat-stable and occur as both the ortho- and para-hydroxyphe-nols. Some examples of these antagonists are caffeic acid, chlorogenic acid, and tannic acid. These compounds interact with the thiamine to oxidize the thiazole ring, thus rendering it unable to be absorbed. Two flavonoids, quercetin and rutin, have also been implicated as thiamine antagonists.

Estimation of thiamine

Materials

- Sodium hydroxide (NaOH) 15%
- Potassium ferricyanide 1%
- Isobutyl alcohol

- Sodium sulphate (anhydrous)
- Sulphuric acid 0.1 N
- Standard thiamin hydrochloride – Dissolve 50.0 mg of thiamin hydrochloride in 500 ml of 0.1 N H_2SO_4 containing 25% alcohol (100 µg/ml). Store this solution in a brown bottle in a refrigerator.
- Working standard solution – Dilute 5.0 ml of stock solution to 100 ml with 0.1 N H_2SO_4 and again dilute 5.0 ml of the second solution to 100 ml with 0.1 N H_2SO_4 (0.25 µg/ml). This is prepared just before use.

Procedure

- Take 5.0 g of finely grounded plant sample in a 250 ml flask and slowly add 100.0 ml of 0.1 N H_2SO_4 without shaking and let it stand overnight.
- After overnight, shake vigorously and filter through Whatman filter No. 1. Discard the first 10 – 15 ml of filterate.
- Take 10.0 ml of filterate in a 100 ml separating funnel and add 3.0 ml of 15% NaOH followed by four drops (0.2 ml) of ferricyanide solution. Shake well for 30 seconds.
- Add 15.0 ml of isobutanol quickly and stopper immediately, shake vigorously for 60 seconds and allow the layer to separate. Drain the lower layer and discard it.
- Add few grams of sodium sulphate directly to the separating funnel, put the stopper swirl gently to clarify the extract. If the solution is not clear, put few more grams of Na_2SO_4 and clarify the solution.
- Collect the clear extract from top using pipette into a test tube.
- The blank is also prepared by taking 10.0 ml of rxtract and following the above procedure, but omitting addition of fericyanide.
- Select a suitable primary (366 nm) and secondary filter as per the make of fluorimeter.
- Set the fluorimeter by initially adjusting the standard blank to 0 reading and standard to 100 reading. Then read the sample blank and sample. Since the light intensity is low so it is desirable to take 5 – 6 readings.

Calculation

Thiamin (μg) content in 100 g sample = ((0.25 x 10) / (a – a')) x (((x – x') x 100) / 10) x (10/5)

Where a = reading of standard = 100

a' = reaing of standard blank = 0

x = reading of standard sample and

x' = reading of standard sample blank

Alternative method for estimation of thiamine

Thiamin estimation (flourimetric analysts): Two gramms (2.0 g) of the sample (fresh and dried *Hibiscus sahdarriffa* and *Lactuca sativa* are grinded in a mortar and 30.0 ml of hydrochloric acid was added by portions. The content was thoroughly mixed and filtered using filter paper.

This analysis is of three (control, test sample and standard), for each sample both fresh and dried. To the first (control), 5.0 ml of hydrochloric acid solution is transferred, the second (sample). To 1.0 ml of the extract of the sample and 4.0 ml of hydrochloric acid was transferred. To the third (standard) 5.0 ml of thiamin solution is transferred. 1.5 ml of oxidizing mixture was poured into each phial and mixed, with gentle shaking to homogeneity. Then 5ml of butanol is added to each phial and stopper-using flask shaker and shaken vigorously for 5 minutes. The phials were then allowed to stand untill the contents are separated in to two layers. Then 0.5 ml of ethanol is added cautiously to further clarify the butanolic phase. The cleared butanolic layer is cautiously decanted into a flourimeter cell and successive measurement of the fluorescence intensity for the three (test sample, control and standard) solutions of both fresh and dried leaves of *Hibiscus sabdarriffa* and *Lactuca sativa* were recorded.

18.3. Riboflavin (Vitamin B_2)

Riboflavin, also known as vitamin B_2 and is the vitamin formerly known as C, is an easily absorbed coloured micronutrient with a key role in maintaining health in humans and other animals. It is the central component of the cofactor FAD and FMN, and is therefore required by all flavoproteins. As such, vitamin B_2 is required for a wide variety of cellular processes. It plays a key role in energy metabolism, and for the metabolism of fats, ketone bodies, carbohydrates, and proteins. Milk, cheese, leaf vegetables, liver, kidneys, legumes, yeast, mushrooms, and almonds are good sources of vitamin B_2.

The name "riboflavin" comes from "ribose" (the sugar whose reduced form, ribitol, forms part of its structure) and "flavin", the ring-moiety which imparts the yellow colour to the oxidized molecule (from Latin *flavus*, "yellow"). The reduced form, which occurs in metabolism along with the oxidized form, is colourless. Riboflavin is best known visually as the vitamin which imparts the orange colour to solid B-vitamin preparations, the yellow colour to vitamin supplement solutions, and the unusual fluorescent-yellow colour to the urine of persons who supplement with high-dose B-complex preparations. Riboflavin can be used as a deliberate orange-red food colour additive, and as such is designated in Europe as the E number E101.

Flavin mononucleotide (FMN) and flavin adenine dinucleotide (FAD) function as cofactors for a wide variety of oxidative enzymes and remain bound to the enzymes during the oxidation-reduction reactions. Flavins can act as oxidizing agents because of their ability to accept a pair of hydrogen atoms. Reduction of isoalloxazine ring (FAD, FMN oxidized form) yields the reduced forms of the flavoproteins ($FMNH_2$ and $FADH_2$).

Riboflavin is yellow or yellow-orange in colour and in addition to being used as a food colouring, it is also used to fortify some foods. It is used in baby foods, breakfast cereals, pastas, sauces, processed cheese, fruit drinks, vitamin-enriched milk products, and some energy drinks. Regarding occurrence and sources of vitamin B_2, yeast extract is considered to be exceptionally rich in vitamin B_2, and liver and kidney are also rich sources. Wheat bran, eggs, meat, milk, and cheese are important sources in diets containing these foods. Cereals contain relatively low concentrations of flavins, but are important sources in those parts of the world where cereals constitute the staple diet. The milling of cereals results in considerable loss (up to 60%) of vitamin B_2, so white flour is enriched in some countries such as USA by addition of the vitamin. The enrichment of bread and ready-to-eat breakfast cereals contributes significantly to the dietary supply of vitamin B_2. Polished rice is not usually enriched, because the vitamin's yellow colour would make the rice visually unacceptable to the major rice-consumption populations. However,

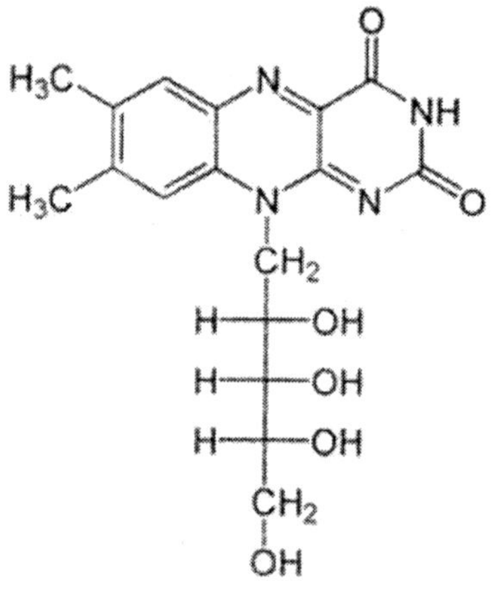

Ribiflavin (7,8-Dimethyl-10-[(2*S*,3*S*,4*R*)-2,3,4,5-tetrahy-droxypentyl] benzo[*g*] pteridine-2,4- dione)

most of the flavin content of whole brown rice is retained if the rice is steamed (parboiled) prior to milling. This process drives the flavins in the germ and aleurone layers into the endosperm. Free riboflavin is naturally present in foods along with protein-bound FMN and FAD. Bovine milk contains mainly free riboflavin, with a minor contribution from FMN and FAD. In whole milk, 14% of the flavins are bound noncovalently to specific proteins. Egg white and egg yolk contain specialized riboflavin-binding proteins, which are required for storage of free riboflavin in the egg for use by the developing embryo. Other sources of riboflavin include leafy green vegetables, fish, cottage cheese, avocados, tangerines, prunes, asparagus, broccoli, mushrooms, beef, salmon, turkey.

It is difficult to incorporate riboflavin into many liquid products because it has poor solubility in water, hence the requirement for riboflavin-5'-phosphate (E101a), a more expensive but more soluble form of riboflavin. Riboflavin is generally stable during the heat processing and normal cooking of foods if light is excluded. The alkaline conditions in which riboflavin is unstable are rarely encountered in foodstuffs. Riboflavin degradation in milk can occur slowly in dark during storage in the refrigerator.

Estimation of riboflavin (Vitamin B_2)

Sample preparation

- Weigh one riboflavin tablet on an analytical balance, record the mass.
- Grind the tablet using a clean, dry mortar and pestle.
- Accurately weigh about one fifth of the tablet and record the mass.
- Quantitatively transfer the sample to a 250 ml volumetric flask.
- Add 3 ml 1 M NaOH to the flask and gently swirl to dissolve the tablet.
- Add 3 ml glacial ethanoic acid to the flask and make up to the mark with water.
- Take a 25 ml aliquot from the 250 ml flask and makeup to 100 ml in a volumetric flask.
- Make a further 1 in 10 dilution of the unknown.

Standard preparation

- Using the 100mg/l Riboflavin stock solution make 100 ml of the following working solutions, diluting with 1% ethanoic acid, 2.5, 5.0, 7.5 and 10 mg/l.

- Using the 1% ethanoic acid as the blank, scan the 5.0 mg/l standard and the sample between 240 nm and 550nm.
- Confirm the sample contains riboflavin by comparing the two spectra.
- Select the wavelength that gives the highest absorbance value for a clearly defined peak.
- Read all standards and the sample.

Calculations

- Plot the calibration graph and determine the concentration of riboflavin in the diluted sample
- Use the dilution factor to determine the mass of riboflavin in the original analytical sample
- Determine the percentage of riboflavin in the analytical sample
- Compare the value obtained to the published value for the tablet, by calculating the mass in the original tablet

Alternative method for estimation of riboflavin

Riboflavin fluoresces at wavelength 440 – 500 nm. The intensity of fluroscence is proportional to the concentration of riboflavin in dilute solutions. The riboflavin is measured in terms of differences in fluroscence before and after chemical reduction.

Materials

- Sulphuric acid (H_2SO_4) 0.1 N
- Sodium acetate 2.5 M
- Potassium permanganate 4%
- Hydrogen peroxide 3%
- Riboflavin standard - Riboflavin is first dried over P_2O_5 in a desiccatorfor 24 hours, Dissolved 50.0 mg in 1500 ml water and 2.4 ml glacial acetic acid in a two liter flask. Warm, cool and make up to volume. Store under toluene in amber coloured bottle in a refrigerator. This solution represents a concentration 25 µg/ml riboflavin.
- Dilute 20.0 ml of the above riboflavin standard solution to 50.0 ml with water. Dilute 10.0 ml of the second riboflavin solution to 100.0 ml with water. This gives a solution of 1.0 µg/ml riboflavin.

- Sodium hydrosulphite (solid).

Procedure

- Take 2 – 5 g of the thoroughly mixed and ground plant sample into a 250 ml flask and add 75.0 ml of 0.1 N H_2SO_4 and mix.
- Autoclave the flask at 15 lbs for 30 minutes or keep the flasks in boiling water for 30 minutes. Cool it to room temperature.
- Add 5.0 ml of 2.5 M sodium acetate solution, mix and allow it to stand for 60 minutes and adjust the pH at 4.5.
- Transfer to a 100 ml volumetric flsk and make the volume with water. And filter through Whatman No. 2 or No. 4 discarding the first 10 – 15 ml.
- Add 0.5 ml of potassium permanganate solution sample and blank tubes, shake well till foaming is almost stopped.
- Set the flurimeter so that glass standard or sodium flurescein solution gives suitable galvanometer deflection. Determine fluroscence of blank (A) and sample (B) solutions. Do not expose the solutions more than 10 seconds.
- To the sample tube add 20 mg sodium hydrosulphite, stir and note the blank fluroscence (C). Do not use reading (C) taken after colloidal sulphur began to form.

Calculation

Riboflavin µg/100 g = ((B-C) / (A – C)) x (R/S) x (V/V_1) x 100

Where A = reading of the sample plus riboflavin standard, B = reading of sample plus water, C = reading after adding sodium R = standard riboflavin added = µg/V_1 of sample solution, V = original volume of sample solution = ml, V_1 = volume of sample solution taken for measurement (10 ml), S = sample weight (g).

Reading

Americal Association of Cereal Chemists (1976). (Complied-approved methods committee) AACC Inc. St. Paul Minn, USA 2.

18.4. Niacin (Vitamin B_3)

Niacin (also known as vitamin B_3 and nicotinic acid) is an organic compound with the formula $C_6H_5NO_2$ and, depending on the definition used, one of the 20 to 80 essential human nutrients. Insufficient niacin in the diet can cause nausea, skin and mouth lesions, anemia, headaches, and tiredness. Chronic niacin deficiency leads to a disease called pellegra. The lack of niacin may also be observed in pandemic deficiency disease which is caused by a lack of five crucial vitamins: niacin, vitamin C, thiamin, vitamin D and vitamin A, and is usually found in areas of widespread poverty and malnutrition. Niacin has been used for over 50 years to increase levels of HDL in the blood and has been found to decrease the risk of cardiovascular events modestly in a number of controlled human trials.

This colourless, water-soluble solid is a derivative of pyridine, with a carboxyl group (-COOH) at the 3-position. Other forms of vitamin B_3 include the corresponding amide, nicotinamide ("niacinamide"), where the carboxyl group has been replaced by a carboxamide group (CONH2), as well as more complex amides and a variety of esters. Nicotinic acid and niacinamide are convertible to each other with steady world demand rising from 8,500 tonnes per year in 1980s to 40,000 in recent years.

Niacin cannot be directly converted to nicotinamide, but both compounds are precursors of the coenzymes nicotinamide adenine dinucleotide (NAD) and nicotinamide adenine dinucleotide phosphate (NADP) *in vivo*. NAD converts to NADP by phosphorylation in the presence of the enzyme NAD^+ kinase. NADP and NAD are coenzymes for many dehydrogenases, participating in many hydrogen transfer processes. NAD is important in catabolism of fat, carbohydrate, protein, and alcohol, as well as cell signaling and DNA repair, and NADP mostly in anabolism reactions such as fatty acid and cholesterol synthesis. High energy requirements (brain) or high turnover rate (gut, skin) organs are usually the most susceptible to their deficiency. Although the two are identical in their vitamin activity, nicotinamide does not have the same pharmacological effects (lipid modifying effects) as niacin. Nicotinamide does not reduce cholesterol or cause flushing. Niacin is involved in both DNA repair, and the production of steroid hormones in the adrenal gland.

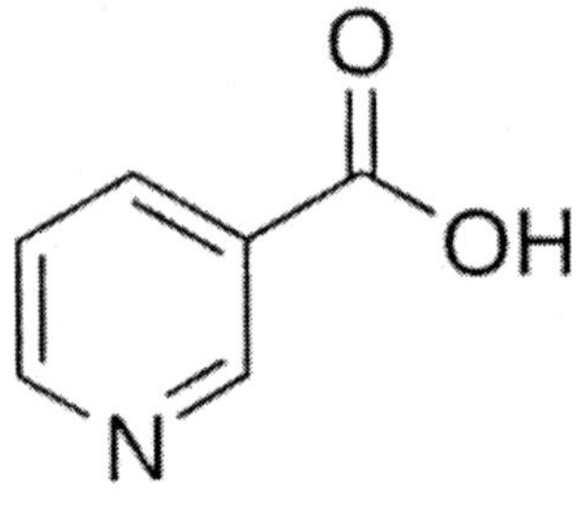

O
OH
N

Niacin (pyridine-3-carboxylic acid)

Niacin is found in variety of foods, including liver, chicken, beef, fish, cereal, peanuts and legumes, and is also synthesized from tryptophane, an essential amino acid found in most forms of protein *e.g.*, liver, heart and kidney, chicken, chicken breast, beef, tuna, salmon halibut, eggs, avacadoes, dates, tomatoes, leaf vegetables, broccoli, carrots, sweet potatoes, asaparagus, nuts, whole grain products, legume.

Estimation of Niacin (Vitamin B_3)

Niacin reacts with cyanogens bromide to give a pyridium complex which undergo's rearrangement yielding derivatives. These derivatives couple with aromatic amines to give yellow coloured pigment. Under proper conditions, the intensity of the yellow colour produced is proportional to the amount of niacin present.

Materials

- Cyanogen bromide – Keep a bottle containing bromine in an ice bath for a few minutes and then using a vacupet take 25 ml of bromine into a beaker containing 500 ml distilled water. Take a 10% solution of sodium cyanide and deliver slowly into the bromine water with constant shaking until solution is clear.
- Redistilled 4% aniline in absolute alcohol.
- Standard – Dissolve 10 mg niacin and make the volume to 100 ml with water. One ml of the standard contains 100 µg of niacin.

Procedure

- Grind the sample (5 g) in 30.0 ml of 4.0 N sulphuric acid and steam it for 30 minutes. Cool make the volume to 50 ml with water. Filter through Whatman No. 1 filter paper Add 60% basic lead acetate to the filterate (5 ml to 25 ml filterate). Adjust the pH to 9.5 with a pH meter or thymol blue indicating paper using 10 N NaOH. Centrifuge and to the supernatant add 2.0 ml of conc. H_2SO_4 and allow to stand for 60 minutes and again centrifuge. Collect the supernatant and add 5.0 ml of $ZnSO_4$ and adjust the pH to 8.4 with 10 N NaOH. Centrifuge and adjust the pH of the supernatant to 7.0. Use the supernatant as the source of niacin.
- Take the standard of niacin in varied volumes (0.1 – 0.5 ml) along with the sample(s).
- Make the volume to 6.0 ml with water and add 3.0 ml of cyanogens bromide using a burette and shake the mixture and after 100 minutes add 1.0 ml of 4% aniline solution to the tubes.

- Allow the tubes to then measure the optical density at 420 nm against reagent blank.

18.5. Pantothanic acid (Vitamin B_5)

Pantothenic acid, also called pantothenate or vitamin B_5 (a B vitamin), is a water-soluble vitamin. For many animals, pantothenic acid is an essential nutrient. Pantothenic acid is the amide between pentoic acid and β-alanine. Its name derives from the Greek *pantothen*, meaning "from everywhere", and small quantities of pantothenic acid are found in nearly evry food, with high amounts in whole grain cereals, legumes, eggs, meat, royal jelly, avacado, and yogurt. It is commonly found as its alcohol analog, the provitamin panthenol, and as calcium pantothenate. Pantothenic acid is an ingredient in some hair and skin care products.

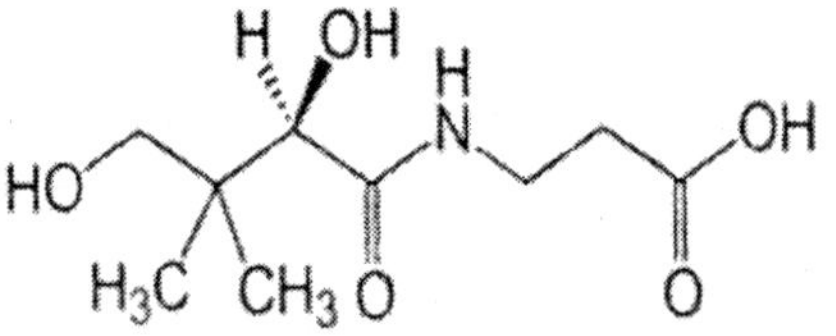

H OH H N OH HO H_3C CH_3 O O

3-[(2,4-Dihydroxy-3,3-dimethylbutanoyl)amino]propanoic acid

Only the dextrorotatory (D) isomer of pantothenic acid possesses biologic activity. The levorotatory (L) form may antagonize the effects of the dextrorotatory isomer.

Pantothenic acid is used in the synthesis of coenzyme A (CoA). Coenzyme A may act as an acyl group carrier to form acetyl-CoA and other related compounds; this is a way to transport carbon atoms within the cell. CoA is important in energy metabolism for pyruvate to enter the tricarboxylic acid cycle (TCA cycle) as acetyl-CoA, and for α-ketoglutarate to be transformed to succinyl-CoA in the cycle. CoA is also important in the biosynthesis of many important compounds such as fatty acids, cholestrol, and acetylcholine. CoA is incidentally also required in the formation of ACP, which is also required for fatty acid synthesis in addition to CoA.

Pantothenic acid in the form of CoA is also required for acylation and acetylation, which, for example, are involved in signal transduction and enzyme activation and deactivation, respectively. Since pantothenic acid participates in a wide array of key biological roles, it is essential to all forms of life. As such, deficiencies in pantothenic acid may have numerous wide-ranging effects, as discussed below.

18.6. Pyridoxine (Vitamin B_6)

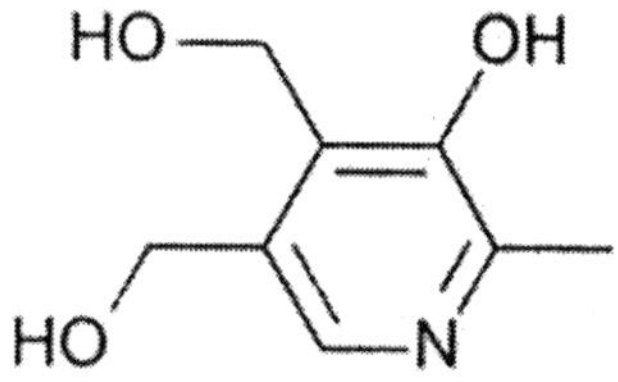

Pyridoxine is one of the compounds that can be called vitamin B_6, along with pyridoxal and pyridoxamine. It differs from pyridoxamine by the substituent at the '4' position. Its hydrochloride salt pyridoxine hydrochloride is often used. It is on the World Health Organisatio's List of Essential Medicines, a list of the most important medication needed in a basic health system. It is based on a pyridine ring, with hydroxyl, methyl, and hydroxymethyl substituents. It is converted to the biologically active form puridoxal-5-phosphate.

Pyridoxine (4,5-Bis(hydroxymethyl)-2-methylpyridin-3-ol)

Vitamin B_6 assists in the balancing of sodium and potassium as well as promoting red blood cell production. It is linked to cardiovascular health by decreasing the formation of homocysteine. Pyridoxine may help balance hormonal changes in women and aid the immune system. Lack of pyridoxine may cause anemia, nerve damage, seizures, skin problems, and sores in the mouth. It is required for the production of the monoamine neurotransmitters serotonin, dopamine, norpinephrine and epinephrine, as it is the precursor to puridoxal phosphate: cofactor for the enzyme aromatic acid decarboxylase. This enzyme is responsible for converting the precursors 5-hydroxytryptophan (5-HTP) into serotinin and melatonin, and levodopa (L-DOPA) into dopamine, noradrenaline and adrenaline. As such it has been implicated in the treatment of depression and anxiety. Fish, soybeans, avocado, lima beans, chicken, bananas, cauliflower, green peppers, potatoes, spinach, raisins, brewer's yeast, blackstrap molasses, liver, kidney, heart.

18.7 Estimation of Vitamins B-Complex (B_2, B_3, B_5 and B_6) by HPLC Method

As the chemicals structures are not related, a considerable number of methods have been reported describing different physical, chemical and biological methods to analyze the vitamins. The simultaneous determination of several water-soluble vitamins is difficult and often many different analyses have to be performed. Different instrumental methods have been used for the determination of B-group vitamins, including electrochemical methods, *spectrophotometry*, derivative UV spectrophotometry, spectrofluorimetry, normal phase and reversed phase TLC and HPLC procedures as well as capillary electrophoresis. The determination of B-complex mainly in tablets using HPLC methods is most widely used methods for the determination of B-group vitamins are reversed-phase HPLC, using a C18 column and aqueous–organic mobile phases, in acidic media. This HPLC method for B-vitamin profile *viz.*, riboflavin (vitamin B_2), niacin (vitamin B_3), pantothenic acid (vitamin B_5) and pyridoxine (vitamin B_6) in leaves of *Lagenaria vulgaris, Amaranthus viridis, Amaranthus gangeticuss, Basella alba,* and *Momordica charantia* are given below.

Materials

- High Performance Liquid Chromatographic system equipped with an auto sampler and UV-Visible detector is used for the analysis. The data were recorded using LC-solutions software.
- Extraction solution is made by mixing 50 ml of acetonitrile with 10 ml of glacial acetic acid and the volume is finally made up to 1000 ml with double distilled water.
- To prepare buffer, 1.08 g of hexane sulphonic acid sodium salt and 1.36g of potasium dihydrogen phosphate were dissolved in 940 ml HPLC water and 5 ml of triethylamine is added to it and the pH is adjusted to 3.0 with orthophosphoric acid.
- For the preparation of mobile phase, buffer and methanol were mixed in a ratio of 96:4 and filtered through 0.22 μM membrane filter and sonicated for degassing in an ultrasonic bath.
- Standard stock solution for vitamin B_2 (riboflavin) is prepared by dissolving 6.9 mg of riboflavin in 100 ml of extraction solution. Standard stock solution for vitamin B_3 (niacin) is prepared by dissolving 41.5 mg of niacin in 25 ml of double distilled water. Standard stock solution for vitamin B_5 (calcium salt of pantothenic acid) is prepared by dissolving 21.4 mg of calcium d-pantothen-ate in 25 ml of double distilled water. Standard stock solution for vitamin B_6 (pyridoxine) is prepared by dissolving 20.8 mg of pyridoxine hydrochloride in 25 ml of double distilled water.

Procedure

- Fresh samples are collected from field and in laboratory the samples are washed with distilled water and wiped with filter paper. About 10 g sample are chopped out into small pieces and homogenized for the estimation of B-vitamins. Some freshly chopped samples were preserved in refrigerator.
- 10 g of each sample is weighed and made into homogenized in mortar with pestle and transferred into conical flask and 25 ml of extraction solution is added, kept on shaking water bath at 70°C for 40 min. Thereafter, the sample is cooled down, filtered and finally the volume is made up to 50

ml with extraction solution. Then sample filtered through 0.45 μm filter tips and aliquots of 20 μL from this solution is injected into the HPLC by using auto-sampler.

- Analytical reversed phase C-18 column (ODS column, 250 × 4.6 mm, 5 μm) is used for the separation.
- Mobile phase consisting of a mixture of buffer and methanol in the ratio of 96:4 (v/v) is deliv-ered at a flow rate of 1 ml/min with UV detection at 210 nm.
- The mobile phase is filtered through 0.22 ìm membrane filter, sonicated and degassed before use.
- Analysis is performed at room temperature (~26°C) temperature. All the prepared sample solutions were first chromatographed to ensure that interfering peaks were not present. 20 μL aliquots of the standard solutions and sample solutions were injected.

Reading

Md. Nazmul Hasan1, M. Akhtaruzzaman1, Md. Zakir Sultan (2013). Estimation of vitamins B-complex (B_2, B_3, B_5 and B_6) of some leafy vegetables indigenous to Bangladesh by HPLC method. *Journal of Analytical Sciences, Methods and Instrumentation*, **3:** 24-29

18.8. Folic acid (Vitamin B_9)

Folic acid (also known as vitamin M, vitamin B_9, vitamin B_c (or folacin), pteroyl-L-glutamic acid, and pteroyl-L-glutamate) is a form of the water-soluble vitamin B_2. Folate is a naturally occurring form of the vitamin, found in food, while folic acid is synthetically produced, and used in fortified foods and supplements. Folic acid is itself not biologically active, but its biological importance is due to tetrahydrofolate and other derivatives after its conversion to dihyroxyfolic acid in the liver.

Vitamin B_9 (folate converted from folic acid) is essential for numerous bodily functions. Humans cannot synthesize folate *de novo*; therefore, folate has to be supplied through the diet to meet their daily requirements. The human body needs folate to synthesize DNA, repair DNA, and methylate DNA as well as to act as a cofactor in certain biological reactions. It is especially important in aiding rapid cell division and growth, such as in infancy and pregnancy. Children and adults both require folate to produce healthy red blood cells and prevent anemia.

Folate and folic acid derive their names from the Latin word *folium*, which means "leaf". Folate occurs naturally in many foods and, among plants, are especially plentiful in dark green leafy vegetables. A lack of dietary folates can lead to folate deficiency. A complete lack of dietary folate takes months before deficiency develops as normal individuals have about 500–20,000 μg of folate in body stores. This deficiency can result in many health problems, the most notable one being neutral tube defects in developing embryos. Common symptoms of folate deficiency include diarrhoea, macrocytic anemia with weakness or shortness of breath, nerve damage with weakness and limb numbness (peripheral neuropathy), pregnancy complications, mental confusion, forgetfulness or other cognitive deficits, mental depression, sore or swollen tongue, peptic or mouth ulcers, headaches, heart palpitations, irritability, and behavioral disorders. Low levels of folate can also lead to homocysteine accumulation. DNA synthesis and repair are impaired, which could lead to cancer development.

Folate naturally occurs in a wide variety of foods, including vegetables (particularly dark green leafy vegetables), fruits and fruit juices, nuts, beans, peas, dairy products, poultry and meat, eggs, seafood, grains, and some beers. Spinach, Liver, yeast, asparagus, and Brussels sprouts are among the foods with the highest levels of folate. Folic acid is added to grain products in many countries, and in these countries, fortified products make up a significant source of the population's folic acid intake. Because of the difference in bioavailability between supplemented folic acid and the different forms of folate found in food, the dietary folate equivalent (DFE) system was established. 1 DFE is defined as 1 μg of dietary folate, or 0.6 μg of folic acid supplement. This is reduced to 0.5 μg of folic acid if the supplement is taken on an empty stomach.

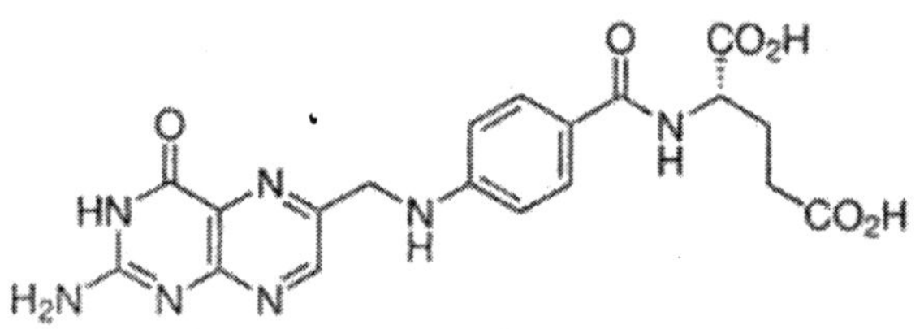

Folic acid
(2*S*)-2-[(4-{[(2-amino-4-hydroxypteridin-6-yl)methyl] amino} phenyl)formamido] pentanedioic acid

Folate naturally found in food is susceptible to high heat and ultraviolet light, and is soluble in water. It is heat-labile in acidic environments and may also be subject to oxidation. Some meal replacement products do not meet the folate requirements as specified by the RDAs. Folate (B_9) can also be processed from the pro-vitamin Pterolmonoglutamic acid (Vitamin B_{10}).

Estimation of folic acid

Materials

- Standards - Standard stock solutions are prepared by dissolving folic acid and 5-methyltetrahydrofolate separately in 0.01 mol/l NaOH to a concentration of 20 μmol/l. Concentrations were determined in pH 7.0 buffered solutions, using UV absorption at 280 nm for folic acid and a molar extension coefficient of 27,600 L · mol^{-1} · cm^{-1} and at 290 nm for 5-methyltetrahydrofolate and a molar extension coefficient of 32,000 L · mol^{-1} cm^{-1}. Following spectral determinations, the standard stock solutions were stored in small aliquots in 1% sodium ascorbate at -20° C. These standards were further purified by affinity column chromatography before use.

Procedure

- Extraction of folates from fortified food - Samples were thawed and were suspended (1 g/10 ml final volume) in a 0.026 mol/ Tris-HCl extraction buffer (pH 7.4) containing 1% (w/v) sodium ascorbate and 0.02 mCi/l [^{3}H] folic acid tracer (specific activity: 69 Ci/m mol). The samples were capped and heated in an autoclave for 15 min at 120°C (1.034 bar), cooled in an ice bath, then homogenized with a polytron homogenizer for 30 s at medium setting.

- The homogenates are incubated for 4 h at 37°C with 1.25 ml of α-amylase solution and 0.2 ml of rat plasma folate conjugase. Subsequently, the samples were treated with 1 ml protease for an additional 1 h at 37°C, then heated for 5 min in a boiling water bath, cooled in ice, and centrifuged for 20 min at 36,000 × *g* and 4°C. The supernatants were filtered through a microfilter and radioactivity in an aliquot of the sample was measured. If analysis is within 12 h, the filtrates are stored at 4°C until folate analysis. For longer storage, the folate extracts were stored at -20° C until analysis.

- HPLC measurement of folic acid and 5-methyltetrahydrofolate - The folates were separated on an ODS-Hypersil (5 μm, 4.6 × 250 mm i.d.) analytical column. A flow rate of 1 ml/min is used. The mobile phase program consisted of 3 min with 100% A (28 mmol/l dibasic potassium phosphate and 60 mmol/l phosphoric acid in water) followed by a linear gradient of 10 min to 70% A:30% B (28 mmol/l dibasic potassium phosphate and 60 mmol/l phosphoric acid in 200 ml/l acetonitrile and 800 ml/l water). A second linear gradient from 70% A:30% B to 45% A:55% B is then run over the next 17 min, followed by a third linear gradient to 43% A:57% B over the next 15 min. At 45 min, the column is equilibrated for 5 min in the initial conditions and another sample analysis could be initiated immediately.

- The absorbance of folic acid was monitored with a diode array detector set at 280 nm, and of 5-methyltetrahydrofolate with a fluorescence detector set at 295 nm excitation and 360 nm emission wavelengths. Peak identification was based on a combination of the retention time and the spectral characteristics.
- Quantification - Quantification was based on an external standard method in which peak areas for both UV (folic acid) and fluorescence (5-methyltetrahydrofolate) were plotted against concentration. Calibration plots using linear regression analysis were prepared for 5-methyltetrahydrofolate and for folic acid. The standard solutions were purified through FBP–Affigel 10 affinity columns, and the calibration curves were prepared from those using similar standard levels as expected in the samples. Both standards and samples were adjusted for tritium folic acid recovery.

Reading

Rosalia Póo-Prieto, David B. Haytowitz, Joanne M. Holden, Gail Rogers, Silvina F. Choumenkovitch, Paul F. Jacques, and Jacob Selhub. (2006). Use of the Affinity / HPLC Method for Quantitative Estimation of Folic Acid in Enriched Cereal-Grain Products. *Journal of Nutrition*, **136** (12): 3079 - 3083.

Alternative method for estimation of folic acid

A colourimetric method has been developed for estimating folic acid. The method is about to reduce folic acid to 2,4,5-triamino-6-hydroxy-pyrimidine (TAHP), which is treated with ninhydrin to yield a stable purple complex. Participation of 5-amino and 6-hydroxy groups of TAHP in the colour reaction with ninhydrin is suggested on the basis of experimental evidence. The colour obtained has an absorbance maximum at 555 nm and obeys Beer's law in the concentration range of 4.5 to 45 µg folic acid/ml. The method has been successfully used for assaying folic acid in samples.

Reading

Rao GR, Mahajan SN, Kanjilal G, Mohan KR, (1977). New colourimetric method for folic acid assay in dosage forms. *Journal - Association of Official Analytical Chemists*, **60(3):** 531 - 535.

18.9. Cobalamine (Vitamin B_{12})

Vitamin B_{12}, also called cobalamin, is a water-soluble vitamin with a key role in the normal functioning of the brain and nervous system, and for the formation of blood. It is one of the eight B-vitamins. It is normally involved in the metabolism of every cell of the human body, especially affecting DNA synthesis and regulation, but also fatty acid synthesis (especially odd chain fatty acids) and energy production. Neither fungi, plants, nor animals are capable of producing vitamin B_{12}. Only bacteria and archaea have the enzymes required for its synthesis, although many foods are a natural source of B_{12} because of bacterial symbiosis. The vitamin is the largest and most structurally complicated vitamin and can be produced industrially only through bacterial fermentation-synthesis.

Vitamin B_{12} consists of a class of chemically related compounds (vitamins), all of which have vitamin activity. It contains the biochemically rare element cobalt sitting in the center of planar tetra-pyrrole ring called a Corrin ring. Biosynthesis of the basic structure of the vitamin is accomplished only by bacteria (which usually produce hydroxycobalamine), but conversion between different forms of the vitamin can be accomplished in the human body. A common semi-synthetic form of the vitamin, cyanocobalamine, does not occur in nature, but is produced from bacterial hydroxocobalamin and then used in many pharmaceuticals and supplements, and as a food additive, because of its stability and lower production cost. In the body, it is converted to the human physiological forms methylcobalamine and 5'-deoxyadenosylcobalamine, leaving behind the cyanide, albeit in minimal concentration. More recently, hydroxocobalamin, methylcobalamin, and adenosylcobalamin can be found in more expensive pharmacological products and food supplements. The extra utility of these is currently debated.

Vitamin B_{12} was discovered from its relationship to disease pernicious anaemia, which is an autoimmune disease in which parital cells of the stomach responsible for secreting intrinsic factor are destroyed, the same cells responsible for secreting acid in the stomach. Intrinsic factor is crucial for the normal absorption of B_{12}, so a lack of intrinsic factor, as seen in pernicious anaemia, causes a vitamin B_{12} deficiency. Many other subtler kinds of vitamin B_{12} deficiency and their biochemical effects have since been elucidated.

The names vitamin B_{12}, vitamin B_{12}, or vitamin B_{12}, and the alternative name cobalamin, generally refer to all forms of the vitamin. Some medical practitioners have suggested that its use be split into two categories.

- In a broad sense, B_{12} refers to a group of cobalt-containing vitamer compounds known as cobalamins: these include cyanocobalamine (an artifact formed from using activated charcoal, which always contains trace cyanide, to purify hydroxycobalamin), hydroxycobalamine (another medicinal form, produced by bacteria), and finally, the two naturally occurring cofactor forms of B_{12} in the human body: 5'-deoxyadenosylcobalamin (adenosylcobalamine—AdoB_{12}), the cofactor of Methylmalonyl Coenzyme A mutase (MUT), and methylcobalamine (MeB_{12}), the cofactor of enzyme Methonine synthase, which is responsible for conversion of homocysteine to methionine. Methionine synthase also needs Methyl Tetra-hydro folate as a cofactor along with B_{12}.
- The term B_{12} may be properly used to refer to cyanocobalamine, the principal B_{12} form used for foods and in nutritional supplements. This ordinarily creates no problem, except perhaps in rare cases of eye nerve damage, where the body is only marginally able to use this form due to high cyanide levels in the blood due to cigarette smoking; it thus requires cessation of smoking or B_{12} given in another form, for the optic symptoms to abate. However, tobacco amblyopia is a rare condition, and it is yet unclear whether it represents a peculiar B_{12} deficiency that is resistant to treatment with cyanocobalamin.

Finally, so-called pseudovitamin-B_{12} refers to B_{12}-like analogues that are biologically inactive in humans and yet found to be present alongside B_{12} in humans, many food sources (including animals), and possibly supplements and fortified foods. In most cyanobacteria, including *Spirulina*, and some algae, such as dried Asakusa-nori (*Porphyra tenera*), pseudovitamin-B_{12} is found to predominate.

Estimation of cobalamine

The qualitative and quantitative analysis of cobalamine and its derivatives is of great importance in pharmaceutical analysis. Several methods including paper chromatography, paper electrophoretic, and column chromatography has been used for the separation and quantitative determination of the compounds and the quantitation of the separated compounds have been accomplished.

Procedure

- The separation is performed on precoated Kieselgel 60 chromaplate, 20 x 20 cm. The samples are transferred to the plate in a 1.0 cm narrow line and it is eluted upto 15 cm distance in a presaturated chamber preotected from light.

- After the run, the plate is dried in air stream at room temperature.
- The separated spots are evaluated by densitometry using chromatogram-spectrophotometer.

Reading

Szepesi, G. And Molnar, J. (1981). Improved quantitative thin layer chromatographic method of separation of cobalamine derivatives. *Cromatographia*, **14 (12):** 709

18.10. Estimation of Vitamins B_1, B_3, B_6, B_9 and B_{12} by HPLC

Materials

Standard solutions

- Standard stock solution of vitamin B_1 is prepared by dissolving 25.0 mg of thiamine chloride hydrochloride in 50.0 ml of water.
- Standard stock solution of vitamin B_3 is prepared by dissolving 25.0 mg of nicotinamide in 50.0 ml of water.
- Standard stock solution of vitamin B_6 is prepared by dissolving 25.0 mg of pyridoxine hydrochloride in 50.0 ml of water.
- Standard stock solution of folic acid is prepared by dissolving 25.0mg of folic acid in 50.0 ml of water. 1 ml of the standard stock solution of folic acid is diluted to 50 ml with 15 % methanol solution.
- The working standard solution of B_1, B_3, B_6 and folic acid is obtained by diluting 0.4 ml of standard stock solution of B_1, 0.8 ml of standards stock solution of B_3, 0.2 ml of standard stock solution of B_6 and 0.8 ml of standard stock solution of folic acid to 10 ml with 15 % methanol solution.
- Standard stock solution of vitamin B_{12} is prepared by dissolving 10.0 mg of cyanocobalamine in 100.0 ml of water.
- The working standard solution of B_{12} is obtained by diluting 1 ml of the standard stock solution of vitamin B_{12} to 10 ml with water. 1 ml of this solution is diluted to 10.0 ml with the same solvent.

Procedure

Sample preparations

- Sample preparation of B_1, B_3, B_6 and folic acid. Twenty tablets were weighed and triturated to a fine powder. The average mass of one tablet is transferred into a 50 ml volumetric flask and 15 % of methanol solution is added. The mixture is sonicated (15 min) and diluted to the mark with the same solvent. 1 ml of this solution is transferred into a 10 ml volumetric flask, diluted to the mark with the same solvent and filtered through a 0.2 m Millipore filter.
- Sample preparation of B_{12}. Twenty tablets were weighed and triturated to a fine powder. The average mass of two tablets is transferred into a 100 ml volum etric flask and water is added. The mixture is sonicated (20 min), diluted to the mark with the same solvent and filtered through a 0.2m Millipore filter.

Apparatus, mobile phase and chromatographic conditions

- A chromatographic LKB 2150 HPLC System, equipped with a Waters M 484 UV/VIS detector, is connected with a computed integrator Maxima 820 Work Station. The detection wavelength is adjusted to 290 nm with a sensitivity of 0.05 AUFS for the determination of B_1, B_3, B_6 vitamins and folic and to 550 nm with a sensitivity of 0.05 AUFS ·for the determination of vitamin B_{12}.
- A column (15 cm 4.6 mm; particle size 5m) is used for the determination of B_1, B_3, B_6 vitamins and folic acid. Another column (15 cm 4.6 mm; particle size 5m) is used for the determination of vitamin B_{12}.
- The estimations are done at 35° C for B_1, B_3, B_6 vitamins and folic acid and at 25 °C for vitamin B_{12}.
- The mobile phase for the determination of B_1, B_3, B_6 vitamins and folic acid is of methanol – 5 mM heptanesulphonic acid sodium salt / 0.1 % TEA (25:75 V/V). The pH 2.8 is adjusted with orthophosphoric acid. The flow rate is 1 ml/min and the injected volume 10 l. The mobile phase for the determination of vitamin B_{12} is methanol–water (22:78 V/V). The flow rate is 0.8 ml/min and the injected volume 100 l.
- The prepared mobile phases were filtered through a 0.2 m Anotop filter and degassed with an ultrasonic bath.

HPLC procedure

Prior to injection into the chromatographic system, all analytical solutions were degassed by sonication. All the prepared sample solutions were first chromatographed to ensure that interfering peaks were not present. 101 and 100 l aliquots of the standard solutions and sample solutions were injected. Triplicate injections were made for each solution.

Reading

Amidzic, R., Brboric, J., Udina, O. and Vladimirov, S. (2005). RP-HPLC Determination of vitamins B1, B3, B6, folic acid and B12 in multivitamin tablets. *J. Serb. Chem. Soc.* **70(10):** 1229–1235.

18.11. Biotin

Biotin, also known as vitamin H or coenzyme R, is a water-soluble B-vitamin (vitamin B_7). It is composed of a ureido (tetrahydroimidizalone) ring fused with a tetrahydrothiophene ring. A valeric acid substituent is attached to one of the carbon atoms of the tetrahydrothiophene ring. Biotin is a coenzyme for carboxylase enzymes, involved in the synthesis of fatty acid, isoleucine, and valine, and in gluconeogenesis. The only human health condition for which there is strong evidence of biotin's potential benefit as a treatment is biotin deficiency.

Biotin is necessary for cell growth, the production of fatty acids, and the metabolism of fats and amino acids. Biotin assists in various metabolic reactions involving the transfer of carbon dioxide. It may also be helpful in maintaining a steady blood sugar level. Biotin is often recommended as a dietary supplement for strengthening hair and nails, though scientific data supporting this outcome are weak. Nevertheless, biotin is found in many cosmetics and health products for the hair and skin.

Biotin deficiency is rare because, in general, intestinal bacteria produce biotin in excess of the body's daily requirements. For that reason, statutory agencies in many countries, for example the USA and Australia, do not prescribe a recommended daily intake of biotin. However, a number of metabolic disorder exist in which an individual's metabolism of

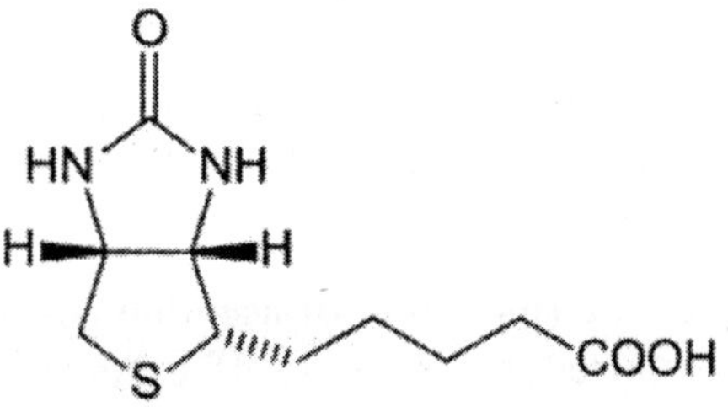

Biotin (5-[(3a*S*,4*S*,6a*R*)-2-oxohexahydro-1*H*-thieno[3,4-*d*]imidazol-4-yl]pentanoic acid)

biotin is abnormal, such as deficiency in the holocarboxylase synthetase enzyme which covalently links biotin onto the carboxylase, where the biotin acts as a cofactor. Biotin is consumed from a wide range of food sources in the diet, but few are particularly rich sources. Foods with a relatively high biotin content include Swiss chard, raw egg yolk (however, the consumption of avidin-containing egg white with egg yolks minimizes the effectiveness of egg yolk's biotin in one's body), liver, Saskatoon berries, and leafy green vegetables. The dietary biotin intake in Western populations has been estimated to be 35 to 70 μg/d (143--287 nmol/d). Biotin is also available in supplement form and can be found in most pharmacies. The synthetic process developed by Leo Sternbach and Moses Wolf Goldberg in the 1940s uses fumaric acid as a starting material.

18.12. Ascorbic Acid (Vitamin C)

Vitamin C or L-ascorbic acid, or simply ascorbate (the anion of ascorbic acid), is an essential nutrient for humans and certain other animal species. Vitamin C refers to a number of vitamers that have vitamin C activity in animals, including ascorbic acid and its salts, and some oxidized forms of the molecule like dehydroascorbic acid. Ascorbate and ascorbic acid are both naturally present in the body when either of these is introduced into cells, since the forms interconvert according to pH.

Vitamin C is a cofactor in at least eight enzymatic reactions, including several collagen synthesis reactions that, when dysfunctional, cause the most severe symptoms of scurvy. In animals, these reactions are especially important in wound-healing and in preventing bleeding from capillaries. Ascorbate may also act as an antioxidant against oxidative stress. However, the fact that the enantiomer D-ascorbate (not found in nature) has identical antioxidant activity to L-ascorbate, yet far less vitamin activity, underscores the fact that most of the function of L-ascorbate as a vitamin relies not on its antioxidant properties, but upon enzymic reactions that are stereospecific. "Ascorbate" without the letter for the enantiomeric form is always presumed to be the chemical L-ascorbate.

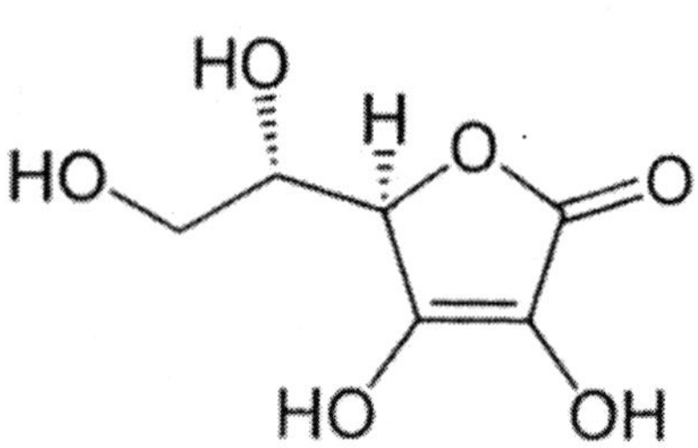

Ascorbic acid (2-Oxo-L-threo-hexono-1,4-lactone-2,3-enediol)
or
(*R*)-3,4-dihydroxy-5-((*S*)- 1,2-dihydroxyethyl)furan-2(5*H*)-one

Ascorbate (the anion of ascorbic acid) is required for a range of essential metabolic reactions in all animals and plants. It is made internally by almost all organisms; the main exceptions are most bats, all guinea pigs, capybaras, and the Anthropodea (*i.e.*, Haplorrhini, one of the two major primate suborders, consisting of tarsiers, monkeys, and humans and other aoes). Ascorbate is also not synthesized by some species of birds and fish. All species that do not synthesize ascorbate require it in the diet. Deficiency in this vitamins causes the disease scurvy in humans. Ascorbic acid is also widely used as a food additive, to prevent oxidation.

The biological role of ascorbate is to act as a reducing agent, donating electrons to various enzymatic and a few non-enzymatic reactions. The one- and two-electron oxidized forms of vitamin C, semidehydro-ascorbic acid and dehydroascorbic acid, respectively, can be reduced in the body by glutathione and NADPH-dependent enzymatic mechanisms. The presence of glutathione in cells and extracellular fluids helps maintain ascorbate in a reduced state. In plants, aAscorbic acid is associated with chloroplasts and apparently plays a role in ameliorating the oxidative stress of photosynthesis. In addition, it has a number of other roles in cell division and protein modification. Plants appear to be able to make ascorbate by at least one other biochemical route that is different from the major route in animals, although precise details remain unknown.

Estimation of ascorbic acid (volumetric method)

Ascorbic acid reduces by 2,6-dichlorophenol indophenols dye to a colourless leuco-base. The ascorbic acid get oxidised to dihydroascorbic acid. Though the dye is a blue colourd compound, the end point is the appearance of pink colour. The dye is pink coloured is acid medium. Oxalic acid is used as the titrating medium.

Materials

- Oxalic acid (4%).
- Dye solution – take 42 mg of sodium bicarbonate in a small volume of water. Dissolve 52 mg of 2,6-dichloro phenol indophenols is sodium bicarbonate solution and make the volume to 200 ml with water.
- Stock solution – Take 100 mg of ascorbic acid in 100 ml of 4% oxalic acid solution in a standard flask (1 mg/ml).
- Working standard solution – Dilute 10 of the stock solution to 100 ml with 4% oxalic acid solution and this gives a concentration of 100 μg/ml.

Procedure

- Extract the sample (0.5-5.0 g) in 4% oxalic acid solution and make up the volume to 100 ml and centrifuge. The supernatant is used as the source of ascorbic acid.
- Take out 5.0 ml of the supernatant and add 10.0 ml of 4% oxalic acid solution and titrate against the dye (V2 ml).
- Take 5.0 ml of working standard solution in a 100 ml flask and add 10 ml of 4% oxalic acid solution and titrate against the dye (V1 ml). Ens point would be the appearance of pink which persist for few minutes. The amount of dye solution used is equal to the amount of ascorbic acid.

Calculation

Amount of ascorbic acid mg/100 g sample = ((0.5 mg / V1 ml) x (V2 / 5.0 ml) x (100 ml / Wt. of sapmle)) x 100

Reading

Harris, L.J. and Ray, S.N. (1935). Ascorbic acid, *Lancet*, **1:** 462.

Alternative volumetric method for estimation of ascorbic acid

This method determines the vitamin C concentration in a solution by a redox titration using iodine. As the iodine is added during the titration, the ascorbic acid is oxidised to dehydroascorbic acid, while the iodine is reduced to iodide ions.

$$\textbf{ascorbic acid} + I_2 \rightarrow 2\, I^- + \textbf{dehydroascorbic acid}$$

Due to this reaction, the iodine formed is immediately reduced to iodide as long as there is any ascorbic acid present. Once all the ascorbic acid has been oxidised, the excess iodine is free to react with the starch indicator, forming the blue-black starch-iodine complex. This is the endpoint of the titration. The method is suitable for use with fresh or packaged fruit juices and solid fruits and vegetables. This method is more straight forward than the alternative method using potassium iodate, but as the potassium iodate solution is more stable than the iodine as a primary standard, the alternative method is more reliable.

Materials

- Iodine solution: (0.005 mol^{-1}). Take 2.0 g of potassium iodide into a 100 ml beaker. Take 1.3 g of iodine and add it into the same beaker. Add a few ml of distilled water and swirl for a few minutes until iodine is dissolved.

Transfer iodine solution to a 1 liter volumetric flask, making sure to rinse all traces of solution into the volumetric flask using distilled water. Make the solution up to the 1 liter mark with distilled water.

- Starch indicator solution: (0.5%). Take 0.25 g of soluble starch and add it to 50 ml of near boiling water in a 100 ml conical flask. Stir to dissolve and cool before using.

Procedure

- For packaged fruit juice: This may also need to be strained through cheesecloth if it contains a lot of pulp or seeds.
- For fruits and vegetables: Cut a 100 g sample into small pieces and grind in a mortar and pestle. Add 10 ml portions of distilled water several times while grinding the sample, each time decanting off the liquid extract into a 100 ml volumetric flask. Finally, strain the ground fruit/vegetable pulp through cheesecloth, rinsing the pulp with a few 10 ml portions of water and collecting all filtrate and washings in the volumetric flask. Make the extracted solution up to 100 ml with distilled water.
- Alternatively the 100 g sample of fruit or vegetable may be blended in a food processor together with about 50 ml of distilled water. After blending, strain the pulp through cheesecloth, washing it with a few 10ml portions of distilled water, and make the extracted solution up to 100 ml in a volumetric flask.

Titration

- Pipette a 20 ml aliquot of the sample solution into a 250 ml conical flask and add about 150 ml of distilled water and 1 ml of starch indicator solution.
- Titrate the sample with 0.005 mol l^{-1} iodine solution. The endpoint of the titration is identified as the first permanent trace of a dark blue-black colour due to the starch-iodine complex.
- Repeat the titration with further aliquots of sample solution until you obtain concordant results (titers agreeing within 0.1 ml).

Calculations

- Calculate the average volume of iodine solution used from your concordant titers.
- Calculate the moles of iodine reacting.

- Using the equation of the titration (below) determine the number of moles of ascorbic acid reacting.

$$\text{ascorbic acid} + I_2 \rightarrow 2\,I^- + \text{dehydroascorbic acid}$$

- Calculate the concentration in mol l^{-1} of ascorbic acid in the solution obtained from fruit/vegetable/juice. Also, calculate the concentration, in mg/100ml or mg/100g of ascorbic acid, in the sample of fruit/vegetable/juice.

Alternative method for ascorbic acid estimation (Colourimetric method)

Ascorbic acid is first dehydrogenated by bromination. The dehydroascorbic acid is then reacted with 2,4 dinitrophenyl hydrazine to form osazone and dissolved in sulphuric acid to give an orange-red colour solution which can be measured ar 540 nm.

Materials

- Oxalic acid solution (4%)
- Sulphuric acid (0.5 N)
- 2,4 dinitrophenyl hydrazine (DNPH) reagent (2%) – Dissolve 2.0NPH in 100.0 ml of 0.5 N H_2SO_4 by heating. Filter and use.
- Thiourea solution (10%)
- Sulphuric acid (80%)
- Bromine water – Dissolve 1 – 2 drops of liquor bromine in approximately 100ml of cool water.
- Ascorbic acid stock solution

Procedure

- Grind 0.5 – 5.0 g of sample material either mechanically or using a pestle and mortar in 25 – 50 ml of 4% oxalic acid solution. Centrifuge and collect the supernatant.
- Transfer an aliquot (~10.0 ml) to a conical flask and add bromine water dropwise with constant mixing. When the extract is orange yello blow some air. Make the volume upto 25 – 50 ml with 4% oxalic acid solution.
- Repeat this bromination with stock solution also.

- Take 10 – 100 μg standard dehydroascorbic acid solution into a series of tubes and similarly take different amount of aliquot (0.1 – 2.0 ml) of brominated sample solution.
- Make the volume in each tube to 3.0 ml by water.
- Add 1.0 DNPH reagent followed by 1 – 2 drops of thiourea to each tube.
- Set the blank as above but water in place of ascorbic acid solution.
- Mix the contents of the tubes thoroughly and incubate at 37^0 C for 180 minutes.
- After incubation period,dissolve the orange-red osazone crystals by adding 7.0 ml of 80% sulphuric acid.
- Observe the optical density at 540 nm against reagent blank.
- Calculate the amount of ascorbic acid by comparing the optical density from standard graph.

18.13. Cholecalciferol (Vitamin D)

Vitamin D is a group of fat-soluble secosteroids responsible for enhancing intestinal absorption of calcium, iron, magnesium, phosphate and zinc. In humans, the most important compounds in this group are vitamin D_3 (also known as cholecaliferol) and vitamin D_2 (ergocalciferol). Cholecalciferol and ergocalciferol can be ingested from the diet and from supplements. The body can also synthesize vitamin D (specifically cholecalciferol) in the skin, from cholesterol, when sun exposure is adequate (hence its nickname, the "sunshine vitamin"). Although vitamin D is commonly called a vitamin, it is not actually an essential dietary vitamin in the strict sense, as it can be synthesized in adequate amounts by most mammals exposed to sunlight. A substance is an essential vitamin when it cannot be made in sufficient quantities by an organism, and must be obtained from its diet. In common with other compounds commonly called vitamins, vitamin D was nevertheless discovered in an effort to find the dietary substance lacking in a disease, namely rickets, the childhood form of oesteomalacia. Additionally, like other compounds called vitamins, in the developed world, vitamin D is added to staple foods, such as milk, to avoid disease due to deficiency. Synthesis from exposure to sunlight, as well as intake from the diet, generally contribute to the maintenance of adequate serum concentrations. Evidence indicates the synthesis of vitamin D from sun exposure is regulated by a negative feedback loop that prevents toxicity, but, because of uncertainty about the cancer risk from sunlight, no recommendations are issued by the Institute of Medicine, USA, for the amount of sun exposure required to meet vitamin D requirements.

Beyond its use to prevent osteomalacia or rickets, the evidence for other health effects of vitamin D supplementation in the general population is inconsistent. The best evidence of benefit is for bone health. The effect of vitamin D supplementation on mortality is not clear, with one meta-analysis finding a decrease in mortality in elderly people, and another concluding there is no clear justification for recommending vitamin D.

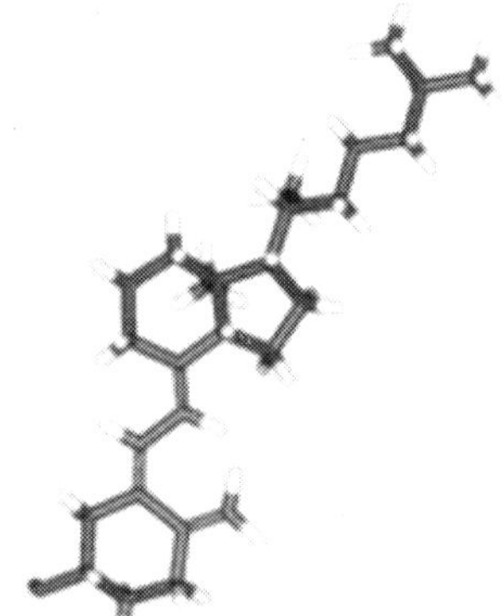

Cholecalciferol (D_3)

18.14. Tocopherol (Vitamin E)

Vitamin E refers to a group of eight fat-soluble compounds that include both tocopherols and tocotrinols. Of the many different forms of vitamin E, γ-tocopherol is the most common in the North American diet. γ-Tocopherol can be found in corn oil, soybean oil, margarine, and dressings. α-tocopherol, the most biologically active form of vitamin E, is the second-most common form of vitamin E in the diet. This variant can be found most abundantly in wheat germ oil, sunflower, and safflower oils. As a fat-soluble antioxidant, it stops the production of reactive oxygen species formed when fat undergoes oxidation. Amounts over 1,000 mg (1,500 IU) per day are called Hypervitaminosis E, as they may increase the risk of bleeding problems and vitamin K deficiency.

The eight forms of vitamin E are divided into two groups; four are tocopherols and four are tocotrienols. They are identified by prefixes alpha- (α-), beta- (β), gamma- (γ-), and delta- (Δ-). Natural tocopherols occur in the RRR-configuration only. The synthetic form contains eight different stereoisomers and is called 'all-rac'-α-tocopherol.

α-Tocopherol

α-Tocopherol is an important lipid-soluble antioxidant. It performs its functions as antioxidant in the glutathione peroxidase pathway, and it protects cell-membranes from oxidation by reacting with lipid radicals produced in the lipid peroxidation chain reaction. This would remove the free radical intermediates and prevent the oxidation reaction from continuing. The oxidized α-tocopheroxyl radicals produced in this process may be recycled back to the active reduced form through reduction by other antioxidants, such as ascorbate, retinol or ubiquinol. However, the importance of the antioxidant properties of this molecule at the concentrations present in the body are not clear and the reason vitamin E is required in the diet is possibly unrelated to its ability to act as an antioxidant. Other forms of vitamin E have their own unique properties; for example, γ-tocopherol is a nucleophile that can react with electrophilic mutagens.

Tocotrienols

Compared with tocopherols, tocotrinols are sparsely studied. Less than 1% of papers on vitamin E relate to tocotrienols. The current research direction is starting to give more prominence to the tocotrienols, the lesser known but more potent antioxidants in the vitamin E family. Some studies have suggested that tocotrienols have specialized roles in protecting neurons from damage and cholesterol reduction by inhibiting the activity of HMG-CoA reductase; δ-tocotrienol blocks processing of sterol regulatory element binding proteins (SREBPs). Oral consumption of tocotrienols is also thought to protect against stroke-associated brain damage *in vivo*. Until further research has been carried out on the other forms of vitamin E, conclusions relating to the other forms of vitamin E, based on trials studying only the efficacy of α-tocopherol, may be premature.

Estimation of vitamin E (Paper chromatographic Method)

Preparation of solution for chromatography: Since lipids, sterols and carotenoids interfere with running of tocopherols on the chromatograms, it was necessary to remove them from extracts of natural products. Lipids can be readily removed by saponification, and the vitamins and carotenoids are removed by the adsorption on the Floridin Earth. Sterols are removed through crystallization from methanol solution at -10° C. After centrifugation and removal of mother liquor the crystals are redissolved in warm methanol and the solution is again cooled to -10° C. The process is repeated till all the tocopherols are removed and mother liquor are the pooled.

Standard solutions: These were prepared by dissolving the synthetic tocopherols in the absolute ethanol. The solution is stored in amber coloured bottle at 2° C. Solutions containing more that one tocopherols (*e.g.*, α and β solutions) were most conveniently prepared from the stock solution.

Chromatography: Whatman No. 1 filter paer (30 x 12 cm) coated with Vaseline and 75% (v/v) ethanol is used as mobile phase. Ethanol is distilled from KOH and $KMnO_4$ before use. Two spots each of unknown and known solution are placed at 3 cm apart. After developing the chromatogram for 16 hours the sheet is dried for 30 minutes at room temperature. The chromatogram is sprayed with 2:2'-dipyridyl solution (0.25% v/v in ethanol) and then with $FeCl_3$ solution

(0.1% w/v, in ethanol). This gives a clear definition of the positions occupied by individual tocopherols.

Reading

Brown, F. (1952). Estimation of vitamin E. 2. Quantitative analysis of tocopherol mixtures by paper chromatography. *Biochem. J*, **52 (3):** 523-526.

Alternative method for vitamin E estimation (colourimetric method)

Tocopherols can be estimated after Emmerie-Engel reaction. This is based on the reduction of ferric to ferrous ion by tocopherols, which then forms a red colour with 2,2'-dipyridyl reagent. Tocopherol is first extracted into xylene and read at 460 nm and a correction is made after addition of ferric chloride and read at 520 nm.

Materials

- Absolute alcohol
- Xylene
- 2,2'-dipyidyl – Take 1.2 g of the chemical in one liter of n-propanol.
- Ferric chloride solution – Take 1.2 g of $FeCl_3.6H_2O$ in one liter of ethanol and store in a brown bottle.
- Standard DL-α-tocopherol (10 mg/l) – Take 10 mg of the chemical and dissolve in one liter of ethanol. 91.0 mg of α-tocopherol is equivalent to 100 mg of tocopherol acetate.

Procedure

- Take 0.5 g of homogenized tissue into a stopper tube and slowly add 10.0 ml of 0.1 N sulphuric acid. Stopper and allow the content to stand overnight. After overnight, shake the tubes vigorously and filter through Whatman filter No. 1. Use aliquot as the source of vitamin E.
- Take 1.5 ml of tissue extract (test) and in other tubes take 1.5 ml of standard solution (standard) and 1.5 ml of water (blank), and cap the tubes.
- Add 1.5 ml of ethanol to test and blank tubes and 1.5 ml of water in standard tube.
- Add 1.5 ml of xylene in each tube close the tubes and centrifuge ansd transfer the xylene layer in other stopper tubes.

- Add 1.0 ml of 2,2'-dipyridyl reagent to each tube and put the stopper.
- Take 1.5 ml of this mixture solution and measure the OD at 460 nm against blank.
- Add 0.33 ml of ferric chloride solution in all the tubes, mix well and incubate for 15 minutes at room temperature for colour to develop.
- Measure the ODs of test and standard tubes at 520 nm against reagent blank.

Calculation

Amount of tocopherol (µg/g) = ((OD of test at 520 nm) – (OD of test at 460 nm) / (OD of standard at 520 nm)) x 0.29 x 15 x ((volume of homogenate) / (volume used x Weight of tissue))

Reading

Rosenberg, H.R. (1992). In "*Chemistry and Physiology of Vitamins*", Interscience Publisher, INC, New York, pp. 452-453.

18.15. Phylloquinone (Vitamin K)

Vitamin K is a group of structurally similar, fat-soluble vitamins that the human body needs for modification of certain proteins that are required for blood coaqualation, and in bone and other tissue. The modification of the proteins allows them to bind calcium ions. Chemically, the vitamin K family are 2-methyl-1,4-napthoquinone (3-) derivatives.

Vitamin K includes two natural vitamers: vitamin K_1 and vitamin K_2. Vitamin K_2, in turn, consists of a number of related chemical subtypes, with differing lengths of carbon side chains made of isoprenoid groups of atoms. Vitamin K_1, also known as phylloquinone, phytomenadione, or phytonadione, is synthesized by plants, and is found in highest amounts in gree leafy vegetables because it is directly involved in photosynthesis. It may be thought of as the "plant form" of vitamin K. It is active as a vitamin in animals and performs the classic functions of vitamin K, including its activity in the production of blood-clotting proteins. Animals may also convert it to vitamin K_2.

Vitamin K_2, the main storage form in animals, has several subtypes, which differ in isoprenoid chain length. These vitamin K_2 homologues are called menaquinones, and are characterized by the number of isoprenoid residues in their side chains. Menaquinones are abbreviated MK-n, where M stands for menaquinone, the K stands for vitamin K, and the n represents the number of isoprenoid side chain residues. For example, menaquinone-4 (abbreviated as MK-4) has four isoprene residues in its side chain. Menaquinone-4 (also known as menatetrenone from its four isoprene residues) is the most common type of vitamin K_2 in animal products since MK-4 is normally synthesized from vitamin K_1 in certain animal tissues (arterial walls, pancreas, and testes) by replacement of the phytyl tail with an unsaturated geranylgeranyl tail containing four isoprene units, thus yielding menaquinone-4. This homolog of vitamin K_2 may have enzyme functions that are distinct from those of vitamin K_1. Bacteria in the colon (large intestine) can also convert K_1 into vitamin K_2. In addition, bacteria typically lengthen the isopreneoid side chain of vitamin K_2 to produce a range of vitamin K_2 forms, most notably the MK-7 to MK-11 homologues of vitamin K_2. All forms of K_2 other than MK-4 can only be produced by bacteria, which use these forms in anaerobic respiration. The MK-7 and other bacteria-derived form of vitamin K_2 exhibit vitamin K activity in animals, but MK-7's extra utility over MK-4, if any, is unclear and is presently a matter of investigation.

Three synthetic types of vitamin K are known: vitamins K_3, K_4, and K_5. Although the natural K_1 and all K_2 homologues have proven nontoxic, the synthetic form K_3 (menadione) has shown toxicity. Synthetic K_4 and K_5 are also non-toxic.

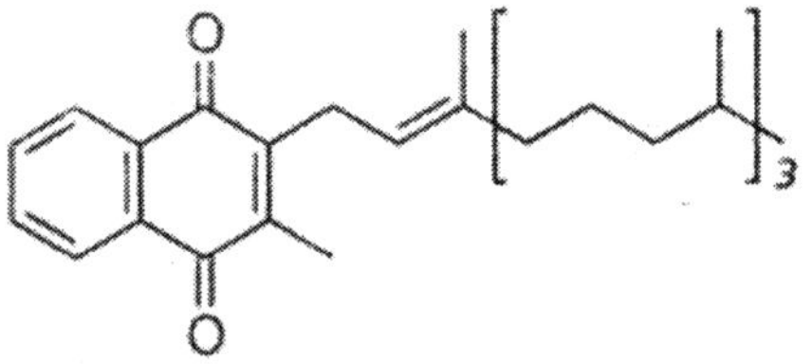

Vitamin K_1 (phylloquinone) - both forms of the vitamin contain a functional napthoquinone ring and an aliphatic side-chain. Phylloquinone has a phytyl side-chain

Vitamin K_1, the precursor of most vitamin K in nature, is a steroisomer of phylloquinone, an important chemical in green plants, where it functions as an electron acceptor in photosystem I during photosynthesis. For this reason, vitamin K_1 is found in large quantities in the photosynthetic tissues of plants (green leaves, and dark green leafy vegetables such as romaine lettuce, kale and spinach), but it occurs in far smaller quantities in other plant tissues (roots, fruits, *etc.*). Iceberg lettuce contains relatively little. The function of phylloquinone in plants appears to have no resemblance to its later metabolic and biochemical function (as "vitamin K") in animals, where it performs a completely different biochemical reaction.

Estimation of vitamin K using HPLC

Vitamin K occur in a number of single and complex preparations, however analytical standards rarely forsee quantitative separation, because of the small content and difficulty in quantitative separation of this vitamin from a multicomponent mixture.

To determine vitamin K use can be made of the great chemical reactivity of the napthoquinone system and particularly of its capability towards colour reaction which enables the application of colourimetric method. By spectrophotometric determination of vitamin K (the absorption spectrum of UV light at demonstrated maximal values of 243, 248, 261, 270 and 328 nm) it is necessary to separate the vitamin in pure form from amongst the other quinines. But shortcoming of these methods is the low sensitivity. For this reason chromatographic method, especially the high-performance liquid chromatogram-phy (HPLC) method has been chosen by many.

Standards

Phytonadione (vitamin K) USP reference standard is used.

Materials

- Dimethylsulphoxide
- Methanol
- n-Hexane
- HPLC grade water
- Anhydrous sodium sulphate.

Chromatographic parameters

- A colum Luna C18, 250 x 4.6 mm, 5 μm from Phenomenex
- Column temperature – 30^0 C.
- Detection – 248 nm
- Injection volume – 100 μl
- Flow rate – 1.2 ml/minute
- Mobile phase – Methanol.

Calibration curve

- Concentration range could be 0.5 – 10 μg/ml – Vitamin K_1 solution at the concentration of 0.5, 2.5. 5.0, 7.5, and 10.0 μg/ml is prepared.
- Each solution is injected in the column three times.
- Chromatograms obtained from the standard vitamin K_1 solutions and their peaks are recorded.

Procedure

- Samples containing 25.0 μg/ml of vitamin K_1 is prepared and to it 15.ml of a mixture of dimethyl sulphoxide (DMSO) and H_2O (3:1) is added and heated at 55°C in a waterbath for 15 minutes.
- Samples are sonicated for 10 minutes followed by shaken mechanically for 40 minutes with 25.0 ml of n-hexane.
- The n-hexane layed is washed three times with 15.0 ml of water.
- After the phases is separated, the hexane layer is filtered through a layer of anhydrous sodium sulphate.
- From the filterate a 10.0 ml aliquot is taken and after evaporation over a stream of nitrogen, the residue is redissolved in 2.0 ml of methanol and injected in column.
- Standard solution of vitamin K1 is also prepared in the range of 5.0 μg/ ml and run in HPLC

Reading

Klaczkow, G.A. and Anuszewska, E.L. (2002). Determination of the content of vitamin K, in complex vitamin preparations by the HPLC method. *Actu. Poloniac Pharmaccitica – Drug Research,* **59 (2):** 93 - 97.

18.16. Estimation of Water and Fat Soluble Vitamins by HPLC

Traditionally, vitamins have been classified as fat-soluble (A, D, E, and K) and water-soluble (C and the B complexes B_1, B_2, B_3, B_5, B_6, B_8, B_9, and B_{12}). Because these vitamins are not naturally synthesized in the human body, a balanced diet is mandatory in order to maintain the quantity of vitamins at the required level. However, at times, dietary habits can create a deficiency of these vitamins. Therefore, the quality control is very important, and the widespread use of multivitamin preparations has stimulated research into the development of accurate, efficient, and easy methods for quality control.

The standard and official analytical methods for vitamin analysis are tedious and involve a pre-treatment of the sample through complex chemical, physical, or biological reactions in order to eliminate interference of other constituents in the vitamin, which is typically followed by individual methods for each different vitamin. Many analytical methods, such as those based on UV-Vis spectrophotometry, fluorimetry, chemiluminiscence, capillary electrophoresis, microbiology, and high-performance liquid chromatography have been proposed only for the determination of water-soluble or fat-soluble vitamins and are tedious, sometimes nonspecific, and time-consuming. Among them, both normal and reversed-phase HPLC techniques are most widely used and provide rapid, sensitive, and accurate vitamin analysis and are often used with detection modes, including UV spectrometry with a diode array, fluorometry, and electrochemical detection.

However, few methods using HPLC have been developed for the simultaneous determination of water- and fat-soluble vitamins. A rapid, reliable, and reproducible method based on microemulsion electro kinetic chromatography for the simultaneous determination of 13 types of water- and fat-soluble vitamins is developed. Separate methods based on reversed-phase HPLC for the determination of eight water-soluble vitamins and seven fat-soluble vitamins were proposed. A method for the simultaneous determination of 12 water- and fat-soluble vitamins using gradient HPLC with diode array detection (DAD) is also developed. This method can be applied to the determination of vitamins in pharmaceutical preparations, food supplements, fortified drinks, and foods with relatively complicated matrices.

The development and optimization of a high-performance liquid chromatographic method using fluorescence detection (FLD) and DAD for the determination of four fat-soluble and seven water-soluble vitamins in a single run has successfully applied to the determination of vitamins in multivitamin tablets, and the separation of the vitamins is achieved on a reversed-phase C18 110A column. The method provides a high degree of chromatographic resolution, sensitivity, and reproducibility, and enables quantitative determination.

Samples and chemicals

- HPLC-grade acetonitrile and methanol
- Sodium hydroxide and trifluoroacetic acid
- Ultra-pure water
- All solutions were sonicated and filtered through 0.45 μm filters prior to use.
- Retinol acetate, cholecalciferol, α-tocopherol, phytonadione, thiamine, riboflavin, niacin, cyanocobalamin, ascorbic acid, folic acid, and *p*-aminobenzoic acid vitamin standards (purity > 99.0%), α-tocopherol (purity > 98.5%).

Standard solutions

- *Water- and fat-soluble vitamin solutions* - The water-soluble vitamins and the fat-soluble vitamins were diluted in deionized water and methanol, respectively, in order to obtain stock solutions (approximately 1000 mg l^{-1}). Each solution is freshly prepared before use and stored in the dark at 4 °C. Working standard solutions in the 0.05 - 20 mg l^{-1} range were daily prepared by step-wise dilution with a water-methanol solution (50:50, v/v).
- *Mixed standard solution* - A mixed standard solution containing the 11 vitamin standards at different concentrations is weekly prepared by using a water - methanol solution (50:50, v/v).

Instrumentation

- The HPLC system used throughout this study consisted of a quaternary pump, an auto-sampler, a degasser, a column thermostat, a diode array detector, and a fluorescence detector. A reversed-phase C18 110A column (150 × 4.6 mm, 3.0 μm particle size), is used.
- The column temperature is held at 30 °C.
- Chromatographic separations were performed under gradient conditions by using solvent A (0.01% TFA, pH = 4) and solvent B (100% methanol) at a flow rate of 0.7 ml min^{-1}.
- A 35-min-gradient program is used in the study, beginning with 0.01% TFA and methanol (95:5, v/v) for 4 min, followed by ramping down to 5% TFA over 6.0 min, holding at that ratio for 20 min, and then ramping up to 95% TFA again.
- The results were monitored with two different detectors (DAD and FLD) and after optimization studies, it is determined which detector is suitable for each vitamin.

Optimization of the chromatographic conditions

- In order to determine the most suitable absorption wavelength for the simultaneous detection of the vitamins, the spectra of standard solutions (10 mg l^{-1}) are measured by using DAD for retinol acetate, cholecalciferol, phytonadione, thiamine, niacin, cyanocobalamin, ascorbic acid, folic acid, and *p*-aminobenzoic acid, and an FLD detector is for α-tocopherol and riboflavin in the range of 190 - 600 nm.

- The wavelengths that had the highest intensity are accepted as the optimum values.
- Studies are then performed with two different columns in order to determine the best column for separation of the vitamins.
- To evaluate the separation, the resolution (R_s), symmetry value, and *S/N* ratio are taken into account.
- After column selection, the most appropriate carrier phase composition and flow rate are optimized. The flow rate optimization studies were carried out using five different flow rates (0.3, 0.5, 0.7, 0.9, 1.1 ml min^{-1}) and 0.01% TFA and MeOH solutions were used to define the carrier phase composition.
- To determine the effect of temperature and pH on the separation of multivitamins, studies were performed at 20, 25, 30, 35, and 40 °C and pH values of 2, 3, 4, 5, and 6, respectively. In addition, the separation procedure is performed using the three different gradient profiles and the most suitable gradient elution program is chosen for all of further studies.

Selection of the extraction solution ratio

- In order to extract the water- and fat-soluble vitamins from the multivitamin tablets, various extraction solutions containing 0.01% TFA and methanol in different ratios are used.
- During the extractions, the samples are weighted according to the amount of active substance, and the solution ratios used in the extraction treatment are as follows: 100:0, 80:20, 50:50, 20:80, and 0:100.

Simultaneous determination of the fat- and water-soluble vitamins in multivitamin tablets

- Dried samples are accurately weighed and then finely powdered.
- A portion of the powder is transferred into a beaker. After adding a 0.01% TFA-methanol (10 ml; 80:20, v/v) solution, the mixture is vortexed for 15 min and then centrifuged for 15 min at 5000 rpm.
- The supernatant is filtered through a 0.45 μm nylon membrane. Next, a 5 μL aliquot of the sample solution is injected into the HPLC system.
- The chromatographic procedure is performed using the optimized conditions for gradient elution program III.

- The retention time, symmetry, and resolution factors were used as criteria for the determination of the separation efficiency. The concentrations of the fat- and water-soluble vitamins were calculated from the integrated areas of the sample and the corresponding standards.

Reading

Kucukkolbasi, S., Bilber, O., Ayyildiz, HF. A. Kara, H. (2013). Simultaneous and accurate determination of water- and fat-soluble vitamins in multivitamin tablets by using an RP-HPLC method. *Quím. Nova.*, **36 (7):** (http://dx.doi.org/10.1590/S0100-40422013000700020)